Relationships Between Perception and Action

Current Approaches

Edited by

O. Neumann and W. Prinz

With Contributions by

P. Bieri · B. Bridgeman · H. Cruse · J. Dean · C.-A. Hauert
H. Heuer · D. G. MacKay · D. W. Massaro · P. Mounoud
O. Neumann · W. Prinz · E. Scheerer · R. A. Schmidt
A. H. C. van der Heijden · A. Vinter · P.-G. Zanone

Springer-Verlag Berlin Heidelberg New York
London Paris Tokyo Hong Kong

Dr. ODMAR NEUMANN
Abteilung für Psychologie
Universität Bielefeld
Postfach 8640
4800 Bielefeld, FRG

Prof. Dr. WOLFGANG PRINZ
Abteilung für Psychologie
Universität Bielefeld
Postfach 8640
4800 Bielefeld, FRG

With 28 Figures

ISBN-13:978-3-642-75350-3 e-ISBN-13:978-3-642-75348-0
DOI: 10.1007/978-3-642-75348-0

Library of Congress Cataloging-in-Publication Data. Relationships between perception and action: current approaches / edited by O. Neumann and W. Prinz; contributors, P. Bieri ... [et al.]. p. cm. Includes bibliographical references. ISBN-13:978-3-642-75350-3 1. Perceptual-motor processes. I. Neumann, Odmar. II. Prinz, Wolfgang. III. Bieri, Peter, 1943– . [DNLM: 1. Behavior. 2. Perception. BF 311 R382] BF295.R44 1990 152.1 – dc20 DNLM/DLC 90-9414

Softcover reprint of the hardcover 1st edition 1990

2126/3130 (3011)-543210 – Printed on acid-free paper

Preface

This book is the fruit of a study group on perception and action that worked at the Center for Interdisciplinary Research (ZiF) of the University of Bielefeld, FRG in the academic year 1984-1985. We express our gratitude to the ZiF for hosting the group and for providing financial and organizational support for its scientific activities, including a meeting of the authors of the present volume that took place at the ZiF in July 1986.

This is the study group's last common product, and it took considerable time to give the book its final shape. Most of the editing was done while one of us (O. N.) was a Fellow at the Netherlands Institute for Advanced Study in the Humanities and Social Sciences (NIAS) during the academic year 1987-1988. Thanks are due to NIAS for its generous support. We also thank all our friends and colleagues who contributed to the book.

Bielefeld, March 1990

ODMAR NEUMANN
WOLFGANG PRINZ

Contents

Contributors

Dr. P. BIERI, Universität Bielefeld, Abteilung für Philosophie, Postfach 8640, 4800 Bielefeld, FRG

Dr. B. BRIDGEMAN, Program in Experimental Psychology, University of California at Santa Cruz, Santa Cruz, CA 95064, USA

Dr. H. CRUSE, Universität Bielefeld, Fakultät für Biologie, Postfach 8640, 4800 Bielefeld, FRG

Dr. J. DEAN, Universität Bielefeld, Fakultät für Biologie, Postfach 8640, 4800 Bielefeld, FRG

Dr. C.-A. HAUERT, Université de Genève, Faculté de Psychologie et des Sciences de l'Education, 24, rue Général Dufour, 1211 Genève, Switzerland

Dr. H. HEUER, Philipps-Universität, Fachbereich Psychologie, Gutenbergstr. 18, 3550 Marburg/Lahn, FRG

Dr. D.G. MACKAY, Department of Psychology, University of California at Los Angeles, Los Angeles, CA 90024, USA

Dr. D.W. MASSARO, Program in Experimental Psychology, University of California at Santa Cruz, Santa Cruz, CA 95064, USA

Dr. P. MOUNOUD, Université de Genève, Faculté de Psychologie et des Sciences de l'Education, 24, rue Général Dufour, 1211 Genève, Switzerland

Dr. O. NEUMANN, Universität Bielefeld, Abteilung für Psychologie, Postfach 8640, 4800 Bielefeld, FRG

Dr. W. PRINZ, Universität Bielefeld, Abteilung für Psychologie, Postfach 8640, 4800 Bielefeld, FRG

Dr. E. SCHEERER, Universität Oldenburg, Fachbereich Psychologie, Institut für Kognitionsforschung, Birkenweg 3, 2900 Oldenburg, FRG

Dr. R.A. SCHMIDT, Department of Psychology, University of California at Los Angeles, Los Angeles, CA 90024, USA

Dr. A.H.C. VAN DER HEIJDEN, University of Leiden, Faculty of Social Sciences, Wassenaarseweg 52, P.O. Box 9555, 2300 RB Leiden, The Netherlands

Dr. A. VINTER, Faculté des Lettres et Sciences Humaines, 30, rue Mégevand, 25030 Besancon Cedex, France

Dr. P.-G. ZANONE, Université de Genève, Faculté de Psychologie et des Sciences de l'Education, 24, rue Général Dufour, 1211 Genève, Switzerland

Introduction

W. PRINZ and O. NEUMANN

The French historian Jules Michelet used to begin his lectures on Great Britain with the statement, "Gentlemen, Britain is an island." Presumably, he did not suppose that this fact was unknown to his students. He cited it in order to point out that many of the peculiarities of British history and political life can be better understood if one takes into account that country's geographical situation.

Similarly, an appropriate opening sentence for this book could have been, "The brain is a device for the control of the bodily apparatus." Everybody knows, of course, that this is true in a descriptive sense. Less trivially, it may also be true in the functional sense in which Michelet interpreted Britain's insular geographical condition. If the brain has evolved as a control instrument for the guidance of physical organisms, then an analysis of its functions should profit from an adequate understanding of the way in which these functions contribute to this ultimate task.

While such an approach is commonplace in biology, it is much less prevalent in present-day psychology. Further, while it may suggest itself readily for fields such as motivation or the mechanisms of motor control, it seems, at first glance, less obvious that it can foster our understanding of sensory, perceptual, and cognitive functions.

Indeed, modern psychology has until recently tended to regard perception and cognition as fields that can, and in fact should, be investigated without reference to action control. The highly successful approach to psychology that started as the information processing approach and is now usually called cognitive psychology (for recent overviews and retrospects see, e.g., Anderson, 1985; Gardner, 1985; Massaro, 1986; Neumann, 1985; Scheerer, 1988) has had a strong preference for studying perceptual and cognitive processes as decoupled from action control. One of the founding fathers of this approach, Ulric Neisser, has aptly described its tenet as "attempting to trace the fate of the input" (Neisser, 1967, p. 4) - a pursuit that usually stopped short of the control of overt action. (Neisser's book nicely illustrates this: Although cognitive psychology was purportedly "concerned with all human activity" [p. 4], the journey through its realm ended with a chapter on memory and thought).

There are signs that this self-imposed abstinence is beginning to taper off. Although the often-heard dictum "out is in" may overstate the situation, there are indications of a renewed interest in the motor aspect of human activity and its relationships to sensory, perceptual, and cognitive aspects. In part, this is probably due

Relationships Between Perception and Action
Edited by O. Neumann and W. Prinz

to the challenge from ecological realism with its insistence on the inseparability of perception and action (e.g., Gibson, 1979; Turvey, 1977; Turvey & Carello, 1986). In part, it may have been a logical result of the progression of psychomotor research, which has begun to embrace complex movements and their planning, thus bridging or at least reducing the gap to cognitive psychology (for surveys see, e.g., Heuer & Fromm, 1986; Magill, 1983). Moreover, the upsurge of robotics, one of the latest branches of artificial intelligence, is likely to have buttressed the novel interest in motor performance. And, finally, it seems that all this takes place within a more general reorientation of theoretical psychology, favorable to a reconsideration of the perception-action relationship.

This reorientation started only a few years ago. Its character and impact are still difficult to assess. Its most predominant manifestation has been the advance of the connectionist movement (e.g., Hinton & Anderson, 1981; McClelland & Rumelhart, 1987; Rumelhart & McClelland, 1987), which, according to Schneider (1987), bears the hallmark of a paradigm shift in the sense put forward by Kuhn (1962), i.e., a scientific revolution in which the basic assumptions within a field are being reconsidered and redefined. However, the success of connectionism may not be a singular event. There are signs that it is just one facet of a broad transition that may involve a reconsideration of psychology's place in the geography of the sciences. The information processing approach tended to view psychology as a close kin (if not an offspring) of computer science. The new orientation that may be emerging places it closer to the biological sciences.

Unlike technological computation, biological computation involves the control of a complex effector apparatus. The biological sciences have always been aware of the close functional linkage between information uptake and action control. This is why, in our view, the general theoretical climate that is presently developing is propitious to the exploration of relationships between perception and action.

In this volume the reader will find some of the flavor of this novel theoretical climate. The book has a prologue, two epilogues, and ten chapters in between. The first three chapters consider basic issues involved in the interaction between information uptake and action control: Bridgeman stresses the reciprocal nature of this interaction, drawing upon Gibson's concept of perceptual systems. In perceptual systems there is action in the service of information uptake, and at the same time the acquired information is used for the control of further action. This reciprocity characterizes the operation of perception and action as well as their development. In the next chapter, Cruse, Dean, Heuer, and Schmidt provide a thorough conceptual analysis of modes of utilizing information for action control. They propose a unified taxonomy for types of motor control structures, based on a distinction between three basic ways in which sensory information is used for motor control. Dean's chapter gives an account of the neuroethology of the relationships between perception and action. As this approach relies on study material drawn from the entire animal kingdom, it stresses both comparative and ecological aspects in the study of behavior and its underlying neural mechanisms.

In the next two chapters the emphasis is shifted from acting animals to reacting subjects, i.e., to human subjects who are instructed to perform certain movements under certain conditions.The reaction time paradigm which has become so popular in the past three decades has usually not been viewed as a means for studying action and action control. It has rather been used as a method for studying processing of input information, and especially for decomposing this processing into steps and stages. Though this has usually included a step of linking the internal representations of stimuli to responses, the exact nature of this operation has not received much consideration. Massaro's chapter seems to be the first comprehensive treatment of the linkage problem from an information processing point of view. This chapter elaborates the notion of translation (between perception and action) as the central theoretical metaphor for conceptualizing the linkage problem. In the following chapter, one of us (W. P.) proposes a different approach, which is based on the notion of common coding. Its basic assumption is that a common representational medium exists for perception and action, so that no translation is required.

In the subsequent chapters, the authors turn from subjects who react to subjects who select and attend. Van der Heijden gives a critical account of two solutions to the problems of the mind's presumed limited processing capacity, viz. encoding and selection. In a thorough evaluation of the evidence he argues that there is no basic capacity problem in the system so that encoding and selection cannot be considered solutions to capacity problems. Rather they form part of the organism's solution to problems of action planning and action control. In the next chapter, one of us (O. N.) starts from a similar functional view of attentional phenomena and the underlying mechanisms and applies it to the area of visual attention. MacKay then presents what he calls a three-body problem, arguing that the interactions between perception and action can be fully understood only if problems of awareness are included as a third party. His chapter gives an outline of his theory of this threefold interaction.

The two concluding chapters return to the issue of development that was touched upon in Bridgeman's chapter. Vinter gives an account of the development of modes of action control in early human development, and Hauert, Zanone, and Mounoud present evidence that there are different modes of information uptake for action control in the child, related to different stages of development. Both chapters demonstrate convincingly that from the developmental perspective, perception and action are so intricately interwoven that one cannot be studied without the other.

The ten chapters are framed by a prologue and two epilogues. The prologue gives an outline of perception and action in the history of experimental psychology. In the first epilogue Bieri presents a philosopher's sceptical reflections on the varieties of informational accounts of cognitive functioning that are contained in the book's chapters. How could we reconcile the informational and the intentional vocabulary? Could we ever? Should we ever? In the second epilogue, Scheerer offers a psychologist's critical reflections on the diversity of the approaches in the book, but also on the possible merits of such diversity - and on some hidden unity behind it as well. His concluding remark is in defence of diversity: "Given that perception-action relations are polymorphous, all we can do is to apply the analytical tools and conceptual structures provided in this book to the study of individual instances of sensorimotor coordination."

References

Anderson, J.R. (1985). *Cognitive psychology and its implications*. New York: Freeman.

Gardner, H. (1985). *The mind's new science: A history of the cognitive revolution*. New York: Basic Books.

Gibson, J.J. (1979). *The ecological approach to visual perception*. Boston: Houghton Mifflin.

Heuer, H., & Fromm C. (1986). *Generation and modulation of action patterns*. Berlin, Heidelberg, New York, Tokyo: Springer.

Hinton, G.E., & Anderson, J.A. (1981). *Parallel models of associative memory*. Hillsdale, NJ: Erlbaum.

Kuhn, T. (1970). *The structure of scientific revolutions*. Chicago: University of Chicago Press.

Magill, R.A. (1983). *Memory and control of action*. Amsterdam: North-Holland.

Massaro, D.W. (1986). The computer as a metaphor for psychological inquiry: Considerations and recommendations. *Behavioral Research Methods, Instruments, and Computers, 18*, 73-92.

McClelland, J.L., Rumelhart, D.E. (Eds.). (1987). *Parallel distributed processing: Explorations in the structure of cognition: Vol. 2. Psychological and biological models*. Cambridge, MA: MIT Press.

Neisser, U. (1967). *Cognitive psychology*. New York: Appleton-Century-Crofts.

Neumann, O. (1985). Informationsverarbeitung, Künstliche Intelligenz und die Perspektiven der Kognitionspsychologie [Information processing, artificial intelligence, and the perspectives of cognitive psychology]. In O. Neumann (Ed.), *Perspektiven der Kognitionspsychologie* (pp. 3-37). Berlin, Heidelberg, New York, Tokyo: Springer.

Rumelhart, D.E., & McClelland, J.L. (Eds.). (1987). *Parallel distributed processing: Explorations in the structure of cognition: Vol.1. Foundations*. Cambridge, MA: MIT Press.

Scheerer, E. (1988). Towards a history of cognitive science. *International Science Journal, 115*, 7-19.

Schneider, W. (1987). Connectionism: Is it a paradigm shift for psychology? *Behavioral Research Methods, Instruments, and Computers, 19*, 73-83.

Turvey, M.I. (1977). Preliminaries to a theory of action with reference to vision. In R. Shaw & J. Bransford (Eds.), *Perceiving, acting and knowing: Toward an ecological psychology* (pp. 211-275). Hillsdale, NJ: Erlbaum.

Turvey, M.T., & Carello, C. (1986). The ecological approach to perceiving-acting: a pictorial essay. *Acta Psychologica, 63*, 133-155.

Prologue: Historical Approaches to Perception and Action

O. NEUMANN and W. PRINZ

CONTENTS

Introduction

Relationships between perception and action were a prominent theme in the early days of experimental psychology. The theoretical systems that dominated nineteenth-century psychology addressed both fields as well as connections between them. In this sense[1] it may be said that experimental psychology embarked as a science of perception and action. It was only relatively late, and for particular historical reasons, that the two fields separated. This chapter presents a brief review of some stations of this historical development.

Perception and Action in Nineteenth-century Psychology

Textbooks of psychology often depict mainstream nineteenth-century psychology as "structuralism," describing it as a mainly introspective endeavor, aimed at the analysis of consciousness into sensations. This is a one-sided picture at best. The term "structuralism" was coined by Titchener (1898, 1899) towards the end of the classical period in defense against the upcoming functionalism (whose legitimacy as an alternative, and complementary, approach Titchener acknowledged, how-

[1] In another sense, it was not. As Scheerer (this volume) points out, the fields of perception and cognition carried much more weight in nineteenth-century psychology than action and motor control did.

Relationships Between Perception and Action
Edited by O. Neumann and W. Prinz

ever). Titchener was a pupil of Wundt, but his radical sensualism and introspectionism was his particular brand of nineteenth-century theorizing.

Most nineteenth-century theorists - and certainly the most prominent ones, Wilhelm Wundt and William James - were not pure sensualists; nor did they believe in introspection as the only true method.[2] They were interested in action and motor behavior as well as in perception and sensation, and they usually employed objective methods where these were available, even though conscious experience (or "immediate experience," as Wundt called it) was believed to be the ultimate topic of psychological inquiry.

Relationships between perception and action were explored in several areas. At the most general level, there was the question of how overt behavior is related to conscious experience. At more detailed levels, there were investigations of how sensory information enters into action and how motor processes influence perception. Let us consider three of these topics: levels of sensorimotor control, ideomotor action, and space perception.

Levels of Sensorimotor Control

Most classical psychological texts included chapters on the control of motor movements. For example, Wilhelm Wundt discussed motor control under the heading "Volitional processes" (*Willensvorgänge*) (e.g., Wundt, 1903, chap. 16, 1914, chap. 14). William James (1890) had three chapters on this topic, one about "The production of movement" (chap. 22), one about "Instinct" (chap. 23), and one about "Will" (chap. 25). Theodor Ziehen, one of the sharpest minds among nineteenth-century theorists, had a full-fledged action theory in which his account of the interplay between mental representations was embedded (Ziehen, 1920).

Although differing in detail and their general theoretical orientation, all three approaches may be considered characteristic of the multilevel conception of sensorimotor control that had emerged in the second half of the nineteenth century. Dissents were mainly of a metatheoretical nature. For Wundt, all sensorimotor processes belonged to the realm of volition. For James, voluntary movements were functionally secondary, based on more primitive functions such as reflexes and instincts. Ziehen rejected the concept of volition altogether.

But there was consensus among these three - and many other nineteenth-century - authors that there is a hierarchy of levels of sensorimotor control, and that higher mental processes have their place within this hierarchy. They were thought to be linked to lower levels in a twofold manner: First, higher mental processes genetically derive from simpler forms of sensorimotor linkage; second,

[2] Take, for example, Wundt's (1882) paper on the methodology of psychology. It lists mainly (a) the classical methods of sensory psychophysics (method of limits, of constant stimuli, etc.) and (b) methods for the "psychological measurement of time" (essentially RT measurement), which Wundt considered the "most important topic of experimental psychology, which also meets with the greatest theoretical interest" (Wundt, 1882 p. 25; translation ours). Almost all the methods Wundt mentioned are still in use.

they serve to achieve basically the same end - adjusting action to the environment - in a more complex manner.

The most explicit formulation of the latter point can be found in James:

> The whole neural organism, it will be remembered, is, physiologically considered, but a machine for converting stimuli into reactions; and the intellectual part of our life is knit up with the middle or 'central' portion of the machine's operation (James, 1890, p. 372).

Wundt's and Ziehen's standpoints were similar, though perhaps less articulate. Ziehen described the association of ideas (his term for higher mental processes) as "the sum of all those mental processes ..., which, starting from sensations, eventually give rise to action" (Ziehen, 1920, p. 27, translation ours). Wundt distinguished between three levels of action control: "drive actions" (*Triebhandlungen*), which arise from a single motive, "voluntary actions" (*Willkürhandlungen*), which result from a competition between antagonistic motives, and "choice actions" (*Wahlhandlungen*), where the competition is clearly represented in consciousness, leading to explicit deliberation and decision before the action is executed.

Within these multilevel frameworks, all three authors offered detailed, largely empirically based analyses of the different levels and modes of sensory, perceptual, and cognitive control of action. For example, Ziehen argued that there were three different modes of control: reflexes, where the only function of the stimulus is to trigger the reaction (e.g., a tactual stimulus causes a decorticated frog to jump); deflexes, where stimulus information in addition shapes the course of execution (a decorticated frog with the thalamus spared will circumvent obstacles, while a totally decorticated frog will not); finally, actions, where behavior is determined not only by present stimulation but also by memory images from previous experience.

Perhaps the best-known empirical attack on different modes of linkage between perception and action was Wundt's (1903) distinction between sensorial and muscular reactions, based on the reaction time (RT) research carried out in his laboratory (Lange, 1888): In a simple RT experiment, the subject can take either a sensorial attitude (attending to the expected stimulus) or a muscular attitude (attending to the response to be performed). The two cases differ both subjectively and objectively. In sensorial reactions, the subject experiences first perceiving the stimulus and then initiating the response, whereas in muscular reactions the response is initiated without any prior conscious awareness of the stimulus. Objectively, muscular reactions are faster than sensorial reactions and show a more skewed RT distribution with a smaller standard deviation. Wundt's interpretation was that only sensorial reactions are mediated via a conscious representation (apperception) of the stimulus and a subsequent voluntary impulse to react, whereas apperception or even - after prolonged training - both apperception and the will impulse are absent in muscular reactions. In the latter case the linkage between stimulus and reaction has become automated.

Reaction time research is an example of the most obvious and best-investigated aspect of the perception-action relationship: sensory stimulation can trigger a motor response. But there are other aspects. Sensory information does not only determine *whether* an action is performed; it may, in addition, specify its course of execution, i.e., *how* it is carried out (cf. Ziehen's distinction between reflex and deflex, mentioned above). Further, not only does perception affect action, but action may also have its effects on perception. Among the classical examples for the investigation of these two aspects are, respectively, ideomotor action and space perception.

Ideomotor Action

In the second half of the nineteenth century the ideomotor principle was a popular account of how action is guided by perception and imagery. The term "ideomotor action" was introduced by Carpenter (1874) and later popularized by James (1890), but it seems that the underlying idea goes back to Lotze's (1852) account of imitative and voluntary action (for details see Prinz, 1987). Lotze started from the observation that when we perceive a motion (e.g., that of a bowling ball), the perception is often accompanied by slight bodily movements (e.g., of the arms and hands), or at least we feel impulses to move. There is a tendency to imitate or copy perceived motions (*"Nachahmungsbewegungen"*; Lotze 1852, p. 293).

Though this seems to be a rather special form of movement control at first glance, Lotze was convinced that imitative movements were manifestations of a very basic principle. This principle says that mental representations of motion or action possess the intrinsic power to induce corresponding movements or actions in the perceiver.

It is important to note that Lotze regarded this as a general principle in two respects. First, he believed that it applied not only to the triggering of movements and actions, but also to the guidance of their course of execution. Second, the ideomotor principle was not restricted to perception proper, but encompassed all mental representations (*"Vorstellungen"*), i.e., all kinds of mental contents arising from perception, imagery, or thought. Based on this broad notion of *"Vorstellungen"*, the category of imitative movements could be extended to the case where there is no external stimulus, i.e., where the mere thought of an action or movement is sufficient to induce its execution.

This enabled Lotze to apply the ideomotor principle to all voluntary action (Lotze, 1852, p. 302): Not only does the mental representation of a motion lead to bodily movement; we must postulate that such a *Vorstellung* is always present as a prerequisite to voluntary action. As soon as there is a mental representation of a movement that is accepted and approved by the will, the corresponding action will be executed. There is no more basic principle to which this can be reduced: Here we have, according to Lotze, a fundamental principle of man's psychophysical machinery.

What *can* be explained is how the mind knows which particular *Vorstellung* has to be generated in order to induce a particular movement or action: This is the result of lifelong learning of the correlations between mental representations and bodily movements. However, this does not explain how it is possible at all that a mental content can get control over a physical event. For a dualist like Lotze, there could be no answer to this question; it was just an ultimate fact of psychophysical functioning.

Perhaps it was in recognition of this that William James' (1890) restatement of Lotze's notion was so solemnly worded (and highlighted by italics):

> We may ... lay it down for certain that every representation of a movement awakens in some degree the actual movement which is its object; and awakens it in a maximum degree whenever it is not kept from so doing by an antagonistic representation present simultaneously to the mind (James, 1890, p. 526). Movement is the natural immediate effect of feeling, irrespective of what the quality of the feeling may be. It is so in reflex action, it is so in emotional expression, it is so in voluntary life (p. 527).

Space Perception

While the ideomotor principle related movement to an antecedent perception or thought, nineteenth-century psychology was also interested in the opposite relationship. The idea that perception not only determines action, but is in turn affected by movements that the perceiver executes, dates back to the early eighteenth century, especially to Berkeley and Condillac (cf. Neumann, 1972; Scheerer, 1984). In the second half of the nineteenth century it had become commonplace. The question was not *whether* this is the case, but *how* movements enter into what is perceived.

Basically, there were two problems, and two kinds of solutions for each of them. The first problem was how space can be perceived at all. The other problem referred to what is nowadays called space constancy: How does it come about that we perceive a spatially stable environment in spite of our own movements (of the eyes, the head, the whole body) that continuously change the sensory stimulation which arrives at the sensory surface?

Space perception as such was a theoretical problem because of the doctrine of sensations that had come to be almost universally accepted in the first half of the nineteenth century (see Neumann, 1972). Sensations, the building blocks of all experience, were believed to be tied to the excitations of single nerve fibers. From this it was usually concluded that they had to be punctiform, possessing quality and intensity, but no spatial extension. Space was not contained in the sense data; it had to be contributed from elsewhere.

This produced two interrelated questions: First, how can extension be perceived at all as a *quale*, different from quality and intensity? Second, how does spatial *order* arise; i.e., how is the spatial location of a given stimulus determined? With respect to the first problem, most nineteenth-century theorists were Kantians; i.e., they believed that spatiality is an innate, irreducible attribute of perception (*Anschauungsform*), not open to further explanation (cf. Boring, 1942, pp. 233 ff). Different opinions existed, however, on the second problem. There were the nativists, most notably Hering (1879), who assumed that spatial order was innate and available without any experience. The second position belongs into our present context. The theory of "local signs" [often attributed to Lotze (1852), though the idea had been around since the beginning of the nineteenth century (see

Neumann, 1980)] postulated that space perception is mediated by effector movements.[3]

There were many versions of this idea. Lotze (1852) started from the observation that extrafoveal stimuli tend to elicit a reflexive eye movement that brings them on the fovea. Each retinal position is affiliated with a particular pattern of activity of the eye muscles; hence, its pattern of kinesthetic feedback unequivocally characterizes the retinal location and, by becoming associated with it through repetition, can serve as its local sign. Once established, these local signs will also be generated by movement impulses that are not actually executed.

Wundt, among others, extended the local sign idea to tactile sensations, arguing that the intensity of the muscular effort that is needed to attain a given position varies as a function of location; hence, space can be internally coded on an intensity dimension, in accordance with the doctrine of sensations (e.g., Wundt, 1914, chap. 10).

Obviously the theory of local signs was a direct consequence of this doctrine. It was needed for systematic reasons, but there was very little direct empirical support for it. It is therefore not surprising that it vanished with the downfall of classical experimental psychology (see Scheerer, 1984). Its basic assumption was indeed wrong; modern brain physiology has definitely established that retinal information is spatially represented in the primary visual cortex independently of efferent processes (see, e.g., Neumann, this volume). Yet one may wonder whether the baby was not thrown out with the bathwater when the idea of local signs was completely abandoned. Eye movements are not needed to spatially order visual sensations, but they may play an important part in interpreting them veridically, i.e., in perceptual learning (see, e.g., Wolff, 1985, 1986, 1987).

The significance of the classical work on the second problem is less equivocal. The constancy of the perceived spatial environment (direction and position constancy) despite movements of the eyes, the head, etc. requires that retinal information is somehow evaluated against information about egomotion. One possibility is that this information is provided by a corollary discharge or efference copy (von Holst & Mittelstaedt, 1950). Both concepts refer to a neural representation of an active movement against which the movements's sensory consequences are matched. As Scheerer (1984, 1987) has pointed out, this idea dates back to at least Herrmann von Helmholtz' (1866) *Physiological Optics* (Scheerer, 1984, pp. 80f; Scheerer, 1987, pp. 186ff; see also Grüsser, 1986).

In the second half of the nineteenth century, the usual term for the internal representation of an active movement was "innervation feeling". The idea that sensory information is merged with innervation feelings to produce spatial perception was one of two competing classical explanations of space constancy. The rival conception (among whose most ardent advocates we find William James) maintained that the only source of information about one's own movements was

[3] The term "local signs" was used in a broad and in a narrow sense. In the narrow sense it referred to Lotze's (1852) and similar empiricist concepts. In its broad meaning it included the nativist version according to which sensations as such carry information about their location (see Neumann, 1980 for details). In the following we refer only to the empiricist theory of local signs.

provided by kinesthetic feedback. Thus, the nineteenth century already had the "inflow" vs "outflow" debate that still dominates the present discussions (for a recent overview of the state of the art on sensorimotor interactions in space perception see Bouwhuis, Bridgeman, Owens, Shebilske, & Wolff, 1986).

These examples may suffice to demonstrate that relationships between perception and action were a central topic of experimental psychology in its first decades. Though sometimes motivated by self-generated difficulties of contemporary theorizing, this research as a whole was empirically well founded and theoretically sound. Yet it came to an abrupt end in the period around the First World War. The reason was not that all problems had been solved, or that trying to solve them had proven futile. Perception/action research simply became outmoded.

The Twentieth-century Schism Between Perception and Action

Behaviorism and Gestalt Psychology

Neisser (1976) has suggested that classical experimental psychology vanished because it had become sterile and had proven inept to answer any questions of practical relevance. This was certainly one factor, especially in the American intellectual climate with its pragmatic background. But there were other factors as well.

One of the main shortcomings of the classical approach had been the special status that it claimed for psychology. In Wundt's terms, all other sciences were concerned with "mediate experience" - the outer physical world -, whereas the ultimate topic of psychology was "immediate experience". This meant that, by its very definition, psychology had isolated itself from the rest of the sciences. The movements that succeeded the classical approach, behaviorism and Gestalt psychology, corrected this. For behaviorism, psychology was a biological science (e.g., Watson, 1917, p. 349). Gestalt psychology sought the explanation of psychological phenomena in principles from physics, particularly those of field theory (e.g., Köhler, 1940). Thus, both were eminently modern movements in their time because they brought psychology in close contact with booming natural sciences. This may have been the main reason why they could virtually wipe out the classical approach within not much more than a decade, roughly between 1910 and 1920.

One of the prerequisites of this success was that, programmatically, each of the two movements aimed at an integrative theory that was to cover the whole field of psychology. Gestalt psychology was not just a theory of perception and thought, and behaviorism was not just a theory of learning. Both claimed to offer an account that included, in principle, all topics that had been investigated by classical psychology. The early programmatic writings of the behaviorists were full of claims that, in principle, all fields of experimental psychology could be incorporated into behaviorism:

There is hardly a problem in human vision is not also a problem in animal vision ... Every one can be worked out by behavior methods ... All of the work upon the senses can be consistently carried along the lines we have suggested for vision (Watson, 1914, p. 15). The result of our examination into the nature of both image and affection seems to indicate that after all the behaviorist can bring them into his general scheme of work without in any way weakening his position (p. 26).

Thus, behaviorism was clearly intended - at least by Watson - to include perception and thought.

Similarly, Gestalt psychology was intended to include behavior. In his outline of the tasks of psychology from the Gestalt point of view, Koffka (1935, p. 67) explicitly named the following two issues: first, how the structure of a person's environmental field and the forces that derive from this structure give rise to bodily movements; second, how forces that may arise between the environmental field and the ego produce behavior in all of its forms.

Neither of the two movements lived up to these claims. Indeed, their contributions to the investigations of relationships between perception and action fell far behind those from nineteenth-century psychology. Most of the earlier research problems were abandoned.

As we have seen, the classical questions had been how sensory information enters into action control, and how perception is influenced by efferent processes. Behaviorism had hardly anything to say about the latter problem, and it tended to dispose of the former problem by invoking some basic learning principle (e.g., stimuli elicit responses in a reflex-like manner; responses become associated with stimuli via reinforcement; "emitted" response are under the control of discriminative stimuli). The aim was to predict *which* stimuli lead to *which* responses, not so much to analyze *how* perception and action are functionally interrelated.

An analogous deficiency can be seen in Gestalt theorizing. There were several among the Gestalt psychologists who discussed problems of action and action control at some length, especially Koffka (1935) and Lewin (e.g. 1938, 1954). However, a closer look at these analyses reveals that they did not really refer to action, but rather to a kind of action-encapsulated-in-perception.

Take, for instance, the discussion in Koffka (1935). Koffka starts with the question of how movements arise and how behavior is produced. In his view, actions are changes within the total psychophysical field, and their occurrence can be accounted for in terms of forces acting towards a better organization of that field. This is the case both for the environmental field and for the total psychophysical field (which includes not only the environment, but also the ego). As an example Koffka mentions eye movements, where the eye moves to fixate the figure rather than some location on the ground, because this leads to a stable organisation of the environmental field (Koffka, 1935, p. 311). Similarly, organisational processes within the total psychophysical field are involved when we sidestep a hurtling stone that is approaching us (Koffka 1935, p. 372). In both cases a certain amount of stress arises somewhere in the psychophysical field, and this stress is relieved by movement.

Thus, Koffka suggests that actions have their functional basis in the same types of organizational processes that, according to the Gestalt psychologists, are also at work in perception and thinking. This is a unified perception/action theory, but precisely because of this unity it is not really a theory of action at all. What this approach can account for at best is action-as-perceived or action-as-intended, i.e., which action comes to mind under given circumstances. Only this is an event at the level of the Gestalt psychologists' "psychophysical field". What this kind of approach cannot account for is action-as-executed, i.e., how the intended action's physical realization is put into effect and controlled. Indeed, there are indications that Koffka (1935) may have been aware of this, although he seems not to have regarded it as a limitation. The book contains more than 100 pages about action-as-perceived (two whole chapters), but only the following two sentences about action-as-executed:

> ... all we have to assume is that such a stress [i.e., between the ego and the object] may be in communication with the motor centres in the brain or a lower part of the central nervous system. In this case the motor system would be thrown into action, and the form which the ensuing behavior takes is thereby determined (p. 373).

The parallel to behaviorist nonexplanations of perception-action relationships is fairly obvious: There is a "communication," by means of which the motor system is "thrown into action," and this is "all we have to assume." This is not far from the behaviorist assertion that stimuli simply "evoke" or "elicit" the responses that are attached to them.

Lewin (1938, 1954) worked out in much greater detail what Koffka (1935) had presented in a nutshell. Without going into the intricacies of Lewin's system, it seems fair to say that he was no more interested in action-as-executed than Koffka had been. Lewin developed a complex and detailed model of the structures and operations that determine action, a model that has influenced the psychology of motivation until the present (e.g., Heckhausen, 1987). Similar to Koffka (1935), the basic idea is that of a field (called "psychological field" by Lewin) in which an interplay of forces takes place. The field determines action, and action changes the field's internal organization. However, the field determines action only in the sense of deciding *what* the person does. The model has nothing to say about *how* the action is executed or how perception is involved in this process. The problem of action control is reduced to (or rather substituted by) an intraphenomenal (or at least intrarepresentational) problem.

In brief, just as perception became lost in action in behaviorism, action became lost in perception in Gestalt psychology. During the dominance of these two movements, empirical research was split up correspondingly. For several decades - roughly between 1920 and the 1950s - research on cognitive processes and research on the control of behavior usually took place in different laboratories (often even on different continents) under different, if not antagonistic theoretical auspices.

Information Processing and Cognitive Psychology

The early 1960s saw a theoretical reunification of experimental psychology. The Gestalt movement had not survived its founding generation, and behaviorism gradually receded to its due field - the study of animal behavior. The vacuum was filled by a new theoretical movement, the information processing approach. It was the result of a number of interacting influences. Among them were the human factors research that had evolved during the Second World War (e.g., Sanders, 1971), Shannon and Weaver's (1949) mathematical theory of communication (e.g., Broadbent, 1958; Miller, 1956), cybernetics (e.g., Miller, Galanter, & Pribram, 1960), and, perhaps most important, the developing computer technology (e.g., Moray, 1967). As Sanders mentioned in the foreword to one of the first comprehensive documentations of the movement (Sanders 1967), there was also the rediscovery of many problems that were first investigated in the times of Wundt and James. All this merged into a remarkably productive amalgam, which was to become the dominant theoretical approach in experimental psychology between the mid-1960s (Neisser, 1967) and the mid-1970s (e.g., Massaro, 1975).

Information processing psychology has not been a "school" in the style of the earlier theoretical schools. As has been pointed out repeatedly (e.g., Carello, Turvey, Kugler, & Shaw, 1984; Gentner & Grudin, 1985; Massaro, 1986; Neumann, 1985; Scheerer, 1988), it is based less on a theory than on a metaphor, or rather on a whole set of metaphors that have continuously changed as technology has progressed - from the stores and switches of early, hardware-oriented models (e.g., Broadbent, 1958) to the encoding, addressing, comparison etc. operations of the 1970s (e.g., Posner, 1973), to the AI analogies that emerged in the late 1970s and have dominated the 1980s (e.g., Simon, 1979; Pylyshyn, 1984). Common to all these variants has been the conviction that there is a close resemblance between information processing in humans and in technical devices, notably computers, a resemblance at least close enough to justify using similar basic concepts to describe and analyze them.[4]

In the early days of the information processing approach there was a strong interest in perception-action relationships. The very name of one of its root sciences, cybernetics, refers to control. The subtitle of Wiener's (1948) book on cybernetics, i.e., *Communication and Control in the Animal and the Machine*, explicitly included this aspect. Further, much of the early experimental work that initially stemmed from applied research was concerned with human movement control. Examples are the tracking task (overview, e.g., in Poulton, 1984) and research into the psychological refractory period (overview, e.g., in Welford, 1967). One of the most influential publications series in the field was entitled *Attention*

[4] Norman (1986) has objected to the term "computer metaphor" with the argument that the early theoretical work on von Neumann machines in the 1940s and 1950s was itself influenced by contemporary ideas about the mode of operation of the mind and the brain. True enough. The point is not that there has been a strictly one-way influence (though psychology has probably taken over more from computer science than vice versa), but that the two fields were viewed as being similar enough to warrant a common theoretical framework. See, e.g., Pylyshyn (1978, 1980, 1984) for a clear exposition of this conviction.

and Performance, and indeed its first volumes (e.g., Sanders, 1967; Koster, 1969) carried sections on topics such as reaction processes and eye movements, and generally conveyed a strong interest in the "performance" aspect.

However, these fields did not shape the information processing approach as it unfolded and turned into cognitive psychology. Cybernetics is an example. With one notable exception (Miller et al., 1960), it had little influence on theoretical psychology. Attempts to employ cybernetic concepts in the study of empirical phenomena, e.g., sensorimotor adaptation (e.g., Hein & Held, 1962), usually took place in a more biological than psychological research context. Biocybernetics became a flourishing field (for its contribution to perception-action research see Cruse et al., this volume), but a corresponding psychocybernetics did not emerge. Similarly, psychomotor research took place largely outside the mainstream of information processing psychology (see, e.g., Semjen, 1978).

Thus, history repeated itself: Just as behaviorism and Gestalt psychology had set out as approaches to both perception and action, but in fact restricted themselves to one of these fields, the information processing approach became essentially a psychology of perceptual and cognitive processes.

The reasons for this development are much less obvious than in the earlier cases. Gestalt psychology had subscribed to the phenomenological method, which made perception and thought its natural topics. For similar methodological reasons, the study of overt action was the natural field of behaviorism. No such bias existed for the information processing approach.

So why this self-imposed restriction? First the metaphors that were provided by communication technology may have played a role. Transmission channels do not move or act, and the early computers were too slow for the on-line control of complex physical processes. Thus, both the channel metaphor and the computer hardware metaphor initially carried with them a bias against studying efferent processes. Similarly, when AI research began in the 1960s and 1970s, it was concerned mainly with the analysis and interpretation of input information.[5]

Second, the cognitive orientation of the information processing approach was probably due in part to the existing fields of research that it absorbed. Research in traditional cognitive areas (perception, thinking, attention, etc.) had not vanished under the reign of behaviorism (for attention, this has been documented by Lovie, 1983). But, with the exception of relatively short-lived movements such as the "New Look" school (e.g., Bruner, 1957) and transactionalism (e.g., Ittelson, 1962), it lacked a common theoretical framework. Such a general framework was provided by the information processing approach. It did not primarily create new research topics; rather it boosted research on already existing (sometimes temporarily forgotten) subjects by offering a new theoretical perspective on them.[6] These subjects happened to be mainly from perceptual and cognitive areas.

---

[5] Meanwhile, process control has, of course, become an important field of computer science, and AI has moved toward robotics. Updated versions of the computer metaphor should reflect this progress.

[6] For example, an early, influential reader on information processing approaches to visual perception (Haber, 1969) contained sections on topics such as visual masking, simultaneity, reaction time, microgenesis, and attention - all fields with a long research history.

Third, there has been a change of focus within the information processing approach that has moved it even farther away from the incorporation of output aspects. Originally, the emphasis was on sensory and perceptual processes. In the mid-1970s, interest began to shift to higher mental structures and processes, such as those underlying representation, understanding, and thought.[7] In part, this was due to the influence (and challenge) from AI. In part, it was the result of the mutual approach between psychology and linguistics (e.g., Chomsky, 1965; Miller & Johnson-Laird, 1976). Perhaps most important, it seems to have been a reaction to what was felt to be the sterility of previous research with its emphasis on elementary operations (e.g., Neisser, 1976).

This growing interest in higher cognitive processes has turned much of the research away from the consideration of perception-action relationships. Among the earlier, more elementary information flow models were the stage models, intended to trace the complete sequence of processing steps between stimulus and response, including S-R translation as well as response programming and response execution (for a review see Sanders, 1980). This is where the information processing approach came closest to an analysis of perception-action relationships (see Massaro, this volume, for a comprehensive review of this research as well as an apt defense of the information processing approach). When interest began to focus on higher mental processes, much of this perspective on the ensemble of information processing operations was lost. Of course, research on higher cognitive processes has continued to make extensive use of response latency measures - but it is not really the latency of *responses* that has been of interest within this research tradition. As a rule, motor output has been used simply as a convenient means for accessing central processes.

This was the situation until mid-1980s. More recently, there has been the upsurge of connectionism (e.g., McClelland & Rumelhart, 1987; Rumelhart & McClelland, 1987; Schneider, 1987; Smolensky, 1988). There are signs that we are in the midst of a theoretical revolution comparable to the Gestalt and behaviorist revolutions earlier in the century. What will be its impact on perception-action research?

As we have argued elsewhere (Prinz & Neumann, this volume), one aspect of the present theoretical development seems to be a rapprochement between psychology and the biological sciences. To the degree that connectionism enhances this trend, it is likely to at least indirectly promote perception-action research. Further, the "subsymbolic" (Smolensky, 1988) style of connectionist theorizing makes it equally applicable to the sensory/perceptual and the action/motor areas (Indeed, one of the earliest connectionist models by Dell & Reich, 1980 was a model of speech production.)

Thus, connectionism certainly has the potential to foster the study of relationships between perception and action. On the other hand, we have seen that other

[7] Compare, for example, the monographs on cognitive psychology by Neisser (1967) and Anderson (1985), each of them intended to cover the whole field of cognition. In Neisser (1967), the proportion of text devoted to higher mental processes was about 10% (Chap. 11). Two decades later, in Anderson (1985), this proportion had increased to almost 80% (Chaps. 4-14).

approaches had this potential, too, but did not really use it. In its present state of development, the connectionist approach consists of a plausible set of ideas about how computation in the brain works, plus a powerful set of formalisms to model such processes. This framework can be filled with content in many different ways. The issues whose history has been sketched in this chapter are among the possibilities.

Conclusion

It has often been asked to which degree psychology is a "cumulative" science which adds new findings to earlier insights rather than replacing them. Our look back into the history of perception-action research has revealed little evidence of cumulativeness in this sense. Scientific psychology started as a science of perception and action. None of the approaches and schools that replaced classical experimental psychology has explicitly abandoned this notion. Yet, none of them has systematically built upon the earlier perception-action research either. For various reasons, a schism between perceptual (sensory, cognitive) and motor (behavioral, action) research has been the rule, up to and including the cognitive approach that has dominated the past few decades.

Outside these dominating approaches, there have of course been theoretical movements that have stressed perception-action relationships, in particular Gibson's ecological realism (e.g., Gibson, 1979; Michaels & Carello, 1981) and Piaget's genetic epistemology (see Hauert & Zanone, this volume; Vinter, this volume). We have not included these approaches in this brief historical review, because their contributions are well known. Moreover, their influence can be seen in various other chapters of this volume.

Acknowledgements. This chapter was drafted while the first author was a Fellow at the Netherlands Institute for Advanced Study in the Humanities and Social Sciences (NIAS) at Wassenaar, The Netherlands. Thanks are due to the NIAS for its splendid hospitality. The second author was supported by an Akademie-Stipendium from the Volkswagenstiftung.

References

Anderson, J.R. (1985). *Cognitive psychology and its implications.* New York: Freeman.

Boring, E.G. (1942). *Sensation and perception in the history of experimental psychology.* New York: Appleton-Century-Crofts.

Bouwhuis, D.G., Bridgeman, B., Owens, D.A., Shebilske, W.L., &Wolff, P. (Eds.). (1986). *Sensorimotor interactions in space perception and action [Special issue Acta Psychologica] (Amsterdam), 63.*

Broadbent, D.E. (1958). *Perception and communication.* Oxford: Pergamon.

Bruner, J.S. (1957). On perceptual readiness. *Psychological Review, 64,* 123-152.

Carello, C., Turvey, M.T., Kugler, P.N., & Shaw, R.E. (1984). Inadequacies of the computer metaphor. In M.S. Gazzaniga (Ed.), *Handbook of cognitive neuroscience* (pp. 229-248). New York: Plenum.

Carpenter, W.B. (1874). *Principles of mental physiology, with their applications to the training and discipline of the mind and the study of its morbid conditions.* New York: Appleton.

Chomsky, N. (1965). *Aspects of the theory of syntax.* Cambridge, MA: MIT Press.

Dell, G.S., & Reich, P.A. (1980). Toward a unified model of slips of the tongue. In V.A. Fromkin (Ed.), *Errors in linguistic performance* (pp. 273-286). New York: Academic.

Gentner, D., & Grudin, J. (1985). The evolution of mental metaphors in psychology: A 90-year retrospective. *American Psychologist, 40,* 181-192.

Gibson, J.J. (1979). *The ecological approach to visual perception.* Boston: Houghton Mifflin.

Grüsser, O.-J. (1986). Interaction of efferent and afferent signals in visual perception. A history of ideas and experimental paradigms. *Acta Psychologica (Amsterdam), 63,* 3-21.

Haber, R.N. (1969). *Information processing approaches to visual perception.* New York: Holt, Rinehart & Winston.

Heckhausen, H. (1987). Vorsatz, Wille und Bedürfnis: Lewins frühes Vermächtnis und ein zugeschütteter Rubikon (Intention, will and need: Lewin's early legacy and a filled-up Rubikon). In H. Heckhausen, P.M. Gollwitzer, & F.E. Weinert (Eds.), *Jenseits des Rubikon: Der Wille in den Humanwissenschaften* (pp. 86-96). Berlin, Heidelberg, New York, Tokyo: Springer.

Hein, A., & Held, R. (1962). A neural model for labile sensorimotor coordinations. In E.E. Bernard & M.R. Kare (Eds.), *Biological prototypes and synthetic systems: Vol. 1.* New York: Plenum.

Hering, E. (1879). Der Raumsinn und die Bewegungen des Auges (The spatial sense and the movements of the eye). In L. Herrmann (Ed.), *Handbuch der Physiologie (Vol. 3,* pp. 343-601). Leipzig: Vogel.

Ittelson, W.A. (1962). Perception and transactional psychology. In S. Koch (Ed.), *Psychology: A study of a science: Vol. 4.* New York: McGraw-Hill.

James, W. (1890). *The principles of psychology.* New York: Holt.

Koffka, K. (1935). *Principles of Gestalt psychology.* London: Routledge & Kegan Paul.

Köhler, W. (1940). *Dynamics in psychology.* New York: Liveright.

Koster, W.G. (Ed.). (1969). *Attention and performance: Vol. 2.* Amsterdam: North-Holland.

Lange, L. (1888). Neue Experimente über den Vorgang der einfachen Reaktion auf Sinneseindrücke (New experiments on the process of the simple reaction to sensory impressions). *Philosophische Studien 4,* 479-510.

Lewin, K. (1938). *The conceptual representation and the measurement of psychological forces.* Durham, NC: Duke University Press.

Lewin, K. (1954). Behavior and development as a function of the total situation. In L. Carmichael (Ed.), *Manual of child psychology (2nd ed).* New York: Wiley.

Lotze, H. (1852). *Medicinische Psychologie oder Physiologie der Seele* (Medical psychology, or physiology of the mind). Leipzig: Weidmann.

Lovie, A.D. (1983). Attention and behaviorism - fact and fiction. *British Journal of Psychology, 74,* 301-310.

Massaro, D.W. (1975). *Experimental psychology and information processing.* Chicago: Rand-McNally.

Massaro, D.W. (1986). The computer as a metaphor for psychological inquiry: Considerations and recommendations. *Behavior Research Methods, Instruments, and Computers, 18,* 73-92.

McClelland, J.L., & Rumelhart, D.E. (Eds.). (1987). *Parallel distributed processing: Explorations in the structure of cognition: Vol. 2. Psychological and biological models.* Cambridge, MA: MIT Press.

Michaels, C.F., & Carello, C. (1981). *Direct perception.* Englewood Cliffs: Prentice-Hall.

Miller, G.A. (1956). The magical number seven, plus or minus two: Some limits on our capacity for processing information. *Psychological Review, 63,* 81-97.

Miller, G.A., & Johnson-Laird, P.N. (1976). *Language and perception.* Cambridge University Press.

Miller, G.A., Galanter, E., & Pribram, K.H. (1960). *Plans and the structure behavior.* New York: Holt, Rinehart & Winston.

Moray, N. (1967). Where is capacity limited? A survey and a model. *Acta Psychologica (Amsterdam), 27,* 84-92.

Neisser, U. (1967). *Cognitive psychology.* New York: Appleton-Century-Crofts.

Neisser, U. (1976). *Cognition and reality.* San Francisco: Freeman.

Neumann, O. (1972). Empfindungen (Sensation). In J. Ritter (Ed.), *Historisches Wörterbuch der Philosophie (Vol. 2,* pp. 464-474). Basel: Schwabe.

Neumann, O. (1980). Lokalzeichen (Local sign). In J. Ritter & K. Gründer (Eds.), *Historisches Wörterbuch der Philosophie (Vol. 5,* pp. 527-530). Basel: Schwabe.

Neumann, O. (1985). Informationsverarbeitung, künstliche Intelligenz und die Perspektiven der Kognitionspsychologie (Information processing, artificial intelligence, and the perspectives of cognitive psychology). In O. Neumann (Ed.), *Perspektiven der Kognitionspsychologie* (pp. 3-37), Berlin, Heidelberg, New York, Tokyo: Springer.

Norman, D.A. (1986). Reflections on cognition and parallel distributed processing. In J.L. McClelland & D.E. Rumelhart (Eds.) *Parallel distributed processing: Explorations in the structure of cognition: Vol. 2. Psychological and biological models* (pp. 531-546). Cambridge, MA: MIT Press.

Posner, M.I. (1973). *Cognition. An introduction*. Glenview, Il: Scott, Foresman.
Poulton, E.C. (1974). *Tracking skill and manual control*. New York: Academic.
Prinz, W. (1987). Ideomotor action. In H. Heuer & A.F. Sanders (Eds.), *Perspectives on perception and action* (pp. 47-76). Hillsdale, NJ: Erlbaum.
Pylyshyn, Z.W. (1978). Computational models and empirical constraints. *Behavioral and Brain Sciences, 1*, 93-127.
Pylyshyn, Z.W. (1980). Computation and cognition: Issues in the foundations of cognitive science. *Behavioral and Brain Sciences 3*, 111-169.
Pylyshyn, Z.W. (1984). *Computation and cognition: Toward a foundation for cognitive science*. Cambridge, MA: MIT Press.
Rumelhart, D.E., & McClelland, J.L. (Eds.). (1987). *Parallel distributed processing: Explorations in the structure of cognition: Vol. 1. Foundations*. Cambridge, MA: MIT Press.
Sanders, A.F. (Ed.). (1967). *Attention and performance*. Amsterdam: North-Holland.
Sanders, A.F. (1971). *Psychologie der Informationsverarbeitung*. (Information processing psychology). Bern: Huber.
Sanders, A.F. (1980). Stage analysis of reaction time. In G.E. Stelmach & J. Requin (Eds.), *Tutorials in motor behavior* (pp. 331-354). Amsterdam: North-Holland.
Scheerer, E. (1984). Motor theories of cognitive structure: A historical review. In W. Prinz & A.F. Sanders (Eds.), *Cognition and motor processes* (pp. 77-98). Berlin, Heidelberg, New York, Tokyo: Springer.
Scheerer, E. (1987). Muscle sense and innervation feelings: a chapter in the history of perception and action. In H. Heuer & A.F. Sanders (Eds.), *Perspectives on perception and action* (pp. 171-194). Hillsdale, NJ: Erlbaum.
Scheerer, E. (1988). Towards a history of cognitive science. *International Science Journal, 115*, 7-19.
Schneider, W. (1987). Connectionism: Is it a paradigm shift for psychology? *Behavioral Research Methods, Instruments, and Computers, 19*, 73-83.
Semjen, A. (1978). From motor learning to sensorimotor skill acquisition. *Journal of Human Movement Studies, 3*, 182-191.
Shannon, C., & Weaver, W. (1949). *The mathematical theory of communication*. Urbana: The University of Illinois Press.
Simon, H.A. (1979). Information processing models of cognition. *Annual Review of Psychology, 30*, 363-396.
Smolensky, P. (1988). On the proper treatment of connectionism. *Behavioral and Brain Sciences, 11*, 1-74.
Titchener, E.B. (1898). Postulates of a structural psychology. *Philosophical Review, 7*, 449-465.
Titchener, E.B. (1899). Structural and functional psychology. *Philosophical Review, 8*, 366-376.
von Helmholtz, H. (1866). *Handbuch der physiologischen Optik*. (Handbook of physiological optics). Leipzig: Voss.
von Holst, E. & Mittelstaedt, H. (1950). Das Reafferenzprinzip (Wechselwirkungen zwischen Zentralnervensystem und Peripherie). [The reafference principle. (Interactions between central nervous system and periphery)]. *Naturwissenschaften, 37*, 464-476.
Watson, J.B. (1914). *Behavior. An introduction to comparative psychology*. New York: Holt.
Watson, J.B. (1917). An attempted formulation of the scope of behavior psychology. *Psychological Review 24*: 329-352.
Welford, A.T. (1967). Single-channel operation in the brain. *Acta Psychologica (Amsterdam), 27*, 5-22.
Wiener, N. (1948). *Cybernetics. Communiation and control in the animal and the machine*. New York: Wiley.
Wolff, P. (1985). Wahrnehmungslernen durch Blickbewegungen. (Perceptual learning through eye movements). In O. Neumann (Ed.), *Perspektiven der Kognitionspsychologie* (pp. 63-111). Berlin, Heidelberg, New York, Tokyo: Springer.
Wolff, P. (1986). Saccadic exploration and perceptual-motor learning. *Acta Psychologica (Amsterdam), 63*, 263-280.
Wolff, P. (1987). Perceptual learning by saccades: A cognitive approach. In H. Heuer & A.F. Sanders (Eds.), *Perspectives on perception and action* (pp. 249-271). Hillsdale, NJ: Erlbaum.
Wundt, W. (1882). Über psychologische Methoden. (On psychological methods) *Philosophische Studien, 1*, 1-38.
Wundt, W. (1903). *Grundzüge der physiologischen Psychologie, 5th ed. (Fundamentals of physiological psychology)*. Leipzig: Engelmann.
Wundt, W. (1914). *Grundriss der Psychologie* (An outline of psychology). Leipzig: Kröner.
Ziehen, T. (1920). *Leitfaden der physiologischen Psychologie* (A manual of physiological psychology). Jena: G. Fischer.

The Physiological Basis of the Act of Perceiving

B. BRIDGEMAN

CONTENTS

Introduction

The close interaction of perception and action, an organizing theme of this book, can be studied with a variety of methods. This chapter examines the physiological organization of the interaction, drawing material principally from the oculomotor system where the interplay of perception and action is particularly clear and easy to study. Analyzing the oculomotor system requires a reinterpretation of sensory physiology: the passive sensory systems must be replaced by active perceptual systems, so that the organism can combine sensory and motor sources to interpret the visual world.

The Concept of Perceptual Systems

Aristotle's classical list of five senses - sight, hearing, touch, taste and smell - is incomplete; vestibular and kinesthetic senses do not appear on the list. And there are dozens of types of receptor cells. Further, sensory experience fails to mirror the anatomy of the sensory systems. The complex and subtle sense to taste, for instance, is mediated largely by olfaction, for if the olfactory cavities are blocked, foods taste flat and discriminations are difficult. Much of what is experienced as taste is mediated by the olfactory anatomy. The pleasures and distinctions of fine wines, for instance, are almost entirely a matter of odor.

Relationships Between Perception and Action
Edited by O. Neumann and W. Prinz

Further, some seemingly unimodal experiences are compounded of neural signals that originate from many types of receptors. The experience of wetness, for example, is a combination of temperature and pressure that is difficult to analyze in experience. The fact that introspection cannot distinguish among the many types of receptors shows the tenuous link between receptors and experience. The Müllerian doctrine of specific nerve energies applies only to the coding of signals entering the CNS, not to the experience that the stimulation of a receptor evokes in combination with other stimulations. The nature of the receptors must be examined with physiology and anatomy, not only with psychology.

Physiologically it would seem that Aristotle's list of five senses is hopelessly obsolete; yet the idea persists because, though there are many types of receptors, there are only five modes of sensory attention. Gibson (1966) clarified this relationship by distinguishing between receptors and perceptual systems, but he had an idiosyncratic way of analyzing the modes. Here, Gibson's idea will be modified to relate active perceptual systems directly to the Aristotelian senses. They are best described as the five *activities* of looking, listening, touching, tasting, and smelling. Perception is inherently active; passive anatomical or physiological analyses are inherently incomplete and inadequate.

Using these systems, one is generally aware of external events rather than physiological processes. Perception consists of information in the formal sense (Shannon & Weaver, 1949) about the outside world, not about the status of the receptors. When perceiving a flower I experience the smell and look and feel of the flower, not of olfactory, visual, and tactile sensations. According to Gibson (1966), sensation is the experience of energy falling on receptors, while perception is the pickup of information from the environment. Sensation and perception are separate phenomena, and perceptions need not be based on conscious, analyzable sensations. In short, the senses evolved to provide information about the world, not about their own states.

All of the perceptual systems can direct action for the pickup of information. Touching uses the whole body to investigate objects, employing muscle systems that are also used in locomotion and in other activities. Chemoreception also uses multipurpose muscle systems, making it difficult to isolate the contributions of the muscles to perception. Listening can use the uniquely sensory middle ear muscles, but these are poorly understood and, in any case, play only a minor role in the hearing process. Seeing, however, requires a specialized system of extraocular muscles whose only role is in the service of information gathering. Therefore, the oculomotor system is the best source of information about the physiology of the interaction of perception and action.

The Oculomotor System

The arrangement of the extraocular muscles in the shark represents the basic vertebrate pattern. The six muscles are arranged in three opposing pairs: two rotate the eye vertically (superior and inferior rectus), two turn it horizontally (lateral and medial rectus), and two cyclorotate it with the line of sight as an axis (superior and inferior oblique). These muscles serve to compensate for the shark's locomotion; as the shark moves, its eyes track the visual world to stabilize it on the retina. When the eye nears the end of its normal range, it snaps back in a saccade and begins tracking again. If the shark turns to the left or right (yaw), the two eyes move conjunctively to continue stabilizing the retinas. These are the roles of the horizontal recti. The vertical recti have analogues functions for the shark's vertical locomotion and roll, while the obliques compensate for changes in pitch. Together these reflexes provide optokinetic compensations for the six dimensions of movement (three rotational and three translational).

The oculomotor system uses vestibular as well as visual signals to minimize image movement. The advantage of vestibulo-ocular reflexes (VOR) is their short latency. Compensation can continue in darkness, so that the eyes are stabilized as soon as light is available after an occlusion. In the vertical direction the compensation is called the "doll reflex" because in humans the eyes, like those of a doll, roll downward when the head tilts back.

The system also predicts image motions, by monitoring motor commands sent to the neck muscles. When the head turns, an efference copy (a copy of the innervation to the muscles) drives extraocular compensating movements with no latency. The system cannot be very accurate because like the VOR it is a feed-forward. It predicts retinal image motions rather than measuring them. The predictions are never completely accurate, but together with the optokinetic system they reduce errors. Visual feedback error correction can then proceed with smaller error signals and corresponding greater stability.

Humans have the same six muscles as the shark, but their arrangement and functions are more subtle and flexible. During evolution the orbits have rotated to the front of the head to enable the processing of distance information with binocular vision. The orbits never rotated completely into parallel positions, however; the axes of the orbits still diverge slightly so that the vertical recti turn the eye somewhat inward as well as vertically (Fig. 1). To move the eye vertically the vertical muscles have been linked neurologically with the obliques, so that an oblique and a rectus muscle are innervated together (Fig. 2). This arrangement increases the rotational range of the eye while maintaining a linear relationship between muscle force and eye rotation.

The three dimensions of space are built into the human oculomotor control system. Consider the double-eye *(Doppelauge)* as a single organ, handled by a single neurological control system (Hering, 1868). The double-eye controls the point of sight, the intersection of the lines of sight from the two eyes. In addition to the horizontal and the vertical systems, the point of sight is moved toward the head by the two medial recti, while the two lateral recti move the point of sight further away. These vergence movements have an independent control system, with slower dynamics than the saccadic system that controls vertical and horizontal conjugate movements.

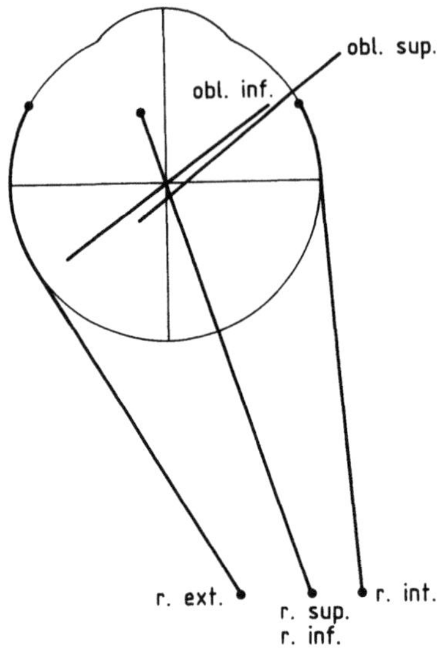

Fig. 1. Anatomical locations of the six human eye muscles; *r. ext.*, lateral rectus; *r. sup.*, superior rectus; *r. inf.*, inferior rectus; *r. int.*, medial rectus; *obl. inf.*, inferior oblique; *obl. sup.*, superior oblique. All muscles are aligned to exert forces slightly lateral to a fully straight-ahead orientation. (From Hering, 1868)

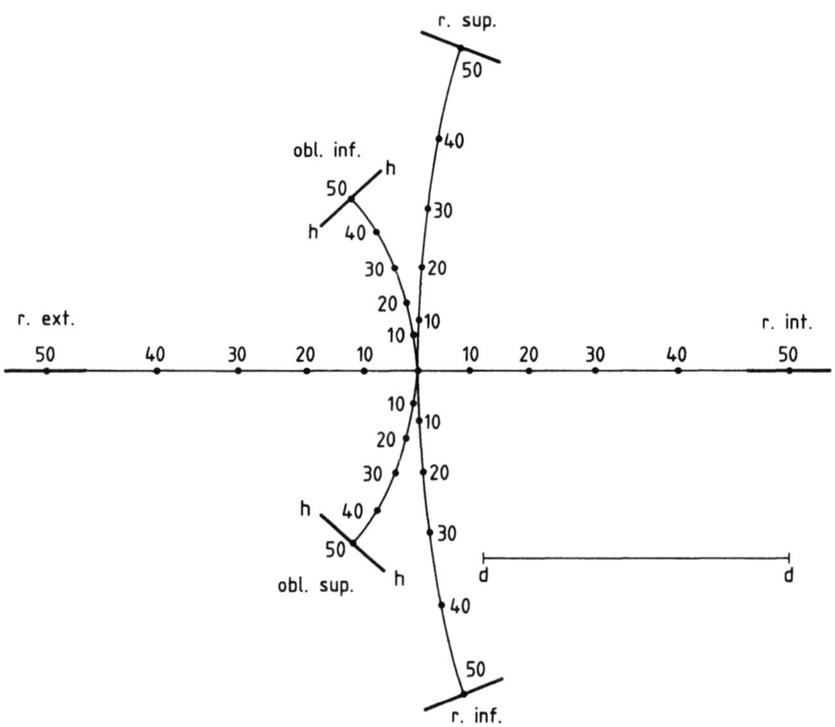

Fig. 2. Directions of action of the six extraocular muscles. Abbreviations as in Fig. 1. Each *light line* represents the position of the eye with a given tension applied to the corresponding muscle. The *heavy line* at the end of each muscle's representation shows the orientation of the horizontal meridian of the eye at the corresponding eye position if only the illustrated muscle is active. To move the eye in a straight vertical line, the superior rectus and inferior oblique muscles must be innervated together in the ratios suggested in the figure. (From Hering, 1868)

Thus each orthogonal direction of visual space is represented by a pair of extraocular muscles operating in various combinations. Rectangular coordinates were not invented by mathematicians; they have always been present in the oculomotor control system.

The Oculomotor System in Perception

The oculomotor system provides a stable platform to facilitate visual transduction. The shark accomplishes this by alternating saccades with slow movements compensating for locomotion. Humans do the same, with two complications introduced by the presence of a fovea. First, pursuit movements track an object of interest, regardless of the motion of the rest of the image; second, exploratory saccades jerk the fovea from one part of the visual world to another.

Pursuit Eye Movements

In humans, a phylogenetically old optokinetic system similar to that in the shark stabilizes the foveal image with negative feedback. If the image begins to drift, the extraocular muscles move the eye to cancel the drift and to restore the image to its original location on the retina. This system is involuntary and is actively inhibited during saccades (Robinson, 1981).

There is a conflict between pursuit and optokinetic systems when the eye tracks an object that is moving relative to the background. In this case the pursuit system takes over, suppressing the optokinetic system and tracking the foveal image despite the movement created on the peripheral retina.

The resulting situation seems paradoxical. Pursuit prevents the image from moving on the fovea, while the background drifts across the peripheral retina, yet we perceive just the opposite - the fixated object undergoes apparent motion while the background remains perceptually stable. When we track a moving train, for example, the background moves on the retina and the train does not, yet we perceive the train as moving and the tracks as fixed.

It is possible to describe this distinction between stable background and moving target in terms of the retinal stimulus conditions alone (Gibson, 1966), for the background is always large, projects mostly on the peripheral retina, and moves as a whole on the retina, while the target is small and projects mostly on the fovea. A figure/ground distinction cannot explain the perception of target motion, however, for there are counterexamples where the background is perceived as moving and the figure as standing still. Induced motion (Duncker, 1929) is a perception of figure motion induced by the opposite motion of a background; if a real motion is added to the figure such that it exactly cancels the induced motion, then the figure will be perceived as standing still and the background as moving (Bridgeman, Kirch, & Sperling, 1981). This case cannot be explained by assuming that the visual system perceives a large background as stable and a small target in relative motion as moving, for here the relative motion relations of the array and even the motions of the projected objects on the retina are the same as in the example of tracking a train, yet the perceived motions are the opposite. Only the dynamics of oculomotor innervation are different. Thus the innervation to the pursuit system must contribute to motion perception.

If this were the complete description of oculomotor control, pursuit would lock vision onto a single object forever. The saccadic system breaks the feedback loop while jumping the eye to a new position.

Saccadic Eye Movements

Saccades are time-optimal; that is, the eye accelerates at the maximum physically possible rate (Clark & Stark, 1975). The saccade is an oculomotor sprint to a new fixation point, with no concern for vision during the movement. In fact, saccadic suppression (discovered by Dodge, 1900) reduces visual sensitivity to flashes of light during and just before each saccade.

Saccadic suppression has been explained by an internally generated signal, corollary to the saccade, that inhibits visual sensitivity (Latour, 1962; Zuber & Stark, 1966; Volkman, Riggs, White, & Moore, 1978). Another interpretation, however, holds that masking alone can generate psychophysical results indistinguishable from saccadic suppression even in the absence of eye movements if the background is displaced saccadically near the time of a test flash (MacKay, 1970), and one laboratory found no saccadic suppression with dark or dim backgrounds (Mitrani, Mateeff, & Yakimoff, 1971).

In order to settle the questions brought up by this interpretation, Riggs, Merton, and Morton (1974) measured saccadic suppression quantitatively in experiments designed to eliminate artifacts which might have yielded spurious results in previous experiments. All light was eliminated from the experimental design, with subjects detecting electrically elicited phosphenes. Under these conditions, Riggs et al. obtained a saccadic suppression equivalent to about 0.45 log units of light intensity. Collewijn (1969) separated eye movement and stimulus movement in an animal preparation: he immobilized one eye of a rabbit and occluded the other, normal eye. Optokinetic nystagmus was then induced, and cortical evoked responses to a flash in the immobilized eye were found to be reduced near the time of a saccade in the occluded eye. Thus saccadic suppression was found even in an eye which did not move during a saccade.

The two interpretations, masking vs. internal suppressive signal, can be reconciled by distinguishing between visual environments in texture-rich (usually photopic) vs. texture-poor (usually scotopic) conditions. Masking might be adequate to suppress vision in texture-rich environments, while an internal signal would be required in a texture-poor environment (Bridgeman, 1977).

But the saccadic suppression of brightness paradigm lacks ecological validity: we would not expect the visual system to be concerned with brightness during saccades, for the visual image does not change the brightness from one fixation to the next; rather, it changes in position. It is more reasonable to hypothesize that perception of position of the visual world should be distorted during saccades.

To test this hypothesis, a pattern can be projected through a mirror that is flipped during a saccade, to jump the pattern across a screen. The dependent variable is the magnitude of pattern displacement. Thresholds for the smallest detectable displacement under these conditions are huge. The target must move about 10%-30% as far as the eye does, under optimal conditions, to be detectable (Brune & Lücking, 1969; Mack, 1970; Bridgeman, Hendry, & Stark, 1975). The perceptual system seems to assume that any small displacement of the entire visual world during a saccade is due to the eye movement. The displacement equivalent of MacKay's masking suppression of flash experiment was done by Bridgeman (1981); suppression of the perception of displacement was much larger during saccades than during fixation even when retinal motion conditions were made comparable.

During fixation, displacement thresholds are as small as 10 s of arc, so that the saccadic suppression of displacement effect can be more than 3 log units in magnitude. The saccadic suppression of flash is only about 0.7 log units, and even that is only seen near threshold, making one suspect that flash suppression is an epiphenomenon surrounding displacement suppression. The function of displacement suppression in space constancy is clear – errors in compensation for the new position of the retinal image are suppressed from perception. It is more difficult to hypothesize a function for flash suppression.

Saccadic suppression of image displacement implies that information about spatial location is degraded after a saccade from one point to another. Yet we do not become disoriented, because image displacements during saccades are not suppressed in a second, motor-oriented visual system. Empirically, this spatial or ambient system can be isolated from perception by asking a subject to point to a target open-loop (without being able to see his hand). When abrupt target displacement is masked by saccadic suppression of displacement, the subject still indicates the true position with open-loop pointing (Bridgeman, Lewis, Heit, & Nagle, 1979). Information about displacement is actively removed from perception, where it would only be a distraction, but the resulting change in position is coded in the motor-oriented system.

Neural Mechanisms of Eye Movement Generation

Differences between pursuit and saccadic movements are also seen in the neural machinery that drives them (Robinson, 1981).

Pursuit Mechanisms

Smooth pursuit begins about 100 ms after a target motion begins; if about 55 ms are lost in peripheral visual processing and another 25 ms in the motor system, only 20 ms remain to organize the pursuit movement. In response to this time pressure, the strategy of the pursuit system is to pursue first and ask questions later. The initial portion of a pursuit movement may be only approximately accurate and may move the eye away from the target if, during the eye's latent period, the target has jumped in the direction opposite its smooth motion.

The only cortical neurons known to be concerned exclusively with pursuit are in the parietal lobe (Lynch, Mountcastle, Talbot, & Yin, 1977). The cells fire when a monkey tracks a moving target, the discharge rate depending on the target's direction but not on its speed. This would be expected of neurons concerned more with initiation of pursuit than with its feedback control. Once the neurons initiate the movement, pursuit seems to be controlled mainly in the cerebellum. Not much is known about the anatomical connections between cortical and cerebellar pursuit mechanisms.

Cells directly related to pursuit are found in the flocculus. Some Purkinje cells there fire in proportion to the velocity of gaze in space; their discharge rates are the sum of eye velocity in the head and head velocity in space. Such cells seem necessary to control pursuit, for removing the cerebellum abolishes pursuit in monkeys (Westheimer & Blair, 1973). Gaze Purkinje cells project to brainstem oculomotor centers by unknown pathways, probably through the paramedian pontine reticular formation (PPRF).

Saccade Mechanisms

A saccadic eye movement is ballistic in the sense that once started, it cannot be stopped. At the start of a saccade, the agonist extraocular muscles are activated maximally. The burst duration depends on the size of the intended movement. At the same time the motor neurons innervating the antagonist muscles are completely inhibited. About halfway through the movement, the innervation drops back to the level required to hold the eye in the new position (Bahill & Stark, 1975). This is the pulse-step pattern of innervation. For large saccades, there is a braking pulse in the antagonist muscles near the end of the movement.

The first electrophysiological sign of the organization of a new saccade is in long-lead burst neurons in the PPRF. These cells begin firing 100-150 ms before a saccade, in what Robinson (1981) calls a "ragged, low-frequency preamble" followed by a burst before, during, or after the saccade. Long-lead neurons seem to encode saccade amplitude in retinocentric coordinates rather than the initial or final eye position.

Similar cells are found in the superior colliculus, where Sparks and Pollack (1977) captured the single-cell signs of a monkey making a decision and changing his mind. They trained monkeys to saccade when a small visual target jumped from point A to B. Long-lead burst neurons in the PPRF gave the expected ragged discharge followed by a sharp burst. Now the target was moved from A to B and then jumped again to C before the monkey could make the A-B saccade. Sometimes the monkey would cancel the A-B movement and look directly to C. In these cases, long-lead neurons active for the A-B direction would give a preamble but no burst, as though the plan to make the A-B saccade were being formulated and then discarded.

Medium-lead burst cells generate the pulse of the pulse-step control signal; they excite the motoneurons monosynaptically. Once the pulse is delivered, a tonic innervation must be supplied to the motoneurons to hold the eye in the new position. The innervation comes from tonic cells in the PPRF whose discharge rates are proportional to eye position.

Initiation of Eye Movements

Under normal visual conditions (people looking at things, rather than psychologists introspecting), the only conscious decision involved in oculomotor behavior is the initiation of saccades. Even this is usually accomplished automatically, without the subject thinking about the saccadic system. Accordingly, there are two systems for initiation of saccades, one in the midbrain (superior colliculus) and another in the frontal eye fields, Brodmann's area 8. It was once thought that area 8 neurons played little role in initiating saccades, for most of them fired only after the beginning of a saccade. These recordings were made when the animal was only casually looking about the laboratory. But when a trained monkey makes a deliberate saccade to obtain a reward, many neurons in the posterior part of area 8 fire just after a change in the visual stimulus but before the saccade (Goldberg & Bushnell, 1979).

All of this machinery is designed to acquire a target and to steady the image on the fovea for as much time as possible. What does the visual system do with that image, and how does it decide where to fixate next?

Scanpaths

When a subject encounters a new image, the eyes jump first to the area of greatest interest, where there is the greatest density of information. Then they jump to another spot determined by information content, distance, and the task. A series of saccades generates a scanpath, the pattern of fixational eye movements (Noton & Stark, 1971).

The scanpath is the meeting point of oculomotor physiology and cognitive psychology. Decisions about where to move the eye next are made with great rapidity (up to five per second) and usually without awareness of the decision process, yet they involve sophisticated information processing. Separation of figure and background is important in guiding the scanpath, as Fig. 3 shows; in this rather difficult image, subjects scan the image widely before they identify it. After the subject finds the face, the scanpath changes, narrowing to those areas that are generally the most informative within faces. In ambiguous figures, the scanpath depends on the interpretation of the moment (Stark & Ellis, 1981).

After the identity of a scene is established, higher-level information governs scanning patterns. Yarbus (1967) showed subjects a painting while asking them questions about the people in the painting; the resulting scanpaths depended on the nature of the question being asked. Subjects still scan the most information-rich

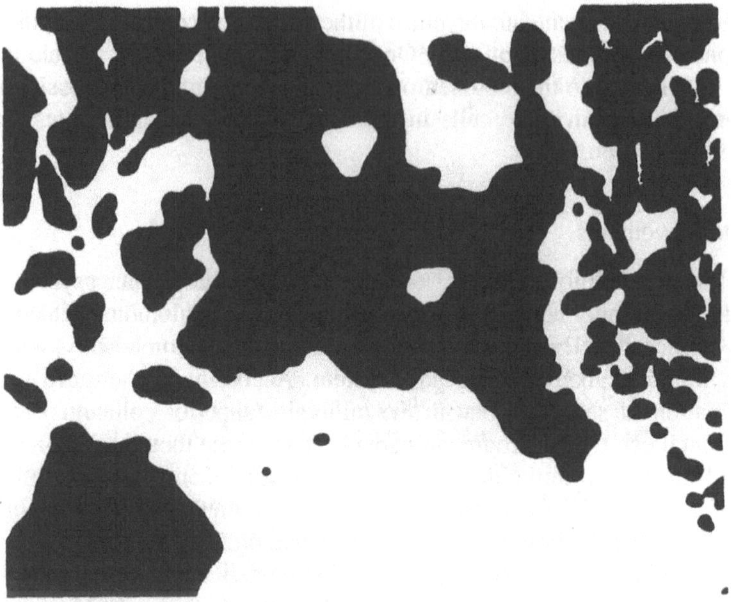

Fig. 3. A fragmented figure, made by high-contrast processing of a photograph. (From Stark & Ellis, 1981)

regions of the painting, but the definition of what provides the most information changes with the question being asked.

Between exploratory saccades, another family of eye movements, physiological nystagmus, appears during fixation.

Physiological Nystagmus

During fixation, the eye generates three kinds of movements: fast (50-80 Hz) oscillations of a few seconds of arc, having no known function; drifts at a few minutes of arc per second; and microsaccades, small jumps of 0.25° or less that occur one to two times per second. The microsaccades have the same dynamics as exploratory saccades, falling on an extension of the main sequence for saccades, but they are made without awareness and are smaller than the smallest possible voluntary saccades. The function of the three kinds of movement is best seen by suppressing them, either by complete paralysis of the body, including the eyes, with synthetic curare analogues (Stevens et al., 1976), or by optically or mechanically compensating for the eye movements (Yarbus, 1967). With either method the result is the same - if all retinal motion is prevented, the image disappears permanently in a few seconds (Gerrits, Stassen, & van Erning, 1984). There has been some methodologically based confusion on this issue, for even slight imperfections of stabilization will allow the image to reappear. The phenomenon is probably related to negative afterimages: with a stabilized retinal image, the afterimage increases in contrast until it exactly cancels the real image (Burbeck & Kelly, 1982).

Which of the components of physiological nystagmus is responsible for maintaining a visible image? The question was addressed with the discovery that microsaccades can be suppressed voluntarily (Steinman, Haddad, Skavenski, & Wyman, 1973), even though they are performed without awareness. In fact, the saccadic eye movement in isolation is by far the most common of all behavioral acts performed by humans. Suppression is quite easy; one need only relax, stare blankly into space, and look at nothing in particular. Though suppression can be maintained for many seconds, the world does not fade and the slow drifts keep fixation centered near the original fixation point. This result has been replicated by the author and Derek Hendry (unpublished observations).

The Function of Microsaccades. If microsaccades are not necessary to compensate for slow drifts and are not needed to keep foveal images from disappearing due to stabilization, they might be useful in tasks requiring high acuity. This hypothesis was tested with ingenious experiments involving needle-threading and rifle-shooting. Reasoning that a high-acuity task might elicit more microsaccades, Winterson and Collewijn (1976) had subjects thread a needle that was electrically insulated except for the inside of the eye. The "thread" was a wire insulated except for the tip. At the instant of success a circuit between thread and needle provided a synchronization pulse. In the rifle-shooting task, the trigger closed a switch to synchronize the recordings.

The results were clear, if disconcerting - instead of increasing in frequency, microsaccades were suppressed for a few seconds before the high-acuity event occurred. But perhaps the motor activity required in the tasks had affected the production of microsaccades, with a phylogenetically older motor-oriented system, incapable of generating microsaccades, temporarily taking over oculomotor control. According to this hypothesis, the reduction should not occur if subjects perform an observational task that does not involve an immediate motor response.

Bridgeman and Palca (1982) tested this hypothesis by asking subjects to observe a needle that moved toward a target and then suddenly disappeared. The task was to predict whether the needle would have hit the target if it had continued moving. The results were similar to those of Winterson and Collewijn (1976), with microsaccade frequency decreasing just before the judgement had to be made. Thus microsaccades are not necessary for observing events with high acuity. In fact, the rate of generating microsaccades decreases just when high acuity is needed most.

A recent hypothesis about the functions of drifts and microsaccades has arisen from studies where the retinal image is stabilized, and then image movements are reintroduced artificially. This is done with a variation of the Yarbus technique: a small optical system is fixed to the eye with a suction contact lens. A fiber-optical bundle allows image motion to be controlled. The stabilized image disappears as expected, but foveal images fade faster than peripheral ones. Artificially induced drifts in the image restore foveal sensitivity, while the periphery remains indistinct. The larger jumps of microsaccades are needed to restore vision in the periphery as well. Since these jumps need not be very frequent, drifts can maintain foveal vision between microsaccades (Gerrits et al., 1984). Thus it seems that both drifts and microsaccades are necessary to maintain complete retinal image visibility. Thus, even at the very lowest level, action is necessary to insure a continuing flow of information.

Active Perception in the Somatosensory System

Samples of sensory input are also integrated in other senses. The somatosensory system, for instance, receives a sequence of deformations of the skin during tactual exploration. Exploring a chair in the dark, one feels only small samples of the chair at any one time, yet one perceives a constant, whole object. Again the information from successive samples is stored in a form of perceptual memory to synthesize a perception of the sensory world. The presently available information and the memory information are integrated so seamlessly that we are ordinarily unaware that all of our perceptions are hybrids of perception and memory.

The physiology of this process has been investigated in awake monkeys by training them to explore a texture with their fingers. While the monkeys were exploring, neurons were recorded in areas 1 and 3b of postcentral somatosensory cortex. Neural discharges were influenced by both the finger movement and the characteristics of the spatial pattern. Because the respective field patterns of the neurons varied so widely, spatial features of the surface can be represented only by responses of a population of cells. The responses of individual cells are ambiguous (Darian-Smith, Sugitani, & Heywood, 1982). As is the case in vision, these cortical neurons do not function as detectors of particular patterns.

In all perceptual systems, much of what we think of as perceptual information comes from top-down hypotheses. The combining of a top-down context with bottom-up sensory information makes a good definition of perception. In this interpretation, perception is the combining of sensory signals with meaning. The act of naming an object,. for instance, requires interaction between sensory characteristics of the object and the definitions in a lexicon (reviewed by Anderson, 1985; see MacKay, this volume). Active interaction with the environment plays another part in perception, in the development of the perceptual systems themselves.

Tuning the Brain to Its Environment

In the early 1950s, it seemed that it would soon be possible to follow the process of development step by step, practically molecule by molecule, until ontogenesis was understood. When the genetic molecular code was finally worked out, and the amount of information in the genes was calculated, there was an unpleasant surprise. The information in the genes was inadequate to specify even the construction of a brain, let alone an entire organism. The problem was worse in higher organisms, for phenotypic complexity increased faster than the genes' information capacity.

Eventually the puzzle was resolved in several ways. First, it was unnecessary for a separate gene to code each neuron or each synapse; using more economical coding schemes, nature could specify a large network by laying down only the rules for network growth. Painting in broad strokes, the genes could control the general structure without determining all the details.

Embryonic Development

Several economical principles of brain growth have emerged from neuroembryology. One is that embryonic neurons grow along chemical or electrical gradients when their processes are developing. An axon develops a growth cone, a set of fingerlike processes on its growing tip that seem to sense the local environment and guide the elongation of the axon behind it. The growth cone undulates as it pushes toward its destination. A target tissue, such as the cortex, may send out chemical signals to guide the axons. If the target tissue is removed, the axons get lost.

The first neurons to innervate a developing nerve tract, the pioneer cells, need genetic information telling them where to go. Later axons growing along the tract need only be informed by the genes that they should follow the paths of the other axons.

The implications of this solution were profound. If synaptic connections were specified in only a general way, the details must be at least partly random. This idea led to efforts to investigate information flow in randomly connected networks, and to study self-organizing systems (Minsky & Papert, 1969). The capabilities of these systems turned out to be very limited, though the new connectionists have shown some useful self-organizing properties of distributed networks if the inputs are specified carefully enough.

A second solution was more successful at supplementing the inadequate genetic capacity, using another source of information available to construct organisms, less easily measured but just as real as the genetic source. The hidden source was the information available in the regularities of the environment.

The world is full of environmental regularities that can be used to help build an organism. Gravity always pulls downward, light appears in a 24-h cycle, etc. Beyond these basic constants are others that can tune a perceptual system, such as the rules of geometric optics. Turning the head or eyes to the right results in a retinal image transfer to the left, and moving forward a highly structured flow field. Other transformations in the visual world specify moving, rotating, or looming objects.

If these regularities of the sensory world could be combined with genetic information to construct a perceptual system, the sensory system could escape the information limitations imposed by genetics. At the same time it could be fine-tuned to peculiarities of the environment in which it found itself. An organism living in a visual world particularly rich in horizontal and vertical contours, for instance, could develop greater sensitivity to these stimuli. In this way, the perception-action relationship could influence not only the activities of the organism, but also the design of the organism itself.

Environmental Influences on Visual Development

Tuning of sensory systems to the environment is a relatively late development in evolution, appearing first in mammals. The evidence against environmental adaptation in other animals comes from experiments in which sensory input is rearranged, either by surgery or by environmental manipulations. A frog whose eyes have been surgically inverted, for instance, will snap at flies on the symmetrically opposite side of the visual field. The frog never learns from its mistakes and will starve even in a room full of buzzing flies.

The situation is similar in young chicks, who will peck accurately at small objects from a few minutes after hatching. The accuracy appears to be genetically determined. A chick fitted with prism goggles to displace the visual world will consistently peck to one side of the target (Hess, 1956). Again no adaptation to the prisms is found. In both the frog and the chick, visuomotor coordination takes place in the optic tectum, a phylogenetically ancient midbrain structure analogous to the superior colliculus of mammals. These experiments show that the tectum lacks the developmental plasticity available to mammals.

In young kittens, active interaction with the environment seems to be necessary to establish effective sensorimotor coordination. Merely restraining a kitten's actions and finding impaired development would not prove the point, for impaired mobility might retard development by restricting sensory input. Held and Hein (1963) exposed two kittens to nearly identical sensory stimulation while giving only one of them sensorimotor feedback, using an ingenious gondola arrangement. Eight pairs of kittens were dark-reared until they were strong enough to move in the apparatus, at about 8-12 weeks old.

After the exposure period, the active kittens showed normal behavior in several visually guided tasks. They blinked at an approaching object, they showed visual placing when carried toward a surface, and they avoided walking over the edges of tables. The passive kittens showed none of these behaviors. At first they appeared to be blind. They quickly developed normal coordination, however, after a few days of visual experience. These experiments can be criticized because the passive kitten learns more than Held and Hein had intended – it not only fails to learn sensorimotor coordination, but it also discovers that none of its limb movements has any consequences for perception or locomotion. Upon testing it must unlearn this fact before it can begin to establish normal coordination. Regardless of the outcome of this controversy, though, the experiments establish the importance of environment in calibrating the young animal's sensorimotor function. Further work has investigated the mechanisms of this influence.

Development of Receptive Fields

The first work on the interaction of environment and development was based on clinical observations. If congenital cataracts are not surgically removed within 1 year after birth, vision never becomes normal even after the optics of the eye are fully corrected.

The phenomenon was studied physiologically in kittens. Neurons recorded from kitten striate cortex just before the time of normal eye opening show "sloppy" receptive fields. The fields are large, are not particularly sensitive, have a wide tuning for orientation and spatial frequency, and often fire with long latencies. The inhibition that is so prominent in adult cats is irregular in kittens.

Visual environments can be restricted during development by dark rearing, or by sewing the eyelids shut so that the kittens get some light (prolonged dark-rearing leads to retinal degeneration). The kittens are then exposed to normal environments after varying periods of deprivation. Those deprived for less than about 12 weeks quickly develop normal visual abilities, and their cortical neurons have normal receptive fields. Those deprived longer show behavioral impairment, and many of their cortical neurons have abnormal receptive fields or none at all. If deprived long enough, the cats appear to be blind (Hubel & Wiesel, 1965: Wiesel & Hubel, 1965). The interval during development when the environment can affect visual organization defines the critical period, about 12 weeks in cats.

Binocular Effects

The mechanisms which normally sharpen the receptive fields of cortical neurons were investigated further by more subtle conditions of deprivation. One such condition is monocular rather than binocular deprivation. A monocularly deprived cat or monkey has almost no binocular neurons; most of them are driven only by the exposed eye. This means that cortical neurons normally innervated primarily by the deprived eye are being driven instead by the exposed eye, as though axons from the left and right lateral geniculate nuclei were competing for access to synaptic sites. The competition hypothesis was later supported by mapping the ocular dominance columns of a monkey with autoradiography. In normal monkeys both eyes have columns (really cortical slabs) of equal thickness, while deprived animals show thicker columns for the exposed eye (Hubel & Wiesel, 1977).

These exciting results opened up the possibility that the mechanisms of tuning by the environment could be probed with ever more specific environmental manipulations, initiating a race for better "environmental surgery." One of the first of such manipulations was alternate monocular rearing. One eye would be patched on odd-numbered days, and the other on even-numbered days. The result was that both eyes drove neurons in striate cortex, but hardly any cells received equal input from both eyes.

A more subtle deprivation is not to close an eye, but merely to misalign it, simulating clinical strabismus. The deviated eye always sees patterns, but the patterns in the two eyes do not match. Again, most neurons became monocular, and stereopsis is poor in these cats (Wiesel & Hubel, 1965).

Can binocular vision recover if the eyes are misaligned temporarily during development? The eyes can be temporarily misaligned with prisms worn continuously over the eyes during development. When young monkeys wore prisms during the 2nd month of life, they became stereoblind; even as adults, the monkeys had abnormally few binocular neurons both in area 17 and in area 18 (Crawford, Smith, Harwerth & von Noorden, 1984). So the answer to the question is "no": abnormal correlation between the images in the two eyes has permanent effects if it begins during the critical period and continues beyond it. Nature performs a similar experiment on humans with congenital strabismus. Once the critical period is past, normal function cannot be restored. The strabismus can be corrected surgically so that the two eyes see corresponding images, but binocular depth perception remains poor (Schor, Bridgeman, & Tyler, 1983) even if acuity in each eye is normal.

It now appears that the competition among incoming axons for access to synaptic sites on neurons in visual cortex gives the more active axons the advantage in establishing connectivity. Most neurons become binocular only because excitation is normally equal to the two eyes.

Stimulus-Specific Receptive Fields

In 1970 Helmut Hirsch, then a graduate student at Stanford, began studies depriving cats of vision in all orientations but one. Each animal could act as its own control because the eye would be exposed to one orientation but deprived of others. After much effort Hirsch was able to raise kittens through the critical period. For a few hours each day, the kittens wore ingenious masks that exposed them to only one pattern, a vertical grating in one eye and a horizontal grating in the other.

Hirsch's plan was to see whether the cats responded differently to the exposed orientation in psychophysical discrimination tasks. Just before testing was to begin, Nico Spinelli suggested that they record from the cortical neurons first. Spinelli, Robert Phelbs, and I had been recording from single neurons with an automated system in K.H. Pribram's laboratory.

Preparations for recording the first kitten began as usual at 8:00 a.m. The first receptive fields were disappointing - they were large and diffuse, revealing no specific patterns.

About 10:00 p.m. we mapped another cell and saw a large streak running most of the way across our $25^\circ \times 25^\circ$ screen. At first we thought that it must be an artifact. Perhaps the cell had put out a long spontaneous burst of firing, or something had gone wrong with the display apparatus. We checked the animal; everything seemed in order. We mapped the receptive field again, this time scanning orthogonal to the original direction. The same streak appeared, larger than any receptive field we had ever seen. We looked at each other, astonished: this was real! We mapped the neuron with the left eye covered, then the right. The field was monocular, a rarity in the layer of cat cortex we were recording. Later that night we recorded several other such fields; all vertical or horizontal, all monocular, and all huge.

After several more cats, it was clear that the receptive fields nearly always matched the orientation of the stripes seen by the corresponding eye. Rather than selectively depriving the cat of some receptive field orientations, the masks had fundamentally reorganized the structure of visual cortex. About one-third of the cells turned out to have the monocular, oriented character. And the large size of the fields implied that they were tuned to the very large stripes in the mask.

Later, Spinelli and Jensen (1979) described a few cortical neurons with even more specific receptive fields; parallel sensitive areas appeared at the same interline spacing as the previous exposure pattern. Could such cells be related to memory for the stimulus?

When Hirsch and Spinelli (1970) published their results, Blakemore and Cooper (1970) rushed their own similar findings into print. They had used a different technique: their cats, raised in long striped tubes, saw only one orientation. The receptive fields were normal except that most of them were oriented near the exposed orientation. Behaviorally the tube-reared cats showed apparent blindness to contours at right angles to their exposure; cats from horizontally striped tubes bumped into table legs, while those from vertically striped tubes walked off tables. [The physiological result has proved controversial; replication of the findings of Blakemore and Cooper has succeeded twice (Mize & Murphy, 1973; Turkel, Gijsders, & Pritchard, 1975) and failed once (Stryker & Sherk, 1975)].

Relating Behavior to Receptive Fields. When Hirsch (1972) completed his dissertation, the behavioral results were paradoxical; the visual abilities of his cats were nearly normal! They distinguished vertical from oblique lines about as well as normal cats, despite the changes in their respective fields. Why were the mask-reared cats nearly normal behaviorally, while the tube-reared cats, with much less drastic cortical changes, became so significantly impaired?

The answer to this question may lie in an analysis of the information flow from the cats' visual world in each experiment. Hirsch's cats received retinal images that were unrelated to environmental information. There was no reduction in uncertainty and thus no information (in the formal sense) in the images. In terms of information, these cats should act like deprived cats. The tube-reared cats, in contrast, continued to receive visual information from their abnormal world. When they moved their heads, image motion gave visual information about the head movement. There was an abnormal distribution of orientations of the stimuli giving this information, though, so that the cats adapted to an abnormal world.

Another reason for the difference relates to the critical period. Hirsch's cats were removed from deprivation just before the end of the critical period. They could develop some normal function, though many of their receptive fields remained changed for years (Spinelli, Hirsch, Phelbs, & Metzler, 1972). The tube-reared cats, deprived beyond the critical period, could not recover when brought into a normal environment.

Once these results had come out, a race began to find other kinds of flexibility in the cortex. Pettigrew and Freeman (1973) used a highly abnormal stimulus, raising "planetarium cats" in a visual world consisting only of points of light in a dome. In these cats, a small point of light excited a cortical cell anywhere in a large area of the visual field. The receptive fields, qualitatively different from those in normal cats, again reflected the exposure conditions.

The area remains controversial, some researchers maintaining that receptive fields are not modified by experience, but require stimulation to maintain their innate organization and to develop normally (Stryker, Sherk, Leventhal, & Hirsch, 1978). The weight of the evidence reviewed above, however, supports the conclusion that environment can substantially restructure the visual cortex under extreme conditions.

These studies showed the enormous flexibility of visual cortex, but did not reveal how the system eventually stabilizes at the end of the critical period. Spinelli and Jensen (1979) were able to renew sensitivity to environmental surgery long after the end of the critical period by injecting the toxin colchicine into the cortex of adult cats. The drug may have disorganized the existing synaptic connections, so that the environment could influence the regrowth of new ones. Kasamatsu and Pettigrew (1979) lengthened the critical period with intracortical infusion of drugs that stimulate beta-noradrenergic systems in the cortex, implying that norepinephrine maintains plasticity in young animals.

Subsequent work has specified the mechanism further, implicating postsynaptic beta-adrenergic receptors on cortical neurons. In one experiment, the cortex of a 6-7-week-old kitten was continuously perfused with a beta-adrenergic blocker for 1 week while one eyelid was sutured shut. The blocker prevented the development of monocular neurons - many normal, binocular neurons could still be recorded from the perfused cortex (Shirokawa & Kasamatsu, 1984).

These studies demonstrate that noradrenergic manipulations can change neuronal plasticity. Do such influences affect animals under naturalistic conditions? To address this question, the noradrenergic neurons of the locus coeruleus were stimulated electrically while normal adult cats were given brief monocular exposure; otherwise, the cats were housed in the dark. After 1 week, recordings showed a lower proportion of binocular cells than normal. Thus plasticity had been restored to striate cortex without direct intervention in that cortex. The effect was clearly noradrenergic, since the plasticity could be prevented if the cortex were perfused beforehand with the catecholamine-related neurotoxin 6-hydroxydopamine (Kasamatsu, Watabe, Schöller, & Heggelund, 1983).

Conclusion

Action cannot be separated from the design and operation of perceptual systems. Their very definition involves action, and both their operation and their development depend on it. The old concept of a passive sensory surface with stimuli impinging on it exists only in laboratories; in nature, the organism must act to gather information.

Perception and the activities that it drives have a reciprocal relationship; a perception leads to a need for more information from the environment, resulting in an action. The action in turn exposes the receptors to new aspects of the environment, resulting in new perceptions. As a result of this relationship, patterns on the receptors are almost always either transitory or in motion. The physiology of each side of the perception-action system is adapted to the physiology of the other side. Again, this is seen most clearly in the oculomotor system. The roughly 200-ms minimum intersaccadic interval, for instance, reflects the time needed to process the visual scene present during a fixation. The speed limit of the pursuit system at full gain, about 30° per second, also reflects limits of the input physiology - beyond this speed, the organism puts itself in the dangerous position of no longer being able to transduce the background sweeping across the retina.

The perception-action reciprocity is particularly important during development, when the sensory channels are tuned by their environments; and these environments are influenced in turn by the developing organism's actions. For instance, the only substantial difference between the exposure conditions of the mask- and the tube-reared cats reviewed above was in the perception-action link. Both groups of kittens saw only patterns of stripes, and both were allowed unrestricted movement. The difference in correspondence between movement and stimulation of the receptors was the only significant cause of the differences in the development and later visual behavior of the two groups of cats.

At this point, it becomes unproductive to attempt to isolate perceptual from motor physiologically; each side of the circle blends imperceptibly into the other.

References

Anderson, J.R. (1985). *Cognitive psychology and its implications*. New York: Freeman

Bahill, A.T., & Stark, L. (1975). The high-frequency burst of motoneuronal activity lasts about half the duration of saccadic eye movements. *Mathematical Biosciences, 26*, 319-323.

Blakemore, C., & Cooper, G.F. (1970). Development of the brain depends on the visual environment. *Nature, 228*, 467-478.

Bridgeman, B. (1977). Reply to Brooks and Fuchs: Exogenous and endogenous contributions to saccadic suppression. *Vision Research, 17*, 323-324.

Bridgeman, B. (1981). Cognitive factors in subjective stabilization of the visual world. *Acta Psychologica, 48*, 111-121.

Bridgeman, B., & Palca, J. (1980). The role of microsaccades in high acuity observational tasks. *Vision Research, 20*, 813-817.

Bridgeman, B. Hendry, D., & Stark, L. (1975). Failure to detect displacement of the visual world during saccadic eye movements. *Vision Research, 15*, 719-722.

Bridgeman, B., Lewis, S., Heit, G., & Nagle, M. (1979). The relationship between cognitive and motor-oriented systems of visual position perception. *Journal of Experimental Psychology: Human Perception and Performance, 5*, 692-700.

Bridgeman, B., Kirch, M., & Sperling, A. (1981). Segregation of cognitive and motor aspects of visual function using induced motion. *Perception and Psychophysics, 29*, 336-342.

Brune, F., & Lücking, C.H. (1969). Oculomotorik, Bewegungswahrnehmung und Raumkonstanz der Sehdinge. *Der Nervenarzt, 40*, 413-421.

Burbeck, C., & Kelly, D.H. (1982). A mechanism in the distal retina that accounts for the fading of stabilized images. *Investigative Ophthalmology and Visual Science, 22 (Suppl.)*, 50.

Clark, M.R., & Stark, L. (1975). Time optimal behavior of human saccadic eye movement. *IEEE Transactions on Automatic Control, 20*, 345-348.

Collewijn, H. (1969). Changes in visual evoked responses during the fast phase of optokinetic nystagmus in the rabbit. *Vision Research, 9*, 803-814.

Crawford, M.L.J., Smith, E.L. III, Harwerth, R.S., & von Noorden, G. (1984). Stereoblind monkeys have few binocular neurons. *Investigative Ophthalmology and Visual Science, 25*, 779-781.

Darian-Smith, I., Sugitani, M., & Heywood, J. (1982). Touching textured surfaces: Cells in somatosensory cortex respond both to finger movement and to surface features. *Science, 218*, 906-909.

Dodge, R. (1900). Visual perception during eye movement. *Psychological Review, 7*, 454-465.

Duncker, K. (1929). Über induzierte Bewegung. *Psychologische Forschung, 12*, 130-259.

Gerrits, H.J.M., Stassen, H.P.W., & van Erning, L.J.T.O. (1984). The role of drifts and saccades for the preservation of brightness perception. In L. Spillman & B. Wooten (Eds.), *Sensory experience, adaptation, and perception (pp 439-459)*. Hillsdale, NJ: Erlbaum.

Gibson, J.J. (1966). *The senses considered as perceptual systems*. Boston: Houghton Mifflin.

Goldberg, M.E., & Bushnell, M.C. (1979). Monkey frontal eye fields have a neuronal signal that precedes visually guided saccades. *Society for Neurosciences Abstracts, 5*, 779.

Held, R., & Hein, A. (1963). Movement-produced stimulation in the development of visually guided behavior. *Journal of Comparative Physiological Psychology, 56*, 872-876.

Hering, E. (1977). *The theory of binocular vision*. (B. Bridgeman, Trans.). New York: Plenum (original work published 1868).

Hess, E.H. (1956). Space perception in the chick. *Scientific American, 195*, 71-80.

Hirsch, H.V.B. (1972). Visual perception in cats after environmental surgery. *Experimental Brain Research, 15*, 405-423.

Hirsch, H.V.B., & Spinelli, D.N. (1970). Visual experience modifies distribution of horizontally and vertically oriented receptive fields in cats. *Science, 168*, 869-871.

Hubel, D. & Wiesel, T. (1965). Binocular interactions in striate cortex kittens reared with artificial squint. *Journal of Neurophysiology, 28*, 1041-1059.

Hubel, D., & Wiesel, T, (1977). Functional architecture of macaque monkey visual cortex. *Proceedings of the Royal Society of London, 198*, 1-59.

Kasamatsu, T., & Pettigrew, J. (1979). Preservation of binocularity after monocular deprivation in the striate cortex of kittens treated with 6-hydroxydopamine. *Journal of Comparative Neurology, 185*, 139-162.

Kasamatsu, T., Watabe, K., Schöller, E., & Heggelund, P. (1983). Restoration of neuronal plasticity in cat visual cortex by electrical stimulation of the locus coeruleus. *Neuroscience Abstracts, 9*, 911.

Latour, P. (1962). Visual threshold during eye movements. *Vision Research, 2*, 261-262.

Lynch, J.C., Mountcastle, V., Talbot, W.H., & Yin, T. (1977). Parietal lobe mechanisms for directed visual attention. *Journal of Neurophysiology, 40*, 362-389.

Mack, A. (1970). An investigation of the relationship between eye and retinal image movement in the perception of movement. *Perception and Psychophysics, 8*, 291-298.

MacKay, D. (1980). Elevation of usual threshold by displacement of retinal image. *Nature, 225*, 90-92.

Minsky, M., & Papert, S. (1969). *Perceptrons*. Cambridge, MA: MIT Press.

Mitrani, L., Mateeff, S., & Yakimoff, N. (1971). Is saccadic suppression really saccadic? *Vision Research, 11*, 1157-1161.

Mize, R., & Murphy, E.H. (1973). Selective visual experience fails to modify receptive field properties of rabbit striate cortical neurones. *Science, 180*, 320-323.

Noton, D., & Stark, L. (1971). Scanpaths in eye movements during pattern perception. *Science, 171*, 308-311.

Pettigrew, J., & Freeman, R.D. (1973). Visual experiences without lines: Effect on developing cortical neurons. *Science, 182*, 599-601.

Riggs, L., Merton, P., & Morton, H. (1974). Suppression of visual phosphenes during saccadic eye movement. *Vision Research, 14*, 997-1011.

Robinson, D.A. (1981). Control of eye movements. In V.B. Brooks (Ed.), *Handbook of physiology Sec 1. The nervous system: Vol. II. Motor control, Part 2*. Bethesda, MD: American Physiological Society.

Schor, C., Bridgeman, B., & Tyler, C.W. (1983). Spatial characteristics of static and dynamic stereoacuity in strabismus. *Investigative Ophthalmology and Visual Science, 24*, 1572-1579.

Shannon, C., & Weaver, W. (1949). *The mathematical theory of communication*. Urbana: University of Illinois Press.

Shirokawa, T., & Kasamatsu, T. (1984). Beta-adrenergic receptor mediates neuronal plasticity in visual cortex. *Investigative Ophthalmology and Visual Science, 25 (Suppl.)*, 214.

Sparks, D., & Pollack, J. (1977). The neural control of saccadic eye movements: the role of the superior colliculus. In B.A. Brooks & F. Bajandas (Eds.), *Eye movements*. New York: Plenum.

Spinelli, D.N., & Jensen, F. (1979). Plasticity: The mirror of experience. *Science, 203*, 75-78.

Spinelli, D.N., Hirsch, H.V.B., Phelps, R., & Metzler, J. (1972). Visual experience as a determinant of the response characteristics of cotical receptive fields in cats. *Experimental Brain Research, 15*, 289-304.

Stark, L.W., & Ellis, S. (1981). Scanpaths revisited: Cognitive models direct active looking. In D. Fisher, R. Monty, & J. Senders (Eds.), *Eye movements: Cognition and visual perception*. Hillsdale, NJ: Erlbaum.

Steinman, R., Haddad, G., Skavenski, A., & Wyman, D. (1973). Miniature eye movement. *Science, 181*, 810-819.

Stevens, J.K., Emerson, R., Gerstein, G., Kallos, T., Neufeld, G., Nichols, C., & Rosenquist, A. (1976). Paralysis of the awake human: visual perceptions. *Vision Research, 16*, 93-98.

Stryker, M., & Sherk, H. (1975). Modification of cortical orientation of selectivity in the cat by restricted visual experience: A reexamination. *Science, 190*, 904-906.

Stryker, M., Sherk, H., Leventhal, A., & Hirsch, H.V.B. (1978). Physiological consequences for the cat's visual cortex of effectively restricting early visual experience with oriented contours. *Journal of Neurophysiology, 41*, 896-909.

Turkel, J., Gijsders, K., & Pritchard, R. (1975). Environmental modification of oculomotor and neural function in cats. *Investigative Ophthalmology and Visual Science, 14 (Suppl.)*, 63.

Volkman, F., Riggs, L., White, K., & Moore, R. (1978). Contrast sensitivity during saccadic eye movements. *Vision Research, 18*, 1193-1199.

Westheimer, G., & Blair, S.M. (1973). Oculomotor defects in cerebellectomized monkeys. *Investigative Ophthalmology, 12*, 618-621.

Wiesel, T., & Hubel, D. (1965). Extent of recovery from the effects of visual deprivation in kittens. *Journal of Neurophysiology, 28*, 1060-1072.

Winterson, B., & Collewijn, H. (1976). Microsaccades during finely guided visuomotor tasks. *Vision Research, 16*, 1387-1390.

Yarbus, A.L. (1967). *Eye movements and vision* (L.A. Riggs, Trans.). New York: Plenum.

Zuber, B., & Stark, L. (1966). Saddadic suppresson: elevation of visual threshold associated with saccadic eye movements. *Experimental Neurology, 16*, 65-79.

Utilization of Sensory Information for Motor Control

H. CRUSE, J. DEAN, H. HEUER, and R.A. SCHMIDT

CONTENTS

Introduction

Interactions between perception and action can be considered on various levels, as stressed by Prinz and Sanders (1984), starting with neurophysiological problems and ending with philosophical ones. The problem of how an animal uses information about its own state and about the environment for controlling movement is among the lower-level problems. It is now generally acknowledged that most behaviors arise from a combination of "central" elements which rely on autonomous neural mechanisms and "peripheral" elements which rely on sensory information. Movements are not simply driven by stimuli in a reflex-like manner: in controlling their movements, organisms have some degree of autonomy, that is, some degree of independence from sensory input (von Holst, 1937). On the other hand, there can be no doubt that most movements are influenced by sensory input, despite some opinions to the contrary (e.g., Jones, 1974). Thus, the problem is to determine how sensory input is combined with central or autonomous control. In the combined system, peripheral information can be seen to have three functions: (a) it influences the decision as to what action is to be performed; (b) it participates in adjusting the parameters of the movements required for the action; and (c) it acts

Relationships Between Perception and Action
Edited by O. Neumann and W. Prinz

to ensure that the movements are carried out correctly in the face of potential disturbances.

In the present chapter, our first goal is to review several important concepts and experimental findings relevant to understanding the control of movement. We will begin by considering several autonomous mechanisms, but our primary focus will be on how information from the outside world is used for motor control. In discussing the role of sensory input and central control, we will present a model for control mechanisms which shows that an experimental discrimination between a peripheral and an endogenous system is not always easy.

Movement forms the basis of behavior, so it is of interest not only for physiologists but also for psychologists, ethologists, and scientists in various other related disciplines to understand how movements are prepared and performed. Thus, the study of movement is a multidisciplinary endeavor including areas of psychology, biology, medicine, and engineering. Unfortunately, the interaction among the different fields has often been limited. Owing to this separation and to differences in the examples studied, it is not surprising that very different terminologies have arisen in the different fields. As a consequence, one and the same object of research is analyzed not only on many different levels but using a multitude of concepts as well. This diversity can lead to unnecessary controversies: a difference in terminology does not necessarily imply a difference of opinion.

Differences in terminology arise not only from the different fields of investigation but also from the complexity of the subject itself and the number of levels on which it can be studied. Motor control can be investigated by studying the behavior as a whole, the physiology of movement control, or the interaction between motor and perceptual events. In turn, each of these subjects can be further partitioned. For example, the physiology of movement involves two distinct entities: a peripheral skeletomotor system which defines the physical constraints and does the actual mechanical work need for movement to occur, and a neural system which does the planning and coordination. This neuronal system selects movements for performance, sends appropriate commands to the peripheral motor system, and receives information on the state of this system via sensory organs located so that some centers are primarily concerned with global features of a movement, and others are concerned with more specific details such as the activity in a single muscle. Motor behavior can be approached on any of these different levels. One subject of central interest, which is also a source of confusion, is the interaction among the different levels of organization. To avoid confusion, one needs to establish at the outset the unit of analysis and the terminology that will be employed.

Thus, the second goal of this article is to point out some of the unnecessary confusion which arises when common terms are used with different meanings, or different terms are used with a common meaning, and to suggest a set of definitions for a less confusing terminology. We will show that terms such as "feedback", "feedforward", "reference input", "open loop", and "closed loop" are often used in an ill-defined way or not clearly distinguished from one other. For example, performing a movement to a target with the eyes closed is often referred to as an "open loop" condition. However, proprioceptive control loops, which may also be involved in the action, remain intact or "closed." Thus the kind of signal (or sensory modality) for which the loop is open should always be made explicit. Furthermore, "open loop" is used in two different ways to describe either a control system which does not use feedback or a system in which a normally existing feedback loop has been removed or "opened," usually through an experimental manipulation. A second example illustrates that the appropriate term also depends on the level of analysis. A sensory signal which is part of a "feedback loop" in one context might function as a "feedforward" or "reference signal" in other contexts. Consider the control of movement in a limb with several joints. When externally applied movement at one joint elicits a resistance reflex at this joint, the corresponding sensory signal would clearly be designated a "feedback signal." Assume that this signal also elicits a movement in another joint of the limb as part of a distributed reflex. Then this signal would be called a "feedforward" or "reference signal" when this second joint is the subject of interest. It could also be labeled a "feedback signal" when the whole limb is considered as the unit. Thus, in trying to avoid some of these problems, we will propose a new set of terms which hopefully can be used with less confusion. In this scheme, we consider sensory information as acting in three different ways which subserve the three functions mentioned above. Sensory information can elicit a movement (advance control, type 3); then, during a movement, sensory information can be used for intermittent control (type 2) or for continuous control (type 1).

Motor Control Structures

As a starting point for our discussion of how sensory information is combined with autonomous mechanisms of control, we introduce the concept of a "motor control structure." A motor control structure is here defined in very general terms to include many different aspects of motor control. It is intended as a superordinate term to designate the functional mechanisms involved in controlling movement. Broadly speaking, a motor control structure embodies both the autonomous activity of the nervous system and the rules by which the nervous system handles sensory information. It operates using neural networks which are specified genetically and may be further modified through development and learning. On this neural foundation, the active control structure is established as a pattern of activity from a set of possible patterns generated in response to situational demands communicated by actual sensory information. Thus, the motor control structure emerges from the properties of both the peripheral sensorimotor apparatus and those parts of the CNS which are involved in the control of a particular movement pattern. The former properties include both physical aspects of the skeletomotor elements and physiological characteristics of sensory and neuromuscular function. The latter include the cellular and network properties of a core of anatomical structures centrally involved in motor control, but activity in other CNS centers may be incorporated temporarily as required by the situation. A motor control structure is considered to be task specific; it is defined as the control mechanism for movements which perform identifiable functions. Thus, individual motor control structures are differentiated at the level of functional movement patterns rather than patterns of motor output in a specific performance. In this sense, one can regard walking or grasping an object or the prey-capture strike of a mantid as under the control of separate motor control structures. In any given performance, the properties of the control structure are instantiated by the task represented by the sensory information about the environment combined, of course, with stored information and autonomous activity. Some elements may be involved in nearly any movement one can think of. The concept is not intended to imply a particular control mechanism: at different levels of analysis a motor control structure may include mechanisms described by more specific concepts such as pattern generator, motor program, coordinative structure, or closed-loop control.

As Poulton (1957) emphasized in his distinction between "closed" and "open" skills, movements differ with regard to the amount of adjustment to the environment that is required. Therefore, different movements require more or less processing of sensory information. Thus, a motor control structure has to subserve two functions: first, the autonomous generation of control signals to the peripheral system; and second, the processing of sensory information. For many years these two functions have been discussed as if they were logical alternatives (e.g., Kelso & Stelmach, 1976). The "open-loop versus closed-loop" debate, however, appears to be fruitless and to distract from the important question of how sensory information is integrated with autonomous control. The main part of this chapter is concerned with this problem. First, however, different conceptions of autonomous control need to be examined.

Autonomous Control

Evidence that autonomous mechanisms, those which operate independently of sensory information, play a role in the control of movements, is available for many species including humans. It has been summarized in several places (e.g., Schmidt, 1982, pp. 237-243; Schmidt, 1988; Heuer, 1988) and will not be repeated here. The major piece of evidence is that limb movements are usually only slightly impaired when sensory information from the limb is not available.

In different fields these autonomous neural mechanisms are conceptualized in different ways. Three of these concepts are reviewed below. They are often seen as logical alternatives, but in our view they can be better described as emphasizing different approaches to the problem of motor control. More complete knowledge of the neural mechanisms should clarify the relationship among these alternative descriptions. For the present, the question should not be whether one concept is "right" and the others are "wrong," but rather which concept is most useful for the purpose at hand.

Central Pattern Generators

The term "central pattern generator" (CPG) is used to describe the control of both rhythmic and nonrhythmic movements. Most often, however, it is used as a synonym for the term "central oscillator"; we will use it in this sense here. This concept comes up whenever the origin of control signals for rhythmic movements such as those of locomotion, respiration, or ingestion is discussed (e.g., Wilson, 1968; Pearson & Iles, 1973; Selverston, 1980; Grillner, 1981). A CPG is thought of as a collection of neurons connected in such a way that they produce an intrinsic oscillation; the neural activity does not depend on sensory information for its pattern, although it may require general excitation from sensory or central sources. Thus the concept of a CPG is intimately related to an analysis of physiological mechanisms on the cellular level. For example, it is applied to neuronal networks that produce simple patterns of alternating activity in antagonistic muscles. The current physiological methodology restricts its use to simple systems and behaviors, like the locomotion of insects or spinal cats, where there is at least a chance of relating behavioral phenomena to details of neuronal networks.

Although we adopt the CPG concept here to designate such neural networks, it should be mentioned that model considerations show that even in very simple networks there may exist a continuum between a pure CPG and a system which is solely driven by sensory signals (Bässler, 1986). Consider the example of the step cycle of a leg. Brown (1911) proposed the simple CPG model shown in Fig. 1a to explain his result with deafferented cats. The two neurons (or sets of neurons) are tonically excited and have mutually inhibitory connections. In addition, they are subject to fatigue. When tonic excitation is present, the less-fatigued neuron will be dominant, and its activity will inhibit the activity of the other. During the inhibition, the inactive neuron recovers from fatigue, while increasing fatigue in the active neuron reduces the strength of the inhibition: these factors combine at some point to allow the previously inactive neuron to become active and rapidly inhibit the previously dominant neuron. The result is a system in which the dominant activity alternates between the two neurons and the switch between the two states is rapid (see Brown, 1911; Wilson, 1968; Pearson & Iles, 1973; Grillner, 1981, p. 1217). In this model, as applied to the example of a walking leg, one neuron might excite the muscles active during the stance phase and the other neuron those active during the swing phase.

The corresponding peripheral oscillator models (Fig. 1b) were based on findings that some sense organs signal the end of each half cycle; in the example of a step cycle (e.g., Bässler, 1977, 1986; Land, 1972; Graham, 1977), some sense organs could signal an anterior leg position appropriate for the end of the swing phase. If one assumes further that these sense organs excite the stance phase muscles, then, upon reaching this anterior position during its swing phase, the leg would switch to stance phase. This switch would not occur if some perturbation during the swing phase prevented the leg from reaching the appropriate end position. An analogous mechanism depending on other sense organs detecting an appropriate posterior position would become active at the end of the stance phase.

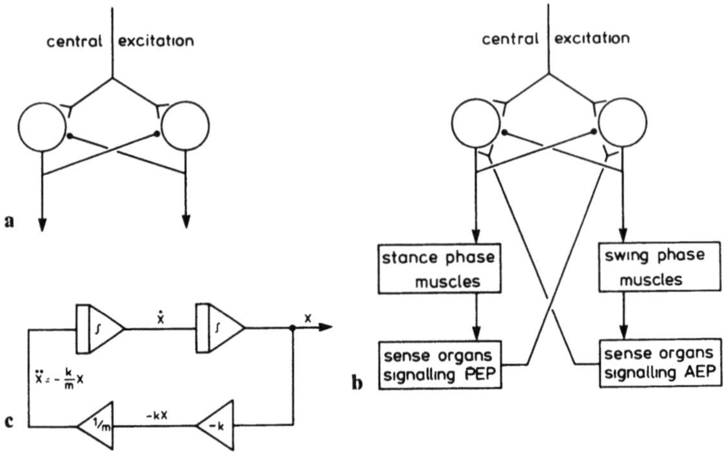

Fig. 1 a-c. Networks illustrating central pattern generators. **a** Two neurons or groups of neurons are tonically excited and mutually inhibit each other. This inhibition is subject to fatigue. **b** A peripheral oscillator model applied to the control of walking legs. It is assumed that sense organs which monitor when the leg reaches the posterior extreme position (*PEP*) at the end of its stance phase excite the swing phase motonenons. Corresponding sense organs detect the anterior extreme position (*AEP*) at the end of the swing phase and excite the stance phase motoneurons. **c** A simple electronic circuit which is able to produce an oscillating output. (**b** After Bässler, 1986)

These components of the stepping motor control structure could therefore be described as a reflex chain. Thus, depending on the relative strengths of central and peripheral signals, the combined system can behave as a more centrally or a more peripherally driven oscillator (Bässler, 1986). Similarly, the required tonic excitation might arise from central neural mechanisms as shown in Fig. 1a; it might also arise from sense organs (Fig. 1b). This example shows that the experimental destruction of the sense organs might artificially change the system from a peripheral to a central oscillator. Thus, a clear distinction between central and peripheral oscillators is experimentally difficult.

Central Motor Programs

Central motor programs can be considered as a more general concept which subsumes central oscillators and CPGs. The central motor program idea is related to the CPG concept but it is applied at a higher level of abstraction: it refers more to the functional sequence of movements and less to the neural network in which the motor commands are generated. In addition, it is applied equally to both rhythmic and nonrhythmic movements. It should be noted, however, that if "CPG" is not used as a synonym for a central oscillator, then it is often used in an abstract manner that is equivalent to a central motor program. It should also be mentioned that sometimes, particularly in the biological literature, the term "motor program" without the qualifier "central" is used in the sense of a process controller to describe mechanisms in which data on the actual state of the process, i.e., sensory signals, influence the performance of the program. Used in this way, "motor program" would correspond closely to "motor control structure" as the term is used here.

Central motor programs are usually illustrated using the analogy of a program for a digital computer, but analog mechanisms have also been used as metaphors. For example, Hoyle (1965, 1983) refers to a "motor tape," Schmidt (1988) uses a phonograph record analogy, while Taylor and Birmingham (1948), writing before the advent of modern computers, use the analogy of a mechanical camshaft controlling timing. These different analogies, referring to digital and analog control devices, reflect different ways of thinking about the nature of a central motor program.

The program of a digital computer consists of a series of commands that are performed in sequence at a rate governed by the computers clock signal. Analogous to this, the "digital" motor program concept considers the content of a motor program as a series of "response commands" that are performed under the control of an internal clock (Rosenbaum, 1985). Setting up a motor program, in this view, involves assigning response commands to clock pulse identifiers; each clock pulse identifier is used to trigger the associated command when the corresponding clock pulse occurs. This digital metaphor is suited to the analysis of movement sequences in which subunits can be meaningfully distinguished. Such subunits, for example, might be the penstrokes in writing, the keystrokes in typing, or the phonemes in speaking. It is, however, also possible to apply the "digital characterization" of a motor program to sequential muscle contractions, e.g., the successive contractions of agonists and antagonists in rapid aiming movements or even to discrete corrections within seemingly continuous movements (see p. 68f).

A rather different way of describing the content of a central motor program is suggested by the analog metaphors (cam control, analog computer, etc.). In very general terms, one can describe a central motor program as a function of time with parameters which are specified by some control signal. The concept of a generalized motor program (Schmidt, 1975, 1988), for example, can be written as:

$$p\ (t) = A \cdot f\ (t/T) \qquad 0 < t < T \qquad (1)$$

The parameters T and A define the movement duration and the amplitude, respectively; $f\ (\vartheta)$, $0 < \vartheta < 1$, constitutes a "prototypical function" (cf. Meyer, Smith, & Wright, 1982). Position, $p\ (t)$, is specified as a function of a dummy variable, ϑ, that, for each single movement, is transformed into real time according to $t = \mathrm{T} \cdot \vartheta$. The prototypical function is the essence of the central motor program. It defines those temporal and spatial relationships that remain invariant if T and A are varied. In the temporal domain, this mechanism for programming movement leads to phase constancy or invariant relative timing.

It should be noted that Eq. 1 serves as an example of how, in principle, the content of a central motor program can be specified. Starting with von Holst (1939), it has been proposed several times that duration and amplitude can be treated as independent parameters for motor control. During recent years evidence has accumulated showing invariant relative timing when movement duration (*T*) is varied (for review see Schmidt, 1985a, 1988), as predicted by the prototypical function concept underlying Eq. 1. Nonetheless, Eq. 1 should be considered as an example only, since invariant relative timing has been shown for some but not all movements.

The relative duration of stance and swing phase in cockroaches, for example, is independent of walking speed (Delcomyn, 1971; see also Carter, 1984) as it is in humans (Shapiro, Zernicke, Gregor, & Diestel, 1981). In contrast, in adult stick insects the absolute duration of swing phase is about the same for all walking speeds, so the relative duration is not constant for these animals (Graham, 1972). This suggests that the hypothesis represented in Eq. 1 may be an appropriate description for some but by no means all movement patterns.

According to the "digital" and "analog" ways of thinking, central motor programs are characterized either as a sequence of commands executed in discrete time intervals or as a signal specified by a continuous function of time. As models for the underlying mechanism, the digital and analog views are not logical alternatives. In the first instance, they are different ways for describing the autonomous functions of a motor control structure with direct reference to the output, i.e., to the spatiotemporal characteristics of the movement. However, they usually differ with regard to the level of description. Analog descriptions are on a low level, referring to the detailed space-time characteristics of a movement. Digital descriptions, on the other hand, are in terms of labels for segments of the space-time characteristics, i.e., for units of the apparently continuous stream of movement (cf. Miller, Galanter, & Pribram, 1960). Initially the definition of units may be arbitrary, but often evidence is sought for their reality (e.g., Monsell, 1986). Reality means that these units are not just units of description, but also units of control (cf. Viviani, 1986).

For both views on central motor programs, the nature of the relationship between kinematic and kinetic characteristics of the movement and the actual parameters of the motor control structure remains unspecified. Motor programs are often discussed as if some kinematic variables, such as movement amplitude or duration, have their counterparts in parameters of the control structure. As a matter of course, the existence of a control mechanism is implied for the kinematic variables that the observer happens to measure or consider important. In general, however, the problem of which kinematic variables are the controlled ones has not been solved (see Stein, 1982). In fact, the controlled parameters may well vary among different types of movements according to the requirements placed on the behavior.

Coordinative Structures

Coordinative structures, as defined by Turvey (1977) and Kugler, Kelso, and Turvey (1980, 1982), consist of a set of muscles and limbs which are constrained to act together as a unit. Like the motor program idea, this concept is applied to both rhythmic and nonrhythmic movements. Another similarity is that the coordinative structure concept is not explicitly concerned with the neural basis; the goal is an abstract formulation of movement characteristics including responses to disturbance. The role of peripheral influences is not explicitly addressed in this theory. Instead, the concept is applied to a state of the organism. Therefore, we have included it under autonomous control, although it seems obvious that proprioceptive mechanisms contribute to the dynamic properties present in specific formulations.

Compared to motor program approaches, more emphasis is placed on physical and dynamic features of movement and less emphasis on representations or information processing in the nervous system. However, the constraints present in a coordinative system must arise from a combination of physical and neural factors. At the physiological level, the muscles require neuronal activity to perform their tasks, so we consider coordinative structure formulations to represent implicitly the properties of those neuronal subsystems which control the set of muscles involved (cf. Kelso, 1986).

One of the important problems that the concept of coordinative structures addresses is the "degrees-of-freedom problem" first discussed by Bernstein (1957, 1975). He postulated that movement can only be efficiently controlled if the nervous systems has some way of mastering the enormous number of variables (each representing a degree of freedom) inherent in the numerous joints, muscles, and motor units making up the motor system and in the many gravitational and dynamic configurations under which it acts. Bernstein applied the word "coordination" to the mechanisms performing this task. Reducing the number of variables is one possible solution. Here, "reduction" is meant in a mathematical sense. It does not mean that a real degree of freedom is simply held constant - for example, by fixation of a joint.

However, if two joints are constrained to flex with proportional angular velocities, only one central control signal is required, and the number of degrees of freedom is thereby reduced. Coordination among the various degrees of freedom is one focus of the coordinative structure approach, although, as discussed below, this feature is also implied in the concept of generalized motor programs.

The notion that the role of coordination is to reduce the number of independently controllable muscles might suggest that coordination is always absolute, but this is not the case. The classic example of "relative" coordination was described by von Holst (1939) in his study of the coordination of fine movements in fish. The fins behave like weakly coupled oscillators; each strives to impose its own rhythm or a specific phase relation on the rhythm of the other fins (the "magnet effect" of von Holst) but does not succeed in establishing a rigid coordination. Relative coordination has also been described for walking legs of insects (Wendler, 1964) and lobster (Clarac, 1981), as well as for human arm movements (Gunkel, 1962).

A magnet effect is also illustrated by more recent data on intermanual coordination. Kelso, Southard, and Goodman (1979) studied aimed movements performed simultaneously with the two hands. When performed separately, the test movements of different amplitudes to targets of different sizes showed different movement times according to the well-known Fitts' law, which predicts a logarithmic relation between the ratio of amplitude and imposed accuracy on the one hand, and movement time on the other. When the two movements were performed simultaneously, however, movement durations for left and right hands became nearly identical. This could be taken to suggest one and the same temporal control signal is used for both hands, and only the amplitude parameter is varied to produce movements of different amplitudes. However, the synchrony of the movements was not perfect. Although the durations became more similar, the movement that was of shorter duration when performed in isolation was still of shorter duration when performed in combination. This residual difference is stressed, for example, by Marteniuk and MacKenzie (1980) and by Marteniuk, MacKenzie, and Baba (1984). Its existence points to the danger of over-simplification in describing complex, coordinated movements in terms of a single coordinative structure (cf. Heuer & Wing, 1984) which neglects a more detailed account of the underlying mechanisms.

Theoretically, coordinative structures can be described using linear or nonlinear differential equations. This theoretical approach is restricted explicitly to the macroscopic level of behavior. No attempt is made to detail the network that generates this behavior. As an example, consider the model of Saltzman (1986) and Saltzman and Kelso (1987) for aimed movements. The starting point of this model is Feldman's equilibrium-point concept (Feldman, 1966, 1974 a,b, 1986) – sometimes referred to as the mass-spring model (see, for example, Schmidt, 1987, 1988). However, this model is applied to an abstract "task space" which is defined by two axes, the one running through the initial and target positions, and the second orthogonal to the first. Movement along both axes is governed by linear second-order differential equations. Thus, the "physical" mass-spring model becomes a "metaphysical" mass-spring model.

Movement along the two axes of the task space, according to this model, is "uncoordinated," that is, independent. Coordination arises only when the real movement is considered. This requires coordinate transformations, which, in moving from the abstract task space to the actual joint movements, are fairly complex. The basic principle, however, by which coordinated joint movements can arise from uncoordinated movements in the abstract task space, is simple. Imaging a movement along the x-axis of a coordinate system; as soon as the coordinate system is rotated, the same movement requires a "coordinated" change of x' and y' values in the new coordinate system. The transformation from task space to joint space has a similar consequence for a two-joint arm; moving the end of the arm along a straight line forces "coordinated" rotations at the two joints.

The simple principle of deriving coordination from coordinate transformations fails as soon as coordination is not completely determined by task demands. This, for example, is the case for planar aiming movements with a three-joint arm. Here the redundancy problem can be solved by assuming additional constraints. In one example, these constraints have been described as cost functions defined at each joint. Adding a constraint that the total cost be minimized provides solution for this redundancy problem (Cruse, 1986). In this case, a description by nonlinear equations without derivative terms was shown to be appropriate. The corresponding kinematic model also shows the property of immediate reorganization after blocking an angle (Cruse & Brüwer, 1987).

A somewhat different approach to the coordination problem is presented by Haken, Kelso, and Bunz (1985) and Schöner, Haken, and Kelso (1986). The phenomenon studied has been described by Cohen (1971): it is easy to oscillate both wrists symmetrically (flexion and extension of the two wrists are in-phase), but it is harder to oscillate them alternately (anti-phase). When the frequency of anti-phase oscillations is increased, it becomes impossible to maintain them, and a shift to symmetric oscillation can be observed (see Kelso, Schöner, Scholz & Haken, 1987). This phenomenon can be conceived as a shift between two modes of coordination, comparable for example to the transitions among walk, pace, trot, and gallop observed in quadrupeds as a function of speed. The wrist system has been modeled using coupled nonlinear oscillators. The interesting feature of the model is that it mimics the observed phase transitions when the frequency of oscillation is increased. Moreover, it duplicates the instabilities occurring in the range of frequencies in which the transition occurs. Here there is nothing like a qualitative shift from one motor program to the other in changing from anti-phase to in-phase; rather, the shift between modes of coordination arises from the intrinsic properties of the motor control structure when frequency is increased.

The "dynamic approach," with its emphasis on an abstract description at a macroscopic level by way of differential equations (e.g., Kelso & Kay, 1987), is not the only approach to the coordinative structure problem. An information-processing or cybernetic approach aims at specifying the computations required within the structure rather than simply describing its global behavior. For example, coordination may be interpreted as arising from information exchange between the various segments of the limb(s) involved. The focus here is on the kinds of signals indicating the states of the various segments or articulators and on how these signals are used to achieve coordinated control of different segments. For example, in locomotion studies, a basic question is how information on the state of one leg is used to "convince" another leg to behave in a certain way. In the crayfish, two such influences have been found (Cruse & Müller, 1986). First, for as long as a leg is in stance phase, the next anterior leg is held in swing phase; second, a leg which is near the end of the stance phase sends a signal to its posterior neighbor to elicit the start of a stance phase in this leg: in this way the swing phase of this leg may be shortened. This type of conditional coordinating influence can more easily be described by means of a digital computer program than by differential equations. A description of the wrist phase transitions by means of the motor program concept has not yet been attempted. As related to the above discussion, the dynamic approach provides an interesting quantitative description of this phenomenon and therefore leads to testable hypotheses.

Multiple Characterizations of the Autonomous Functions of Motor Control Structures

As mentioned above, we consider CPGs, motor programs, and coordinative structures to be complementary concepts that focus on different aspects of the autonomous function of a motor control structure. This position may appear awkward in view of the ongoing controversies. These arise in part from the discrepant theoretical frameworks in which the motor program and coordinative structure concepts in particular are embedded. The motor program concept generally uses an information-processing framework, while the coordinative structure concept (associated with a dynamic description) arose from the ecological framework of Gibson (Turvey, 1977). However, these different associations may simply reflect historical development rather than logical necessity.

It should be stressed that we are not arguing that CPGs, motor programs, and coordinative structures are identical concepts - a number of differences have been described in the preceding sections. What we are arguing is that the concepts are complementary rather than logically exclusive. They refer to the same thing, that which we labeled a "motor control structure," but they focus on different aspects of it. They can be likened to the descriptions of the sea provided by a chemist, a physicist, and a poet, or the description of an airplane provided by an engineer, a pilot, and a passenger.

The view that the concepts are complementary is supported by the fact that the different descriptions applied to one system can often be transformed into one other as can be illustrated by the following simple example. Figure 1c shows a simple electronic circuit that exhibits essentially the same behavior as the neural oscillator of Fig. 1a. In this example we consider the output of the circuit to represent the position of the moving limb rather than neural activity. We can describe this system in three ways. First, the physical components and their connections can be specified: this corresponds to the description of a CPG. Second, we note that Fig. 1c is a wiring scheme for an analog computer to solve the following linear differential equation.

$$m\ddot{x} + kx = 0 \tag{2}$$

Tracing around the circuit, we see that the output value, x, is multiplied by a constant, $-k$, and then by a second constant, 1/m. The result, $-x \cdot$ k/m, is integrated twice to give the value of $\ddot{x}$. The input to the second integrator is labeled the first derivative of x because integration of $\dot{x}$ yields x. By the same argument, the input to the first integrator is $\ddot{x}$. Therefore this circuit corresponds to the equation, $\ddot{x} = -x \cdot k/m$, which is equivalent to Eq. 2. This equation represents a description of the system that corresponds to the dynamic approach to motor control. The equation does not explicitly specify the space-time characteristics of the movement; that is, it does not specify "$x(t)$ = ...". Instead these characteristics are implicitly specified.

The third kind of description makes the space-time characteristics explicit by solving Eq. 2. According to the rules of calculus, the general solution can be written as shown in Eq. 3.

$$x(t) = C \cdot \sin(\omega t - \varphi) \quad \text{for } \omega = \sqrt{k/m} \qquad (3)$$

where φ is a constant determined by the initial conditions. This description of the system corresponds to a motor program for a harmonic movement. The complementarity of a CPG on the one hand and a central motor program on the other is quite obvious: the first emphasizes anatomy and physiology, the second emphasizes function. As soon as the CPG is used in an abstract manner without reference to its physical instantiation, CPGs and central motor programs become virtually equivalent terms.

The complementarity of Eqs. 2 and 3 is obvious as well. There seems to be no reason to ascribe a higher degree of "reality" to one equation or the other, or to consider the one as being a consequence of the other. Both equations describe the system in equivalent ways, but for particular purposes one may be better suited than the other. For example, the wiring diagram can easily be derived from Eq. 2 but not from Eq. 3. On the other hand, what the output, $x(t)$, really looks like can easily be seen in Eq. 3 but not in Eq. 2.

Although a dynamic approach, focusing on the global description of the behavior of coordinative structures, and a motor program approach, focusing on the temporally ordered motor commands and their generation, can be considered as being complementary, major differences appear as soon as these formal descriptions are discussed in the context of possible mechanisms. For example, an obvious question is how systems described by equations like Eqs. 2 or 3 come into being. As pointed out by Schmidt (1987), a motor program is generally considered as some kind of central representation of the movement, while advocates of the dynamic approach deny the existence of such a representation. It is worth noting, however, that a differential equation like Eq. 2 "represents" a movement described by Eq. 3 in that a system described by Eq. 2 produces the output given by Eq. 3. Only if the equations are reified, could one say that Eq. 3 is a representation of a sinusoidal movement but Eq. 2 is not. Reification of functional or abstract descriptions is, of course, completely unjustified. It could be that the general orientation of the dynamic approach differs from more concrete formulations with regard to the representation issue.

The fundamental equivalence of these different ways to describe the autonomous function of motor control structures should not obscure the fact that the concepts are of different value for different purposes. Because of the different levels of abstraction, this is clear for the central pattern generator concept on the one hand and the central motor program and coordinative structures concepts on the other. At present, however, we feel that dynamic descriptions of coordinative structures have some strength in modeling transitions between modes of coordination. However, they are necessarily global and neglect the "mechanisms inside the box." Therefore, motor program descriptions appear to be more powerful as far as the details of motor control structures are concerned.

The Role of Sensory Input in Motor Control Structures

Motor control structures usually contain more than just autonomous central mechanisms; they also incorporate peripheral mechanisms dependent on sensory input. The problem here is to identify how autonomous control mechanisms are integrated with those that depend on sensory input. In what is to follow we will describe various modes of integration. The exposition is organized around what we will consider as three different types of such integration, somewhat arbitrarily defined according to the temporal relation between the sensory information and the motor act it affects. As already mentioned, there is often ambiguity in identifying units of motor output. Similarly, the coding of sensory information in a frequency code and the delays inherent in processing complex information limit the ability to attribute sensory information to discrete, precisely defined points in time. Despite these difficulties, we propose to distinguish the following types.

At one extreme is continuous control or, more accurately, "quasi-continuous" control within the limits of temporal resolution (type 1). Here sensory information is continuously used to influence motor output. (When the sensory information is frequency coded, the control cannot be strictly continuous, since the code itself is extended in time.)

The next type (type 2) refers to intermittent use of sensory information for motor control. In this case, there are clearly identifiable periods where sensory input is not incorporated into control. Although the sensory signal may be continuously monitored, it affects motor commands only when the signal exceeds a certain threshold, at certain times, or at certain intervals. Whether or not a particular input is processed during a particular time interval is a feature of the motor control structure and may depend on the requirements placed on the action.

The final type (type 3) is advance processing. Again, sensory information is not continuously used. In advance processing it is incorporated into motor control before execution of the movement. In this case, sensory information is used to set up a motor control structure and to specify initial values for parameters rather than to modify an ongoing movement as is the case in intermittent and continuous processing.

In discussing motor control, it is particularly important to distinguish signals that indicate the current states of a terminal device or the joints controlling the terminal device from signals that indicate what these states should be. The former are feedback signals, while the latter are reference signals. Although, at first sight, type 1 and type 2 influences seem to refer to feedback signals and type 3 influences to reference signals, it will be shown below that this distinction is only sensible when the system is well defined in the sense that its input and output channels are clearly identifiable (see also "Introduction"). Therefore, although we will partly distinguish between feedback and reference signals, we realize that in practice an unambiguous classification is often impossible.

Type 1: Continuous Incorporation of Sensory Information into Motor Commands

Feedback Signals

Thinking in terms of closed-loop systems dominates the cybernetic approach to motor control. There are two ways to verify that a particular signal is used as a feedback signal: (a) the signal can be eliminated; and (b) it can be distorted. Either manipulation should result in gross malfunctioning of the system if the control system depends only on servo-mechanisms relying on feedback. A remarkable finding about motor control is that elimination of sensory input that presumably acts as a feedback signal has only minor effects on motor behavior, but distortion of the very same signal has drastic effects. We will first consider this evidence and then present a design principle for motor structures that exhibits these characteristics.

In insects and crustaceans, minor effects of eliminating sensory input that presumably serves as feedback have been found. Destroying different sense organs that signal leg position produces only small changes in the movement of a walking leg. The same is true for ablation of strain-sensitive proprioceptors (for review see Bässler, 1983; Graham, 1985).

Findings reported for deafferented mammals are basically similar. After dorsal root section, monkeys show only small motor deficits, provided that they use the deafferented limbs at all (for review see Taub & Berman, 1968; Taub, 1976). Taub, Goldberg, and Taub (1975) investigated the accuracy of aiming by a deafferented limb. Although other sensory input that could replace the missing proprioceptive input was carefully excluded, deafferented animals exhibited better performance in their "good" sessions than intact animals in their "bad" sessions. Thus, while deafferented animals were less accurate on the average, the distributions of session means for intact and deafferented animals showed considerable overlap (see also Polit & Bizzi, 1978). In humans, corresponding results we have been reported by Lashley (1917) and Laszlo (1967).

Minor effects on motor performance are found not only when proprioceptive input is eliminated, but also - as is perhaps more familiar - when other sources of feedback are eliminated. We are able to point to a target without seeing our hand. If the movement takes longer than 100-200 ms, accuracy is slightly reduced compared to conditions where visual information is available (Carlton, 1981; Keele & Posner, 1968; Zelaznik, Hawkins, & Kisselburgh, 1983), but these are not gross disturbance. The view of the hand can be obscured through the first half of the path to the target without any detectable effect on accuracy or movement time (Carlton, 1981). Also, we are able to speak without gross disturbance when we cannot hear our own voice.

Elimination of potential feedback signals, in summary, has astonishingly little behavioral effect. This finding has been the major reason for attributing primary importance to autonomous CNS activity in the control of movement (Delcomyn, 1981).

If, as the evidence above suggests, the various potential feedback signals are not used for feedback, then disturbing them should also have minor effects. This, however, is not the case. Distortion of a sensory signal can result in drastic changes, even when completely eliminating the same signal has little effect. When joint proprioceptors in a walking insect are not ablated but fixed so that they constantly signal a forward leg position, the leg rarely starts a swing movement (Bässler, 1977; Dean & Wendler, 1983). More drastically, when an internal proprioceptor (the femoral chordotonal organ) is operated on in such a way that its position signal is inverted (i.e., extension is reported as flexion and vice versa), the leg is hindered from completing a normal step movement: it remains in the air at the endpoint of the swing phase (Bässler, 1977; Graham & Bässler, 1981). Corresponding effects are found when the strain-sensitive proprioceptors are stimulated so that they always signal the leg to be under load (Bässler, 1977).

Distortion of proprioceptive input has gross effects in humans as well. Goodwin, McCloskey, and Matthews (1972) disturbed proprioceptive input by vibrating muscle tendons at the elbow joint. One of the effects of vibration appears to be a tonic excitation of the muscle spindles. The message contained in this hyperactivation is that the length of the corresponding muscle is greater than its true length. When the biceps tendon is vibrated, the elbow angle should be registered as being too large. In accord with this prediction, the extension of the elbow joint was too small when subjects had to match the position of the vibrated arm to passive movements of the other arm (see also Sittig, 1986).

A nice example involving visual feedback was reported by Nielsen (1963). The subject's task was to move one hand along a vertical line. The gloved hand which the subject could see was not his own but that of the experimenter. The visible hand followed a curved path rather than the straight line. This visual signal was processed to correct the apparent error, causing the subject's hand to deviate from the straight line in the direction opposite to the deviation of the observed hand. In spite of the obvious discrepancies between intended movement and visual feedback, it required large deviations and several trials before subjects gave up the search for magical explanations and realized that the gloved hand they were watching could not be their own. This appears to be a rather dramatic case of "visual capture" (see also Jordan, 1972).

Closing the eyes does not prevent humans from maintaining a vertical stance: body sway may increase, especially if we imagine we are falling, but the mean is still vertical. However, presenting a distorted visual stimulus by moving the visual surrounding induces activity in the leg muscles that would be appropriate if the visual input were a result of body movement (Berthoz, Pavard, & Young, 1975; Lee & Lishman, 1975). If the amplitude of the forward-backward oscillation of the walls is sufficiently large, this distorted feedback about the orientation of the body can induce falling. The effect is stronger in children; they exhibit greater body sway and also a larger effect of lack of visual information (Shumway-Cook & Woollacott, 1985). This particular effect of distorted feedback is exploited for fair attractions like the "haunted swing" (where the walls swing rather than the people). The latencies of compensatory responses to distorted visual information can be as short as 100 ms (Nashner & Berthoz, 1978), less than the usual latencies for responses to visual stimuli (Zelaznik et al., 1983). Thus, a disturbed signal again produces a strong effect, whereas completely removing the input elicits only minor alterations.

In speech production the presence of delayed auditory feedback induces hesitations and repetitions which are similar to stuttering. The maximal effect is found at a delay of about 200 ms (see, for example, MacKay, 1986). This speech disturbance can be avoided at least partly if attempts to neglect the auditory feedback are successful (see Roeck, 1977). Similar effects of delayed auditory feedback have been observed when finger tapping is accompanied by tones and these are delayed (Wing, 1977; see Chase, 1965). Again, although we need to hear neither our own voice speaking nor auditory feedback for tapping, if these signals are present and distorted they have gross effects on behavior.

How can the apparently discrepant results obtained from elimination and distortion of various sensory signals be reconciled? The minor effects of elimination indicate that these signals are not continuously processed as feedback, while the strong effects of distortion indicate that they are. We shall outline a design principle for a motor control structure that can show these different effects of elimination and distortion of sensory signals. This design can be identified with physiological mechanisms on lower levels of the motor system; we suggest that it might be repeated on higher levels.

The basic scheme is shown in Fig. 2a. We will use the muscle spindle system as an illustration. Through the alpha-gamma linkage, the supraspinal movement command is fed onto alpha and gamma motoneurons roughly simultaneously. The direct signal to the alpha motoneurons is amplified by a factor A and sent to the muscle, where it causes a change in muscle length according to the factor M. Muscle length (y) is measured by the muscle spindles, which effectively measure the difference between the setpoint determined by the gamma motoneuron discharge and actual muscle length. The error signal, with gain S, is added to the alpha control signal. The essential feature of this scheme is the parallel arrangement of the open-loop control of muscle length via the alpha motoneurons and the closed-loop control via the muscle spindles. The reference signal for the feedback loop is provided by the gamma motoneuron activity, so it is linked to the open-loop control signal.

With appropriate values for M, A, and S, a system of the kind shown in Fig. 2a will show only minor changes in its output when the feedback loop is opened. The controlled variable, y, can be written as a function of the input signal, x, as follows:

$$y = \frac{MA + MS}{1 + MS} x \tag{4}$$

Opening the feedback loop by cutting the dorsal roots is equivalent to setting $S = 0$, so that the contribution of the loop to the controlled variable, y, becomes zero. If the gain, A, is chosen properly so that $MA = 1$, then $y = x$ for all values of S including $S = 0$. If the gain is not properly chosen, then eliminating the feedback will produce smaller or larger effects depending on the extent to which A deviates from the reciprocal of M.

Although such a system is insensitive to elimination of feedback, it will be affected by distorted feedback. In this case the disturbance, d, is added to y before the error is computed. Then the controlled variable is given by:

$$y = \frac{MA + MS}{1 + MS} x + \frac{1}{1 + MS} d \tag{5}$$

Even if $MA = 1$, so that feedback will be ineffective in the "normal" case, a disturbance will reveal that it is processed. It will induce a change in y that increase according to the gain (MS) of the feedback loop, but always remains smaller than d itself.

Before generalizing the design-principle of Fig. 2a to higher levels of motor control, some remarks on its advantages are appropriate. First, the redundancy in the system makes it less prone to failure. Second, and more important, it combines the merits of closed-loop and open-loop control. The advantage of an open-loop system is that it responds rapidly to an input signal and cannot become unstable like a closed-loop system subject to a delay in the feedback signal (Cruse, 1981, pp. 66-67). The disadvantage, however, is that it is unable to compensate for disturbances. In the example, A has to be properly adjusted to M so that a desired output, y, can be achieved. Even in the absence of outside disturbance, A or M may vary due to fatigue and cause errors in a pure open-loop control. (In the generalized case, where the simple gain factor M has to be replaced by complex transfer characteristics – i.e., when the dynamic properties of the elements are taken into account – this problem becomes even more serious.)

A closed-loop system in parallel with the open-loop system will compensate for the weaknesses of the latter: it will serve to reduce the errors that occur through variation in A or M or through external disturbances. Owing to the operation of the open-loop system, the closed-loop system only has to deal with small error signals, so the gain (S) can be relatively low, which helps avoid instability. The actual gain of the spinal reflex loop does appear low; the system can only partially compensate for gross disturbances (e.g., Bizzi, Polit, & Morasso, 1976; Hopf, Handwerker, & Hausmanns, 1967). In addition proper design of the dynamic properties of S could further reduce stability problems (Cruse, 1981, pp. 69-72). In summary, the design principle of Fig. 2a amounts to adding the advantages of open-loop and closed-loop control while eliminating their disadvantages.

We now have to consider the case of a feedback loop in which the sense organ, unlike the muscle spindles, is not under efferent control. This is shown in Fig. 2b (solid lines); S is considered to be 1 and is therefore omitted in this drawing. In this circuit the interruption of the feedback loop by ablation of the sense organs cuts the loop at another position. It corresponds to $F = 0$ in Fig. 2b. In this case, as long as we consider only linear systems, cutting the feedback loop always produces major differences in the behavior of the system compared to the intact situation.

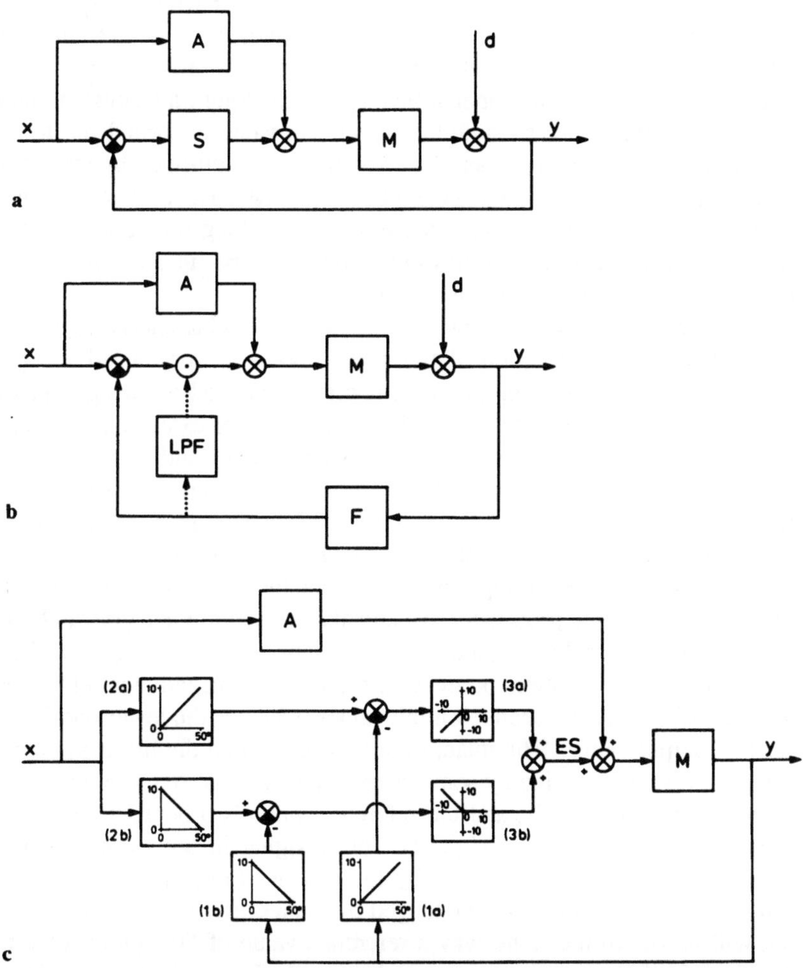

Fig. 2. **a** A basic scheme describing the control of skeletal muscles in vertebrates and the role of the muscle system. The amplication factor of the channel directly influencing the alpha motoneurons is *A*, that of the muscle spindle is *S*, and that of the muscles is *M*. Muscle length, *y*, is controlled via a reference input, *x*, and a disturbance signal, *d*. **b** A feedback system in which the sense organ (*F*) is not efferently controlled. The interruption of the feedback loop by ablation of the sense organ does not significantly influence the behavior of the system with respect to the reference input, *x*, because the nonlinear influence of the low-pass filter (*LPF*) cancels the effect of the error signal. **c** A corresponding result is obtained by another system using antagonistically arranged parallel channels (**a**, **b**). Each channel contains a feedback transducer (*1*), a characteristic producing the reference value (*2*), and a rectifier (*3*). Summation of the outputs of the latter gives the final error signal (*ES*) (Cruse, 1980). The disturbance input is omitted for sake of simplicity. (**a** After Cruse, 1981)

The gain of the system, $y/x = (M + MA) / (1 + MF)$, is different according to whether $F = 0$ or $F \neq 0$.

However, two nonlinear systems can be constructed, which have the property that cutting the feedback does not change the output of the system. As shown in

Fig. 2b by dotted lines, we now add a low-pass filter with a very long time constant (in the order of hours or days) which receives the signals from the sense organs as input. The output of this low-pass filter will provide a sort of long-term mean value of the sensory input which, due to the long time constant, will be a relatively constant value. Assume that this mean value is normalized to have a numerical value of about 1. When the error signal is multiplied by this constant value, the resulting system behaves essentially like the linear system with no low-pass filter. However, if the sensory input is zero for a long time due to destruction of the sense organs, then the output value of the low-pass filter will also decay to zero and the action of the error signal is removed.

Another circuit which shows invariant output characteristics when sensory feedback is destroyed occurs if the sense organs are organized antagonistically (Fig. 2c). Joint position, for example, is often measured by two groups of sensory cells which are increasingly excited by joint flexion or extension, respectively. Assume that the joint angle can adopt values between 0° and 50°, and the excitation can vary between 0 and 10 in arbitrary units which may correspond to spike frequency. The antagonistic responses of the two sense organs are represented in Fig. 2c by the two response characteristics yielding increasing (1a) or decreasing (1b) excitation when the joint angle increases. For the respective comparison with the reference value these proprioceptive signals also require antagonist characteristics (2a, b); then the results are passed through separate rectifiers (3a, b). For positive input values, both rectifiers produce zero output; for negative input values, the upper one (3a) produces negative output values, whereas the lower one (3b) produces positive output values. The intact system behaves like the linear system shown in Fig. 2a, as can be verified by some simple calculations. Assume, for example, that the actual position is 25°, and the reference value is 0 degrees thus demanding a movement with a negative sign. The error signal in the lower channel will be $10 - 5 = 5$ which is made to zero by the rectifier (3b); that in the upper channel will be $0 - 5 = -5$ which is not changed by the rectifier (3a). The total error signal, ES, will be –5. In the same way a reference value of 50°, produces a final error signal of 5. If the sense organs are destroyed, there will be no activity at their output (a value of zero). Therefore the input values to the rectifiers in each pathway will be greater than zero - yielding a zero output. Thus the total ES becomes zero and, if $AM = 1$, it follows that $y = x$ even when the sense organs are ablated. This system shows a normal response immediately after sensory ablation, whereas the system described earlier (Fig. 2b) adapts slowly due to the long time constant of the low-pass filter.

Generalizing the design principle to higher levels of motor control requires some minor changes (Fig. 3) but retains the parallel arrangement of open-loop and closed-loop control. The scheme has as its input two (or more) variables. The one called "autonomous control signal" corresponds to a movement command. The "feedback reference" corresponds to a sensory representation of the expected feedback for the intended movement. There may be different reference signals for different types of feedback. It should be noted that these two inputs are equivalent to

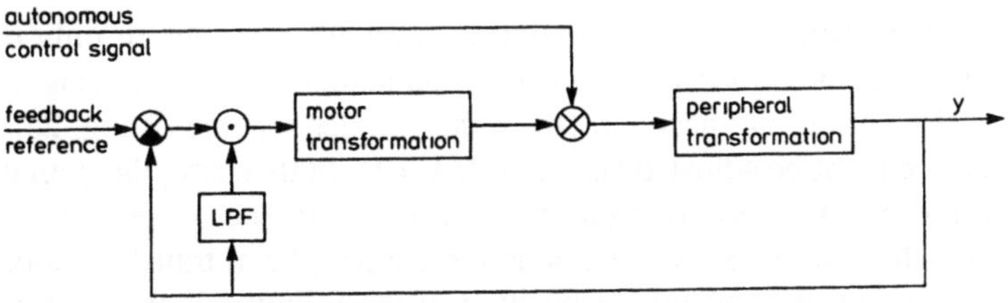

Fig. 3. The control principle shown in Fig. 2b generalized for higher levels of motor control. *LPF*, low-pass filter with a very long time constant. See text for further explanation

the representations postulated by Adams (1971) and Schmidt (1975). Hoyle (1964) proposed a similar distinction in his use of motor and sensory tapes.

The autonomous control signal affects the controlled variable, *y*, via the "peripheral transformation." If, for example, *y* is the position of the hand, then the peripheral transformation includes all transformations that central commands undergo due to mechanical characteristics of the muscles and the limbs (see Partridge, 1979). If *y* is the position of a terminal device, such as a part of a tool, then the peripheral transformation also includes the transformations added by the tool. It should be noted that to adequately control *y*, the generation of the autonomous control signal has to take the peripheral transformation into account (see Heuer, 1983, pp. 15-16).

The sensory feedback is processed in parallel with the open-loop system. The "sensory error" is transformed into a corresponding motor command and added to the autonomous control signal. For this transformation, the motor control structure must select motor commands which are suited to cancel the "sensory error." More generally, the use of sensory error signals for motor control requires that their "motoric meaning" is known. As one way to solve this problem Hein and Held (1962) postulated the existance of a correlation storage which assigns motor commands to sensory signals. Transforming the sensory error into motor commands does not require that the peripheral transformation be taken exactly into account.

What arguments justify this generalization of the design principle of parallel open-loop and closed-loop control to higher levels of motor control? First, the evidence reviewed above for sensory signals processed either at the spinal level or at higher levels shows that distorted input produces stronger effects than complete removal. It should be noted that the case where removal of the feedback signal produces no effect at all is only a limiting case of the design that will be observed if the autonomous control signal is exactly adapted to the peripheral transformation. Whenever the peripheral transformation is not taken into account exactly – and this will be common in biological systems - elimination of the feedback signals will produce some small effects. Second, as will be discussed in the next paragraph, the scheme allows correct predictions as to when the elimination or the distortion of feedback signals will produce greater or lesser effects. Finally, there is evidence that separable motor and sensory representations do indeed exist.

If the peripheral transformation is unusually complex or only partially known, then it may be impossible to generate an autonomous control signal that takes this transformation into account. (If, in the simpler scheme of Fig. 2a, M is unknown, then A cannot be adjusted such that $MA = 1$). In this case, the output, the controlled variable y, varies according to the presence or absence of feedback. Such situations arise when complex tools are added to the peripheral transformation. One example is a tracking task where the position of the follower is the controlled variable. If the only feedback from the controlled variable is visual, then performance should depend more strongly on visual feedback as the peripheral transformation becomes more complex. This dependence was confirmed by varying the peripheral transformation and by measuring the performance decrement when the follower is occasionally extinguished for a few seconds (Heuer, 1988). The increasing importance of visual feedback with increasing complexity of the peripheral transformation is also indicated by the fact that quasi-linear, closed-loop models are reasonably successful with highly complex tracking tasks (Poulton, 1966).

Postulating separable motor and sensory representations of a movement implies that it should be possible to modify one without changing the other. There are some data from prism adaptation experiments which suggest that this may happen. Bailey (1972) examined slow and rapid movements to a target during prism adaptation and also during pre- and posttests. If subjects pointed slowly at the target while wearing the prism, then the after-effect - a systematic bias in pointing after removing the prism - was much larger for this slow movement than for the fast movement. This finding suggests that during prism adaptation the kinesthetic representation of the visual target was changed, giving rise to the bias in slow pointing, but the motor representation, which is the major determinant of rapid aiming movements was unchanged. Also Schmidt, Christenson, and Rogers (1975) showed that delaying knowledge of results (KR) during learning affected the development of memory for movement recognition, but did not affect the memory of movement production. This result again supports the separation of sensory (recognition) and motor (production) representations (see also Schmidt, 1975).

So far our discussion has been simplified by considering only a single controlled variable - e.g., the position or velocity of the terminal device. Control of the terminal device, however, usually requires coordinated control of many muscles and joints. In general, there are many inputs to the box labeled "peripheral transformation" in Fig. 3, so a network has to be postulated that "distributes" the global control signal referring to the controlled variable among the various muscles involved. The function of this distributing network is partly constrained by how the peripheral transformation combines the various inputs to determine the single controlled variable, y. But partly its function seems to be arbitrary or the product of habit (cf. MacNeilage, 1980). Some arbitrariness in the function of the distributing network is possible when various input patterns to the peripheral transformation produce one and the same output value (c.f. "Coordinative Structures").

What is the role of sensory input for the distributing network? As an example, consider lip closure when speaking. A lip closure of amplitude a can be achieved by various combinations of movements by the upper lip (a_u), lower lip (a_l), and jaw (a_j). The task of the distributing network is to insure that $a = a_u + a_l + a_j$. One way of doing so is to use constant weights, setting $a_u = w_u \cdot a$, $a_l = w_l \cdot a$, and $a_j = w_j \cdot a$ with the weights adding to 1. This would specify a rigid pattern of the three articulators, which runs counter to experimental findings. In fact, in a sequence of utterances with constant a the contributions of the various articulators vary, so that sometimes the one and sometimes the other contributes more to lip closure (e.g., Hughes & Abbs, 1976). Thus, instead of setting a rigid distribution, the function of the distributing network is to specify the required covariation. One of the oldest reports of such covariation concerns the task of throwing a ball a certain distance: the initial velocity of the ball and the deviation of the initial trajectory from 45° are negatively correlated, which corresponds to the physical fact that these two variables have to be negatively correlated if distance are to remain constant (Stimpel, 1933).

So far the distributing network could be considered as being a pure output device. Perturbation experiments, however, show that its function depends on sensory input. For example, interfering with the movement of one of the articulators during speech elicits rapid compensatory movements in remote articulators (for review see Abbs, Gracco, & Cole, 1984). If an unexpected state of one of the articulators is signaled via sense organs, then other articulators take over to achieve the desired state of the controlled variable, which in this case is the task-related variable lip closure. The occurrence of remote compensation in the case of perturbations implies that the covariations defined by the distributing network are formed at least partly in sensory terms and not just in motor terms.

Should the sensory inputs to the distributing network be called "feedback"? If one focuses on individual articulators, then proprioceptive signals from the perturbed articulator clearly have feedforward effects when they modify the movement of other articulators, as well as probable feedback effects onto the perturbed articulator to resist the perturbation (see Abbs et al., 1984). But if the total system is considered with respect to the task variable lip closure, or the sensory signals themselves relate to this variable, then it is justified to talk about feedback.

One final aspect of the scheme of Fig. 3 as amended to include a distributing network should be mentioned. If one and the same autonomous control signal can be channeled through different distributing networks to activate different effector groups, then this could explain the phenomenon of motor equivalence. The most common example is the observation that a person's handwriting retains many common features when written small on paper or large on a blackboard. Therefore it is reasonable to think that these common features are part of a higher level control structure, organized above the level of the specific muscle groups. This effectively means that different motor control structures may share the higher levels of control related to their respective controlled variables (e.g., position of a pen on paper or a piece of chalk on a blackboard), but use different distributing networks to translate the control signal into the pattern of neural activity necessary to coordinate the different effector muscles.

Reference Signals

Reference signals indicate what the state of a moving limb should be. Thus they include both the "autonomous control signal" and the "feedback reference" of Fig. 3. Often, however, reference signals are produced not just by autonomous processes but also by sensory input. These sensory inputs may derive either from sense organs which are unaffected by the motor system using the reference signal - in which case the sensory signal serves as feedforward - or they may derive from sense organs which are affected by the system for which they provide reference information. In the latter case it may be difficult to distinguish between feedback and feedforward: a sensory signal relating to the controlled variable can provide feedback as outlined in the scheme of Fig. 3 and at the same time participate on a higher level in the production of the reference signal.

It is a rather trivial observation that sensory reference signals can define the desired positions of a terminal device. For example, in a continuous tracking task, the reference signal provides an exact prescription for the position-time curve of the follower. In a discrete tracking task, where the reference signal consists of a series of steps, the movement of the follower is incompletely specified. (As we will discuss on p. 68f, discontinuous or intermittent processing of sensory information may occur even when the target is continuously present.) In both cases the afferent information is used to update reference values incorporated into autonomous control mechanisms, even if this updating does not occur continuously.

In some movement patterns, a more complex integration of sensory reference signals and autonomous control functions can be observed. For example, consider a central pattern generator producing a harmonic oscillation. On a more abstract level, this mechanism can be described as a motor program with a sine wave as the prototypical function and the amplitude and frequency as parameters. It is possible that a sensory reference signal continuously controls one or the other parameter. An example is the control of wing beat amplitude by a flying insect, in which an optomotor turning response is induced by rotating the insect's visual surroundings. A similar optomotor response can be elicited from the stick insect during walking: the rotation of the visual field induces changes in amplitude or frequency of leg movements or both, causing the insect to walk in curves (Jander, 1985).

The role of this afferent information as a reference signal might also be called feedforward but, as mentioned above, the distinction between feedforward and feedback depends on the definition of the system and on the selection of the variable considered to be the controlled variable. As long as the movement defined as the output does not directly influence the sense organ considered, the afferent activity should be regarded as providing a reference signal.

As outlined above, a central motor program can be described using a prototypical function. So far, we have considered the transformation of the dummy variable ϑ into real time by a simple scaling operation, $t = T \cdot \vartheta$, to obtain an actual movement of duration T, but we have not discussed how this might occur. Lashley (1951) outlined a simple mechanism for generating a temporal series of events from a nontemporal representation. According to his hypothesis, a spatially organized representation is read out sequentially. This is what actually happens in stepping through the "spatially" organized command sequence for a digital computer. If the same mechanism applies to the execution of a central motor program, then the relative timing should remain invariant, as is often the case. This kind of design might require some internal clock mechanism that determines how fast the program is executed. Alternatively, a chain reflex mechanism which allows the movement velocity to be varied would also show invariant relative timing or phase constancy.

Whether something like "metering out of time" does occur in motor control has been a subject of debate (e.g., Schmidt, 1980, 1987; Kelso, Holt, Kugler, & Turvey, 1980; Kelso & Kay, 1987). The experimental results seem to depend on the nature of the movement studied (Schmidt, 1980; Schmidt, McGown, Quinn, & Hawkins, 1986). To answer the question by "always" or "never" probably underestimates the versatility of motor control structures. An internal clock mechanism which can run at different rates would represent an internal time base for the motor program, providing a signal that replaces the dummy variable ϑ. But an autonomous, clock-like signal is not the only candidate for replacing the dummy variable. In fact, the motor control structure may use several different signals: some of them internally generated, others possibly derived from sensory signals. In some sense, these signals can be considered equivalent to "time," but they do not measure time itself and they may represent different functions of real time.

Lee, Young, Reddish, Lough, and Clayton (1983) describe a case where a sensory reference signal seems to fulfill this timing function. The task for the subject is to time a jump so as to punch a falling ball at the apex of the jump. This task requires preparatory flexion of the knees followed by rapid extension to propel the body. The hypothesis explored was that the timing of the preparatory movements is coupled to a stimulus variable called τ. τ is defined as $\alpha/(d\alpha/dt)$, where α is the visual angle of the approaching object and $d\alpha/dt$ is its first derivative with respect to time. This variable is directly related to the time that remains until the object collides with the observer, given that it approaches with constant velocity. Since a falling ball has a constant acceleration, its velocity increases during the approach. When the ball falls from different heights, different functions for τ as a function of time to contact can be obtained. From these functions, specific predictions can be made as to how the anticipatory movement pattern should depend on the height from which the ball is dropped, given the assumption that the movement is temporally coupled to τ.

Lee et al. (1983) confirmed these predictions. The timing of the preparatory movements for striking a ball dropped from different heights varies when knee angle is potted as a function of time, but these differences vanish when the data are plotted as a function of τ. This finding strongly suggests that the dummy variable ϑ of a motor program is replaced by the sensory variable τ. Here we have a case of what Lee and Young (1983) call extrinsic timing; in this case the extrinsic timing is continuous. It is not hard to imagine other instances where a sensory reference signal may be processed in the manner described here for τ.

Type 2: Intermittent Use of Sensory Information

The continuous stream of movement can often be segmented into units. In the computer program analogy, the autonomous control mechanisms for such units are often referred to as subroutines. It is not always clear whether the units used in the description of behavior have a counterpart in separate units of control, but in some instances a division of the apparently continuous flow of movement into subunits seems realistic (e.g., swing phase and stance phase of legs walking, in Grillner, 1981; Bässler, 1983; see also Viviani, 1986; Monsell, 1986; Schmidt, 1985b). Unfortunately, the criteria for defining units are unclear.

Sometimes it can be shown that sensory information for motor control is only used at discrete points in the movement rather than continuously. This could provide one criterion for dividing a complex movement into subunits. In this section we will consider this intermittent processing of sensory information, focusing on the question of when and how sensory information is used. It is again possible to classify signals as feedback and reference signals. However, as mentioned above, the distinction varies according to how the units of control are defined. This problem is particularly acute for intermittent processing, so we will discuss both types together in this section. The problem is acute because the effect of intermittent processing can either be related to the immediately preceding movement segment, in which case it serves as feedback, or to the succeeding segment, in which case it serves as feedforward or as a reference signal.

Updating Parameters

The points in the time at which a sensory signal is processed need not depend on the signal itself; they can be specified either through autonomous mechanisms or other sensory inputs. This, for example, is the case in an intermittent servo-system that samples the error signal at a fixed rate. Such a system served as the model for Craik's (1947) description of human operators in manual control tasks. Craik's model is based on the following observation: when inexperienced subjects have to track a ramp, their movement is uneven and oscillates around the target ramp. The response looks more like a sequence of small, discrete aiming movements than like a smooth ramp movement. Therefore, Craik postulated that the movement consists of a sequence of ballistic corrections occurring approximately every 0.5 s. In Craik's model of manual tracking, the visual stimulus, in this case a deviation from the target, is sampled according to some kind of internal clock and used to adjust the parameters of a motor program for the corrective movements. During programming and execution of the corrections the system is thought to be refractory and insensitive to sensory input. Since the frequency of the intermittent corrections was not affected by amplification of the error signal, the conclusion was that the sampling rate is internally determined and independent of the amplitude of the error signals itself. However, it was shown later that the start of the first correction is triggered by the error signal (Navas & Stark, 1968). Brooks, Cooke, and Thomas (1973) found very similar discontinuities in arm movements of monkeys.

There is another type of internal specification of the times at which a sensory signal is used for motor control. In some behaviors, particular sensory signals need to be processed only at certain points associated with particular states of the movement. Different sensory signals may be used in different parts of a movement cycle. The fact that a certain signal is not incorporated into the control of movement during some part of the movement, of course, does not imply that the sensory input is not monitored during this time.

As an example, consider the human stride length in running. This depends mainly on the flight distance of the trunk (see Lee, Lishman, & Thomson, 1982). If the trajectory of the trunk is approximated as a parabola, then this distance is given by

$$s = \frac{2\, v_x\, v_y}{g} \qquad (6)$$

where v_x and v_y are horizontal and vertical velocity at takeoff, respectively, and g is the acceleration due to gravity. There is evidence that the horizontal velocity is roughly constant during a step cycle. Hence, distance is proportional to the flight duration as given by

$$t = \frac{2\, v_y}{g} \qquad (7)$$

Since v_y is proportional to the vertical impulse, this impulse has to be adjusted according to an estimate of how long it will take - given a certain horizontal velocity v_x - to reach the position for the next footstep. This "time estimate" is available through the variable τ that we have discussed on p. 67 (see also Warren, Young, & Lee, 1986).

In this example, there are "time windows" in the continuous run during which a visual variable is used for the adjustment of certain parameters of a central motor program. The time windows are imposed by the intermittent opportunity for making corrections - i.e., during contact with the substrate. These properties are presumably incorporated into the motor control structure so as to determine when and how different signals are processed in the course of a periodic movement.

Other examples where the use of afferent information for motor control depends on the state of the system are the so-called phase-dependent reflex reversals or "program-dependent reflexes," which occur in walking movements of crustaceans (DiCaprio & Clarac, 1981), insects (Bässler, 1974, 1977; Graham, 1979), and cats (Forssberg, Grillner, & Rossignol, 1975). In these examples, sensory signals lead to different responses depending on when they occur in the step cycle.

Stick insects show a reflex which prevents the hind leg from standing on the tarsus of the adjacent anterior leg. If the hind leg lands on the forward leg, it is briefly lifted and moved to a more posterior position. The trigger for this reaction is a tactile stimulus to the forward leg. Touching this leg with a brush will do, but this stimulus is only effective during part of the hind leg's step cycle (Graham, 1979). The full reaction only occurs for stimuli during the beginning of the stance phase. Stimuli in the second half of the swing phase cause the hind leg to shorten its swing. When the hind leg is in the second half of its stance phase or in the first half of the swing phase, no reaction of the hind leg is observed. Similar effects have been described for cats (Forssberg et al., 1975). Stimulating the dorsal part of the paw causes the leg to flex when it is in swing phase, whereas the same stimulus elicits increased extensor activity when applied during the second half of the stance phase. Z. Hasan (1986, personal communication) has suggested, however, that instead of being program dependent, the response may vary simply as a function of the position of the leg which of course covaries with the phase of the locomotor cycle.

So far we have considered instances in which processing of sensory information for adjustment of motor program parameters is determined internally or by other sensory input. It can, however, also depend on the sensory information itself. This seems to be the case in an experiment by Pew (1966). In this experiment subjects had to center a marker on a screen by pressing two keys alternately. Pressing a key resulted in a constant acceleration of the marker in one or the other direction. Early in practice, key presses were rather slow. It seemed as if subjects pressed a key and then waited until the marker changed and crossed the center line of the screen before they pressed the other key.

In the course of prolonged practice, subjects learned rapid key press sequences that held the marker approximately stationary with only small oscillations. However, the marker remained completely stationary only for identical transition times between the two keys. This the subjects were unable to obtain. Because the ratio of transition times was not equal to 1, the marker drifted slowly across the screen. At this stage of practice subjects seemed to neglect the small oscillations and only to process the drift. Corrections occurred when the drift became too large. These corrections revealed two different strategies. Some subjects changed the durations of a few key presses, bringing the marker quickly back to the center. Other subjects used a more elegant strategy. They changed the ratio of transition times and thus altered the direction of the drift. This suggests that in both groups - but in a different way for each group - the visual signal itself influences the parameters of the motor control structure when a certain threshold was exceeded.

Triggering Transitions

In the following examples the sensory information itself is assumed to be continuously monitored. In these motor control structures, however, only the transitions from one subroutine to the next appear to depend on critical values of the monitored variables and there is no obvious processing of these signals with regard to parameter adjustment within a subunit. Consider the transition from the stance phase to the swing phase in the stick insect. When the movement of a leg in stance phase is prevented, one might still expect that the swing phase would be initiated as soon as the normal duration for the stance phase is reached. This is true for monkeys trained to make oscillating arm movements: obstructing the movement in one direction does not alter the time at which the antagonists become active to initiate the movement in the opposite direction (Conrad & Brooks, 1974). The situation for the stick insect is different: the leg continues its stance phase. During this stance phase continuous feedback systems which apparently control velocity are active. Proprioceptive information on position does not appear to be incorporated into motor commands. Nevertheless, position information is monitored and used to trigger the transition from the stance to swing phase. This transition occurs only if a given position threshold is reached (Cruse, 1985a, 1985b; Dean & Cruse, 1986). In this way a continuous afferent signal is only used for motor control at particular points which are dependent upon the actual position of the leg.

Similar trigger functions have been reported for the variable τ (see, for example, Lee & Young, 1986). The forceful leg extension of ski jumpers is probably initiated when a certain value of τ is reached. Similarly, the gannet which plummets toward the sea seems to streamline its wings when a τ threshold is reached.

To summarize, there are two basic functions for intermittently processed sensory signals: (a) triggering transitions from one movement phase to another; and (b) updating parameters. The triggering function implies that the signals are monitored continuously. The motor control structure defines a threshold for the transition from one subroutine to the next, and the transition occurs as soon as the threshold is exceeded.

The updating function can be performed at internally or externally determined points in time. Externally determined updating again requires continuous monitoring of the afferent signals. Updating occurs as soon as a certain threshold is exceeded. Thus updating is similar to triggering except that in triggering the monitored signals have no influence on the parameters of the triggered subroutine, while in updating there is a systematic relation between at least one sensory signal and the parameter change. Internally determined updating, in contrast, is completely defined by the requirements of the motor control structure. Certain information may be needed only at certain times, so it can be processed for parameter adjustment during preset time windows.

Type 3: Advance Use of Sensory Information

The third role is played by sensory information which is used prior to or at the start of an action. Sensory information of this type is used to elicit movements, selecting motor programs, and specifying parameters. It is hard to make a cogent distinction between type 3 and type 2 functions in the general case, but in practice the distinction is usually clear. The problem is similar to that of distinguishing between feedback and feedforward: it arises in deciding just what entity is taken as the unit for analysis. When an entire tracking task or a complete step cycle is defined as the action, then the error signals eliciting correction movements in the first case or the transition signals in the second are type 2 - intermittent influences occurring within the movement. However, one might also consider each correction movement or the separate swing and stance phase as the units of organization, in which case the same signals would be considered as type 3, advanced processing influences.

It is rarely the case that an external stimulus is the only determinant of a movement. (Exceptions in humans might be eye movements toward a moving peripheral stimulus or withdrawing a hand in response to pain.) The more frequent condition for elicitation of a movement pattern appears to be a conjunction of internal "drive" and external stimulus. This formulation, of course, is reminiscent of Hull's (1952) theory of behavior. A cat or a crayfish standing on a stationary treadmill will normally begin to walk when the belt begins to move, provided that the internal excitation is high enough. This "drive" is an essential part. It may vary with time or developmental stage or from one species to another. For example, in one crayfish species adults could easily be induced to walk on a motor-driven treadmill. In contrast, young animals of the species or adults of another species could not be stimulated to walk on the same treadmill although they were quite active walkers under other conditions (Cruse & Müller, unpublished observation). Stick insects walk on a motor-driven wheel very reluctantly and only with additional tactile stimulation. Similarly, a decerebrated cat needs electrical or chemical stimulation to begin walking (Grillner, 1981).

In humans it may be hard to relate the initiation of movement to specific drives. Rather, one can speak about intentions. Intention can be conceptualized as tuning the CNS to process certain sensory information in particular ways. The state of the nervous system corresponding to an intention bears some similarity to a motor control structure. The main difference is that its function is not to control the execution of a movement but rather its initiation. The intentional state embodies rules for the assignment of particular input patterns to the establishment or the activation of pre-established motor control structures, similar to conditional GOTO statements in a computer program (Heuer & Prinz, 1987). In man, this assignment can be selected in a rather arbitrary way (for example, as a result of instruction of learning), but some assignments allow for faster and more accurate initiation of movement patterns than do other assignments.

Use of sensory information for parameter adjustment in advance of movement can be clearly seen in knowledge of results (KR) experiments, one of the traditional paradigms of motor learning. In a typical KR experiment subjects have to repeatedly perform a specified movement. After each performance or at longer intervals, they are given information about the accuracy of their movements. This information about the error on an previous movement can then be used to adjust parameters for the next performance. Thus, KR plays two roles. First, KR informs subjects about the target movement and thus is a prerequisite for the development of learning (i.e., creating a relatively permanent internal representations of the desired movement). Second, and this has only recently been recognized (Salmoni, Schmidt, & Walter, 1984), KR provides "external guidance" - external information which can be used for temporary parameter adjustment and may not be used to form a part of the relatively permanent (i.e., learned) representation of the action. Subjects seem to rely partly on this external information for parameter adjustment, so that a performance decrement is observed when it is withdrawn.

Conclusion

To summarize, we have described various ways in which sensory information is used for motor control. Before the actual start of the movement, sensory information (type 3) is used for selecting and activating an appropriate motor control structure and specifying the initial parameter values for the autonomous control signal. During movement, sensory information (type 2) can be used for triggering successive units and for intermittent updating of parameters. These actions may depend either on the current state of the control structure, on the state of the monitored signals, or on a combination of both. Finally, sensory information (type 1) can be used continuously either as feedback, as a reference signal, or as a kind of time base for a motor program. In the latter instance, it defines the "time axis" on which the motor program operates. Sensory information used as feedback can be processed in parallel with the autonomous control signal. It serves to reduce or correct any errors due to changes in the peripheral transformation or to inaccurate parameter specification in the direct control path.

The diverse functions of sensory signals appear to make any simple answer to the question of how the motor system works unrealistic. Instead, one may have to be content with a set of specific answers to specific questions. The specificity enters with the requirements posed by each task and with the role of chance in the evolution of biological solutions. Each task appears to favour some control mechanisms over others in the development of motor control structures for performing the required action. This is particularly true with respect to the use of sensory information. The motor control structure itself determines, to a large extent, how and when sensory signals are processed. In this sense it is not only a "motor system" but rather a "sensorimotor system."

So far we have taken our examples from various species based on the implicit assumption that the basic ways of processing sensory information for motor control are invariant. Species, however, do vary. In particular, the variety of motor control structures seems to increase in higher species.

In the frog, for example, it seems to be possible to compile a rather complete list of the motor control problems this animal faces. More importantly, it is possible to identify anatomical structures that subserve these control problems - neural centers which process sensory information in a particular way for a particular movement patterns (see Ingle, 1982). These structures represent the higher-level anatomical substrates in the associated motor control structures. Thus, motor control structures in the frog seem to be "hard-wired." They process in a preestablished way: once they are established little or no plasticity remains. For example, lower animals do not adapt their motor actions to changed visual input such as occurs when the eyes are surgically rotated in their sockets (see Taub, 1968).

The situation is radically different in humans. Humans adapt readily to even severe changes of visual input (see, for example, Welch, 1978), the assignment of output motor patterns to input signals is nearly arbitrary, and it appears impossible to compile an even modestly complete list of the motor control problems. Motor control structures in humans appear to be "soft-wired": they can be readily adjusted to meet changed environmental demands. Hence, attempts to find general rules for processing sensory information or for the parameterization of motor programs appear doomed to failure. The best that can be done is to find rules for particular tasks, that is, rules that hold for a particular motor control structure.

This sounds like a rather pessimistic view, but there is some consolation. Even if there are innumerable motor control structures, there are some that are particularly relevant and deserve study in their own right, such as locomotion. Also, there should be some general rules defining what is possible for motor control structures. Although these rules will probably not suffice to predict how motor control will actually work in particular situation, they may allow a range of possibilities to be specified. Finally, there may be some features that are shared by many motor control structures. For example, the parallel use of autonomous command signals and closed-loop control seems to be such a feature.

Acknowledgements. We appreciate the helpful comments of Don Graham, Bruce Bridgeman, and the other members of the ZiF study group on Perception and Action.

References

Abbs, J.H., Gracco, V.L., & Cole, K.J. (1984). Control of multi-movement coordination: sensorimotor mechanisms in speech motor programming. *Journal of Motor Behavior, 16*, 195-232.

Adams, J.A. (1971). A closed-loop theory of motor learning. *Journal of Motor Behavior, 3*, 111-150.

Baily, J.S. (1972). Adaptation to prisms: do proprioceptive changes mediate adapted behavior without ballistic arm movements? *Quarterly Journal of Experimental Psychology, 24*, 8-20.

Bässler, U. (1974). Vom femoralen Chordotonalorgan gesteuerte Reaktionen bei der Stabheuschrecke *Carausius morosus:* Messung der von der Tibia erzeugten Kraft im aktiven und inaktiven Tier. *Kybernetik, 16*, 213-226.

Bässler, U. (1977). Sensory control of leg movement in the stick insect *Carausius morosus. Biological Cybernetics, 25*, 61-72.

Bässler, U. (1983). *Neural basis of elementary behavior in stick insects*. Berlin, Heidelberg, New York: Springer.

Bässler, U. (1986). On the definition of central pattern generator and its sensory control. *Biological Cybernetics, 54*, 65-69.

Bernstein, N.A. (1975). Einige heranreifende Probleme der Regulation der Bewegungsakte. In N.A. Bernstein (Ed.), *Bewegungsphysiologie*. Leipzig: Barth. (originally published 1957).

Berthoz, A., Pavard, B., Young, L.R. (1975). Perception of linear horizontal self-motion induced by peripheral vision (linearvection). Basic characteristics and visual-vestibular interactions. *Experimental Brain Research, 23*, 471-489.

Bizzi, E., Polit, A., & Morasso, P. (1976). Mechanisms underlying achievement of final head position. *Journal of Neurophysiology, 39*, 435-444.

Brooks, V.B., Cooke, J.D., & Thomas, J.S. (1973). The continuity of movements. In R.B. Stein, K.G. Pearson, R.S. Smith, & J.B. Redford (Eds.), *Control of posture and locomotion*. New York: Plenum.

Brown, T.G. (1911). The intrinsic factors in the act of progression in the mammal. *Proceedings of the Royal Society, 84B*, 308-319.

Carlton, L.G. (1981). Processing of visual feedback information for movement control. *Journal of Experimental Psychology: Human Perception and Performance, 7*, 1019-1030.

Carter, M.C. (1984). *Simultaneous control of two rhythmic behaviors in the cat: locomotion and the paw-shake response*. Unpublished doctoral dissertation, University of California, Los Angeles.

Chase, R.A. (1965). An information-flow model of the organization of motor activity. I. Transduction, transmission and central control of sensory information. *Journal of Nervous and Mental Disease, 140*, 239-251.

Clarac, F. (1981). Decapod crustacean leg coordination during walking. In C.F. Herreid & C.R. Fourtner (Eds.), *Locomotion and energetics in arthropods*. New York: Plenum.

Cohen, L. (1971). Synchronous bimanual movements performed by homologous and non-homologous muscles. *Perceptual and Motor Skills, 32*, 639-644.

Conrad, B., & Brooks, V.B. (1974). Effects of dentate cooling on rapid alternating arm movements. *Journal of Neurophysiology, 37*, 792-804.

Craik, K.J.W. (1974). Theory of the human operator in control system. I. The operator as an engineering system. *British Journal of Psychology, 38*, 56-61.

Cruse, H. (1980). A quantitative model of walking incorporating central and peripheral influences: I. The control of the individual leg. *Biological Cybernetics, 37*, 131-136.

Cruse, H. (1981). *Biologische Kybernetik. Eine Einführung in die lineare und nichtlineare Systemtheorie*. Weinheim: Verlag Chemie.

Cruse, H. (1985a). Which parameters control the leg movement of a walking insect? I. Velocity control during the stance phase. *Journal of Experimental Biology, 116*, 343-355.

Cruse, H. (1985b). Which parameters control the leg movement of a walking insect? II. The start of the swing phase. *Journal of Experimental Biology, 116*, 357-362.

Cruse, H. (1986). Constraints for joint angle control of the human arm. *Biological Cybernetics, 54*, 125-132.

Cruse, H., & Brüwer, M. (1987). The human arm as a redundant manipulator: the control of path and joint angles. *Biological Cybernetics, 57*, 137-144.

Cruse, H., & Müller, U. (1986). Two coupling mechanisms which determine the coordination of ipsilateral legs in the walking crayfish. *Journal of Experimental Biology, 121*, 349-369.

Dean, J., & Cruse, H. (1986). Evidence for the control of velocity as well as position in leg protraction and retraction by the stick insect. In H. Heuer & C. Fromm (Eds.), *Generation and modulation of action patterns*. Berlin, Heidelberg, New York: Springer.

Dean, J., & Wendler, G. (1983). Stick insect locomotion on a walking wheel: interleg coordination of leg position. *Journal of Experimental Biology, 103*, 75-94.

Delcomyn, F. (1971). The locomotion of the cockroach. *Journal of Experimental Biology, 54*, 443-452.

Delcomyn, F. (1981). Insect locomotion on land. In C.F. Herreid & C.R. Fourtner (Eds.), *Locomotion and energetics in arthropods*. New York: Plenum.

DiCaprio, R.A., & Clarac, F. (1981). Reversal of a walking leg reflex elicited by a muscle receptor. *Journal of Experimental Biology, 90*, 197-203.

Feldman, A.G. (1966). Functional tuning of the nervous system with control of movement maintenance of a steady posture. III. Mechanographic analysis of the execution by man of the simplest of motor tasks. *Biophysics, 11*, 766-775.

Feldman, A.G. (1974a). Change of muscle length due to shift of the equilibrium point of the muscle-load system. *Biophysics, 19*, 544-548.

Feldman, A.G. (1974b). Control of muscle length. *Biophysics, 19*, 766-771.

Feldman, A.G. (1986). Once more on the equilibrium point hypothesis (model) for motor control. *Journal of Motor Behavior, 18*, 17-54.

Forssberg, H., Grillner, S., & Rossignol, S. (1975). Phase dependent reflex reversal during walking in chronic spinal cats. *Brain Research, 85*, 103-107.

Goodwin, G.M., McCloskey, D.I., & Matthews, P.B.C. (1972). The contribution of muscle afferents to kinesthesia shown by vibration-induced illusions of movement and by the effects of paralyzing joint afferents. *Brain, 95*, 705-748.

Graham, D. (1972). A behavioral analysis of the temporal organization of walking movements in the 1st instar and adult stick insect. *Journal of Comparative Physiology, 81*, 23-52.

Graham, D. (1977). Simulation of a model for the coordination of leg movement in free walking insects. *Biological Cybernetics, 26*, 187-198.

Graham, D. (1979). Effects of circum-oesophageal lesion on the behavior of the stick insect *Carausius morosus*. II. Changes in walking co-ordination. *Biological Cybernetics, 32*, 147-152.

Graham, D. (1985). Pattern and control of walking in insects. *Advances in Insect Physiology, 18*, 31-140.

Graham, D., & Bässler, U. (1981). Effects of afference sign reversal on motor activity in walking stick insects (*Carausius morosus*). *Journal of Experimental Biology, 91*, 179-193.

Grillner, S. (1981). Control of locomotion in bipeds, tetrapods, and fish. In V.B. Brooks (Ed.), *Handbook of physiology: Sec. 1. The nervous system: Vol. 2. Motor control, part 2*. Bethesda, MD: American Physiological Society.

Gunkel, M. (1962). Über relative Koordination bei willkürlichen menschlichen Gliederbewegungen. *Pflügers Archiv für die gesamte Physiologie, 275*, 472-477.

Haken, H., Kelso, J.A.S., & Bunz, H. (1985). A theoretical model of phase transitions in human hand movements. *Biological Cybernetics, 51*, 347-356.

Hein, A., & Held, R. (1962). A neural model for labile sensorimotor coordinations. In E.E. Bernhard, & M.R. Kare (Eds.), *Biological prototypes and synthetic systems I*. New York: Plenum.

Heuer, H. (1983). *Bewegungslernen*. Stuttgart: Kohlhammer.

Heuer, H. (1988). Psychomotorik. In H. Spada (Ed.), *Lehrbuch der Allgemeinen Psychologie*. Bern: Huber.

Heuer, H., & Prinz, W. (1987). Initiierung und Steuerung von Handlungen und Bewegungen, In M. Amelang (Ed.), *Bericht über den 35. Kongress der Deutschen Gesellschaft für Psychologie, Heidelberg, 1986*. Göttingen: Hogrefe

Heuer, H., & Wing, A.M. (1984). Doing two things at once: Process limitations and interactions. In M.M. Smyth & A.M. Wing (Eds.), *Psychology of human movement*. London: Academic.

Hopf, H.C., Handwerker, H., & Hausmanns, J. (1967). Die rasche Willkürbewegung des Menschen. Untersuchungen zur zentralen Programmierung. *Deutsche Zeitschrift für Nervenheilkunde, 191*, 186-209.

Hoyle, G. (1964). Exploration of neuronal mechanisms underlying behavior in insects. In R.F. Reiss (Ed.), *Neural theory and modelling*. Stanford: Stanford University Press.

Hoyle, G. (1983). *Muscles and their neural control*. New York: Wiley.

Hughes, O.M., & Abbs, J.H. (1976). Labial-mandibular coordination in the production of speech. Implications for the operation of motor equivalence. *Phonetica, 33*, 199-221.

Hull, C.L. (1952). *A behavior system*. New Haven: Yale University Press.

Ingle, D.J. (1982). Organization of visuomotor behaviours in vertebrates. In D.J. Ingle, M.A. Goodale, & R.J.W. Mansfield (Eds.), *Analysis of visual behavior*. Cambridge, MA: MIT Press.

Jander, J.P. (1985). Mechanical stability in stick insects when walking straight and around curves. In M. Gewecke & G. Wendler (Eds.), *Insect locomotion*. Berlin: Parey.

Jones, B. (1974). Is proprioception important for skilled performance? *Journal of Motor Behavior, 6*, 33-45.

Jordan, T.C. (1972). Characteristics of visual and proprioceptive response times in the learning of a motor skill. *Quarterly Journal of Experimental Psychology, 24*, 536-543.

Keele, S.W., & Posner, M.I. (1968). Processing of visual feedback in rapid movements. *Journal of Experimental Psychology, 77*, 155-158.

Kelso, J.A.S. (1986). Pattern formation in speech and limb movements involving many degrees of freedom. In H. Heuer & C. Fromm (Eds.), *Generation and modulation of action patterns*. Berlin, Heidelberg, New York: Springer.

Kelso, J.A.S., & Kay, B.S.A. (1987). Information and control. In H. Heuer & A.F. Sanders (Eds.), *Perspectives on perception and action*. Hillsdale, NJ: Erlbaum.

Kelso, J.A.S., & Stelmach, G.E. (1976). Central and peripheral mechanisms in motor control. In G.E. Stelmach (Ed.), *Motor control: issues and trends*. New York: Academic.

Kelso, J.A.S., Southard, D.L., & Goodman, D. (1979). On the co-ordination of two-handed movements. *Journal of Experimental Psychology: Human Perception and Performance, 5*, 229-238.

Kelso, J.A.S., Holt, K.G., Kugler, P.N., & Turvey, M.T. (1980). On the concept of coordinative structures as dissipative structures. II. Empirical lines of convergence. In G.E. Stelmach & J. Requin (Eds.), *Tutorials in motor behavior*. Amsterdam: North-Holland.

Kelso, J.A.S., Schöner, G., Scholz, J.P., & Haken, H. (1987). Phase-locked modes, phase transitions and component oscillators in biological motion. *Physica Scripta, 35*, 79-98.

Kugler, P.N., Kelso, J.A.S., & Turvey, M.T. (1980). On the concept of coordinative structures as dissipative structures. I. Theoretical line. In G.E. Stelmach & I. Requin (Eds.), *Tutorials in motor behavior*. Amsterdam: North-Holland.

Kugler, P.N., Kelso, J.A.S., & Turvey, M.T. (1982). On coordination and control in naturally developing systems. In J.A.S. Kelso & J.E. Clarke (Eds.), *The development of movement control and coordination.* New York: Wiley.

Land, M.F. (1972). Stepping movements made by jumping spiders during turns mediated by the lateral eyes. *Journal of Experimental Biology, 57,* 15-40.

Lashley, K.S. (1917). The accuracy of movement in the absence of excitation from the moving organ. *American Journal of Physiology, 43,* 169-194.

Lashley, K.S. (1951). The problem of serial order in behavior. In L.A. Jeffress (Ed.), *Cerebral mechanisms in behavior. The Hixon symposium.* New York: Wiley.

Laszlo, J.I. (1967). Training of fast tapping with reduction of kinesthetic, tactile, visual, and auditory sensations. *Quarterly Journal of Experimental Psychology, 19,* 344-349.

Lee, D.N., & Lishman, J.R. (1975). Visual proprioceptive control of stance. *Journal of Human Movement Studies, 1,* 87-95.

Lee, D.N., & Young, D.S. (1983). Visual timing of interceptive action. In D.J. Ingle, M. Jeannerod, & D.N. Lee (Eds.), *Brain mechanisms and spatial vision.* Dordrecht: Nijhoff.

Lee, D.N., & Young, D.S. (1986). Gearing action to the environment. In H. Heuer & C. Fromm (Eds.), *Generation and modulation of action patterns.* Berlin, Heidelberg, New York: Springer.

Lee, D.N., Lishman, J.R., & Thomson, J.A. (1982). Visual regulation of gait in long jumping. *Journal of Experimental Psychology: Human Perception and Performance, 8,* 448-459.

Lee, D.N., Young, D.S., Reddish, E., Lough, S., & Clayton, T.M.H. (1983). Visual timing in hitting an accelerating ball. *Quarterly Journal of Experimental Psychology, 35a,* 333-346.

MacKay, D.G. (1986). Self-inhibition and the disruptive effects of internal and external feedback in skilled behavior. In H. Heuer & C. Fromm (Eds.), *Generation and modulation of action patterns.* Berlin, Heidelberg, New York: Springer.

MacNeilage, P.F. (1980). Distinctive properties of speech motor control. In G.E. Stelmach & J. Requin (Eds.), *Tutorials in motor behavior.* Amsterdam: North-Holland.

Marteniuk, R.G., & MacKenzie, C.L. (1980). A preliminary theory of two-hand co-ordination control. In G.E. Stelmach & J. Requin (Eds.), *Tutorials in motor behavior.* Amsterdam: North-Holland.

Marteniuk, R.G., MacKenzie, C.L., & Baba, D.M. (1984). Bimanual movement control: information processing and interaction effects. *Quarterly Journal of Experimental Psychology, 36a,* 335-365.

Meyer, D.E., Smith, J.E.K., & Wright, C.E. (1982). Models for the speed and accuracy of aimed movements. *Psychological Review, 89,* 449-482.

Miller, G.A., Galanter, E., & Pribram, K.H. (1960). *Plans and the structure of behavior.* New York: Holt.

Monsell, S. (1986). Programming of complex sequences: evidence from the timing of rapid speech and other productions. In H. Heuer & C. Fromm (Eds.), *Generation and modulation of action patterns.* Berlin, Heidelberg, New York: Springer.

Nashner, L., & Berthoz, A. (1978). Visual contribution to rapid motor responses during postural control. *Brain Research, 150,* 403-407.

Navas, F., & Stark, L. (1968). Sampling or intermittency in hand control system dynamics. *Biophysics Journal, 8,* 252-302.

Nielsen, T.I. (1963). Volition: a new approach. *Scandinavian Journal of Psychology, 4,* 225-230.

Partridge, L.D. (1979). Muscle properties: a problem for the motor controller physiologist. In R.E. Talbott & D.R. Humphrey (Eds.), *Posture and movement.* New York: Raven.

Pearson, K.G., & Iles, F.J. (1973). Nervous mechanisms underlying intersegmental co-ordination of leg movements during walking in the cockroach. *Journal of Experimental Biology, 58,* 725-744.

Pew, R.W. (1966). Acquisition of hierarchical control over the temporal organization of skill. *Journal of Experimental Psychology, 71,* 764-771.

Polit, A., & Bizzi, E. (1978). Processes controlling arm movements in monkey. *Science, 201,* 1235-1237.

Poulton, E.C. (1957). On prediction in skilled movements. *Psychological Bulletin, 54,* 467-478.

Poulton, E.C. (1966). Tracking behavior. In E.A. Bilodeau (Ed.), *Acquisition of skill.* New York: Academic.

Prinz, W., & Sanders, A.F. (1984). Preface. In W. Prinz & A.F. Sanders (Eds.), *Cognition and motor processes.* Berlin, Heidelberg, New York: Springer.

Roeck, E. (1977). Verzögerte auditive Rückkopplung (VAR) der Lautsprache. In E. Roeck (Ed.), *Verzögerte auditive Rückkopplung (VAR).* Bern: Huber.

Rosenbaum, D.A. (1985). Motor programming: a review and a scheduling theory. In H.Heuer, U. Kleinbeck, & K.-H. Schmidt (Eds.), *Motor behavior. Programming, control, and acquisition.* Berlin, Heidelberg, New York: Springer.

Salmoni, A.W., Schmidt, R.A., & Walter, C.B. (1984). Knowledge of results and motor learning: a review and critical reappraisal. *Psychological Bulletin, 95*, 355-386.

Saltzman, E. (1986). Task dynamic coordination of the speech articulators. A preliminary model. In H.Heuer & C. Fromm (Eds.), *Generation and modulation of action patterns*. Berlin, Heidelberg, New York: Springer.

Saltzman, E., & Kelso, J.A.S. (1987). Skilled actions: a task-dynamic approach. *Psychological Review, 94*, 84-106.

Schmidt, R.A. (1975). A schema theory of discrete motor skill learning. *Psychological Review, 82*, 225-260.

Schmidt, R.A. (1980). On the theoretical status of time in motor program representations. In G.E. Stelmach & J. Requin (Eds.), *Tutorials in motor behavior*. Amsterdam: North-Holland.

Schmidt, R.A. (1982). *Motor control and learning: a behavioral emphasis*. Champaign: Human Kinetics.

Schmidt, R.A. (1985a). The search for invariance in skilled movement behavior. *Research Quarterly for Exercises and sport, 56*, 188-200.

Schmidt, R.A. (1985b). Identifying units of motor behavior. *The Behavioral and Brain Sciences, 8*, 163-164.

Schmidt, R.A. (1987). Motor and action perspectives on motor behavior. In O. Meijer & K. Roth (Eds.), *Complex movement behavior: the motor-action controversy*. Amsterdam: North-Holland.

Schmidt, R.A. (1988). *Motor control and learning: a behavioral emphasis* (2nd ed.). Champaign: Human Kinetics.

Schmidt, R.A., Christenson, R., & Rogers, P. (1975). Some evidence for the independence of recall and recognition in motor behavior. In D.M. Landers, D.V. Harris, & R.W. Christian (Eds.), *Psychology of motor behavior and sport II*. State College, PA: Penn State HPER Series.

Schmidt, R.A., McGown, C., Quinn, J.T., & Hawkins, B. (1986). Unexpected inertial loading in rapid reversal movements: violations of equifinality. *Human Movement Science, 5*, 263-273.

Schöner, G., Haken, H., & Kelso, J.A.S. (1986). A stochastic theory of phase transitions in human hand movements. *Biological Cybernetics, 54*, 247-257.

Selverston, A.I. (1980). Are central pattern generators understandable? *Behavioral and Brain Sciences, 3*, 535-571.

Shapiro, D.C., Zernicke, R.F., Gregor, R.J., & Diestel, J.D. (1981). Evidence for generalized motor programs using gait pattern analysis. *Journal of Motor Behavior, 13*, 33-47.

Shumway-Cook, A., & Woollacott, M.H. (1985). The growth of stability: postural control from a developmental perspective. *Journal of Motor Behavior, 17*, 131-147.

Sittig, K. (1986). *Kinesthesis and motor control*. Unpublished doctoral dissertation, University of Utrecht.

Stein, R.B. (1982). What muscle variable(s) does the nervous systems control in limb movements? *Behavioral and Brain Sciences, 5*, 535-577.

Stimpel, E. (1933). Der Wurf. *Neue Psychologische Studien, 9*, 105-138.

Taub, E. (1968). Prism compensation as a learning phenomenon: a phylogenetic perspective. In S.J. Freedman (Ed.), *The neuropsychology of spatially oriented behavior*. Homewood, IL: Dorsey.

Taub, E. (1976). Movements in nonhuman primates deprived of somatosensory feedback. *Exercise and Sport Sciences Reviews, 4*, 335-374.

Taub, E., & Berman, A.J. (1968). Movement and learning in the absence of sensory feedback. In S.J. Freedman (Ed.), *The neuropsychology of spatially oriented behavior*. Homewood, IL: Dorsey.

Taub, E., Goldberg, I.A., & Taub, P. (1975). Deafferentation in monkeys: pointing at a target without visual feedback. *Experimental Neurology, 46*, 178-186.

Taylor, F.V., & Birmingham, H.P. (1948). Studies of tracking behavior. II, The acceleration of quick manual corrective responses. *Journal of Experimental Psychology, 38*, 783-795.

Turvey, M.T. (1977). Preliminaries of a theory of action with respect to vision. In R. Shaw & J. Bransford (Eds.), *Perceiving, acting, and knowing*. Hillsdale, NJ: Erlbaum.

Viviani, P. (1986). Do units of motor action really exist? In H. Heuer & C. Fromm (Eds.), *Generation and modulation of action patterns*. Berlin, Heidelberg, New York: Springer.

von Holst, E. (1937). Vom Wesen der Ordnung im Zentralnervensystem. *Naturwissenschaften, 25*, 625-631, 641-647.

von Holst, E. (1939). Die relative Koordination als Phänomen und als Methode zentralnervöser Funktionsanalyse. *Ergebnisse der Physiologie, 42*, 228-306.

Warren, W.H., Young, D.S., & Lee, D.N. (1986). Visual control of step length during running over irregular terrain. *Journal of Experimental Psychology: Human Perception and Performance, 12*, 259-266.

Welch, R.B. (1978). *Perceptual modification: adapting to altered sensory environments*. New York: Academic.

Wendler, G. (1964). Laufen und Stehen der Stabheuschrecke: Sinnesborsten in den Beingelenken als Glieder von Regelkreisen. *Zeitschrift für vergleichende Physiologie, 48*, 198-250.

Wilson, D.M. (1968). The flight-control system of the locust. *Scientific American, 218*, 83-90.

Wing, A.M. (1977). Perturbations of auditory feedback delay and the timing of movement. *Journal of Experimental Psychology: Human Perception and Performance, 3*, 175-186.

Zelaznik, H.N., Hawkins, B., & Kisselburgh, L. (1983). Rapid visual feedback processing in single-aiming movements. *Journal of Motor Behavior, 15*, 217-236.

The Neuroethology of Perception and Action

J. DEAN

CONTENTS

Introduction

Neuroethology is the study of the physiological mechanisms underlying behavior (Hoyle, 1970). Its conceptual framework is biological in two important senses. First, it emphasizes a broadly comparative approach based on study material drawn from the entire animal kingdom. Second, the neuroethological analysis of behavior and neural mechanisms includes consideration of their evolutionary development and their value to the animal in its natural environment. Thus, the focus is on behavioral units of adaptive significance to the animal - functional units or actions in the language of the psychologist, and on mechanisms for identifying meaningful stimuli - perception in a broad sense.

This biological orientation of the field may make it appear far removed from the more human-centered approaches of psychology and medicine. For the psychologist interested in the relationship between perception and action, the notion that one should consider studying insects, snails, and other invertebrates may still seem strange. At first glance, the world of the invertebrates appears far removed from our own (see Griffin, 1982; Gould & Gould, 1982). Different sensory capacities, seemingly mechanistic behavior subject to the whims of immediate stimuli, and the apparent absence of strong affective responses (Eisemann et al., 1984) are a few of the obvious differences. If insight into our own conscious experience pro-

Relationships Between Perception and Action
Edited by O. Neumann and W. Prinz

vides little guidance for interpreting the actions of even closely related primates (Griffin, 1982), then it surely fails completely when applied to the invertebrates.

Nevertheless, invertebrates and vertebrates face the same basic challenges for survival. In order to reproduce they must locate suitable habitats, find food, seek mates, and protect themselves from predators. This selection for similar performance levels has two sources. On the one hand, all organisms are presented with the same abiotic cues and conditions. On the other hand, coevolution of vertebrates and invertebrates, especially through predator-prey interactions, selects for similar performance levels in terms of speed, force, and agility of movement.

Thus, all animals face similar problems in controlling behavior: they must select suitable actions and generate appropriate movement patterns. As a result, there are many functional similarities. For the most part, invertebrates rely on the same basic sensory modalities that we do: they see, hear, touch, taste, and smell. With few exceptions, differences in the morphology and physiology of the sense organs do not solve the problem of incorporating sensory information into adaptive behavior. For this purpose both vertebrates and invertebrates employ similar integrative mechanisms ranging from simple reflexes through servocontrol mechanisms to pattern recognition and abstract representation of the environment. At each moment only a limited number of behaviors are possible, so both vertebrates and invertebrates have mechanisms for selecting among the potential actions and for adjusting that selection according to the current needs of the animal. Selection of behavior can often be improved when experience is stored and used to modify the responses to future stimuli. This learning through experience takes a variety of forms, most of which have been identified in invertebrates as well as in vertebrates (Sahley, 1984). These common features serve as the basis for the expectation that the study of simple organisms can aid in understanding the behavior of more complex organisms.

Behaviors studied by neuroethologists range from the extremely simple to the very complex. An example of the very simple is the chemotaxis of bacteria (Koshland, 1980). Like many higher organisms, the bacteria *Escherichia coli* is able to orient in chemical gradients present in its environment, moving toward higher concentrations of nutrients and away from noxious substances. Both the receptor mechanisms for several classes of chemical stimuli and the flagellar motor have been described at the biochemical level. In this single-celled organism the chemical stimulus and the motor apparatus are not connected by a specialized nervous system; the two processes are linked directly by the electrical potential across the cell membrane, which makes these bacteria a useful model of chemotransduction in general.

Multicellular organisms rely on a specialized nervous system to provide adaptive links between environmental stimuli and behavior. Qualitatively, the complexity of the behavioral repertoire increases with increasing numbers of neurons. *Caenorhabditis elegans,* a nematode with 302 neurons (White, Southgate, Thomson, & Brenner, 1983, 1986), still belongs to the very simple organisms; nevertheless, it locomotes, shows chemo-, thermo-, and osmotaxes, and performs selective mating behaviors (Brenner, 1974). Its small size facilitated a complete description of the synaptic connectivity and the ontogeny of all 302 neurons (Sulston, 1983; White et al., 1986). *Aplysia,* marine gastropod mollusks with approximately 10 000 neurons, and *Drosophila,* fruit flies with approximately 100 000 neurons, show more complex behaviors including several forms of associative and nonassociative learning (Kandel, 1979; Quinn, 1984; Tully & Quinn, 1985; Cook & Carew, 1986). The honey bee, with nearly a million neurons (Menzel & Erber, 1978), and higher mollusks, such as the octopus (Young, 1971) with millions of neurons, are comparable in complexity to simpler vertebrates. These more advanced invertebrates show still more complicated behaviors including pattern recognition, learning, and some forms of cognitive representation (review Sanders, 1975; Gould, 1984). This diversity among invertebrates provides ample material for comparative study of behaviors of almost any complexity.

The great promise in using invertebrates is that the physiological mechanisms underlying behavior are more readily accessible (Maynard, 1972; Hoyle, 1975). The reduced number of neurons involved in many behaviors of interest and the ability to identify individual neurons having a typical form and carrying out specific functions simplify the study of the neural mechanisms. Historically, the invertebrates have often provided the most productive subjects for the methodology of the day. Thus, they have served as the testing ground for new techniques. At the same time, many physiological mechanisms first discovered in invertebrates were subsequently identified in vertebrates as well.

Although this historical pattern is comforting, one need not assume that results from invertebrates will generalize to vertebrates in order to appreciate the analytical advantages of comparative studies: such an approach refines functional analysis by providing a way to identify functionally significant features of neural organization. Different present-day species often illustrate diverse solutions to common functional problems. This is particularly true at more complex levels of organization whereas cellular structure and physiology have been largely conserved throughout the animal kingdom. Diversity arose because functional factors are not the only constraints affecting evolution; historical, developmental, and structural influences also play important roles (Dumont & Robertson, 1986). Through study of this diversity, comparative analysis provides the main hope for correctly distinguishing functionally important features from historically determined similarities.

In recent years, this comparative analysis has been extended to the level of individual neurons and neural networks. Initial success was achieved with respect to the generation of motor patterns, a subject of intensive investigation in the past 25 years. More recently, increasing sophistication in neurobiological techniques and in behavioral analysis has allowed more complex features of motor control to be investigated in animal models. Mechanisms related to such higher functions as behavioral choice and plasticity, motivation, and learning - processes in the domain of traditional psychology - can now be investigated at the level of individual neurons in some animals.

The present chapter discusses some results from this comparative, neuroethological approach. It focuses on selected physiological results that bear on the relationship between perception and action mechanisms. Here examples are drawn from research on both vertebrates and invertebrates, although most examples of cellular mechanisms necessarily concern the latter. The chapter is organized according to a hierarchy of increasing complexity of control. It is not intended as a comprehensive review but rather as an introduction for the nonphysiologist to selected results from the cellular analysis of behavior. These examples serve to illustrate basic concepts applied to motor control. The result of the survey is a picture of adaptive diversity rather than uniformity in control mechanisms.

The Functional Unit of the Nervous System

A necessary foundation for theoretical studies of neural processing is a list of the elemental operations available to the nervous system. This list has been considerably extended in the past decade. Since the turn of the century, the neuron has been regarded as the basic functional unit - an element which receives excitatory and inhibitory inputs onto its dendrites and cell body, performs a spatial and temporal summation over all active inputs, compares the sum with a threshold, and transmits the result in the form of action potentials along an axon to an output region. The action potential was found to be an all-or-nothing depolarization of the neuron followed by a refractory period in which the neuron is less responsive and, in some cases, a rebound depolarization in which it is more excitable. Together, this strict convergence of multiple inputs onto a single output and the characteristics of the action potential (McCulloch & Pitts, 1943; Kandel & Spencer, 1968; Lewis, 1968) were taken as the basis for evaluating information transmission and modeling neural circuits (e.g., Perkel, 1970; Friesen & Stent, 1977). More generally, they tend to influence thinking about the function of the nervous system as a whole (e.g., MacKay, this volume).

The past decade has seen a revision of this view. Morphological evidence showed that in both invertebrates (review Wilson & Phillips, 1983) and vertebrates (review Shepherd, 1979) many neurons have no strict separation of input and output synapses. Dendrites from two or more cells can form reciprocal or serial synapses (review Shepherd, 1979; Pearson, 1979), so neural processing can occur without involvement of the spike-generating mechanism. In fact, many neurons have been discovered which function entirely without action potentials (review Pearson, 1976; Wilson & Phillips, 1983; Hisada, Takahata, & Nagayama, 1984). In the absence of an intermediate spike code, the potential changes are attenuated during transmission within a neuron. Since the synaptic output is a continuous graded function of the local membrane potential, this graded transmission means that each input synapse will have a stronger effect on the output synapses closer to it than on those farther away (Graubard & Calvin, 1979). Thus, a single neuron can differentially influence its follower neurons. This nonspiking mode appears suited to converting a complex pattern of input into a suitably weighted pattern of output, such as might be needed to enable a limb to perform coordinated movements to many different targets (e.g., Burrows & Siegler, 1982; Burrows, 1985).

The ability to transmit information without action potentials is just one feature in which neurons have been found to be very complex. The Hodgkin-Huxley description of the action potential in the squid axon relied on just two voltage-dependent ion conductances (Hodgkin & Huxley, 1952). Meanwhile, at least six different types of voltage-sensitive channels differing in ion selectivity, voltage sensitivity, and activation kinetics are known (Adams, Smith, & Thompson, 1980; Kandel & Schwartz, 1986). Chemically gated ion channels provide an additional set of variables. These channels can be combined in different ways to give each neuron or different regions within a neuron different membrane characteristics. Furthermore, these bioelectrical and biochemical inputs can set in motion complex biochemical reactions which affect the synaptic output, protein synthesis, and even the morphology of the cell. This complexity underlies cellular properties such as spike-broadening, autoinhibition, intrinsic rhythmicity, postinhibitory rebound, and bursting activity, which determine the response of individual neurons; it gives some neurons the characteristics postulated in Hebb's model of learning (Cotman & Iversen, 1987).

The characteristics of the synaptic output are equally varied. For example, homosynaptic facilitation and depression, each with variable time course, can increase and decrease transmission in response to repetitive activation. Nonlinear heterosynaptic facilitation underlies some forms of learning (Hawkins, Abrams, Carew, & Kandel, 1983). The number of chemical transmitters has greatly expanded (Snyder, 1980), and neurons releasing more than one transmitter have been identified (Pulst, Gusman, Rothman, & Mayeri, 1986; Sigvardt, Rothman, Brown, & Mayer, 1986). The distinction between transmitter and hormone has been found to be graded.

The result is a bewildering number of possible response characteristics (see Bullock, 1976, 1979 and Calvin & Graubard, 1979 for more complete lists) with several implications for the construction of models. The first is that the number of processing units might vastly exceed the number of neurons. Single synapses or synaptic clusters may suffice to perform simple integrative functions, such as mutual inhibition or feedback modulation. Second, these findings show that individual neurons and neural networks as a whole possess a much broader range of processing mechanisms than was originally assumed. For theoretical models which take demonstrated neural processes as their starting point, this vastly increases the number of elements and reduces the chance that theory alone in the absence of experimental work can provide a satisfying description of network function.

Levels of Sensorimotor Integration

The past 15 years have seen tremendous progress in analyzing the neural basis of motor control. On the one hand, the investigation of cellular properties has revealed neurons to be much more complex than originally thought. On the other hand, neural networks of increasing complexity have been described. Progress in this area is evident in a comparison of the simple oscillator models proposed 20

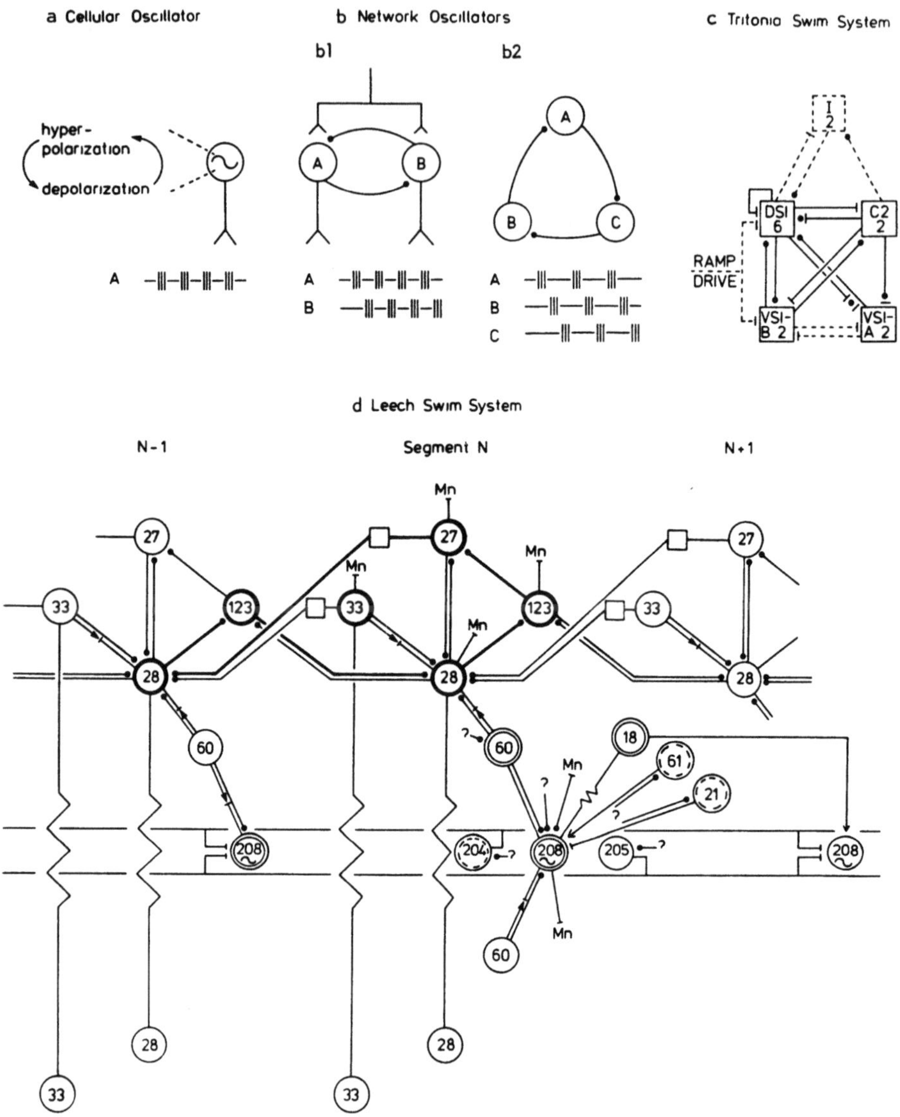

Fig. 1 a-d. Comparison of simple oscillator models and actual neural networks. Early oscillator models (**a, b**) which produce rhythmic spike outputs are compared with well-described neural networks (**c, d**). *Circles* represent single cells, the *boxes* in **c** represent cell groups. Cells marked with a *sine wave*, as in **a**, possess membrane properties which can lead to intrinsic rhythmicity. Connections are indicated as follows: monosynaptic excitatory (–⊣) and inhibitory (–•) chemical synapses, polysynaptic excitatory (– –⊣) chemical synapses, electrical (-\/\/\/-) and rectifying electrical (–⇥–) synapses. For pathways with multiple functions present in the *Tritonia* oscillator (**c**), the temporal sequence begins at the end of the line. The RAMP drive represents a prolonged depolarization resulting from cellular properties. In **d** the *boxes* denote delays; the five *thickly outlined cells* and their connections form the oscillator based on recurrent inhibition as originally described; cells with *double rings* are also formally part of the oscillator - those with *one dashed ring* show rhythmic activity, but rhythmicity is not necessary for rhythmic motor output. (Part **c** is from Getting and Dekin 1985; part **d** is based upon Friesen, Poon, & Stent, 1978, Weeks, 1982a, 1982b, Friesen, 1985, and Nusbaum & Kristan, 1986)

years ago with the complexity of the best-described neural circuits (Fig. 1). However, despite this progress, the search for simple universal principles has proved disappointing. Even in simple animals the neural circuits may be complex, and general principles to account for this complexity and the diversity across species have not yet been found. Thus, the following sections can only illustrate the range of possible control mechanisms rather than provide simple generalities.

Motor control involves an interplay of sensory, motor, and central elements. Central elements may transform sensory input for use in motor output, or they may produce intrinsic "autonomous" activity independent of sensory input, which in turn may be modified by sensory signals. In the following subsections I first consider various oscillator mechanisms and then discuss interactions between sensory and motor processes.

Peripheral Versus Central Models of Rhythmic Behavior

For technical reasons, the greatest progress has been made in elucidating neural mechanisms underlying rhythmic motor outputs - an important subset of motor behavior which includes locomotion, respiration, circulation, ingestion, and many forms of communication. Mechanisms for generating rhythms have been placed in two categories according to the role of the central nervous system and the peripheral neuromuscular system.

For one kind, the peripheral neuromuscular system is central to rhythm generation. In the original reflex theories, each state of the periphery provided the stimulus for the ensuing action. According to the chain reflex hypothesis, rhythmicity results from a series of reflexes joined together to form a closed cycle in which each reflex action generates the stimulus for the subsequent reflex. This notion was applied to both intra- and interlimb coordination by the zoologists von Holst (1935), Roeder (1937), Gray (1950), and Weiss-Fogh (1956) in their studies of locomotion in invertebrates and lower vertebrates. It bears a formal resemblance to the notion of "reaction chains" as used in the early ethological literature to describe open-ended behavioral interactions between individuals (e.g. Tinbergen, 1951; Eibl-Eibesfeldt, 1970). Thus the higher level mechanisms postulated for eliciting behavior in response to external stimuli were similar to those lower level mechanisms postulated for guiding its performance in response to proprioceptive stimuli.

More recent versions of this hypothesis, which will be designated "peripheral oscillator models" following Graham (1977), deviate from a rigid reflex chain scheme. The formal definition is simply that the actual state of the peripheral system serves as a parameter in pattern generation. Conceptually, these models go beyond a rigid chain reflex model in that unpatterned central activity is allowed. For example, central activity can take the place of reflex activity to provide the excitation for leg retraction during walking. However, pattern generation arises from the interaction of this activity with numerous excitatory and inhibitory influences which depend on the state of the skeletomuscular system.

The second hypothesis places the origin of rhythmicity in autonomous neural processes which drive the motor output. These mechanisms are designated "central oscillators" or, more generally, "central pattern generators" (CPG; review Delcomyn, 1980; Pinsker & Ayers, 1983; see Cruse, Dean, Heuer, and Schmidt, this volume). In a pure CPG the state of the peripheral motor apparatus merely reflects the state of central processes but has no influence on it. Brown (1914), and later von Holst (1948) and Weiss (1940), championed this hypothesis for explaining rhythmic stepping in vertebrates and invertebrates.

These notions derived from behavioral observations received a decisive impetus from Wilson's (1961, 1966) demonstration that the isolated nervous system of the locust is able to produce rhythmic efferent activity with many of the characteristics of the flight motor pattern. In the subsequent 25 years this deafferentation method has been applied to numerous rhythmic behaviors, and in virtually all cases a basic rhythm can be produced by the isolated nervous system (review Delcomyn, 1980; Pinsker & Ayers, 1983). As a result, these central processes came to be viewed as the primary source of rhythmic motor output.

This view has been revised in the past 15 years. For several locomotor patterns, the afferent information has been shown to affect the pattern in ways that formally satisfy the criteria used to identify interneurons participating in a CPG (Getting, Lennard, & Hume, 1980). Imposed movement of a limb or specific sensory stimuli may reset or entrain an ongoing rhythm in the same way that intracellular stimulation or inhibition of appropriate interneurons do (Wendler, 1974; Grillner, McClellan, & Perret, 1981; for reviews Pearson, 1981; Grillner & Wallén, 1985; Wendler, 1985; Bässler, 1985).

The relative importance of central and peripheral factors varies greatly in different systems. At one extreme, some oscillatory events are produced by peripheral mechanisms alone, with no rhythmic contribution from the nervous system. For example, the individual sound pulses making up a cricket chirp result from a mechanical mechanism that is equivalent to the pendulum-driven escapement of a clock. The neuromuscular system serves only as the spring, providing the power but not determining the timing (Elliot & Koch, 1985). The myogenic flight rhythm of many insects is a better-known example (Pringle, 1957). At the other extreme, many rhythms can be generated by the isolated nervous system or are unaffected by changes in the actual movement (review Delcomyn, 1980).

In most cases both central and peripheral mechanisms are present, and the actual pattern results from their interaction. Thus, the current approach replaces the 'either peripheral or central' question of the original debate with a more differentiated view: each specific behavior is seen as subject to more or less interference from unpredictable environmental influences. When adaption to unpredictable conditions is functionally more important, as is often the case for walking, then peripheral mechanisms are important. When overall speed of execution is important, as in prey-catching behavior and escape reactions, or when substrate variation is limited, as in flying or swimming through homogeneous media, then central mechanisms dominate.

Types of Central Oscillatory Mechanisms

Neurophysiological investigations have revealed two principle types of neural oscillators. In cellular oscillators the rhythm arises in individual neurons as a result of membrane properties which cause a neuron to alternate between periods of activity and inactivity (Fig. 1a; Mendelson, 1971; Maynard, 1972). The periodic burster R15 in *Aplysia* is a classic example of such a pacemaker: even when isolated from the rest of the nervous system this cell depolarizes periodically to threshold, produces a burst of action potentials, and then hyperpolarizes (Alving, 1968). Similar intrinsic oscillations occur in the pacemaker cells of vertebrate and many invertebrate hearts; in many cases their biophysical origin is well understood (Noble, 1985).

In network oscillators, the second type, the rhythm arises from the pattern of connections among two or more functional groups of neurons (e.g., Brown, 1914; Wilson & Waldron, 1968): a simple example for a two-phase oscillation is given in Fig. 1b - b1. The basic principle is reciprocal inhibition between neurons or groups of neurons, which ensures that when one group becomes active it turns off the activity of the other. To obtain rhythmic alternation, an additional process must be added to counter the inhibition and enable the inactive group to become active. This can take the form of adaptation or delayed self-inhibition of the active group (Brown, 1914; Wilson & Waldron, 1968) or delayed reciprocal excitation (Getting, 1983; Getting & Dekin, 1985).

In order to explain patterns with more than two phases, Szekely (1968) proposed that neurons form ring structures in which each functional group inhibits its neighbor on one side (Fig. 1b - b2). The result is sequential activity traveling around the ring. Conceptually, the activity in each group can activate different muscle groups; Szekely used this model to account for the four phases in the step cycle of newts.

An early description of the neural oscillator controlling swimming in the leech corresponded nicely to Szekely's theoretical models. The basic unit consisted of five interneurons located in adjacent ganglia and forming a ring with recurrent inhibition (Fig. 1d; Friesen, Poon, & Stent, 1976, 1978; Friesen & Stent, 1977). Connections from this ring to motoneurons determine the alternation of dorsal and ventral flexion, the timing of individual muscles, and the phase shift between adjacent body segments. However, the connections among the interneurons are not limited to those proposed in the model; additional connections reciprocally connect neighbors in the ring or even cross the ring to form rings within rings. Moreover, the early model is now known to be incomplete (Weeks, 1981); the swim rhythm can be produced by a single isolated ganglion, so the five-neuron ring normally is only one of several mechanisms contributing to rhythm generation.

Many biological oscillators contain both cellular and network elements, as in the leech example (Fig. 1d). The output of simple pacemaker cells is obviously of limited complexity, but incorporation of several interconnected neuron pools driven by the two-phase oscillation of pacemakers is an alternative to ring networks for generating multiphase rhythms (Kater, 1974; Benjamin, Elliot, & Ferguson, 1985). Conversely, the fatigue and rebound attributed to each hemicenter in Brown's 1914 network model may represent intrinsic cellular properties. Moreover, the "plateauing" shown by some cells, in which depolarization to threshold triggers a prolonged, self-maintaining depolarization, is important in determining the timing and duration of movement phases in many network oscillators (Russell & Hartline, 1978; Getting & Dekin, 1985; Miller & Selverston, 1985; Sillar & Elson, 1986).

The analysis of biological oscillators is further complicated by the fact that similar functions may rest on entirely different mechanisms. For the *Tritonia* swim network (Fig. 1c) to oscillate, the VSI interneurons must receive a delayed excitation (Getting, 1983; review Getting & Dekin, 1985). For the VSI-A subgroup this influence arises from a dual-action synaptic connection from C2 interneurons, which is inhibitory at first and then excitatory. The VSI-B subgroup receives only excitation, but the membrane properties of the cells are such as to counteract this excitation at first and thus introduce a delay.

Similar functional equivalences occur at the synaptic level. In several systems neurotransmitters, designated as neuromodulators, alter cellular properties underlying intrinsic bursting or plateauing characteristics and thereby cause a qualitative change in system output (e.g., Miller & Selverston, 1985). In other cases, the function of a synaptic connection can depend upon the electrical activity of the cells. Two identified cells in the leech are connected by a rectifying electrical synapse and by inhibitory chemical synapses (Friesen, 1985). Depending upon the relative potentials of the two cells, either the electrical excitation dominates and enforces in-phase activity, or the chemical inhibition dominates and enforces antiphase activity.

The presence of many contributing mechanisms, some of which appear redundant, is characteristic of biological oscillators. Redundancy may provide additional stability. It can also provide a substrate for further evolutionary change in which elements differentiate to perform separate functions or the network as a whole assumes new functions. This evolution may leave connections which appear superfluous (e.g., Wine, 1984; Dumont & Robertson, 1986) or even maladaptive in the sense that they provide a synchronizing influence on cells that normally alternate or vice versa. More detailed analysis may reveal that such elements actually do have a functional role, but in other cases such connections may simply represent evolutionary ballast. This complexity has led some investigators to question whether such mechanisms can ever be unraveled, even in relatively simple invertebrates (Selverston, 1980).

Knowledge of oscillatory mechanisms on the cellular level is less advanced for vertebrates (Grillner, 1985). The isolated spinal cord of the lamprey provides a preparation with many of the advantages of invertebrate preparations (review Grillner et al., 1987), but the anatomy is similar to that of higher vertebrates and therefore encourages the hope that the findings will be of general relevance for vertebrates. As in the invertebrate examples, both cellular and network properties are important for pattern generation in swimming. The former include intrinsic rhythmicity in interneurons and plateauing properties contributing to burst formation; the latter include inhibitory interconnections of the interneurons.

In summary, pattern generation in both vertebrates and invertebrates involves both cellular and network properties. The important cellular properties are based on complex combinations of ionic conductances which allow a cell to do more than simply produce action potentials in response to external inputs: they include intrinsic rhythmicity, bursting activity, or time-dependent synaptic modulation. The important network properties include reciprocal inhibition, which imposes alternating activity between cells or groups of cells, and loops with recurrent inhibition, which provide more activity phases. In contrast, electrical synapses and reciprocal excitatory synapses tend to synchronize activity.

The Role(s) of Sensory Information

Although in most behaviors a basic rhythm can be generated by central mechanisms alone, this autonomous activity is augmented by a variety of peripheral influences. The many levels on which sensory information can interact with efferent motor activity are illustrated in Fig. 2, using the framework of the motor control structure as discussed by Cruse et al. (this volume). For descriptive purposes the different functions incorporated in a motor control structure define a hierarchy of control levels, with generation of activity in specific motor units at the bottom and task selection at the top. This functional hierarchy rests on neural networks of increasing complexity, but, as will become evident in the following review of some of the underlying physiological mechanisms, there is no simple 1:1 relationship between functional and neural levels.

The lowest structural level is the monosynaptic reflex in which activity in sensory fibers is passed directly onto motoneurons. The function of such connections ranges from selecting a response to modifying details of motor activity. These direct connections provide fast responses but offer limited complexity, and their action is often complemented by connections through interneurons (see below).

Direct connections from sense organs to motoneurons which trigger actions are common in withdrawal reflexes where a fast response is advantageous for avoiding injury. In both *Aplysia* (review Kandel, 1979) and the leech (review Kuffler & Nicholls, 1976), nociceptors in the skin synapse directly onto motoneurons in order to initiate a protective withdrawal from aversive stimuli. The same sensory fibers also synapse onto interneurons which provide longer-lasting excitation of the motoneurons to maintain the response (Kandel, 1979). Viewing withdrawal as the task, the sensory information performs a type-III function: it is used to trigger the action and to set parameters such as the intensity and duration of contraction.

Other monosynaptic connections mediate negative feedback. In a negative feedback network, sensors and effectors are connected in such a way that a change in a parameter monitored by the sensors leads to effector activity counteracting the change. Some simple proprioceptive responses include monosynaptic connections. The cockroach leg provides a good example (Pearson, Wong, & Fourtner, 1976).

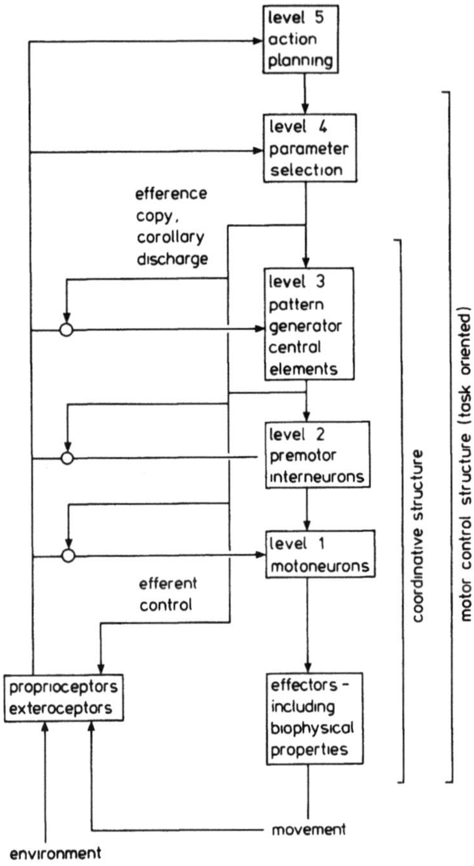

Fig. 2. A schematic illustration of the interactions between efferent and afferent information in motor control. The figure is cast in the manner of the motor control structure proposed in the chapter by Cruse et al. (this volume). It illustrates the many structural/functional levels on which sensory and motor pathways interact. As discussed in the text, the hierarchical representation of the control functions does not necessarily correspond to separate structural levels in the nervous system. In the scheme of Cruse et al., type-III feedback acts primarily at levels 4 and 5 whereas type-I and type-II feedback predominate at levels 1-3

An imposed flexion of the coxa-trochanter joint excites mechanoreceptive hairs which monosynaptically excite motoneurons of the extensor muscles. The increased activity in these muscles serves to straighten the joint and resist the imposed flexion (Wong & Pearson, 1976). This afferent information has a continuous effect (type I) on motoneuron activity. Similar resistance reflexes have been demonstrated behaviorally and physiologically in numerous vertebrates and invertebrates (review Wright, 1976 for invertebrates, Hunt & Perl, 1960, and Henneman, 1980 for vertebrates), although direct monosynaptic pathways have been confirmed in only a few cases. These negative feedback circuits aid in maintaining body posture by compensating for muscle fatigue or differences in loading (e.g., Wendler, 1964). A direct, monosynaptic connection reduces the delay in the feedback loop and therefore allows the system to cope with perturbations of higher frequencies without becoming unstable.

In Fig. 2 premotor interneurons are introduced as an intermediate structural level between the oscillator or pattern generator and the motoneurons. This level is not present in all systems - oscillator neurons may themselves synapse directly onto motoneurons (Poon, Friesen, & Stent, 1978; Getting & Dekin, 1985) or motoneurons may participate in the oscillator (e.g., Heitler 1978; Miller & Selverston, 1985) - but it is useful for illustrating some kinds of temporal and spatial integration in elaborating motor patterns originated by central or sensory sources.

Premotor interneurons can participate in temporal integration by prolonging a response triggered by brief sensory inputs, as mentioned above, and by transforming sensory input in various ways. Such a transformation is evident in the vertebrate "clasp-knife reflex" - a sudden release of muscle tension if a critical threshold is exceeded. This reflex is mediated by strain receptors which provide a graded measure of the forces acting at a joint (Houk & Rymer, 1981). This information is routed through separate interneuron pathways, one of which generates the sharp threshold for the "clasp-knife reflex" while others contribute in a graded manner to tension regulation.

In spatially distributed reflexes, which involve several limbs or limb segments, interneurons abstract inputs from distant sense organs and distribute the activity to the relevant motor pools. In both vertebrate and invertebrate resistance reflexes, monosynaptic excitation of one muscle group is usually accompanied by polysynaptic inhibition of antagonistic muscles. Monosynaptic pathways for both excitation and inhibition appear to be ruled out because they would either violate the constraints on transmitter and receptor distributions in one cell or require separate transmitter systems for each behavior. The introduction of interneurons is a simple way to increase the number of response combinations.

One example involving several limbs is the crossed extensor response in vertebrates, in which the stimulus-induced flexion of one leg is accompanied by an extension of the contralateral limb - an automatic preparation to assume the weight previously carried by the flexing limb (Sherrington, 1906; review Grillner, 1981). Similar coordination responses are seen in insects (C. Walther in Bässler, 1983). In fact, many distributed reflexes contribute to load compensation or posture control (Pearson, 1972; Seyfarth, 1978; Klärner & Barnes, 1986; reviews by Evoy & Ayers, 1982 and Roberts, 1967).

The presence of such distributed reflexes, some of which contribute stabilizing negative feedback and some of which elicit positive feedback within certain limits, formed the basis for chain reflex models of locomotion (Gray, 1950; see also Fraser, 1982). Currently, such distributed reflexes are viewed less as hard-wired, inflexible responses than as part of flexibly programmable states set up to meet task-related criteria, e.g., as an expression of a coordinative structure. In rhythmic or voluntary behavior, sensory information processed at this intermediate level between pattern generator and motoneurons can serve to modulate the amplitude of muscle activation without affecting its timing; this might correspond to setting the amplitude parameter in a prototypic function (Meyer, Smith, & Wright, 1982; Cruse et al., this volume) or to performing "phase-dependent responses" which correct a disturbance within a movement phase without resetting the rhythm (Forssberg, 1979).

An example of phase-specific modulation where the physiology is better understood is provided by the locust flight system (Reichert & Rowell, 1985; Reichert, Rowell, & Griss, 1985). Here sensory information from the simple and the complex eyes is used to determine the deviation of the locust from its desired flight attitude. This information is transmitted to the flight motor centers in the thoracic ganglia by interneurons which form weak synapses onto motoneurons and strong synapses onto premotor interneurons. The premotor interneurons are divided into two groups. Under the control of a thoracic oscillator these two groups are active during the wing upstroke and the downstroke, respectively, and they in turn drive the corresponding motoneurons. The deviation signal is able to modulate the activity of each group during its active phase but is unable to excite the premotor interneurons during their inactive phase. This phase-specific modulation apparently suffices to produce subtle changes in wing movement which are sufficient to alter the forces and correct the deviation in flight attitude (Thüring, 1986). Directing the sensory information through the premotor interneurons apparently ensures that the continuous deviation signal is split into two commands for type-I corrections appropriate to each phase of wing movement.

The pattern generator, the next structural level, may incorporate complicated central networks (see previous section) as well as sensory influences. In many behaviors, sensory inputs can affect a rhythm in ways that formally make the sensory organs and the whole peripheral skeletomotor apparatus part of the oscillator. Entrainment effects are found in walking and swimming of both vertebrates and invertebrates (Grillner, 1981, 1985; Grillner, McClellan, & Perret, 1981; Williamson & Roberts, 1986; Sillar, Skorupski, Elson, & Bush, 1986). In the locust flight system the deviation detectors and the premotor interneurons described above do not entrain or reset the flight rhythm, but other sensory inputs do. Movement of the whole wing has this effect (Wendler, 1974), as does appropriate stimulation of individual proprioceptors which directly monitor the movement of the wing (review Wendler, 1985) and wind-sensitive hairs on the head which detect wing motion indirectly through its effect on the airstream past the head (Horsmann, Heinzel, & Wendler, 1983). The function of both circuits is presumably to stabilize the flight rhythm against external disturbances.

The neural mechanisms are partly known for the stretch receptor and for the wind hairs: both involve multiple pathways. The stretch receptor at the base of the wing has weak monosynaptic connections to motoneurons (Burrows, 1975a) which contribute to a resistance reflex opposing wing elevation (type-I effect). The delay in this pathway is such that this connection can promote the switch from wing elevation to depression at normal wing-beat frequencies (Burrows, 1975a) and thus takes on type-II, intermittent processing characteristics. Two factors besides the delay contribute to this transition. First, the phasic-tonic characteristics of the stretch receptor mean that it produces only a short burst of action potentials, concentrated at the transition from upstroke to downstroke in each wing-beat. Second, the membrane potentials of the motoneurons show large oscillations imposed by the CPG, so the monosynaptic influence from the stretch receptor is effective only when the potential is already above or near threshold. To explain the behavioral effects of stretch receptor activity, the weak monosynaptic connections must be augmented by additional synapses onto premotor and pattern generator interneurons. Similarly, information from wind hairs is transmitted along several pathways to wing motoneurons (Bacon & Möhl, 1983).

In some cases, the effects of proprioceptors on a pattern generator are limited to a particular phase. This is particularly true when sensory information is used to determine the successful completion of a movement (type-II effect). In the cockroach, campaniform sensilla sensitive to loading at leg joints can alter the timing of the step oscillator by prolonging the stance phase - a type-II effect - while at the same time increasing the muscle force in this phase - a type-I effect (Pearson, 1972). A variety of sensory influences related to leg position and to load affect either the force development or the termination of step phases in the stick insect (review Bässler, 1983; Graham, 1985).

Formally, these sensory influences on pattern generation are evidence for peripheral or mixed oscillator models. They introduce the state of the skeletomuscular system as a parameter in the control system. As a direct consequence, the characteristics of the skeletomuscular system must be included in any analysis, as Bernstein (1967) emphasized, because physiological parameters of muscle force development together with the physical parameters of the moving body combine to influence movement. These characteristics may affect movement either directly, in accord with physical laws, or indirectly, by way of feedback through sense organs monitoring the actual movement.

At higher levels of motor organization, sensory information is used to select appropriate motor control structures and to determine their parameters. This function depends on the extraction of appropriate sensory information about the environment. It requires sensory processing of variable complexity and may even involve internal maps or representations of the animal and its relation to its environment.

In invertebrates the automaticity of much behavior implies a relatively straightforward connection between sensory inputs and motor centers. Nevertheless, these connections may be very complex and counterintuitive. For example, in the prey-capture strike of the praying mantis the forelegs can be aimed at prey anywhere in the frontal field (Mittelstaedt, 1962). Information is available from the eyes on the deviation of the target from the center of the visual field and from proprioceptors at the neck on the deviation of the eyes from the midline of the prothorax and the front legs. However, these values cannot be internally summed in a flexible manner to direct the strike. Rather the head movement as the animal directs its gaze toward the target is opposed by a negative feedback system involving the neck proprioceptors: the direction of the strike is determined by the residual deviation of the visual image at the point where the opposing tendencies to fixate the prey and to keep the neck straight are in balance. Such a network might provide a useful model for the direct control of motor output by sensory input which is postulated by various hypotheses derived from Gibson's ecological optics.

Considerable information on the cellular level is available for neural circuits involved in rapid escape responses because the large size of many of these neurons facilitates neurophysiological study. In crayfish (Wine, 1984), cockroaches (Camhi, 1984), and aquatic vertebrates (Eaton & Hackett, 1984), mechanical stimuli portending an attack activate a small set of large intersegmental interneurons which exercise a command function (see section on Variability in Behavior, below). These giant fibers trigger a rapid stereotyped response (type III). Their receptive fields and the orientation of the responses they elicit are matched to ensure that the response is appropriate to the threat. Some of these specialized pathways are also active in normal locomotion, but most are turned off and control is assumed by more complex but slower networks which can better adapt behavior to the current conditions (review Camhi, 1985).

In summary, the functions of sensory information are as varied as the control levels on which it acts; it is used to select a motor control structure, to provide general excitation for motor patterns, to stabilize a basic pattern, or to adapt the motor output to environmental conditions through feedforward or feedback or through learning. As we have seen, there is only a rough correlation between functional and structural levels.

Modification of Sensory Input During Movement: Implications for Control Mechanisms

The previous section illustrated how sensory information is integrated into all levels of motor control. The peripheral motor apparatus provides a pathway through which the movement in turn directly modifies the sensory input from both proprioceptors and exteroceptors. In the present section I will consider this pathway briefly and then turn to other pathways by which motor activity modifies the afferent input. Depending upon whether or not the sensory information in turn modifies the movement, these pathways can form loops in the motor control structure. Naturally, these pathways also modify the form in which information is available to higher centers for perceptual processes.

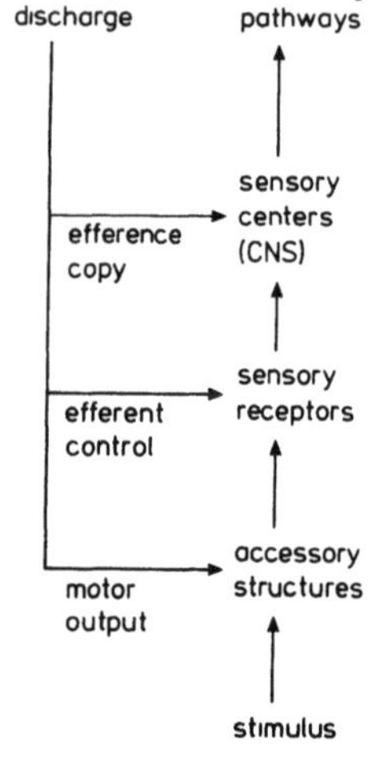

Fig. 3. Levels at which efferent activity can influence afferent signals

Descending motor activity can phasically modulate the afferent information at three levels (Fig. 3). Beginning with the most peripheral, these levels comprise (a) motor mechanisms acting on auxilliary structures involved in sensory transduction, (b) synaptic mechanisms acting at the receptors themselves (efferent innervation), and (c) synaptic mechanisms in the CNS. Efferent fibers are a common feature of sensory systems in vertebrates but not in invertebrates; the first and third mechanisms are found in both.

The ability to modulate the afferent information is particularly important when identical stimulus patterns can arise either through environmental change or through the animal's own actions. Because the functional implications for the animal are different, the afferent information must be processed in the context of the motor commands. In many cases it is assumed that sensory networks receive information on the outgoing motor commands. Here a distinction is made between an "efference copy" - a signal quantitatively representing the reafference, the afference predicted to arise from the animal's own actions (von Holst & Mittelstaedt, 1950), and a "corollary discharge" - a more general term for activity which is associated with motor output and alters sensory processing (Sperry, 1950; see also the discussions in McCloskey, 1981 and Bell, 1984). The function of these mechanisms in thc behaviors for which they were originally proposed has since been challenged, but there is good physiological evidence for both corollary discharges and efference copies.

The three basic functions are illustrated in a weak electric fish (Bell, 1982). The regular discharge of the animal's own electric organ excites three types of electroreceptive organs. Central afferent stations in each system receive corollary discharges appropriate to the function it performs. The CNS area for the ampullary organs, which are primarily involved in detecting weak external current sources, receives an efference copy which cancels out the response to the electric organ discharge. The sign and amplitude of this efference copy are varied to follow changes in the actual reafference signal due to slow changes in the surroundings: in this way effective cancellation is maintained and the system remains highly sensitive to the phasic signals of interest. The mormyromast system, which evaluates the electric field generated by the animal's own discharge, and the knollenorgan system, which detects discharges from other fish for communication purposes, receive sensitizing and desensitizing corollary discharges, respectively (review Bell, 1984).

Proprioceptive input itself can be used to modulate afferent pathways in the same way as corollary discharges. In the grasshopper both central activity (corollary discharge) and proprioceptive input from the wings, the sound-producing organ, combine to turn off an auditory interneuron when the animal stridulates (Wolf & von Helversen, 1986).

The contribution of motor behavior as such to sensory function ranges from facilitating basic sensory function to making additional information available. At the lowest level, the former involves the control of movable sense organs, such as eyes, antennae of many invertebrates, specialized tactile organs, and ears. The control may involve movements serving active exploration of objects of interest or stabilization of sense organs during movement of the body.

Motor behavior can also make novel information available to sense organs. For example, movement can be used to detect gradients, and it has been suggested that some animals detect magnetic fields indirectly by using electroreceptors to sense the currents induced by the animal's motion in the earth's magnetic field (review Kalmijn, 1984). On a more complex level, animals can use self-generated stimuli to extract information about their surroundings. The blind cave fish passively acquires information on moving objects from its mechanosensitive lateral line receptors. By gliding past objects, the fish is able to extract information about non-moving objects (Hassan, 1986). The weak electric fish use self-generated discharges to gain information about their surroundings. The echolocation of bats and marine mimals is a better-known example of extending a passive system by registering the return of an actively produced signal. In all these active systems one might expect a combination of desensitizing and sensitizing corollary discharges which decrease sensitivity during sending and enhance it for the returning information (Bell, 1984).

Motion can also aid in performing more complex feature analysis; i.e., motion induces reafference which contains more information than that present in a static image (Bell, 1984). According to a very detailed theory of visual perception in flies, relative motion of objects resulting from the fly's own motion is necessary for the perception of figure-ground differences (Reichardt, 1979). This discrimination in turn guides locomotion. Another example is the determination of time to contact, which is important for many behaviors. In the context of Gibson's theory of ecological optics, Lee proposed the parameter tau, a measure of the rate of expansion of the visual flow field, as a source of direct information for controlling behavior (Lee, 1976; Lee & Young, 1986).

Corollary discharges, apart from their effect on perceptual processes, are particularly significant for motor control when they interact with peripheral control mechanisms. As discussed in the previous section, some sensory inputs elicit a particular response in a feedforward manner and are not affected by the response. In other cases, the response in turn alters the sensory input, establishing a feedback pathway which can act to stabilize a state or a movement pattern: this reafference can involve both proprioceptors and exteroceptors.

A simple example of the latter is the case of a sensory input serving to trigger a movement. If the movement in turn reexcites this sensory input, then the activity may continue indefinitely. An escape response, such as that of the crayfish, is a good illustration (Wine, 1984). Mechanoreceptors on the abdomen of the crayfish detect water movements generated by approaching predators. Their activity triggers first a stereotyped tailflip and then normal swimming, which together propel the animal away from an approaching predator. However, swimming itself produces relative water movement which strongly stimulates these same mechanoreceptors: this could reelicit the stereotyped tailflip and interfere with directed swimming. A corollary discharge prevents this from occurring: the output synapses of the sensory fibers receive presynaptic inhibition from the interneurons controlling the tailflip and active swimming. This inhibition turns off these outputs once they have elicited the escape response. It also serves to protect the sensory synapses from the homosynaptic depression which would otherwise occur during the strong, self-induced stimulation. When the animal stops swimming and the inhibition is removed, the synapses are immediately ready to transmit information on a new threat. In vertebrates efferent fibers innervating sensory receptors may play a similar protective role.

The type-I reflexes mentioned above, the negative feedback 'resistance' reflexes and positive feedback 'assistance' reflexes, pose a more complex problem. Fixed reflexes can stabilize particular states, but at times they conflict with the need for flexible control. During active movement these reflexes may continue unchanged, in which case the central movement commands must allow for the action of the reflexes (Mittelstaedt, 1951; Wendler, 1964; Möhl, 1985), they may act as servomechanisms to follow a variable reference signal, or they may be turned off in various ways. The provision of an efference copy as a reference signal is one way of 'turning off' the reflex during undisturbed movement.

Studies of exteroceptive systems affecting movement reveal both mechanisms. In many vertebrates and invertebrates, for example, optomotoric responses provide negative feedback serving to stabilize the animal's orientation with respect to its optic surroundings. These responses would oppose voluntary changes of direction. In flies, the optomotoric response continues to function during slow voluntary turns and thereby reduces the amplitude of the actual turn in relation to that performed under open-loop conditions (Collett, 1980). Quantitative findings suggest a similar situation for vertebrates (Sperry, 1950). During fast turns, however, the optomotoric response is apparently turned off, which is in accord with an "efference copy" hypothesis (Heisenberg & Wolf, 1979), although simple filter characteristics of sensory interneurons, which selectively block typical reafference signals, are sufficient to explain much of the behavior (Egelhaaf, 1987). This phenomenon is analogous to the suppression of visual perception during saccades in primates (see Bridgeman, this volume).

Proprioceptive type-I reflexes show similar variability. In the stick insect resistance reflexes arising in position-sensitive proprioceptors at the basal joint of the leg resist both passive movements imposed on the leg and active movements originating in the animal itself (Wendler, 1964). In contrast, the chordotonal organ at the femur-tibia joint contributes to a negative feedback resistance reflex when the animal is quiescent but may provide positive feedback when the animal is active (Bässler, 1976).

A more elegant use of negative feedback circuits in movement control is to incorporate their ability to counter outside disturbance. In technical servomechanisms this control is introduced by adding a variable reference signal which specifies the desired state of the system. The negative feedback causes the system to track the reference signal. As a rule, the function of technical feedback pathways is invariant. However, in the presence of a disturbance a change in feedback gain causes a change in the output.

Again, behavioral evidence indicates that some proprioceptive networks are controlled in this way. At one joint in the crab leg, certain temporal parameters allow reflex activity to be distinguished from other efferent motor activity. Here, resistance reflexes present in the quiescent animal are not evident when the animal itself makes a normal leg retraction, but they do appear if the leg slips or external factors cause the movement to depart from the normal trajectory (Barnes, Spirito, & Evoy, 1972; Barnes, 1975). One interpretation is that in the crab the reflex continues to be active, but the reference value takes the form of an efference copy corresponding exactly to the normal step trajectory. More generally, if a corollary discharge does not exactly cancel the movement-induced afference, it can cause the reflex circuit to show either resistance or assistance characteristics. For example, Pearson (1972) described a positive feedback response for load compensation in the cockroach, but a physiological study of a closely related system revealed connections forming a negative feedback servomechanism (Zill & Moran, 1981 a,b; Zill, Moran, & Varela, 1981). Such a system can perform load compensation if it serves to counteract deviations induced by the load. The reflex reversal discovered by Bässler could also represent a stimulus-induced (in this case, the movement is the stimulus) change in the reference value for a proprioceptive servomechanism.

The physiology underlying such control systems has been studied in both vertebrates and invertebrates. In vertebrates this research has concentrated on the muscle spindles. This system contains efferent channels for both reference signals and gain control. Theoretical interest has focused on the former because it places the comparison between a reference signal and the actual state in the periphery, where it can be readily studied. The reference signal is contained in the gamma motor activity to the intrafusal muscle fibers. These fibers adjust the length of the receptor spindle without measurably changing the length of the whole muscle; hence, they serve to bias the receptor's measurement of muscle length. By its actions on the alpha motoneurons and the extrafusal muscle fibers, which generate the main force in the muscle, the spindle organ in turn provides negative feedback for controlling muscle length.

Based on these pathways, several authors suggested that the activity in the gamma motoneurons controls movement by setting the reference value for a follow-up servomechanism formed by the spindle receptors, the alpha motoneurons, and the extrafusal fibers (Merton, 1952; Wagner, 1960). This idea is inadequate in several respects (review Stein, 1982). First, it would lead to time delays of 50-100 ms before alpha motoneurons were activated (Stein, 1982). In fact, voluntary movements often involve joint activation of alpha and gamma motoneurons (Vallbo, 1971), and both can be activated directly from the cortex. Second, the gain around the muscle spindle loop appears too low to provide effective error correction (review Mathews, 1972; Stein, 1974, 1982).

The notion of a follow-up servomechanism was replaced by the hypothesis of alpha and gamma coactivation (Granit, 1970; Mathews, 1972). The direct alpha activation decreases the movement latency (Vallbo, 1971), whereas the gamma activity controls the reference for the feedback control as before. The low feedback gain is less of a problem because the servosystem is called on only to reduce any position deviations left by the alpha system, rather than to carry out the entire position change. Some recordings from spindle Ia fibers show that during voluntary contraction, spindle discharge increases (Vallbo, Hagbarth, & Torebjörk, 1979). Here the gamma activity is stronger than it need be for a proper efference copy and the feedback loop provides some, albeit limited, servo-assistance for the movement. However, the amount of alpha-gamma linkage is quite variable in different systems (Prochazka, 1985).

In fact, the monosynaptic spinal stretch reflex is only part of the response to disturbance. The springlike properties of an active muscle provide an instantaneous force change for small length changes. A major part of the neural compensation arises through longer latency pathways involving either more slowly conducting group-II spindle afferents (Mathews, 1985) or transcortical and cerebellar circuits (review Wiesendanger & Miles, 1982). This has led to the suggestion that the spinal stretch reflex serves primarily to linearize the muscle characteristics by regulating muscle stiffness over a greater range of muscle lengths and thus simplifies the control task for centers involved in forming movement commands (Nichols & Houk, 1976; Houk & Rymer, 1981).

This simplistic description considers the stretch receptor a detector of length: it does not do justice to the complexity of this organ (review Boyd & Smith, 1984; Mathews, 1981). In fact, the organ contains two kinds of afferent fibers, the group Ia fibers from the primary endings and the group II fibers from the secondary endings. In different combinations these fibers innervate three types of intrafusal fibers differing in fine structure, histochemistry, and electrical and mechanical properties. Similarly, the efferent gamma fibers can be devided into dynamic fibers and two types of static fibers. In addition, many muscles also have beta motoneurons, which jointly innervate intrafusal and extrafusal muscle fibers, and efferent terminals from the autonomic nervous system. The net effect is that the response of the group Ia and the group II fibers depends not just upon the length, but also, in differing degrees, upon the rate of length change: the different classes of motoneurons differentially affect the bias on the receptor response at a given muscle length or the gain of the phasic response to a length change. Behavioral results show that stretch reflexes are modified during active movement (Gottlieb & Agarwall, 1980) but the mechanism is not yet known. Thus, despite intensive study, the precise function in movement is not yet clear (Prochazka, 1981; Boyd & Smith, 1984).

The muscle spindles of vertebrates, with their motor innervation distinct from that of the main muscle, find parallels in the muscle receptor organs of invertebrates (review Fields, 1976; Bush & Cannone, 1985). Functionally, these organs permit a follow-up servocontrol analogous to the alpha-gamma control of vertebrates (Fields, Evoy, & Kennedy, 1967; Bräunig & Hustert, 1985). However, these organs are less common than vertebrate muscle spindles, and the bulk of the proprioceptive information is not subject to efferent control at the periphery.

The physiological basis for an activity-dependent reflex reversal has been uncovered in the crayfish. The reflex is governed by the muscle receptor organ at the base of each leg. The S and T fibers, respectively, respond to dynamic and static parameters of a stretch. In the quiescent animal both fibers contribute to a resistance reflex (review Bush and Cannone, 1985). This reflex can also be demonstrated in an isolated preparation consisting of the receptor organ and the attached ganglion (Sillar & Skorupski, 1986). However, some preparations exhibit assistance reflexes in which T-fiber activity elicits positive feedback (Skorupski & Sillar, 1986). Occasionally, the isolated ganglion produces a spontaneous motor output analogous to that of forward walking. In this active condition an applied stretch elicits resistance or assistance reflexes, depending on the phase of the ongoing rhythm. The phase dependence is produced by a corollary discharge from the pattern generator onto the sensory fibers (Sillar & Skorupski, 1986). This discharge is antiphasic for the two fibers and serves to modulate the gain of the respective reflex pathways. The fact that these sensory inputs can entrain the central rhythm (Sillar et al., 1986), as mentioned above, emphasizes the complexity of circuits for generating patterns and stabilizing them against outside disturbances.

In summary, both vertebrates and invertebrates have several mechanisms available for modifying the afferent information arising from self-induced movement. This action modifies the working of peripheral servomechanisms and trigger stimuli. As discussed by Cruse et al. (this volume), the context-specific processing of particular peripheral information may be an invariant part of a motor control structure or it may be subject to modification on the basis of experience. Modification of type-I reflexes often varies with movement speed, as shown in the optomotor examples. Similarly, reflexes elicited by strain detectors in the leg of the cockroach aid in load compensation during slow walking (Pearson, 1972; Zill & Moran, 1981b). As walking speed increases, the relative timing of muscle and proprioceptive activity shifts so that the reflex effects are no longer appropriate and therefore presumably are turned off centrally.

Model-Reference Control: A Higher-Order Control Using Internal Simulation of External Feedback Loops

In previous sections I have considered central control mechanisms and interactions between afferent and efferent information on several levels. In the last section motor control structures involving modulation of afferent and reflex activity were discussed. In some examples, the change in reflex activity appears to reflect the process of comparison between afferent activity and descending motor commands. In this section I will consider how this interaction of afferent and efferent activity has been elaborated by higher vertebrates into a more sophisticated control system. This system incorporates the spinal reflex loops and lower-level mechanisms for generating basic movement patterns such as locomotion, breathing, and chewing.

In the previous section evidence was presented suggesting that control signals passed to lower centers may take the form of reference signals for servocontrol loops. These reference signals can be considered expected values for the afferent parameters, a notion implied in Hoyle's (1964) use of the term "sensory tape". However, the result of this comparison is apparently also used to modify the afferent information returned to higher centers involved both in perception and in motor control. Thus, an animal is able to differentiate between exafference, which originates in external changes, and reafference, which arises through the animal's own actions, by simply comparing all afferences with expected values for afferences resulting from the animal's own motor behavior. Differences that remain can be attributed to external stimuli. For perceptual processing, this solves the main problem. For the purpose of motor control, this mechanism still suffers from delays, particularly those inherent in the peripheral skeletomuscular system. The efference copy signal itself is subject to delays only through neural conduction and integration, but it must be further delayed for comparison with the corresponding afferent information.

In a model-reference control system this delay is omitted, and the computed expected values are used to provide feedforward information for movement control. In his discussion of these idea, McMahon (1984) uses the model-reference control system of modern airplanes as an illustration. The commands from the pilot are passed both to the control systems and to a simulator of the airplane. The simulation is used to calculate the effect the commands should have under undisturbed conditions. With respect to higher levels, the function of peripheral control loops is internalized into the reference model to provide predictive, feedforward information about the commanded behavior. On lower levels, discrepancies between the behavior of the model and the actual behavior of the plane can be used in follow-up servomechanisms to bring the behavior of the plane into accord with that of the model. The advantage for higher control levels is that the delay times around the loop are reduced: the pilot can see the effect of his commands, or automatic systems can intervene to avoid potential danger even as servosystems translate the commands into actual control signals for the airplane and before the plane actually enters the dangerous state.

The introduction of model-reference elements increases the complexity of the control system in several ways. The model can be envisioned as performing four roles (Fig. 4). First, the simulation can provide feedforward predictions of the results of motor commands and return this information to higher centers. (Confusion is added by the fact that this computation of the expected results of motor commands returned to higher motor centers has also been designated as "internal feedback" (Wiesendanger, 1986a), in the sense that the loop provides a signal to higher motor centers which can be compared with output commands in the same way as those from peripheral feedback loops.) Second, the output information can be directly relayed to activators or follow-up servomechanisms which ensure that the actual behavior of the body matches that of the model. In animals this servocontrol is surely not the sole pathway to lower centers: direct pathways are equally important. Third, the output can be delayed and used as one source of an efference copy for comparison with afferent information. Fourth, the model may be an intermediate station in ascending pathways for information on the static and dynamic state of the body which is not directly available to higher centers.

This more sophisticated control principle combines the advantages of peripheral and central systems. When the environment or the consequences of particular motor patterns are entirely unpredictable, then a control of the sucess of the plan-

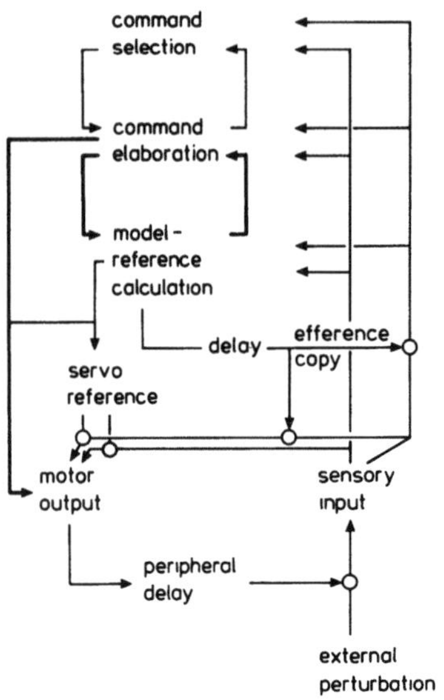

Fig. 4. Flow diagram for a model-reference system. The central feature is a loop (*thick lines*) between a simulation of the peripheral response, the model-reference calculation, and the command elaboration. This loop returns information equivalent to the reafference when the planned movement is carried out undisturbed; the loop is parallel to the peripheral loop through the motor output and sensory input but lacks the delay inherent in the latter

ned action by means of afferent information is necessary. The disavantage of imposing such a control on each movement is that it imposes a penalty on speed. The time for the muscular conversion of activity to force and movement and for the neural transmission and processing of the resulting afferent activity is considerable. The delay between pyramidal cell activity in the motor cortex and the EMG activity in hand muscles can be 50 ms (Evarts, 1966, 1968), and the development of full twitch tension can require anywhere between 30 and 80 ms or longer, depending upon the muscle (e.g., Collatos, Edgerton, Smith, & Botterman, 1978). Visual inputs require 35-40 ms to reach the visual cortex in the monkey (Glickstein, 1972). These peripheral delays make up a substantial part of typical reaction times to visual triggers, which may range between 120 and 190 ms for different tasks and experimental animals, and to somatosensory stimuli, which may be up to 80 ms shorter (e.g., Evarts & Tanji, 1976; Flanders, Cordo, & Ansom, 1986). Especially for large vertebrates, the long conduction distances and the large inertial masses of limbs and body may lead to a significant speed disadvantage in a world where selection is solely on the basis of performance.

When the environment is predictable and the effects of efferent activity patterns invariant, then an open-loop, centrally programmed control network provides sufficient accuracy and offers the advantage of speed. In biological systems such a control network could be predetermined, as it is in an airplane. This would form a very sophisticated central control model: the advantages of the model-reference control obtain as long as the model accurately represents the system's performance. In biological systems, such invariant, preprogrammed mechanisms become adapted to the environmental conditions in the course of evolutionary selection.

However, unlike machines, animals grow and change. They are subject to injury and may face a wide range of environmental conditions. For the model-reference control to be of use, the model must be adapted to current operating conditions. Evolutionary selection may provide preprogrammed solutions for recurring situations, but it cannot accurately allow for individual variability. For many animals the necessary adaptation is provided by learning. Particularly in higher vertebrates, learning mechanisms permit an individual animal to adapt to constant or slowly changing features of its interaction with the environment, regardless of whether the change is in the environment or in the capabilities of the body. The modulation of stimulus-response processing on the basis of experience is just one form. Motor learning through experience provides a means for predicting the effect of various motor commands under different circumstances. Thus, an important feature of biological model-reference control is the capacity to modify the model on the basis of experience: the development of the model becomes a process of self-organization.

In acquiring a novel movement, learning allows a motor control structure to shift from a dependence on peripheral mechanisms to a reliance on a centrally generated motor program. The role of sensory information shifts from controlling the success of movements in a continuous feedback manner to setting parameters in advance in a feedforward mode and to adjusting the model to the current state of the body and the environment. The scheme presented by Cruse et al. (this volume) incorporates these different possibilities: experimentally induced distortions in sensory inputs signal an important error in the current model and shift the control back to a continuous peripheral control mode. Complete absence of afferent input allows the central control to continue more or less undisturbed on the basis of the existing model.

In summary, such a model-reference control system in a biological context can augment the preexisting control systems in two ways. First, it uses a central representation of the body's performance in order to reduce the dependence on the momentary peripheral feedback. Second, it relies on peripheral information and learning to ensure that this representation is an accurate model of the body's performance

The rest of this section will concentrate on the first point - the nature of the control system. It illustrates the way in which movement control can rely on internal representations of the body and the environment. The example chosen is the mass-spring hypothesis for the control of joint angle.

The mass-spring model of movement control is a hypothesis that limb position is specified by the nervous system as a relation between the activities in antagonistic muscle groups (Feldman, 1974a, 1974b). As a first approximation, the muscles are treated as springs in which force increases as a function of muscle length, and this function varies according to the activity in the muscle (review Partridge & Benton, 1981). The equilibrium point for a joint is determined by the intersection of the spring functions for the antagonistic muscle groups plus any external forces acting on the limb.

Evidence to support this notion has been obtained from human beings (Kelso & Holt, 1980; Schmidt & McGown, 1980) and primates. The clearest example is provided by experiments with deafferented monkeys (Polit & Bizzi, 1979; Bizzi,

Chapple, & Hogan, 1982b). After a monkey learns a pointing task, the arm is deafferented to remove proprioceptive information. The monkeys are still able to point correctly at targets in the absence of information on the arm's true position. They specify arm position by activating the antagonistic muscle groups in the proper proportion. When an external force opposing the movement is applied to the arm, it shifts the equilibrium point because the added force is equivalent to adding a constant force to the antagonist spring function: the deafferented arm stops short of the target. When the external force is removed, the arm moves to the proper target position. This delayed correction is significant because it demonstrates that the activity pattern in the nervous system specifies a particular co-contraction which, in the absence of external forces, determines a particular limb position. Similar effects of loading occur in head movements which are incidental to a visual fixation task (Bizzi, Polit, & Morasso, 1976), so this form of control also seems to be a feature of natural movements.

The absolute amplitude of the muscle activity specifies the slope of the force-position function, the stiffness, and thereby determines the resistance to deflection by an external force. Thus, even in the absence of position feedback, the monkey could theoretically set the stiffness in accord with the probable load so as to achieve any desired minimum deviation from the target position. In the intact animal, the altered movement caused by an increased load elicits reflex activity which increases the excitation of the agonists: the result is a new force-length curve with increased stiffness to compensate for the load. In this context it is interesting that for the head movements mentioned above the co-contraction after deafferentation is greater than before the operation (Bizzi et al., 1976): it appears that the absence of information about either the current load or the actual head position leads the monkey to play safe and increase the stiffness. This finding shows that the monkey can specify limb position according to a mass-spring kind of representation, but the monkey may use more efficient schemes when additional information is available.

Further observations by Polit and Bizzi (1979) suggest that learning may play an important role in updating the internal representation of the characteristics of the body, its interactions with gravity, and the effect of particular motor commands. Slightly changing the configuration of the arm is sufficient to cause large pointing errors in a trained, deafferented monkey. An intact monkey rapidly corrects for the altered configuration, but the deafferented monkey does not.

Besides this reduced dependence on position feedback, a further attractive feature of the mass-spring formulation is its suggestion that a movement can be specified simply in terms of its endpoint (Feldman, 1974a,b; Kelso & Holt, 1980; Sakitt, 1980). Simulations of joint movement have shown that a step change in the relative activity of the antagonistic muscles suffices to produce a movement that has many of the kinematic characteristics of simple target movements (Cooke, 1980). These characteristics include the typical dome-shaped velocity profile with a maximum near the middle of the movement (Morasso, 1981). Different movement velocities can be obtained by varying the relative stiffness of agonists and antagonists.

However, this attractive notion is not supported by further experiments with deafferented monkeys (Bizzi, Accornero, Chapple, & Hogan, 1982a). The results show that the arm movement is specified in detail as a trajectory of positions leading from the starting position to the end position. Thus, the nervous system is not spared the problem of calculating continuous movement parameters. This is not surprising, given the evidence for separate control of end position and movement velocity (Sittig, Denier van der Gon, & Giellen, 1985) and the everyday experience that, depending on the task, additional parameters such as force, acceleration, and stiffness can be controlled (Stein, 1982).

Specification of movement as a trajectory of positions is a much more complex control problem. The kinematics are complex because they involve transformations between the rectilinear Cartesian coordinates, which describe the movements of effectors such as the hand or finger tip most simply, and the angular coordinates, which describe the corresponding positions of the limb joints. Similarly, the dynamics are complex because the torques needed at basal joints depend upon the limb's moment of inertia and hence on the configuration of distal joints. The theory of robotics has provided exact solutions for the necessary transformations (see Brady, Hollerbach, Johnson, Lozano-Perez, & Mason, 1982), but even using fast modern computers, the calculations require too much time for efficient real-time control of a moving arm. Presumably, the parallel processing arrangement of the nervous system more than compensates for the slower speed of neural processing elements. Just how it manages is not known.

That the nervous system is capable of calculating movement parameters in advance is evident in more rapid movements. In a true ballistic movement the motor command takes the form of a pulse: the agonist motoneuron activity ends before the movement is well underway (Desmedt & Godaux, 1978). Considered in the framework of the mass-spring formulation, the initial burst of agonist activity can be regarded as transiently setting the equilibrium position to an extreme joint position beyond the target. This pulse-type control works well only when the possibility of external perturbation is small, as in saccadic eye movement (von Holst & Mittelstaedt, 1950; see Bridgeman, this volume), or when the movement is well practiced.

Thus, the behavioral evidence shows that higher vertebrates plan and control multiple movement parameters according to the task at hand. Furthermore, they can maintain spatially coordinated movements in the absence of afferent information as long as an adequate internal representation is available. As discussed by Cruse et al. (this volume), this capability shows that any efference copy is not an invariant feature of movement control because it would lead to systematic deviations after deafferentation. It does not rule out the possibility that this information is used in the intact animal.

Variability in Behavior: Behavioral Choice and Motivation

At any one moment an animal generally is exposed to stimuli that allow many different behaviors, but only a limited number of behaviors can be performed simultaneously. The investigation of physiological mechanisms for behavioral choice involves searching for the neural correlates of this decision process. The respective neurons should operate as threshold detectors. Their input should be proportional to the afferent information, whereas the output should show discrete changes correlated with the presence or absence of a particular behavior.

Some of the early work with insects indicated that behaviorally significant stimulus classes were already created at the receptor level by receptors which respond selectively to a limited set of stimuli and trigger the appropriate behavior. Roeder's (review, 1967) description of the two sensory cells in the ears of noctuid moths is a prime example. The two cells respond at different intensity thresholds to echolocation signals of bats. Under natural conditions, these thresholds are reached at different distances between bat and moth. These two distances correspond to the distances at which free-flying moths begin to show first a negative phonotaxis and then an erratic "emergency" avoidance flight. The suggestion is that the successive activation of the two cells triggers first the far-range and then the near-range avoidance behavior. In addition, intensity differences at the two ears are encoded in the activity of the left and right low-threshold cells and guide the negative phonotaxis.

Detection systems relying on such "specialist" receptors are the exception. Most sensory systems in both vertebrates and invertebrates encode stimuli as a pattern of activity in an array of sensory receptors with broad response characteristics and in the temporal features of that activity. The nervous system must extract usable information from this pattern in order to guide behavior. Nervous systems apparently analyze complex stimuli in terms of simpler parameters or features present in the complex pattern. Neurons which show more specificity than the primary receptors are generally referred to as "feature detectors." Camhi (1984) distinguishes between the recognition of a complex stimulus relevant for the behavior of the animal and the detection of simpler constituent features. He uses "recognition neurons" for the first and restricts "feature detectors" to the second case. For the moth, the receptors themselves appear to recognize the relevant parameters: "bat approaching, alarm level 1 or 2." As the features become more abstract it is more difficult to determine how they guide action. At the same time, they permit a wider range of responses: the threshold detectors or recognition networks are removed from the periphery and several can act in parallel on the incoming afferent information.

Nevertheless, it may be misleading to view these recognition networks in isolation from the motor networks for the responses they elicit. Classical cross-mating experiments have shown an intriguing genetic link between stimulus detection and motor mechanisms, both in vertebrates and in insects. Male crickets and grasshoppers produce mating calls with species-specific sound patterns. When closely related species are crossed, the resulting hybrids produce a mating call which is intermediate to those of the parental species. The surprise is that the female hybrids, which themselves do not sing and therefore cannot experience the altered pattern, nevertheless prefer the calls of the hybrids (Hoy & Paul, 1973). Essentially the same result occurs for fruit flies (Kyriacou & Hall, 1986) and for hybrids between cricket frog populations with different songs (Doherty & Gerhardt, 1983). This behavioral linkage suggests that the detection and the production mechanisms are determined by the same genes: possibly they utilize many of the same neural circuits. Recently, physiological evidence in birds and in frogs has demonstrated that neural centers previously thought to be primarily motor are activated by species-specific acoustic calls (Williams & Nottebohm, 1985). This intimate link between perceptual and motor functions has been postulated in the motor theory of perception (review Kelly, 1986), whereby activity in the 'motor' centers should not necessarily be viewed simply as a repressed performance of the corresponding behavior.

Central to the discussion of physiological mechanisms underlying motivation and behavioral selection is the notion of command neurons (review Kupferman & Weiss, 1978). First introduced in the invertebrate literature by Wiersma and Ikeda (1964), the term has since been widely used for both invertebrates and vertebrates. Wiersma and Ikeda applied the term to fibers in the abdominal connective of the crayfish in which tonic stimulation releases a rhythmic movement of the swimmerets. This provides an operational definition: command neurons are neurons which, when stimulated, elicit a complex behavior pattern which is not determined by the pattern of command neuron activity.

The functional implication of the word command is the notion that the performance of the behavior under natural conditions also involves activating these neurons, but this evidence is generally difficult to obtain. Alternatively, laboratory experiments need to show that activity in these neurons is both necessary and sufficient for the behavior (Kupfermann & Weiss, 1978). The sufficiency condition is met when the behavior is elicited by stimulation of a neuron at a natural frequency. To test the necessity, it must be shown that the behavior disappears when the putative command neurons are destroyed or temporarily inactivated. Both tests are clearcut only when the command function is consolidated in one or a few neurons, as will be seen below. Also, it should be noted that any essential part of a pattern generator will pass the test for necessity.

Implicit in the original notion was a hierarchy of control levels (Kennedy, 1966; Kennedy, 1976; Grillner, 1977). The command neurons at the top of the hierarchy turn on a particular behavior by setting up and activating a motor control structure; Kennedy (1976), citing Hughlings Jackson, states that they "control movements, not muscles." They do so by exciting neurons on the next level which constitute a pattern generator and create the temporal structure. These in turn drive premotor and motoneurons on lower levels.

The tenor of more recent findings is that the structure of neural networks does not usually show this strict hierarchical organization. The flow of information is not just in one direction from command to motor levels: command elements may show rhythmic activity arising from peripheral feedback, as discussed above, or from corollary discharges. They may even form part of the oscillator. Neither is the flow of information restricted to communication between adjacent levels. This is evident in invertebrates: identified neurons which fit the operational definition of command neurons may have outputs directly onto motoneurons or even muscles (e.g., Kupfermann & Weiss, 1978; Davis, 1985). In vertebrates, too, command information flows over several parallel pathways to interact with motor centers at various levels. For example, the supplementary motor area, the medial part of area 6 located just rostral to the primary motor cortex, is thought to be concerned primarily with the planning of movements (Wiesendanger, 1986b). Its major output is into primary motor cortex, but it also has direct connections to spinal motor areas, in particular those controlling trunk muscles which provide the necessary basis for the fine movements of the extremities controlled by the primary motor cortex.

Two different patterns of command activity can be differentiated (Willows & Hoyle, 1969; Stein, 1978; Kupfermann & Weiss, 1978). The form first described by Wiersma and Ikeda (1964) acts like an on-off switch: activity in the command pathways turns on a behavior which continues as long as the command activity continues. Such neurons are now called gating neurons (Willows & Hoyle, 1969; Stein, 1978), although this is a bit of a misnomer. "Gating" implies opening and closing a connection between a source of continuous activity and the motor output. While gating may apply to some systems (Burrows, 1975b), more often the command itself activates the pattern generator. Besides this on-off function, the activity in command systems may also specify global parameters such as the rate of a rhythm.

The midbrain locomotory center in vertebrates is of this kind: stimulation of the center induces a decorticate cat placed on a treadmill to locomote for the duration of the stimulus (review Grillner, 1981). The preferred speed of locomotion varies with the intensity of stimulation. In the lamprey the analogous descending command system formed by reticulospinal pathways appears to release glutamate or other excitatory amino acids onto spinal motoneurons and interneurons (Dale & Grillner, 1986; Grillner et al., 1987). Bath application of these substances suffices to elicit locomotory rhythms from isolated spinal cords. In both insects and the leech, individual intersegmental command neurons have been identified which turn on motor output for locomotion (Pearson, Heitler, & Steeves, 1980; Gynther & Pearson, 1985; Weeks & Kristan, 1978; Weeks, 1982a) and may also modulate its frequency.

Detailed physiological information is also available for the feeding system of various mollusks (Davis & Kovac, 1981; Benjamin, 1983; Benjamin et al., 1985). The paracerebral (PCN) cells, a group of 16 cells in the brain of *Pleurobranchaea*, form a command system for feeding (review Davis, 1985). Depolarization, either as a result of chemical feeding stimulants or directly through intracellular stimulation, turns on a feeding rhythm in the motor centers. Depolarization of a single cell can elicit cyclic feeding bursts, and these cells are normally active during feeding in the intact animal (Gillette, Kovac, & Davis, 1978; Gillette, Gillette, & Davis, 1982). Plateauing properties of the ventral white cells are important for their command function in maintaining feeding (Gillette, Gillette, & Davis, 1980). This command system mediates several kinds of behavioral choice. Satiated animals are less willing to feed and the PCN cells in such animals show inhibition instead of excitation in response to normal feeding stimulants: this effect apparently depends on sensory input from mechanoreceptors signaling distension of the gut. Similarly, the PCN cells show inhibition rather than excitation in response to a test stimulant after the animals are aversively conditioned.

Besides these cells which are clearly identifiable as gating neurons, several cells have been identified which are better described as exerting modulatory functions. In *Lymnaea*, activity in the unpaired SO cell can induce feeding rhythms and is necessary for obtaining normal frequencies, but it is not necessary for the feeding rhythm to occur. In both *Lymnaea* (McCrohan & Benjamin, 1980) and *Pleurobranchaea* (Gillette & Davis, 1977) the metacerebral giant cell exerts widespread modulatory effects on the pattern generator, on motoneurons, and on muscles. These modulatory functions could represent part of a command function which is distributed among many parallel elements and therefore makes a precise distinction experimentally difficult (Kupfermann & Weiss, 1978).

The second form of command activity acts like a trigger (Willows & Hoyle, 1969): brief activity in the command cells initiates a behavior which is not dependent on continuing activity in the command cells. Examples are found in many escape responses of both vertebrates and invertebrates (review Eaton, 1984), some of which have already been mentioned. Escape is a behavior with high priority for the animal, whereby the speed of the initial response is more important than its adaptation to specific conditions. Consequently, this initial response is relatively stereotyped and the command function is concentrated in one or a few specialized neurons. In most cases, control passes to slower-conducting networks which maintain the response and adapt it to specific requirements of the environment.

The seven pairs of giant fibers in the cockroach are a good example (review Camhi, 1984). These intersegmental interneurons receive inputs from mechanoreceptors at the tip of the abdomen which register air movements associated with an approaching predator. If excited to threshold they conduct this information along a large-diameter axon to the motor centers in the thorax. Each giant fiber has a different receptive field, so the pattern of activity in the giants encodes the direction of the approaching threat and directs an appropriate evasive turn. The three largest pairs are the first to excite leg motoneurons and trigger turning in a strictly feedforward manner; then they are turned off. The four smaller pairs transmit similar information more slowly but they remain active during running (Daley & Delcomyn, 1980). Presumably, both sets contribute excitation to additional centers which maintain and coordinate running.

The rapid escape swimming of the crayfish is organized similarly. Activity in the lateral or the medial giant interneurons, respectively, triggers the appropriate all-or-none tailflip to begin rapid escape forwards or backwards. Activity in the lateral giant interneurons has been shown to be both sufficient and necessary for the corresponding tailflip to occur. Their nonlinear response converts graded inputs from mechanoreceptors into the decision to flee (Olson & Krasne, 1981). Both the ensuing swimming and normal swimming are controlled by a second, slower-acting system (review Wine, 1984).

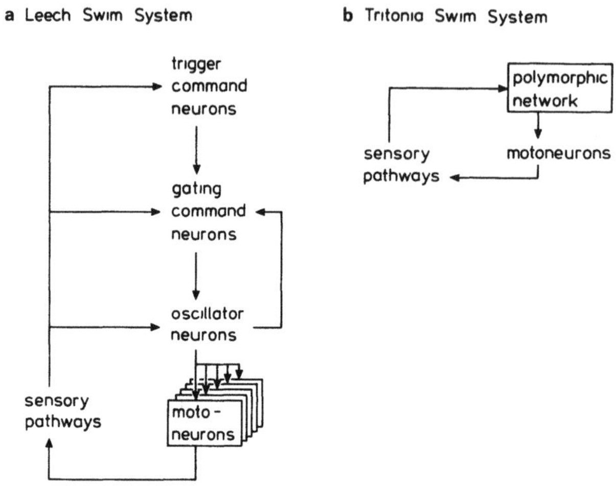

Fig. 5 a, b. Cellular organization of command structure. In the leech (**a**) the gating command neurons but not the trigger command neurons receive recurrent signals from lower levels. In *Tritonia* (**b**) command and oscillator functions are combined in a polymorphic network: sensory excitation above a threshold level acts on nonlinear, cellular properties to set up and turn on the swim state. In both diagrams the peripheral motor transformation is abbreviated. (**a** after Brodfuehrer & Friesen, 1986d; **b** from Getting & Dekin, 1985)

Both gating and trigger neurons have been identified in the swim system of the leech (Figs. 1d, 5a). Gating neurons linked to swimming include a single interneuron (cell 205) with extensive projections throughout the segmental nerve cord, unpaired interneurons in ganglion 10 and more posterior ganglia (cell 204), and unpaired interneurons in each ganglion (cell 208). Several serotonin-containing cells (cells 21 and 61) in each ganglion have local actions restricted to the two adjacent ganglia in each direction. These abdominal interneurons receive inputs from identified tactile cells; they contribute to swim initiation and they remain rhythmically active during swimming. They also excite the Retzius cells to release serotonin, which generally facilitates swimming in a hormonal fashion. These abdominal gating neurons receive inputs from several trigger neurons (cells Tr 1 and Tr 2, Brodfuehrer & Friesen, 1986a,b,c) located in the subesophageal ganglion in the head. These trigger neurons can account for swim initiation following strong mechanical stimuli.

Despite the complexity of the identified pathways among sensory, command, and pattern generator neurons, they do not suffice to explain all the physiological and behavioral findings, so as yet undescribed pathways must contribute. Nevertheless, the current scheme exemplifies a modified hierarchy (Fig. 5a) in which the gating neurons but not the trigger neurons receive recurrent excitation from the pattern generator. Some of these gating neurons fulfill the criteria for inclusion in the pattern generator but the functional importance is unclear: rhythmic activity in the gating neuron is normally present but is not required for pattern generation (Weeks, 1981; Nussbaum & Kristan, 1986). Hence, the rule might be that when recurrent pathways exist, the pattern generator may return rhythmic activity but the main effect is to provide net excitation to gating neurons as a contribution to sustaining their activity and net inhibition to trigger neurons. In *Pleurobranchaea*, Davis, Siegler & Mpitsos (1973) have interpreted the rhythmic input to cells having a command function as providing an efference copy for coordination of multiple oscillators, but their description does not fit the definition of efference copy used here.

The swim network in *Tritonia* (Fig. 5b) conforms even less well to a hierarchical organization. Command and oscillator functions are combined in the same neurons (review Getting & Dekin, 1985). Here one network produces two different patterns in the course of the escape reaction. Initially, the system produces withdrawal because the resting state is such that sensory input activates both the dorsal and the ventral swim interneuron groups (the DSI and the VSI, respectively; Fig. 1c), causing a co-contraction of the dorsal and ventral flexor muscles they control. If the sensory input is sufficiently strong, then its interaction with membrane and network properties of the pattern generator induces a prolonged depolarization of the major oscillator neurons and switches the oscillator into a second functional state in which the DSI and VSI groups alternate and produce rhythmic dorsoventral flexions. Once established, this second state is initially self-sustaining, but a slow increase in inhibitory interactions and a decrease in recurrent excitation eventually return the system to the resting mode. In this compact network, the plateau depolarization of the second state performs the gating function, and the positive-feedback threshold process by which it is established performs the trigger function. Getting and Dekin (1985) propose the term "polymorphic" network to designate such networks which are defined on the basis of the constituent neurons rather than a particular function and which can assume two or more functional states.

Polymorphic output patterns are also produced by the stomatogastric pattern generator under the control of neuromodulators and their effects on cellular properties (review Marder & Hooper, 1985) and possibly by the network controlling ingestion and egestion in mollusks (McClennan, 1982; Getting, 1984). The stomatogastric system of the lobster controls three teeth which reduce food to a size for swallowing. They do this using several different chewing motions (Heinzel, 1988). These patterns are all produced by a set of only 30 neurons, of which 14 form one oscillator network. The oscillation of this network can be altered by modulatory peptides which shift the pattern from that for one type of chewing to that for another (review Marder & Hooper, 1985). Here, the command for a particular pattern is apparently exercised through chemical transmitters from as yet unidentified neurons.

When the command function is solely a yes/no decision, then a few command neurons may suffice, but as discussed above, the command system often includes specification of spatial and temporal response parameters. When the behavior is highly variable, then the number of elements in the command system must be increased. As in the case of sensory coding, the parameter information is then conveyed in the pattern of activity rather than as yes/no signals in a few neurons. The anatomy of vertebrate nervous systems suggests a similar principle. Even in the simpler networks of invertebrates, such complex command networks are apt to be the rule and individual command neurons the exception. As described above, the swim system of the leech contains a hierarchy of command interneurons which mediate sensory motor interactions at ever more global levels. The control of movement in the crayfish abdomen shows a similar organization (Jellies & Larimer, 1986). Stimulation of single interneurons in a large set of interneurons produces a specific local movement of the abdomen. Each interneuron receives specific sensory input from abdominal proprioceptors and so forms a local reflex network. Complex movements require coordinated activity in the whole set. For such movements, the local integration of sensory and motor activity presumably is augmented by complex central control networks.

The foregoing discussion has illustrated mechanisms for initiating and maintaining particular behaviors. This role is just one part of command function: the command must also suppress other command systems which might compete for the motor system. Initial physiological findings with invertebrates suggest that this competition is resolved with the help of reciprocal inhibition among command systems for the conflicting activities. This question has been most intensively studied in *Pleurobranchaea* (review Davis, 1985). The relative priorities have been determined for escape swimming, egg-laying, feeding, mating, withdrawal, and righting behaviors (Kovac & Davis, 1980a). Some of these behaviors exclude others, whereas some behaviors can coexist. For example, egg-laying and righting behaviors can occur simultaneously but feeding suppresses righting. In the latter case, a corollary discharge pathway from the feeding network inhibits specific command interneurons for righting behavior (Kovac & Davis, 1980b).

Similar networks inhibiting competing behaviors are evident in other animals. In the crayfish the lateral giant fiber exerts inhibition at several control levels to suppress motor pathways conflicting with the tailflip motor pattern (Wine, 1984). In fish and larval amphibians, reciprocal inhibition between the contralateral Mauthner cells prevents a counterproductive co-contraction of left and right sides (Furakawa & Furshpan, 1963).

In contrast, positive feedback often is found among the elements within a command system. Positive feedback ensures that all members of the system become active together. In *Aplysia* the release of a defensive ink is controlled by a set of three electrically coupled motoneurons (review Kandel, 1979). The coupling ensures that the ink release is an all-or-nothing behavior. Plateau potentials of some command cells are another kind of positive feedback which can contribute to the maintenance of a behavior (e.g., Getting & Dekin, 1985), as can the release of humoral messengers or neuromodulators with prolonged activity (e.g., Willard, 1981).

Modulatory effects like those seen in the last examples weaken the distinction between specific synaptic interactions and hormonal effects. One tends to attribute behavior to synaptic activity and physiological changes to hormonal action, but some messengers have both effects. In *Aplysia* a whole complex of actions making up the egg-laying behavior is elicited by a set of peptides released by the bag cells (Mayeri & Rothman, 1985). One of the peptides, the egg-laying hormone, exerts both hormonal effects on the gonads and modulatory effects on neurons of the respiratory centers. One peptide alone is not sufficient for the full pattern: the complete mixture is required. This, then, is a case in which the command is exercised not just through synaptic interactions, although these play a part, but also through a complex chemical releaser. The leech swim system also includes nonspecific effects mediated through a general release of serotonin, as well as specific synaptic effects.

For technical reasons, the inputs to command fibers have generally received less attention than the outputs. Hence, the nonlinear response characterizing a decision process has been studied in only a few cases. The analysis is particularly difficult when the sensory inputs are complex, as in chemosensory systems. However, in *Pleurobranchaea* (review Davis, 1985) there is evidence that numerous chemosensory inputs differentially affect excitatory and inhibitory elements in the command network and that the net effect of these inputs, together with the interaction between the command elements themselves, determines whether feeding occurs.

In summary, the physiological evidence shows that the functional hierarchy implicit in many behavioral descriptions - behavioral selection, planning and global parameter setting, execution - is not directly reflected in neural elements. A rough control hierarchy of trigger elements, gating elements, pattern generator, and motoneurons can be proposed but there are many interconnections in both directions and across several levels. Several cellular and network properties involved in pattern generation, such as plateauing properties, reciprocal inhibition or excitation, and the time course of synaptic and membrane actions, are also important for behavioral choice and command functions. In addition, pharmacological agents, including neurohormones and neuromodulators of various kinds, play an important role in setting a behavior in motion and modifying the activity of the motor control centers. Many of these mechanisms are functionally interchangeable.

Mechanisms of Development and Plasticity

The preceding discussion of the functional organization of the nervous system has taken the anatomical structure as given. How this structure is generated is an area receiving ever more attention.

In higher vertebrates it is apparent that the development of the nervous system cannot be entirely programmed in the genetic information. As Bridgeman (this volume) points out, the information capacity of the genes is too small to completely specify the structure of the nervous system, let alone the vast number of connections among the neurons. Moreover, it is difficult to imagine an open-loop process achieving the specificity of connections evident in the resolution of the visual system. The solution is a process of self-organization, whereby interactions among the existing elements and feedback arising from the function of the system at each level of achieved organization are used to guide further development. Abnormal function at intermediate stages disturbs the course of further development. Some of the mechanisms underlying this plasticity are beginning to be understood (e.g., Bear & Singer, 1986). However, the complexity of the vertebrate nervous system still poses a formidable obstacle to the attempt to unravel the interactions of genetic and environmental effects.

The genetic factors are more accessible in various invertebrate systems. Therefore, this section will consider primarily the extent to which invertebrates express plasticity similar to that of vertebrates and thus might provide useful models for studying these phenomena.

Several facts lead to the conclusion that genetic specification is of prime importance in invertebrates. On the behavioral level, the life cycle of invertebrates generally lacks a period of developmental plasticity and learning like that occurring in the period of prolonged parental care present in many vertebrates. Many complex behaviors show no evidence of adaptive change: they are performed precisely the first time. On the anatomical level, the number of neurons is generally much smaller, so a complete specification of their number and their connections appears more feasible. The individuality expressed in the unique morphology of many invertebrate neurons appears to make such a specification more necessary.

It is no surprise, therefore, that genetic and anatomical studies of simple invertebrates provide evidence for genetic control exerted on the differentiation of individual neurons and even on synaptic connectivity. The nervous system of the nematode *C. elegans* has been examined electronmicroscopically to obtain a complete description of the cell lineages and the connections of all 302 neurons (White et al., 1986). Mutations have been isolated which effect specific changes in the behavior and in the neural structure. Mutants leading to specific deficits in behavior and neural structure have also been identified in fruit flies: one mutant, for example, is missing a prominent synaptic connection between an identified giant fiber and large motoneurons important for a startle reflex (review: Wyman, Thomas, Salkoff, & Costello, 1985). Genetic specification of neural structure at this level would explain the perfection with which even complex behaviors are performed without the need for practice or a behavioral model.

For more advanced invertebrates, the premise that the genetic information is sufficient to specify the connectivity in detail may no longer hold: interactive mechanisms may be necessary to appropriately use all the available neurons. The recent report of deficient visual behavior in flies raised without visual experience (Mimura, 1986) shows obvious parallels with vertebrates. How widespread such developmental plasticity may be remains to be seen.

After development is completed the nervous systems of invertebrates retain some structural and functional plasticity. As in vertebrates, the structural changes following abnormal afferent input may reflect processes participating in normal development. In crickets the loss of one ear leads to a rearrangement in the morphology of several important auditory interneurons (Hoy, Nolan, & Casady, 1985; Schildberger, Wohlers, & Schmitz, 1986). This is correlated with changes in the phonotactic orientation of the animals. A similar rearrangement occurs in the receptive fields of the giant fibers in cockroaches. When input from one cercus is lost, which can occur through natural causes, the giant fibers no longer correctly signal the direction of an approaching predator and the animal turns incorrectly (Comer & Camhi, 1984). However, over the course of several weeks, the inputs to the giant fibers are rearranged to again provide the proper spatial characteristics. This rearrangement requires that the intact cercus be exposed to normal wind stimuli, such as occur when the animal runs.

When injury results in the complete loss of sensory inputs, there is no basis for a functional rearrangement. In some cases, the disturbance of motor output is small at first but increases with time. Such a time course indicates that sensory information is essential for maintaining the long-term accuracy of the motor pattern, even when an immediate dependence on sensory feedback is not obvious. Song production in grasshoppers provides an example. The acridid grasshoppers use their hind legs for singing. Immediately following loss of one or both hind legs, the muscles on the remaining proximal segment continue to produce a normal pattern of activity, but over a period of weeks the pattern loses features characteristic of the intact animal (Lindberg & Elsner, 1977).

A third outcome is characteristic of some of the most dramatic rearrangements in behavior - the changes in leg coordination after leg amputation. Leg autotomy is a normal defensive reaction of many arthropods. In insects, spiders, and crustaceans, loss of a leg produces an immediate adaptive change in leg coordination, which does not show clear evidence of learning. The motor control structure governing leg coordination, as expressed in the neural control network, is constructed with sufficient generality that it operates appropriately both in the intact animal and in animals with a reduced number of legs. The coordination change can be attributed both to intact peripheral feedback loops acting under altered conditions and to the effect of altered afferent inputs on central activity. Loss of a leg appears to change not the procedures but rather the conditions under which they operate.

The above examples of adaptive or developmental change represent relatively inflexible, preprogrammed responses to changed conditions. Such mechanisms evolve within a species in the course of natural selection. In contrast, learning provides the individual with a flexible strategy for responding to changed conditions.

With respect to motor control, two functions of learning can be discriminated. One is to provide a quicker and more accurate link between external stimulus patterns and the behavioral response, i.e., a better definition of type-2 and type-3 effects. This formation of links between stimulus and response falls in the province of traditional learning studies. Descriptions of cellular mechanisms are well advanced for some model systems. The second function is to adjust the motor control structure to optimize performance. This can involve tuning peripheral control loops, i.e., modulating type-1 and type-2 effects, or adjusting central networks, such as a model-reference control. Here, little information on a cellular level is available.

Forms of stimulus-response learning which have been described in invertebrates include habituation, sensitization, and both instrumental and classical conditioning. Specialized forms of classical conditioning such as food aversion and higher order mechanisms such as masking and secondary conditioning have also been demonstrated (Sahley, 1984; Carew & Sahley, 1986). The accessibility of the relevant neurons has allowed an investigation of the physiological, biochemical, and morphological changes correlated with learning.

In *Aplysia*, several associative and nonassociative mechanisms have been studied. Habituation of defensive withdrawal responses results from a homosynaptic depression at the terminals of sensory fibers which decreases the transmitter output in response to recurring, nonaversive stimuli. Dishabituation and sensitization result from a heterosynaptic facilitation of these same terminals, which increases the transmitter output. Together, these two mechanisms serve to modulate the gain of the afferent pathways (Kandel & Schwartz, 1982; Quinn, 1984). They act like a variable filter to attenuate background activity and admit biologically significant signals, i.e., those differing from the background or those containing biologically relevant information.

Associative learning has been studied in several invertebrate models. In each case, learning involves changing the strength of preexisting connections. The results indicate that more than one mechanism may be responsible. One mechanism which promises to be widespread involves a nonlinear interaction of nonspecific, presynaptic facilitation with specific activity-dependent changes in the terminals themselves (Kandel & Schwartz, 1982). The interaction disproportionately strengthens the outputs of those sensory cells active in the period immediately preceding the sensitizing stimulus. This interactive mechanism has been investigated on the biochemical level: formally it resembles the interaction between specific and nonspecific priming effects in the model presented by Mackay (this volume). Depending on the intensity of the training, the short-term biochemical changes may be converted into structural changes involving number and size of synapses (Bailey & Chen, 1983). If these growth mechanisms are not restricted to strengthening preexisting connections but also lead to the formation of new connections, then new responses could be developed and subjected to further testing on the basis of experience.

The second model system (Alkon, 1984; reviews by Quinn, 1984 and Carew & Sahley, 1986) is a phototaxis response in *Hermissenda*, another marine mollusk; it does not show this link between sensitization and associative conditioning. Although the details of the network seem specific to *Hermissenda*, the results suggest a postsynaptic mechanism for amplifying a neuron's response to one synaptic input when contiguous activity in this input and a second input consistently activate the cell. A similar function is performed by the more recently discovered *N*-methyl-*D*-aspartate receptor system, which has been especially well studied in the vertebrate hippocampus (review: Gustafsson & Wigström, 1988). These postsynaptic mechanisms for detecting contiguity of activity in two inputs correspond to the learning rule proposed by Hebb (1949).

Although learning research has concentrated on the stimulus and reinforcement relationships, learning mechanisms may be intimately tied to motor processes under natural conditions. Bees and butterflies both learn to recognize flowers and concentrate their feeding on the most lucrative species; at the same time they learn to extract the nectar from this species more efficiently (Heinrich, 1984; Lewis, 1986). This motor learning has not yet been analyzed at the neuronal level.

It is evident that learning plays a much more important part in the motor control of higher vertebrates than in that of other animals. The experience acquired during the long period of parental care is necessary for the development of skills for survival. Play behavior shown by young mammals appears to aid in developing skills for prey-catching and intraspecific conflicts. Learning also plays a role in refining prey-catching in birds. Through experience young oystercatchers become better both at identifying vulnerable mussels and at performing the motor patterns for opening the shell. As the beak matures and hardens they learn to exploit more efficient strategies for extracting mussels (Goss-Custard & Dusell, 1987). In human beings and several laboratory animals, the behavioral changes involved in motor learning and the influence of various paradigms on performance have been studied in detail. The phenomena studied include learning of both simple movements and complex tasks such as athletic skills, the effects of knowledge of results, and various kinds of automatization including the feedforward control of larger and larger movement units (Schmidt, 1982). However, model systems suitable for studying the underlying neural mechanisms are not well developed.

Summary

The preceding discussion has covered a broad range of physiological findings related to motor behavior. The basic question concerns the neural mechanisms participating in a motor control structure. Present physiological findings show that it is important to distinguish between a hierarchy in the description of the functions belonging to a motor task and the actual structure of the nervous system. In a behavioral description it is usually possible to construct a control hierarchy beginning with task selection, initiation, and maintenance at the top and proceeding to pattern formation and then to control of individual movements on lower levels. However, the neural networks are much more complex in several respects.

First, the hierarchy arrived at through an information processing description of neural function need not coincide with the actual structure of the neural network: individual neurons may act at many levels of the functional hierarchy and contribute to both control and command functions.

Second, information does not flow in just one direction from the sense organs up to a decision center and then down to motor centers: interactions between sensory and motor pathways occur at many different levels. These interactions affect both the outgoing motor activity and the incoming sensory signals. Simple reflexes allow direct modulation of motor output on a local level without passing information up and down the hierarchy. These pathways may form feedback loops acting to stabilize specific parameters or, if the reference value is variable, servomechanisms ensuring that commands from higher centers are faithfully carried out. Both these local control units and distributed reflexes affecting several joints can be viewed as simplifying the behavior of the peripheral motor apparatus by buffering it against disturbances arising from external influences. However, the influence of sensory information active in these circuits is usually not limited to the peripheral level; it is also transmitted to networks involved in higher-level sensory processing and in motor pattern formation. In a similar fashion corollary discharges or efference copies, signals linked to motor output, can interact with incoming signals and thereby modulate the information available to higher centers.

Thus Turvey's (1977) notion of a heterarchy appears appropriate. In contrast to a strictly linear hierarchy, this term denotes a network containing a multiplicity of interactions among neurons and groups of neurons which themselves may be identified with several functional parameters.

In the past decade the peripheral motor apparatus and its mechanical characteristics have assumed an important place in this heterarchy. In all movements motor commands are filtered by the mass of the limbs and body and by the muscle characteristics in accord with the laws of physics. These physical factors in turn affect the afferent sensory information and through it can influence reflex performance and pattern generation whenever sensory feedback is important for higher centers.

With respect to the neural mechanisms controlling movement, the study of invertebrates has revealed a multiplicity of specific motor control networks and few general principles. The neural mechanisms underlying the production of motor patterns may depend in varying degrees on properties of individual neurons, network properties, or both. Neurons themselves are much more complex and variable in their function than was originally thought. Membrane properties govern the form of a neuron's response to a constant input. Complex biochemical and bioelectrical interactions act presynaptically to alter a neuron's output and postsynaptically to vary its response. These factors can cooperate to produce synaptic connections which are excitatory or inhibitory according to the pattern of activity or the potential of the cells concerned. The number of identified neurotransmitters has expanded tremendously. Many of these act as neuromodulators to affect variables ranging from muscle response and pattern generation to motivation and mood.

In invertebrates, a graded, nonspiking mode of neural integration may be an important factor in generating the variety of coordinated movements of which animals are capable. In vertebrates, this function may be subsumed by patterns of activity within large populations of neurons, but even here graded transmission is important for local interactions among neurons.

The example of different mechanisms for similar ends serves as a reminder that the control of movement in animals has developed under different rules than have comparable technical systems. Biological evolution does not necessarily lead to optimal solutions in a technical sense: it can only modify that which is already present in some form. The ability to transfer solutions from one functional area to another is limited, and the development of new solutions depends on a measure of chance. At the same time, selection is exerted only on the function in its broadest sense, and much of this selection takes the form of competition with other organisms. Thus, selection permits and evolutionary history may require that an advance to a higher level of performance be based on different mechanisms in different groups. This fact encourages care in referring to "lower" and "higher" animals: all present-day animals have succeeded in the evolutionary competition and therefore "lower" forms cannot be taken to represent the level achieved by ancestors of more complex animals.

Another consequence of evolution is that nonfunctional components may persist. In analyzing neural circuits it is usually assumed that all connections have a functional purpose even if this is not immediately apparent. Increasing knowledge has led some investigators to suggest that some connections may be truly nonfunctional relicts of the past, a kind of neural appendix (Dumont & Robertson, 1986).

The divergence of evolutionary pathways is even more apparent in a comparison of different groups. One might expect the considerable differences in the size and complexity of the nervous system of insects and vertebrates to correspond to striking differences in behavior but this is obviously not true. Qualitatively, the behavioral differences lie not so much in the complexity and speed of performance as in its variety and plasticity, but the latter are hard to quantify. Movement speed is the easiest parameter to measure, but it does not correlate with neural differences between insects and vertebrates.

However, the similarity in speed may conceal a possible difference in control function. The larger size of vertebrates places them at a disadvantage both in terms of neural conduction distances and inertial masses. In the discussion above it was proposed that higher vertebrates compensate for these disadvantages by incorporating more advance planning into the control structure. For mammals this advance planning is thought to include a model-reference control. Some elements of this control may be present in other vertebrates and in invertebrates, but they are less apparent due to the more compact and anatomically less differentiated form of their nervous systems. Some neural circuits in invertebrates do contain complex loops and functional redundancies reminiscent of mammalian circuits (e.g., Davis, Kovac, Croll, & Matera 1984; Davis 1985). The other advanced vertebrate group, the birds, have not been discussed here. In fact, the forebrain of birds is organized quite differently from that of mammals (Hodos, 1982) and it will be interesting to discover whether internal CNS loops play as large a role as in mammals.

Learning permits an individual to adapt to the environment around it and to track changes in the environment and in the animal's own capabilities. The ability to learn is apparently of such value and can be achieved with such basic neural mechanisms that practically all animals show some forms of learning. Nevertheless, learning does appear best developed in higher vertebrates, and this is correlated with more complex nervous systems and longer life spans. The model-reference control outlined above depends upon the accuracy of the model. Although it might be possible to genetically program the necessary connections, mammals appear to rely on motor learning for fine-tuning the model, but the physiological mechanisms are not well understood. Possibly, the study of invertebrate models could aid here as it has in the study of stimulus-response learning.

In the end, one can take a comparative approach to motor control with two attitudes. One is to acknowledge that many features will be specific to a particular evolutionary history and to study this diversity for its own sake. The second is to assume that functional constraints are sufficiently important in evolution to impose general principles of organization that hold for large taxonomic groups. The realization of these principles may build upon a large not infinite number of basic cellular properties, which appear to be common to large taxonomic groups, and a much larger range of possible circuits. Despite this diversity, cellular analysis carried out in one group can improve both the methodology and the conceptual basis for attacking similar problems in other groups. More important, the comparative approach provides the primary means for distinguishing between control principles enforced by functional criteria and those resulting from evolutionary history.

References

Adams, D.J., Smith, S.M., & Thompson, S.H. (1980). Ionic currents in molluscan soma. *Annual Review of Neuroscience, 3*, 141-167.

Alkon, D.L. (1984). Calcium-mediated reduction of ionic currents: A biophysical memory trace. *Science, 226*, 1037-1045.

Alving, B.O. (1968). Spontaneous activity in isolated somata of *Aplysia* pacemaker neurons. *Journal of General Physiology, 45*, 29-45.

Bacon, J., & Möhl, B. (1983). The tritocerebral commissure giant (TCG) wind-sensitive interneurone in the locust. I. Its activity in straight flight. *Journal of Comparative Physiology, 150*, 439-452.

Bailey, C.H., & Chen, M. (1983). Morphological basis of long-term habituation and sensitization in *Aplysia. Science, 220*, 91-93.

Barnes, W.J.P. (1975). Nervous control of locomotion in Crustacea. In: P.N.R. Usherwood, & D.R. Newth (Eds.), *Simple nervous systems*. (pp. 415-441). London: Arnold.

Barnes, W.J.P., Spirito, C.P., Evoy, W.H. (1972). Nervous control of walking in the crab, *Cardisoma guanhumi*. II. Role of resistance reflexes in walking. *Zeitschrift für Vergleichende Physiologie, 76*, 16-31.

Bässler, U. (1976). Reversal of a reflex to a single motoneuron in the stick insect *Carausius morosus. Biological Cybernetics, 24*, 47-49.

Bässler, U. (1983). *Neural basis of elementary behavior in stick insects*. Berlin, Heidelberg, New York: Springer.

Bässler, U. (1985). Proprioceptive control of stick insect walking. In B.M.H. Bush & F. Clarac (Eds.), *Coordination of motor behavior* (Society for Experimental Biology Seminar No. 24, pp. 271-281). Cambridge: Cambridge University Press.

Bear, M.F., & Singer, W. (1986). Modulation of visual cortical plasticity by acetylcholine and noradrenaline. *Nature, 320*, 172-176.

Bell, C.C. (1982). Properties of a modifiable efference copy in an electric fish. *Journal of Neurophysiology, 47*, 1043-1056.

Bell, C.C. (1984). Effects of motor commands on sensory inflow, with examples from electric fish. In L.Bolis, R.D. Keynes, & S.H.P. Maddrell (Eds.), *Comparative physiology of sensory systems* (pp. 637-646). Cambridge: Cambridge University Press.

Benjamin, P.R. (1983). Gastropod feeding: Behavioral and neural analysis of a complex multicomponent system. In A. Roberts & B.L. Roberts (Eds.), *Neural control of rhythmic movements* (pp. 159-193). Cambridge: Cambridge University Press.

Benjamin, P.R., Elliott, C.J.H., & Ferguson, G.P. (1985). Neural network analysis in the snail brain. In A.I. Selverston (Ed.), *Model neural networks and behavior* (pp. 87-108). New York: Plenum.

Bernstein, N. (1967). *The co-ordination and regulation of movements*. Oxford: Pergamon.

Bizzi, E., Polit, A., & Morasso, P. (1976). Mechanisms underlying achievement of final head position. *Journal of Neurophysiology, 39*, 435-444.

Bizzi, E., Accornero, N., Chapple, W., & Hogan, N. (1982a). Arm trajectory formation in monkeys. *Experimental Brain Research*, *46*, 139-143.

Bizzi, E., Chapple, W., & Hogan, N. (1982b). Mechanical properties of muscles: implications for motor control. *Trends in Neuroscience*, *5*, 395-398.

Boyd, I.A., & Smith, R.S. (1984). The muscle spindle. In P.J. Dyck, P.K. Thomas, E.H. Lombert, & R. Bunge (Eds.), *Peripheral neuropathy* (pp. 171-202). Philadelphia: Saunders.

Brady, M., Hollerbach, J.M., Johnson, T.L., Lozano-Perez, T., & Mason, M.T. (Eds.). (1982). *Robot motion: Planning and control*. Cambridge, MA: MIT Press.

Bräunig, P., & Hustert, R. (1985). Actions and interactions of proprioceptors of the locust hind leg coxo-trochanteral joint. 1. Afferent responses in relation to joint position and movement. *Journal of Comparative Physiology. A, Sensory, Neural, and Behavioral Physiology*, *157*, 73-82.

Brenner, S. (1974). The genetics of *Caenorhabditis elegans*. *Genetics*, *77*, 71-94.

Brodfuehrer, P.D., & Friesen, W.O. (1986a). Initiation of swimming activity by trigger neurons in the leech subesophageal ganglion. I. Output connections of Tr1 and Tr2. *Journal of Comparative Physiology. A, Sensory, Neural, and Behavioral Physiology*, *159*, 489-502.

Brodfuehrer, P.D., & Friesen, W.O. (1986b). Initiation of swimming activity by trigger neurons in the leech subesophageal ganglion. II. Role of segmental swim-initiating interneurons. *Journal of Comparative Physiology. A, Sensory, Neural, and Behavioral Physiology*, *159*, 503-510.

Brodfuehrer, P.D., & Friesen, W.O. (1986c). Initiation of swimming activity by trigger neurons in the leech subesophageal ganglion. III. Sensory inputs to Tr1 and Tr2. *Journal of Comparative Physiology. A, Sensory, Neural and Behavioral Physiology*, *159*, 511-519.

Brodfuehrer, P.D., & Friesen, W.O. (1986d). From stimulation to undulation: A neuronal pathway for the control of swimming in the leech. *Science*, *234*, 1002-1004.

Brown, T.G. (1914). On the nature of the fundamental activity of the nervous centres; together with an analysis of the conditioning of rhythmic activity in progression, and a theory of the evolution of function in the nervous system. *Journal of Physiology*, *48*, 18-46.

Bullock, T.H. (1976). In search of principles in neural integration: Are there rules in the combination of elements in neural circuits? In J.C. Fentress (Ed.), *Simpler networks and behavior* (pp. 52-60) Sunderland, MA: Sinauer.

Bullock, T.H. (1979). Evolving concepts of local integrative operations in neurons. In F.O. Schmitt & F.G. Worden (Eds.) *The neurosciences fourth study program* (pp. 43-49). Cambridge, MA: MIT Press.

Burrows, M. (1975a). Monosynaptic connexions between wing stretch receptors and flight motoneurons of the locust. *Journal of Experimental Biology*, *62*, 189-219.

Burrows, M. (1975b). Coordinating interneurons of the locust which convey two patterns of motor commands: Their connections with flight motoneurons. *Journal of Experimental Biology*, *63*, 713-733.

Burrows, M. (1985). Nonspiking and spiking local interneurons in the locust. In A.I. Selverston (Ed.), *Model neural networks and behavior* (pp. 109-125). New York: Plenum.

Burrows, M., Siegler, M.V.S. (1982). Spiking local interneurons mediate local reflexes. *Science*, *217*, 650-652.

Bush, B.M.H., & Cannone, A.J. (1985). How do crabs control their muscle receptors? In W.J.P. Barnes & M.H. Gladden (Eds.), *Feedback and motor control in invertebrates and vertebrates* (pp. 145-166). London: Croom Helm.

Calvin, W.H., & Graubard, K. (1979). Styles of neuronal computation. In F.O. Schmitt & F.G. Worden (Eds.), *The neurosciences fourth study program* (pp. 513-524). Cambridge, MA: MIT Press.

Camhi, J.M. (1984). *Neuroethology: Nerve cells and the natural behavior of animals*. New York: Sinauer.

Camhi, J.M. (1985). Feedback control of an escape behavior. In W.J.P. Barnes & M.H. Gladden (Eds.), *Feedback and motor control in invertebrates and vertebrates* (pp. 93-111). London: Croom Helm.

Carew, T.J., & Sahley, C.L. (1986). Invertebrate learning and memory: From behavior to molecules. *Annual Review of Neuroscience*, *9*, 435-487.

Collatos, T.C., Edgerton, V.R., Smith, J.L., & Botterman, B.R. (1978). Fiber type compositions of flexors and extensors of elbow joint in cat: Implications for motor control. *Journal of Neurophysiology*, *40*, 1292-1300.

Collett, T.S. (1980). Angular tracking and the optomotor response. An analysis of visual reflex interactions in the hoverfly. *Journal of Comparative Physiology*, *140*, 145-158.

Comer, C., & Camhi, J.M. (1984). Behavioral compensation for altered cercal position in the cockroach. *Journal of Comparative Physiology. A, Sensory, Neural, and Behavioral Physiology*, *155*, 31-38.

Cook, D.G., & Carew, T.J. (1986). Operant conditioning of head waving in *Aplysia. Proceedings of the National Academy of Sciences of the United States of America, 83*, 1120-1124.

Cooke, J.D. (1980). The organization of simple, skilled movements. In G.E. Stelmach & J. Requin (Eds.), *Tutorials in motor behavior* (pp. 199-212). Amsterdam: North-Holland.

Cotman, C.W., & Iversen, L.L. (1987). Excitatory amino acids in the brain - focus on NMDA receptors. *Trends in Neurosciences, 10*, 263-265.

Dale, N., & Grillner, S. (1986). Dual-component synaptic potentials in the lamprey mediated by excitatory amino acid receptors. *Journal of Neuroscience, 6*, 2653-2661.

Daley, D.L., & Delcomyn, F. (1980). Modulation of the excitability of cockroach giant interneurons during walking. I. Simultaneous excitation and inhibition. *Journal of Comparative Physiology, 138*, 231-239.

Davis, W.J. (1985). Neural mechanisms of behavioral plasticity in an invertebrate model system. In A.I. Selverston (Ed.), *Model neural networks and behavior* (pp. 263-282). New York: Plenum.

Davis, W.J., & Kovac, M.P. (1981). The command neuron and the organization of movement. *Trends in Neurosciences, 4*, 73-76.

Davis, W.J., Siegler, M.V.S., & Mpitsos, G.J. (1973). Distributed neuronal oscillators and efference copy in the feeding system of *Pleurobranchaea. Journal of Neurophysiology, 36*, 258-274.

Davis, W.J., Kovac, M.P., Croll, R.P., & Matera, E.M. (1984). Brain oscillator(s) underlying rhythmic cerebral and buccal motor output in the mollusc, *Pleurobranchaea californica. Journal of Experimental Biology, 110*, 1-15.

Delcomyn, F. (1980). Neural basis of rhythmic behavior in animals. *Science, 210*, 492-498.

Desmedt, J.E., & Godaux, E. (1978). Ballistic skilled movements: Load compensation and patterning of the motor commands. In T.E. Desmedt (Ed.), *Cerebral control in man: Long loop mechanisms: Vol. 4. Progress in Clinical Neurophysiology* (pp. 21-55). Basel: Karger.

Doherty, J.A., & Gerhardt, H.C. (1983). Hybrid treefrogs: Vocalizations of males and selective phonotaxis of females. *Science, 220*, 1078-1080.

Dumont, J.P.C., & Robertson, R.M. (1986). Neuronal circuits: An evolutionary perspective. *Science, 233*, 849-853.

Eaton, J.P.C. (Ed.). (1984). *Neural mechanisms of startle behavior*. New York: Plenum.

Eaton, R.C., & Hackett, J.T. (1984). The role of the Mauthner cell in fast starts involving escape in teleost fishes. In R.C. Eaton (Ed.), *Neural mechanisms of startle behavior* (pp. 213-266). New York: Plenum.

Egelhaaf, M. (1987). Dynamic properties of two control systems underlying visually guided turning in house-flies. *Journal of Comparative Physiology. A. Sensory, Neural, and Behavioral Physiology, 161*, 777-783.

Eibl-Eibesfeldt, I. (1970). *Ethology: The biology of behavior*. New York: Holt, Rinehart & Winston.

Eisemann, C.H., Jorgensen, W.K., Merritt, D.J., Rice, M.J., Cribb, B.W., Webb, P.D., & Zalucki, M.P. (1984). Do insects feel pain? - A biological view. *Experientia, 40*, 164-167.

Elliott, C.J.H., & Koch, U.T. (1985). The clockwork cricket. *Naturwissenschaften, 72*, 150.

Evarts, E.V. (1966). Pyramidal tract activity associated with a conditioned hand movement in the monkey. *Journal of Neurophysiology, 29*, 1011-1027.

Evarts, E.V. (1968). Relation of pyramidal tract activity to force exerted during voluntary movement. *Journal of Neurophysiology, 31*, 14-27.

Evarts, E.V., & Tanji, J. (1976). Reflex and intended responses in motor cortex pyramidal tract neurons of monkeys. *Journal of Neurophysiology, 39*, 1069-1080.

Evoy, W.H., & Ayers, J. (1982). Locomotion and control of limb movements. In D.C. Sanderman, H.L. Atwood (Eds.), *The biology of Crustacea: Vol. 4. Neural integration and behavior* (pp. 61-105). New York: Academic.

Feldman, A.G. (1974a). Change of muscle length due to shift of the equilibrium point of the muscle-load system. *Biophysics, 19*, 544-548.

Feldman, A.G. (1974b). Control of muscle length. *Biophysics, 19*, 766-771.

Fields, H.L. (1976). Crustacean abdominal and thoracic muscle receptor organs. In P.J. Mill (Ed.), *Structure and function of proprioceptors in the invertebrates* (pp. 65-114). London: Chapman and Hall.

Fields, H.L., Evoy, W.H., & Kennedy, D. (1967). Reflex role played by efferent control of an invertebrate stretch receptor. *Journal of Neurophysiology, 30*, 859-874.

Flanders, M., Cordo, P.J., & Ansom, J.G. (1986). Interaction between visually and kinesthetically triggered voluntary response. *Journal of Motor Behavior, 18*, 427-448.

Forssberg, H. (1979). "The stumbling correction reaction" - a phase-dependent compensatory reaction during locomotion. *Journal of Neurophysiology, 42*, 936-953.

Fraser, P.J. (1982). Views on the nervous control of complex behavior. In D.C. Sandeman, H.L. Atwood (Eds.), *The biology of the Crustacea: Vol. 4. Neural integration and behavior* (pp. 293-319). New York: Academic.

Friesen, W.O. (1985). Neuronal control of leech swimming movements: Interactions between cell 60 and previously described oscillator neurons. *Journal of Comparative Physiology. A, Sensory, Neural, and Behavioral Physiology, 156*, 231-242.

Friesen, W.O., & Stent, G.S. (1977). Generation of a locomotory rhythm by a neural network with recurrent cyclic inhibition. *Biological Cybernetics, 28*, 27-40.

Friesen, W.O., Poon, M., & Stent, G.S. (1976). An oscillatory neuronal circuit generating a locomotory rhythm. *Proceedings of the National Academy of Sciences of the United States of America, 73*, 3734-3738.

Friesen, W.O., Poon, M., Stent, G.S. (1978). Neuronal control of swimming in the medicinal leech. IV. Identification of a network with recurrent cyclic inhibition. *Journal of Experimental Biology, 75*, 24-43.

Furakawa, T., & Furshpan, E.J. (1963). Two inhibitory mechanisms in the Mauthner cell of goldfish. *Journal of Neurophysiology, 26*, 140-176.

Getting, P.A. (1983). Mechanisms of pattern generation underlying swimming in *Tritonia*. III. Intrinsic and synaptic mechanisms for delayed excitation. *Journal of Neurophysiology, 49*, 1036-1050.

Getting. P.A. (1984). Neural control of behavior in gastropods. In A.O.D. Willows (Ed.), *The mollusca: Vol. 8. Neurobiology and behavior: Part 1* (pp. 269-334). Orlando: Academic.

Getting, P.A., & Dekin, M.S. (1985). *Tritonia* swimming: A model system for integration within rhythmic motor systems. In A.I. Selverston (Ed.), *Model neural networks and behavior* (pp. 3-20). New York: Plenum.

Getting, P.A., Lennard, P.R., & Hume, R.I. (1980). Central pattern generator mediating swimming in *Tritonia*. I. Identification and synaptic interactions. *Journal of Neurophysiology, 44*, 151-164.

Gillette, R., & Davis, W.J. (1977). The role of the metacerebral giant neuron in the feeding behavior of *Pleurobranchaea*. *Journal of Comparative Physiology. A, Sensory, Neural, and Behavioral Physiology, 116*, 129-159.

Gillette, R., Kovac, M.P., & Davis, W.J. (1978). Command neurons in *Pleurobranchaea* receive synaptic output from the motor network they excite. *Science, 199*, 798-801.

Gillette, R., Gillette, M., & Davis, W.J. (1980). Action-potential broadening and endogenously sustained bursting are substrates of command ability in a feeding neuron of *Pleurobranchaea*. *Journal of Neurophysiology, 43*, 669-685.

Gillette, R., Gillette, M.U., Davis, W.J. (1982). Control of feeding motor output by paracerebral neurons in the brain of *Pleurobranchaea*. *Journal of Neurophysiology, 47*, 885-908.

Glickstein, M. (1972). Brain mechanisms in reaction time. *Brain Research 40*, 33-37.

Goss-Custard, J.D., & Durell, S.E.A. le V. dit (1987). Age-related effects in oystercatchers, *Haematopus ostralegus*, feeding on mussels, *Mytilus edulis*. I. Foraging efficiency and interference. *Journal of Animal Ecology, 56*, 521-536.

Gottlieb, G.L., & Agarwall, D.C. (1980). Response to sudden torques about ankle in man. III. Suppression of stretch-evoked responses during phasic contraction. *Journal of Neurophysiology, 44*, 233-246.

Gould, J.L. (1984). Natural history of honey bee learning. In P. Marler & H.S. Terrace (Eds.), *The biology of learning* (pp. 149-180). Berlin, Heidelberg, New York: Springer.

Gould, J.L., & Gould, C.G. (1982). The insect mind: Physics or metaphysics? In D.R. Griffin (Ed.), *Animal mind - Human mind* (pp. 269-297). Berlin, Heidelberg, New York: Springer.

Graham, D. (1977). Simulation of a model for the coordination of leg movement in free-walking insects. *Biological Cybernetics, 26*, 187-198.

Graham, D. (1985). Pattern and control of walking in insects. *Advances in Insect Physiology, 18*, 31-140.

Granit, R. (1970). *The basis of motor control*. London: Academic.

Graubard, K., & Calvin, W.H. (1979). Presynaptic dendrites: Implications of spikeless transmission and dendritic geometry. In F.O. Schmitt & F.G. Worden (Eds.), *The neurosciences fourth study program* (pp. 317-331). Cambridge, MA: MIT Press.

Gray, J. (1950). The role of peripheral sense organs during locomotion in the vertebrates. *Symposia of the Society for Experimental Biology, 4*, 112-126.

Griffin, D.R. (Ed.). (1982). *Animal mind - Human mind*. Berlin, Heidelberg, New York: Springer.

Grillner, S. (1977). On the neural control of movement - A comparison of different rhythmic behaviors. In G.S. Stent (Ed.), *Function and formation of neural systems (Dahlem Konferenzen*, pp. 197-224). Berlin, Heidelberg, New York: Springer.

Grillner, S. (1981). Control of locomotion in bipeds, tetrapods, and fish. In V.B. Brooks (Ed.), *Handbook of physiology: Sec. 1. The nervous system: Vol. 2. Motor control.* (American Physiological Society Series, pp. 1179-1236). Baltimore: Williams and Wilkins.

Grillner, S. (1985). Neurobiological bases of rhythmic motor acts in vertebrates. *Science, 228*, 143-149.

Grillner, S., & Wallén, P. (1985). General pattern generators for locomotion, with special reference to vertebrates. *Annual Review of Neuroscience, 8*, 233-261.

Grillner, S., McClellan, A., & Perret, C. (1981). Entrainment of the spinal pattern generators for swimming by mechanosensitive elements in the lamprey spinal cord in vitro. *Brain Research, 217*, 380-386.

Grillner, S., Wallén, P., Dale, N., Brodin, L., Buchanan, J., & Hill, R. (1987). Transmitters, membrane properties and network circuitry in the control of locomotion in lamprey. *Trends in Neurosciences, 10*, 34-41.

Gustafsson, B., & Wigström, H. (1988). Physiological mechanisms underlying long-term potentiation. *Trends in Neurosciences, 11*, 156-162.

Gynther, I.C., & Pearson, K.G. (1986). Intracellular recordings from interneurones and motoneurones during bilateral kicks in the locust: Implications for mechanisms controlling the jump. *Journal of Experimental Biology, 122*, 323-343.

Hassan, E.-S. (1986). On the discrimination of spatial intervals by the blind cave fish *Anoptichthys jordani. Journal of Comparative Physiology. A, Sensory, Neural, and Behavioral Physiology, 159*, 701-710.

Hawkins, R.D., Abrams, T.W., Carew, T.J., & Kandel, E.R. (1983). A cellular mechanism of classical conditioning in *Aplysia*: Activity-dependent amplification of presynaptic facilitation. *Science, 219*, 400-405.

Hebb, D.O. (1949). *The organization of behavior.* New York: Wiley.

Heinrich, B. (1984). Learning in invertebrates. In P. Marler & H.S. Terrace (Eds.), *The biology of learning* (pp. 135-147). Berlin, Heidelberg, New York: Springer.

Heinzel, H.-G. (1988). Gastric mill activity in the lobster. I. Spontaneous modes of chewing. *Journal of Neurophysiology, 59*, 528-550.

Heisenberg, M., & Wolf, R. (1979). On the fine structure of yaw torque in visual flight orientation of *Drosophila melanogaster. Journal of Comparative Physiology, 130*, 113-130.

Heitler, W.J. (1978). Coupled motoneurons are part of the crayfish swimmeret central oscillator. *Nature, 275*, 231-233.

Henneman, E. (1980). Organization of the spinal cord and its reflexes. In V.B. Mountcastle (Ed.), *Medical physiology* (14th ed., pp. 762-786, Vol. 1). St Louis: Mosby.

Hisada, M., Takahata, M., & Nagayama, T. (1984). Structure and output connection of local non-spiking interneurons in crayfish. *Zoological Science, 1*, 41-49.

Hodgkin, A.L., & Huxley, A.F. (1952). A quantitative description of membrane current and its application to conduction and excitation in nerve. *Journal of Physiology, 117*, 500-544.

Hodos, W. (1982). Some perspectives on the evolution of intelligence and the brain. In D.R. Griffin (Ed.), *Animal mind - Human mind* (pp. 33-55). Berlin, Heidelberg, New York: Springer.

Horsmann, U., Heinzel, H.-G., & Wendler, G. (1983). The phasic influence of self-generated air current modulations on the locust flight motor. *Journal of Comparative Physiology, 150*, 427-438.

Houk, J.C., Rymer, W.Z. (1981). Neural control of muscle length and tension. In V.B. Brooks (Ed.), *Handbook of physiology, Sec. 1. The nervous system: Vol. 2. Motor control* (American Physiological Society Series, pp. 257-323). Baltimore: Williams and Wilkins.

Hoy, R.R., & Paul, R.C. (1973). Genetic control of song specificity in crickets. *Science, 180*, 82-83.

Hoy, R.R., Nolen, T.G., & Casady, G.C. (1985). Dendritic sprouting and compensatory synaptogenesis in an identified interneuron follow auditory deprivation in a cricket. *Proceedings of the National Academy of Sciences of the United States of America, 82*, 7772-7776.

Hoyle, G. (1964). Exploration of neuronal mechanisms underlying behavior in insects. In R.F. Reiss (Ed.), *Neural theory and modelling* (pp. 346-376). Stanford: Stanford University Press.

Hoyle, G. (1970). Cellular mechanisms underlying behavior: Neuroethology. *Advances in Insect Physiology, 7*, 349-444.

Hoyle, G. (1975). Identified neurons and the future of neuroethology. *Journal of Experimental Zoology, 194*, 51-74.

Hunt, C.C., & Perl, E.R. (1960). Spinal reflex mechanisms concerned with skeletal muscle. *Physiological Review, 40*, 538-579.

Jellies, J., & Larimer, J.L. (1986). Activity of crayfish abdominal-positioning interneurons during spontaneous and sensory-evoked movements. *Journal of Experimental Biology, 120*, 173-188.

Kalmijn, A.J. (1984). Theory of electromagnetic orientation: A further analysis. In L. Bolis, R.D. Keynes, & S.H.P. Maddrell (Eds.), *Comparative physiology of sensory systems* (pp. 525-560). Cambridge: Cambridge University Press.
Kandel, E.R. (1979). *Behavioral biology of Aplysia*. San Francisco: Freeman.
Kandel, E.R., & Schwartz, J.H. (1982). Molecular biology of learning: modulation of transmitter release. *Science, 218*, 433-443.
Kandel, E.R., & Schwartz, J.H. (Eds.). (1986). *Principles of neural science*. New York: Elsevier.
Kandel, E.R., Spencer, W.A. (1968). Cellular approaches to the study of learning. *Physiological Review, 48*, 65-134.
Kater, S.B. (1974). Feeding in *Helisoma trivolvis*: The morphological and physiological basis of a fixed action pattern. *American Zoologist, 14*, 1017-1036.
Kelly, D.B. (1986). A motor theory of song perception. *Trends in Neurosciences, 9*, 149-150.
Kelso, J.A.S., & Holt, K.G. (1980). Exploring a vibratory systems analysis of human movement production. *Journal of Neurophysiology, 43*, 1183-1196.
Kennedy, D. (1976). Neural elements in relation to network function. In J.C. Fentress (Ed.), *Simpler networks and behavior* (pp. 65-81). Sunderland, MA: Sinauer.
Kennedy, J.S. (1966). Some outstanding questions in insect behavior. In P.T. Haskell (Ed.), Insect behaviour. *Symposium of the Royal Entomological Society of London, 3*, 97-112.
Klärner, D., & Barnes, W.J.P. (1986). The cuticular stress detector (CSD2) of the crayfish. II. Activity during walking and influence on leg coordination. *Journal of Experimental Biology, 122*, 161-175.
Koshland, D.E., Jr. (1980). *Bacterial chemotaxis as a model behavioral system*. New York: Raven.
Kovac, M.P., & Davis, W.J. (1980a). Reciprocal inhibition between feeding and withdrawal behaviors in *Pleurobranchaea*. *Journal of Comparative Physiology. A, Sensory, Neural, and Behavioral Physiology, 139*, 77-86.
Kovac, M.P., & Davis, W.J. (1980b). Neural mechanism underlying behavioral choice in *Pleurobranchaea*. *Journal of Neurophysiology, 43*, 469-487.
Kuffler, S.W., & Nicholls, J.G. (1976). *From neuron to brain*. Sunderland, MA: Sinauer.
Kupfermann, I., & Weiss, K.R. (1978). The command neuron concept. *Behavioural Brain Science, 1*, 3-39.
Kyriacou, C.P., & Hall, J.C. (1986). Interspecific genetic control of courtship song production and reception in *Drosophila*. *Science 232*, 494-497.
Lee, D.N. (1976). A theory of visual control of braking based on information about time-to-collision. *Perception, 5*, 437-459.
Lee, D.N., & Young, D.S. (1986). Gearing action to the environment. In H. Heuer & C. Fromm (Eds.), *Generation and modulation of action patterns* (Experimental Brain Research Series No. 15, pp. 217-230). Berlin, Heidelberg, New York, Tokyo: Springer
Lewis, A.C. (1986). Memory constraints and flower choice in *Pieris rapae*. *Science, 232*, 863-865.
Lewis, E.R. (1968). Using electronic circuits to model simple neuro-electric interactions. *IEEE Proceedings, 56*, 931-949.
Lindberg, D., & Elsner, N. (1977). Sensory influence upon grasshopper stridulation. *Naturwissenschaften, 64*, 342-343.
Marder, E., & Hooper, S.L. (1985). Neurotransmitter modulation of the stomatogastric ganglion of decapod crustacians. In A.I. Selverston (Ed.), *Model neural networks and behavior* (pp. 319-337). New York: Plenum.
Mathews, P.B.C. (1972). *Mammalian muscle receptors and their central actions*. London: Arnold.
Mathews, P.B.C. (1981). Muscle spindles: Their messages and their fusimotor supply. In V.B. Brooks (Ed.), *Handbook of physiology: Sect. 1. The nervous system: Vol. 2. Motor control* (American Physiological Society Series, pp. 189-228). Baltimore: Williams and Wilkins.
Mathews, P.B.C. (1985). Human long-latency stretch reflexes - A new role for the secondary ending of the muscle spindle? In W.J.P. Barnes & M.H. Gladden (Eds.), *Feedback and motor control in invertebrates and vertebrates* (pp. 431-449). London: Croom Helm.
Mayeri, E., & Rothman, B.S. (1985). Neuropeptides and the control of egg-laying behavior in *Aplysia*. In A.I. Selverston (Ed.), *Model neural networks and behavior* (pp. 285-301). New York: Plenum.
Maynard, D.M. (1972). Simpler networks. *Annals of the New York Academy of Sciences, 193*, 59-72.
McClennan, A.D. (1982). Movements and motor patterns of the buccal mass of *Pleurobranchaea* during feeding, regurgitation and rejection. *Journal of Experimental Biology, 98*, 195-211.
McCloskey, D.I. (1981). Corollary discharges: motor commands and perception. In V.B. Brooks (Ed.), *Handbook of physiology, Sec. 1. The nervous system: Vol. II, Part 2* (American Physiological Society Series, pp. 1415-1447). Baltimore: Williams and Wilkins.

McCrohan, C.R., & Benjamin, P.R. (1980). Synaptic relationships of the cerebral giant cells with motoneurons in the feeding system of *Lymnaea stagnalis*. *Journal of Experimental Biology, 85*, 169-186.

McCulloch, W.S., & Pitts, W. (1943). A logical calculus of the ideas immanent in nervous activity. *Bulletin of Mathematical Biophysics, 5*, 115-133.

McMahon, T.A. (1984). *Muscles, reflexes, and locomotion*. Princeton: Princeton University Press.

Mendelson, M. (1971). Oscillator neurons in crustacean ganglia. *Science, 171*, 1170-1173.

Menzel, R., & Erber, J. (1978). Learning and memory in bees. *Scientific American, 239*, 80-87.

Merton, P.A. (1953). Speculations on the servo-control of movement. In G.E.W. Wolstenholme (Ed.), *The spinal cord* (pp. 247-255). London: Churchill.

Meyer, D.E., Smith, J.E.K., & Wright, C.E. (1982). Models for the speed and accuracy of aimed movements. *Psychological Review, 89*, 449-482.

Miller, J.P., & Selverston, A.I. (1985). Neural mechanisms for the production of the lobster pyloric motor pattern. In A.I. Selverston (Ed.), *Model neural networks and behavior* (pp. 37-48). New York: Plenum.

Mimura, K. (1986). Development of visual pattern discrimination in the fly depends on light experience. *Science, 232*, 83-85.

Mittelstaedt, H. (1951). Zur Analyse physiologischer Regelungssysteme. *Verhandlungen der Deutschen Zoologischen Gesellschaft, 45*, 150-157.

Mittelstaedt, H. (1962). Control systems of orientation in insects. *Annual Review of Entomology, 7*, 177-198.

Möhl, B. (1985). Sensory aspects of flight pattern generation in the locust. In M. Gewecke & G. Wendler (Eds.), *Insect locomotion* (pp. 139-148). Berlin: Parey.

Morasso, P. (1981). Spatial control of arm movements. *Experimental Brain Research, 42*, 223-227.

Nichols, T.R., & Houk, J.C. (1976). The improvement in linearity and the regulation of stiffness that results from the stretch reflex. *Journal of Neurophysiology, 39*, 119-142.

Noble, D. (1985). Ionic mechanisms in rhythmic firing of heart and nerve. *Trends in Neurosciences, 89*, 499-507.

Nusbaum, M.P., & Kristan, W.B. (1986). Swim initiation in the leech by serotonin-containing interneurons, cells 21 and 61. *Journal of Experimental Biology, 122*, 277-302.

Olson. G.C., & Krasne, F.B. (1981). The crayfish lateral giants are command neurons for escape behavior. *Brain Research, 214*, 89-100.

Partridge, L.D., & Benton, L.A. (1981). Muscle, the motor. In V.B. Brooks (Ed.), *Handbook of physiology: Sec. 1. The nervous system: Vol. 2. Motor control* (American Physiological Society Series, pp. 43-106). Baltimore: Williams and Wilkins.

Pearson, K.G. (1972). Central programming and reflex control of walking in the cockroach. *Journal of Experimental Biology, 56*, 173-193.

Pearson, K.G. (1976). Nerve cells without action potentials. In J.C. Fentress (Ed.), *Simpler networks and behavior* (pp. 99-110). Sunderland, MA: Sinauer.

Pearson, K.G. (1979). Local neurons and local interactions in the nervous systems of invertebrates. In F.O. Schmitt & F.G. Worden (Eds.), *The neurosciences fourth study program* (pp. 145-157). Cambridge, MA: MIT Press.

Pearson, K.G. (1981). Function of sensory input in insect motor systems. *Canadian Journal of Physiology and Psychology, 59*, 660-666.

Pearson, K.G., Heitler, W.J., & Steeves, J.D. (1980). Triggering of locust jump by multimodal inhibitory interneurons. *Journal of Neurophysiology, 43*, 257-278.

Pearson, K.G., Wong, R.K.S., & Fourtner, C.R. (1976). Connexions between hair-plate afferents and motoneurons in the cockroach leg. *Journal of Experimental Biology, 64*, 251-266.

Perkel, D.H. (1970). Spike trains as carriers of information. In F.O. Schmitt (Ed.), *The neurosciences second study program* (pp. 587-596). New York: Rockefeller University Press.

Pinsker, H.M., & Ayers, J. (1983). Neuronal oscillators. In W.D. Ellis (Ed.), *The clinical neurosciences* (Vol 5, pp. 203-266). New York: Churchill Livingstone.

Polit, A., & Bizzi, E. (1979). Characteristics of motor programs underlying arm movements in monkeys. *Journal of Neurophysiology, 42*, 725-744.

Poon, M., Friesen, W.O., & Stent, G.S. (1978). Neuronal control of swimming in the medicinal leech. V. Connections between the oscillatory interneurons and the motor neurons. *Journal of Experimental Biology, 75*, 45-63.

Pringle, J.W.S. (1957). *Insect flight*. Cambridge: Cambridge University Press.

Prochazka, A. (1981). Muscle spindle function during normal movement. In R. Porter (Ed.), *International Review of Physiology, 25*, 47-90.

Prochazka, A. (1985). Central control of sense organ excitability: introduction. In W.J.P. Barnes & M.H. Gladden (Eds.), *Feedback and motor control in invertebrates and vertebrates* (pp. 115-121). London: Croom Helm.

Pulst, S.M., Gusman, D., Rothman, B.S., & Mayeri, E. (1986). Coexistence of egg-laying hormone and alpha-bag cell peptide in bag cell neurons of *Aplysia* indicates that they are a peptidergic multi-transmitter system. *Neuroscience Letters, 70*, 40-45.

Quinn, W.G. (1984). Work in invertebrates on the mechanisms underlying learning. In P. Marler & H.S. Terrace (Eds.), *The biology of learning* (pp. 197-246). Berlin, Heidelberg, New York: Springer.

Reichardt, W. (1979). Functional characterization of neural interactions through an analysis of behavior. In F.O. Schmitt & F.G. Worden (Eds.), *The neurosciences fourth study program* (pp. 81-103). Cambridge, MA: MIT Press.

Reichert, H., & Rowell, C.H.F. (1985). Integration of nonphase-locked exteroceptive information in the control of rhythmic flight in the locust. *Journal of Neurophysiology, 53*, 1201-1218.

Reichert, H., Rowell, C.H.F., & Griss, C. (1985). Course correction circuitry translates feature detection into behavioural action in locusts. *Nature, 315*, 142-144.

Roberts, T.D.M. (1967). *Neurophysiology of postural mechanisms*. London: Butterworth.

Roeder, K.D. (1937). The control of tonus and locomotor activity in the praying mantis (*Mantis religiosa* L.). *Journal of Experimental Zoology, 76*, 353-374.

Roeder, K.D. (1967). *Nerve cells and insect behavior*. Cambridge, MA: Harvard University Press.

Russell, D.F., & Hartline, D.K. (1978). Bursting neural networks: A reexamination. *Science, 200*, 453-456.

Sahley, C.L. (1984). Behavior theory and invertebrate learning. In P. Marler & H.S. Terrace (Eds.), *The biology of learning* (pp. 181-196). Berlin, Heidelberg, New York, Tokyo: Springer.

Sakitt, B. (1980). A spring model and equivalent neural network for arm position control. *Biological Cybernetics, 37*, 227-234.

Sanders, G.D. (1975). Octopus learning. In W.C. Corning, J.A. Dyal, & A.O.D. Willows (Eds.), *Invertebrate learning* (pp. 1-101). New York: Plenum.

Schildberger, K., Wohlers, D.W., & Schmitz, B. (1986). Morphological and physiological changes in central auditory neurons following unilateral foreleg amputation in larval crickets. *Journal of Comparative Physiology. A Sensory, Neural, and Behavioral Physiology, 158*, 291-300.

Schmidt, R.A. (1982). *Motor control and learning*. Champaign, IL: Human Kinetics.

Schmidt, R.A., & McGown, C. (1980). Terminal accuracy of unexpectedly loaded rapid movements: Evidence for a mass-spring mechanism in programming. *Journal of Motor Behavior, 12*, 149-161.

Selverston, A.I. (1980). Are central pattern generators understandable? *Behavioral Brain Science, 3*, 535-571.

Seyfarth, E.A. (1978). Lyriform slit sense organs and muscle reflexes in the spider leg. *Journal of Comparative Physiology, 125*, 45-57.

Shepherd, G.M. (1979). Functional analysis of local circuits in the olfactory bulb. In F.O. Schmitt & F.G. Worden (Eds.), *The neurosciences fourth study program* (pp. 129-143). Cambridge, MA: MIT Press.

Sherrington, C.S. (1906). *The integrative action of the nervous system*. New Haven: Yale University Press.

Sigvardt, K.A., Rothman, B.S., Brown, R.O., & Mayer, E. (1986). The bag cells of *Aplysia* as a multi-transmitter system: Identification of alpha bag cell peptide as a second neurotransmitter. *Journal of Neuroscience, 6*, 803-813.

Sillar, K.T., & Elson, R.C. (1986). Slow active potentials in walking-leg motor neurons triggered by non-spiking proprioceptive afferents in the crayfish. *Journal of Experimental Biology, 126*, 445-452.

Sillar, K.T., & Skorupski, P. (1986). Central input to primary afferent neurons in crayfish, *Pacifastacus leniusculus*, is correlated with rhythmic motor output of thoracic ganglia. *Journal of Neurophysiology, 55*, 678-688.

Sillar, K.T., Skorupski, P., Elson, R.C., & Bush, B.M.H. (1986). Two identified afferent neurones entrain a central locomotor rhythm generator. *Nature, 323*, 440-443.

Sittig, A.C., Denier van der Gon, J.J., & Giellen, C.C.A.M. (1985). Separate control of arm position and velocity demonstrated by vibration of muscle tendon in man. *Experimental Brain Research, 60*, 445-453.

Skorupski, P., & Sillar, K.T. (1986). Phase-dependent reversal of reflexes mediated by the thoracocoxal muscle receptor organ in the crayfish, *Pacifastacus leniusculus*. *Journal of Neurophysiology, 55*, 689-695.

Snyder, S.H. (1980). Brain peptides as neurotransmitters. *Science, 209*, 976-983.

Sperry, R.W. (1950). Neural basis of the spontaneous optokinetic response produced by visual inversion. *Journal of Comparative and Physiological Psychology, 43*, 482-489.

Stein, P.G.S. (1978). Motor systems, with specific reference to the control of locomotion. *Annual Review of Neuroscience, 1*, 61-81.

Stein, R.B. (1974). The peripheral control of movement. *Physiological Review, 54*, 215-243.

Stein, R.B. (1982). What muscle variable(s) does the nervous system control in limb movements? *Behavioral Brain Science, 5*, 535-577.

Sulston, J.E. (1983). Neuronal cell lineages in the nematode *Caenorhabditis elegans. Cold Spring Harbor Symposia on Quantitative Biology, 48*, 443-452.

Szekely, G. (1968). Development of limb movements: Embryological, physiological and model studies. In G.E.W. Wolstenholme & M. O'Connor (Eds.), Growth of the nervous system. *Ciba Foundation Symposium*, 77-95.

Thüring, D.A. (1986). Variability of motor output during flight steering in locusts. *Journal of Comparative Physiology, 158*, 653-664.

Tinbergen, N. (1951). *The study of instinct.* London: Oxford University Press.

Tully, T., & Quinn, W.G. (1985). Classical conditioning and retention in normal and mutant *Drosophila melanogaster. Journal of Comparative Physiology. A, Sensory, Neural, and Behavioral Physiology, 157*, 263-277.

Turvey, M.T. (1977). Preliminaries to a theory of action with reference to vision. In R. Shaw & J. Bransford (Eds.), *Perceiving, acting and knowing: toward an ecological psychology* (pp. 211-265). Hillsdale, NJ: Erlbaum.

Vallbo, A.B. (1971). Muscle spindle response at the onset of isometric voluntary contractions in man. Time difference between fusimotor and skeletomotor effects. *Journal of Physiology, 218*, 405-431.

Vallbo, A.B., Hagbarth, K.-E., & Torebjörk, H.E. (1979). Somatosensory, proprioceptive, and sympathetic activity in human peripheral nerves. *Physiological Review, 59*, 919-957.

von Holst, E. (1935). Die Koordination der Bewegung bei den Arthropoden in Abhängigkeit von zentralen und peripheren Bedingungen. *Biological Review, 10*, 234-261.

von Holst, E. (1948). Von der Mathematik der nervösen Ordnungsleistungen. *Experientia, 4*, 374-381.

von Holst, E., & Mittelstaedt, H. (1950). Das Reafferenzprinzip. *Naturwissenschaften, 37*, 464-476.

Wagner, R. (1960). Über Regelung von Muskelkraft und Bewegungsgeschwindigkeit bei der Willkürbewegung. *Zeitschrift für Biologie, 111*, 449-479.

Weeks, J.C.(1981). Neuronal basis of leech swimming: Separation of swim initiation, pattern generation, and intersegmental coordination by selective lesions. *Journal of Neurophysiology, 45*, 698-723.

Weeks, J.C. (1982a). Synaptic basis of swim initiation in the leech. I. Connections of a swim-initiating neuron (cell 204) with motor neurons and pattern-generating 'oscillator' neurons. *Journal of Comparative Physiology. A, Sensory, Neural, and Behavioral Physiology, 148*, 253-263.

Weeks, J.C. (1982b). Synaptic basis of swim initiation in the leech. II. A pattern-generating neuron (cell 208) which mediates motor effects of swim-initiating neurons. *Journal of Comparative Physiology. A, Sensory, Neural, and Behavioral Physiology, 148*, 265-279.

Weeks, J.C., & Kristan, W.B., Jr. (1978). Initiation, maintenance and modulation of swimming in the medicinal leech by the activity of a single neuron. *Journal of Experimental Biology, 77*, 71-88.

Weiss, P. (1940). Self-differentiation of the basic patterns of coordination. *Comparative Psychology Monographs, 17*, 1-96.

Weiss-Fogh, T. (1956). Biology and physics of locust flight. IV. Notes on sensory mechanisms in locust flight. *Philosophical Transactions of the Royal Society of London. Series B: Biological Sciences, 239*, 553-584.

Wendler, G. (1964). Laufen und Stehen der Stabheuschrecke *Carausius morosus*: Sinnesborstenfelder in den Beingelenken als Glieder von Regelkreisen. *Zeitschrift für Vergleichende Physiologie, 48*, 198-250.

Wendler, G. (1974). The influence of proprioceptive feedback on locust flight co-ordination. *Journal of Comparative Physiology. A, Sensory, Neural, and Behavioral Physiology, 88*, 173-200.

Wendler, G. (1985). Insect locomotory systems: control by proprioceptive and exteroceptive inputs. In M. Gewecke & G. Wendler (Eds.), *Insect locomotion* (pp. 245-254). Hamburg: Parey.

White, J.G., Southgate, W.E., Thomson, J.N., & Brenner, S. (1983). Factors that determine connectivity in the nervous system of *Caenorhabditis elegans. Cold Spring Harbor Symposia on Quantitative Biology, 48*, 633-640.

White, J.G., Southgate, W.E., Thomson, J.N., & Brenner, S. (1986). The structure of the nervous system of the nematode *Caenorhabditis elegans. Philosophical Transactions of the Royal Society of London. Series B: Biological Sciences, 314*, 1-340.

Wiersma, C.A.G., & Ikeda, K. (1964). Interneurons commanding swimmeret movements in the crayfish. *Procambarus clarkii* (Girard). *Journal of Comparative Neurology, 116*, 209-228.

Wiesendanger, M. (1986a). Redistributive function of the motor cortex. *Trends in Neurosciences, 9*, 120-125.

Wiesendanger, M. (1986b). Initiation of voluntary movements and the supplementary motor area. In H. Heuer & C. Fromm (Eds.), *Generation and modulation of action patterns* (Experimental Brain Research Series No. 15, pp. 3-13). Berlin, Heidelberg, New York, Tokyo: Springer.

Wiesendanger, M., & Miles, T.S. (1982). Ascending pathway of low-threshold muscle afferents to the cerebral cortex and its possible role in motor control. *Physiological Review, 62*, 1234-1270.

Willard, A.L. (1981). Effects of serotonin on the generation of the motor program for swimming by the medicinal leech. *Journal of Neuroscience, 1*, 936-944.

Williams, H. & Nottebohm, F. (1985). Auditory responses in avian vocal motor neurons: A motor theory for song perception in birds. *Science, 229*, 279-282.

Williamson, R.M., & Roberts, B.L. (1986). Sensory and motor interactions during movement in the spinal dogfish. *Proceedings of the Royal Society of London. Series B: Biological Sciences, 227*, 103-119.

Willows, A.O.D., & Hoyle, G. (1969). Neuronal network triggering a fixed action pattern. *Science, 166*, 1549-1551.

Wilson, D.M. (1961). The central nervous control of flight in a locust. *Journal of Experimental Biology, 38*, 471-490.

Wilson, D.M. (1966). Insect walking. *Annual Review of Entomology, 11*, 102-123.

Wilson, D.M., & Waldron, I. (1968). Models for the generation of the motor output pattern in flying locusts. *IEEE Proceedings, 56*, 1058-1064.

Wilson, J.A., & Phillips, C.E. (1983). Pre-motor non-spiking interneurons. *Progress in Neurobiology, 20*, 89-109.

Wine, J.J. (1984). The structural basis of an innate behavioural pattern. *Journal of Experimental Biology, 112*, 283-319.

Wolf, H., & von Helversen, O. (1986). 'Switching-off' of an auditory interneuron during stridulation in the acridid grasshopper *Chorthippus biguttulus* L. *Journal of Comparative Physiology. A, Sensory Neural, and Behavioral Physiology, 158*, 861-871.

Wong, R.K.S., & Pearson, K.G. (1976). Properties of the trochanteral hair plate and its function in the control of walking in the cockroach. *Journal of Experimental Biology, 64*, 233-249.

Wright, B.R. (1976). Limb and wing receptors in insects, chelicerates and myriapods. In P.J. Mill (Ed.), *Structure and function of proprioceptors in the invertebrates* (pp. 323-386). London: Chapman and Hall.

Wyman, R.J., Thomas, J.B., Salkoff, L., & Costello, W. (1985). The *Drosophila* thorax as a model system for neurogenetics. In A.I. Selverston (Ed.), *Model neural networks and behavior* (pp. 513-535). New York: Plenum.

Young, J.Z. (1971). *The anatomy of the nervous system of Octopus vulgaris*. Oxford: Clarendon.

Zill, S.N., & Moran, D.T. (1981a). The exoskeleton and insect proprioception. I. Responses of tibial campaniform sensilla to external and muscle-generated forces in the American cockroach, *Periplaneta americana*. *Journal of Experimental Biology, 91*, 1-24.

Zill, S.N., & Moran, D.T. (1981b). The exoskeleton and insect proprioception. III. Activity of tibial campaniform sensilla during walking in the American cockroach, *Periplaneta americana*. *Journal of Experimental Biology, 94*, 57-75.

Zill, S.N., Moran, D.T., & Varela, F.J. (1981). The exoskeleton and insect proprioception. II. Reflex effects of tibial campaniform sensilla in the American cockroach, *Periplaneta americana*. *Journal of Experimental Biology, 94*, 43-55.

An Information-Processing Analysis of Perception and Action

D.W. MASSARO

CONTENTS

Metatheory for the Study of Perception and Action

How much behavior can psychological theory be expected to predict or fully explain? A recent New Realist theory of philosophy of science allows for predictability in the laboratory, but not in the naturally varying environment (Manicas & Secord, 1983). Complexity of the natural setting far exceeds what any scientific theory could hope to accommodate. Theories will have predictive power only in the laboratory in which the complexity of everyday life can be simplified, measured, and controlled. Our point of departure is even more pessimistic in that the complexity of the prototypical psychology experiment might exceed any theory's predictive power. Revising the dictum that "data without theory are meaningless" (Coombs, unpublished), laboratory data might exceed the constraints demanded by theory (even reasonably correct theory).

There is considerable basis for the above conjecture in the study of perception and the study of action. My contention is that our experiments in these domains too infrequently have the constraints to inform theoretical alternatives (Massaro,

Relationships Between Perception and Action
Edited by O. Neumann and W. Prinz

1975a, 1987). The relationship between perception and action is a question, at least, a log unit more complex than questions concerning their separate functions. Theoretical enlightenment in this arena will be slow in coming, if it is achieved at all. The behaviorists failed; the stimulus does not enable us to predict the response (or even to postdict it). Other more creative efforts are equally unsatisfactory. An influential example is Gibson's (1979) ingenious development of the concept of affordance to explicate the relation between perception and action. The visual world has meaning for behavior; we detect its invariants and we perceive affordance or the actions that the objects or events make available to the perceiver. Affordance is used to explain why apples tend to be seen and eaten rather than seen and kicked. As observed by Cutting (1982), however, the affordance of a stimulus does not constrain behavior sufficiently. There are simply too many possibilities. (In a later section the neo-Gibsonian dichotomy that the perception of stimuli either specifies or simply indicates action proves somewhat helpful in elucidating some perception-action relationships).

The variety of variables influencing the observed relationships between perception and action precludes any hope for a simple solution. Jenkins (1979) has given a similar view of memory research, because of the context sensitivity of any finding. The "law" of memory that is observed depends on subjects, acquisition conditions, memory materials, and memory tasks. The law will differ under different choices of these four classes of variables. Jenkins catalogued these four classes of variables in a Problem Pyramid or Theorist's Tetrahedron. We might expect that laws relating perception to action will have an analogous context sensitivity, depending on subjects, performance history, stimulus situation, and response requirements. Notwithstanding this complexity, the useful heuristic of information-processing models offers the potential for some progress, and this is the tack taken in this chapter. Analogous to Winston Churchill's view of democracy, information processing is not perfect but it is the best that is currently available.

William James (1890) sets the stage for our inquiry, as he does for most inquiries in psychology. "The question is this: 'Is the bare idea of a movement's sensible effects its sufficient mental cue, or must there be an additional mental antecedent, in the shape of a fiat, decision, consent, volitional mandate, or other synonymous phenomenon of consciousness, before the movement can follow?' I answer: sometimes ..." (p. 522). In an information-processing approach, James' question is addressed in the general concern for how input processing relates to output processing. James used the term "ideomotor action" to describe the situation of movement following unhesitatingly and immediately the notion of it in the mind, without awareness. As James said, "We think the act and it is done" (p. 522). Almost a century later, we realize how inadequate this introspective analysis is, but it sets the stage for the study of the relationship between perception and action. Without a doubt, there is a continuum of perception-action relations, from the automatic and fast striking out at the mosquito that has landed on my arm during a walk in the woods, to the deliberate and slow rising from a warm bed onto a cold floor in a cold room.

The preceding statement of the perception-action problem illuminates for me how much of the problem is empirical and not simply theoretical. It is necessary to measure this perception-action relation across the spectrum of human behavior to provide the constraints on potential theory. Without a systematic measurement of the range of phenomena, we are no better off than was James with his introspective method. Luckily, psychologists have been busy since the time of James and have provided useful results on the relation between perception and action. Interestingly, several results were not acquired in the direct search for perception-action relations. Given that even the simplest behavior is assembled from both perception and action, the purest perception experiment has some action, as the purest motor experiment has some perception. As in other domains of psychological inquiry, experimental and theoretical progress are mutually tied to one another.

An Information-Processing Model

We present a stage model to organize the multiplicity of variables influencing perception, action, and the relation between the two. A stage model operates on the assumption of relative independence in that some stages are hypothesized to be influenced by only some variables. Evidence consistent with this hypothesis allows a more parsimonious description of behavior than highly interactive or nonindependence models. In addition, this independence is what makes a theoretical description possible in that the functioning of some stages can be described relatively independently of others. The model is a useful heuristic and is presented in the spirit of falsification and strong inference strategies of scientific inquiry (Massaro, 1979). Specific alternatives or contrasts, not otherwise transparent, can be formulated within the context of the model.

The serial stage model has come under heavy fire in recent years and some defense is warranted even before we begin. The serial nature of the model is assumed because it is the most parsimonious, in many instances, mathematically equivalent to the nonserial models (McClelland, 1979; Townsend & Ashby, 1983; Schweickert, 1978, 1983; Theios & Amrhein, 1989). In these parallel, cascade, or PERT (critical path) networks, stages are somewhat autonomous subprocesses, each with a unique starting time and a unique finishing time. Although the subprocesses can overlap in time, they each contribute to or affect performance time. As stressed by Theios and Amrhein (1989), our prototypical experiments using mean reaction time (RT) as a dependent measure cannot address the issue of serial versus parallel stages. Luckily, the research is still informative regardless of the eventual outcome of the serial/nonserial questions (Townsend & Ashby, 1983). In both instances, performance can be broken down into its component parts, and a theory must describe the parts and how each part is influenced by the multiplicity of variables available to nature and the experimenter. The current effort builds on the foundation assembled by the thorough reviews of Theios (1973), Teichner and Krebs (1974), Sanders (1980), and Keele (1986). The current goal is to confront more directly the relation between perception and action, something not previously explored within the information-processing framework. The information-processing analysis will prove valuable only if it provides a reasonable conceptual framework for the research on perception and action, particularly for those studies previously believed to be exceptions to typical information-processing explanations.

Six Stages of Processing

Following Sanders (1980), it is necessary to distinguish among six stages of processing, as illustrated in Table 1. Sanders distinguishes between perceptual and motor stages but does not make explicit a stage of processing responsible for the translation of perception into action. Our percept-act translation stage replaces his response-choice stage to make the process of translation more apparent. Sanders assumes that stimulus-response compatibility is the critical variable at this stage. Percept-act compatibility is a better descriptor of this variable, however, because it is the relationship between the percept and act that is critical, not simply the relationship between stimulus and response. For example, the overt naming of a Chinese ideogram would be difficult for English but not Chinese readers. The stimulus and response are equivalent for the two groups of subjects, but the translation of the percept (perceptual recognition) into an act (spoken articulation) differs in difficulty for the two groups. Percept-act compatibility better represents the relation between perception and action than does stimulus-response compatibility.

Table 1. Six stages of processing and the variables influencing each stage

Stages	Variables
1. Sensory transduction	1. Signal intensity, duration
2. Featural registration, evaluation, and integration	2. Clarity of features: spatial frequency of visual sine wave gratings ambiguity of acoustic features in speech
3. Pattern classification	3. Similarity of stimulus alternatives
4. percept/act translation	4. Percept-act compatibility, associative strength between perceptual code and action code
5. Motor programming	5. Complexity of movement, specificity of percept
6. Motor execution	6. Physical components of movement, muscle tension

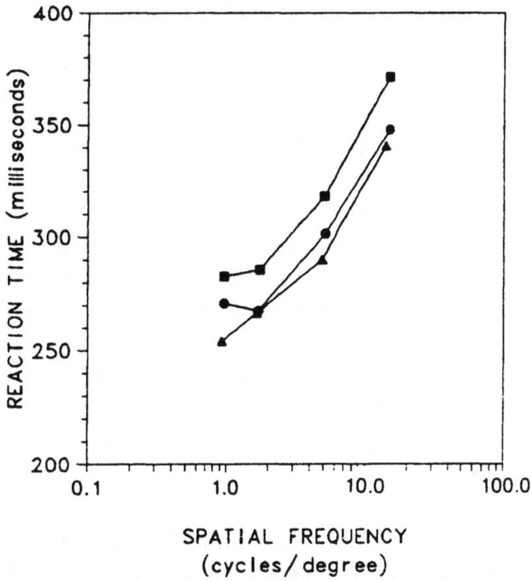

Fig. 1. Simple reaction time to vertical bar gratings of differing spatial frequency and contrast. Patterns of vertical stripes subtending a visual angle of 3.8° by 5.3° were presented to the fovea for 100 ms. The spatial frequency varies from 1.0 to 16.0 cycles per degree (higher frequencies involve more stripes). The contrast between light and dark portions was set 3.0 (*squares*), 4.5 (*circles*), or 6.0 (*triangles*) times greater than the just detectable (threshold) contrast for each spatial frequency. Reaction time is greater for higher frequency gratings but tends to decrease with contrast. (Lupp et al., 1976)

It should be noted that there is a transition from physiological terms to psychological terms between stages 1 and 2 and back to physiological terms at stage 6. This terminology is as it should be in that the algorithmic mental level has proven more appropriate for more central functions and the neural/physiological level more appropriate for more peripheral functions.

Sensory transduction is the first measurable stage and is influenced by signal intensity. As an example, simple RT to detect a stimulus decreases with the intensity of the auditory test signals (McGill, 1961, 1963). Increasing the amplitude of a 1000-Hz test tone from 30 to 100 dB decreases RT by about 100 ms. Analogously, Fig. 1 shows that simple RT to detect visual sine wave gratings decreases with increases in the contrast between the light and dark portions of the pattern (Lupp, Hauske, & Wolf, 1976).

Featural registration, evaluation, and integration is the second stage, which is influenced by the quality of the featural information, the nature of the evaluation, and the type of integration that is required. Our visual acuity varies with detail (spatial frequency) of the pattern, and we might expect spatial frequency also to influence simple RT. As illustrated in Fig. 1, RT changes systematically with spatial frequency, and this change is roughly additive with changes in the contrast between the light and dark portions. This additivity supports the distinction between a sensory-transduction stage and a featural-registration-evaluation stage of processing. In the auditory domain, the second stage will consume more time in speech perception to the extent the featural information is ambiguous, but this effect should be roughly additive with the intensity of the speech signal. Identification RTs increase with increases in the ambiguity of the speech syllable along a synthetic speech continuum (Massaro & Cohen, 1983; Repp, 1981), but the possible interaction between ambiguity and intensity has not been evaluated.

The third stage is pattern classification or the perceptual encoding of a stimulus event relative to various alternatives in memory. If the output of this stage were discrete rather than continuous, we would say that the pattern is categorized as a given alternative. However, this stage of processing makes available the degree to which the signal represents one pattern relative to the degree to which it matches all other patterns. This stage is influenced by the family of alternatives that are relevant to the task at hand. Thus, identification RTs should be shorter to auditory speech syllables if subjects are discriminating between the alternatives /ba/ and /da/ than the alternatives /va/ and /tha/. The latter two are psychophysically much more similar to one another than are the former two. Similarly, time for pattern classification will usually increase with increases in the number of alternatives because of a natural confounding between this variable and similarity among the alternatives. To use another example, literate drivers find it easier to react to the word in the traffic sign "HILL" than to the picture of the international sign depicting a hill. The ambiguity at this stage of pattern classification is greater in the second case relative to the first.

The fourth stage is referred to as "percept-action translation" (more traditionally known as "response selection"). This stage, in many respects the focus of the current chapter, is influenced primarily by what has been covered by the blanket term "stimulus-response compatibility." A more realistic description would be "percept-act compatibility." An important contribution to performance is how readily perception maps into action. Reacting in the direction of a signal is usually easier than reacting away from a signal. It would be a feat of effort to look away from rather than toward someone speaking your name at a cocktail party. Consider the task of identifying pictures of objects by pronouncing either their names or their superordinate categories. A picture of a sparrow would be named as *bird* more quickly than it would be called *sparrow*, whereas the opposite might occur for a picture of a parrot. Given the similarity of sparrow to the prototypical bird and the uniqueness of parrot with respect to the category "bird," we have learned to call sparrows birds and parrots parrots (Rosch, 1978).

The fifth stage is motor programming. Once the action is selected, it has to be programmed before it can be executed. An obvious variable influencing motor programming is complexity of the action that is required. Complexity should increase with the number of discrete actions, and thus this variable should influence programming time. Figure 2 shows that the RT to initiate a response sequence increases with increases in the number of key presses required (Inhoff, Rosenbaum, Gordon, & Campbell, 1984). In addition, the effect of this variable is additive with whether subjects responded in the direction of or away from the test signal. Given that the direction of the response relative to the stimulus has already been described as a percept-act compatibility variable, we have some evidence for distinguish-

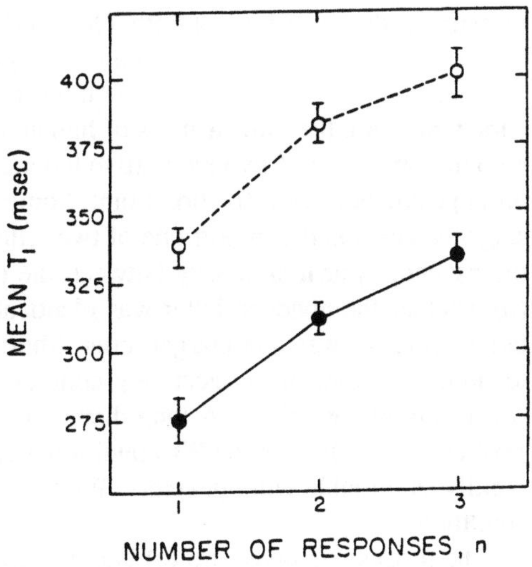

Fig. 2. Mean choice reaction time for response 1 (T_1) of sequences consisting of $n = 1, 2$, or 3 responses tested in compatible (*solid circles*) or incompatible (*open circles*) stimulus-response translation conditions. (About 300 observations contribute to each *point*.) Estimates of -SE are based on between-subjects differences. (Data from Inhoff et al., 1984, experiment 1)

ing between the fourth and fifth stages. The specificity of the percept also appears to be important for motor programming. In some cases, the percept underspecifies the action, whereas, in other cases, the action is inherently specified by the percept. Programming time is longer when preparation and the percept do not specify the required action. Thus, for literate adults, the RT of a naming response to pictures will be longer than the same response to words (all other things being constant, Theios & Amrhein, 1989).

The motor execution stage involves moving the appropriate effectors for the desired action. Needless to say, the physical requirement of the action will influence this stage of processing. It will obviously take longer to move a lever with your foot to one of four locations than to hit one of four buttons with a finger. Muscle tension can also influence execution time of a given response (Sanders, 1980). As noted by Sanders, the motor stages have not been studied as extensively as the perceptual stages. The relation between the two types of stages has been studied even less, but the few dozen studies in this domain are surprisingly informative (Annett, 1983; Keele, 1986; Sanders, 1980).

Accuracy and RT Measures of Performance

An experimenter is fairly limited in dependent measures of performance. A subject's behavior can be described by classification judgements across a set of alternatives in terms of accuracy and/or reaction time. One important recommendation of the current approach is that information.processing models should be developed and tested using both accuracy and RT as dependent measures of performance. Specific theories have been developed too often handle one dependent measure or the other but not both. One working hypothesis explored in this section is that accuracy and RT provide converging measures of human performance.

Santee and Egeth (1982), on the other hand, challenge the idea that RT and accuracy measure the same aspects of human performance. The task they use is letter recognition of one of two alternative letters. Subjects were shown letters presented on opposite sides of a fixation point. Sometime after the onset of the test display, subjects were cued to report one of two letters. The independent variables were (a) the psychophysical similarity between the cued letter and the noncued letter, and (b) whether the noncued letter was identical to the cued letter, was the alternative target letter, or was a nontarget letter. Short durations of the test letters were used to produce errors in the accuracy condition, and we refer to this condition as a "data-limited condition." A long display duration was used to produce essentially perfect performance in the RT condition, but subjects were instructed to respond as rapidly as possible without errors. We therefore call this condition a "time-limited condition."

The process involved are (a) lateral interference of similar letters on letter recognition; and (b) response interference between target and noise items at response selection. Lateral interference occurred in the data-limited task and response interference in the time-limited task. If the noncued letter was identical to the cued letter, accuracy was lower than the in case of different noncued letters in the data-limited task. In the time-limited task, RTs were longer when the noncued test letter was the alternative target. Thus, the results reveal lateral interference in the data-limited task and response interference in the time-limited task. Santee and Egeth used these results to challenge the notion that RT and accuracy are assumed to be converging measures of performance.

However, these results can be explained within a model of information processing. Finding no effect of response interference in the data-limited task is reasonable. These subjects take their time in responding and simply optimize the accuracy of their choice. The stage of response selection should not be influenced by the similarity of the target and background items, and neither accuracy nor RT (which Santee and Egeth do not report) should be influenced by the similarity manipulation in the data-limited task. In the time-limited task, subjects respond as quickly as possible without too many errors. Response interference is reasonable; presentation of the alternative target as the noncued test letter produces response interference analogously to the Stroop color word situation.

One might ask additionally, however, why an effect of similarity of the two test letters on RT is not observed. That is, subjects should be somewhat slower when the noncued letter is identical to the cued letter relative to the case in which the two letters differ. To address the question, however, it is necessary to have a quantitative model of the contribution of lateral interference and response interference. It could be the case that the disadvantage of identical letters in perception is outweighed by an advantage in response selection. For example, the identical background could have slowed down perceptual processing of the target item by 100 ms, but having the background identical to the target could have speeded up response selection by 300 ms. Supporting this interpretation, Shapiro and Krueger

(1983) revealed significant but opposing effects of discriminability and bias due to the similarity of context letters to the test letter.

Given an information-processing description, the Santee and Egeth (1982) experiment does not challenge the common assumption that accuracy and RT are converging measures of human performance. Although it is easy to overlook the differences in the stimulus conditions in the accuracy and RT tasks, these differences could contribute to any differences that are observed. What the results make apparent is that data are meaningless without some kind of model analysis. One needs both information-processing models and fine-grained analyses of performance in the study of perception and action. A good theory will accommodate both accuracy and RT measures of performance.

Our rejection of the Santee and Egeth (1982) conclusion is also supported by Salthouse's (1981) converging evidence for the stage model by comparing accuracy and RT paradigms. Visual information processing in a tachistoscopic task was contrasted with performance in a speed-accuracy RT task. The test stimuli were digits or symbols, and the responses were key presses. In one experiment, for example, the odd integers 1 to 9 were responded to with the right index finger on one key, and the even digits 0 to 8 were responded to with the left index finger on another key. A short stimulus presentation was used to produce errors in the tachistoscopic task, but the stimulus remained in view until the subject responded in the speed-accuracy task. The size of the effect of an independent variable in the two tasks was used to assess whether its influence was at an early stage or a late stage of information processing. As an example, manipulating the number of stimulus alternatives while holding the number of response alternatives constant produced a smaller effect in the tachistoscopic task than in the RT task. This result agrees with earlier research in the RT domain showing that perceptual processing is less influenced by the number of alternatives than is response processing. Another reasonable finding was no effect of stimulus-response compatibility in the tachistoscopic task, but a large effect in the RT task.

An important result for the study of perception and action was finding no significant effect of stimulus quality in the tachistoscopic task along with a small effect of this variable in a speed-accuracy task. A stronger manipulation of stimulus quality would have produced a large effect of this variable in the tachistoscopic task. If the size of the effect remains larger in the speed accuracy task than in the tachistoscopic task, it would provide some evidence that quality influences both perceptual and motor stages of processing. Given the more prominent role of a stage of perception-action translation in the current model, this result is more easily accommodated. Quality of the perceptual code could influence the time needed to translate it into an action code and possibly even the time needed to program the action. Salthouse's comparative-influence method is a promising paradigm for the study of perception and action and should be explored in more detail.

Perception-Action Translation

We have neglected a significant stage of processing in the two decades of information-processing research since Neisser's (1967) book. The stage of processing seldom addressed in inquiry is the translation between perception and action. We have studied one or the other or both but have tended to ignore the translation process. More recently, several researchers have recognized the problem, and their studies will contribute significantly to the present review. Our bias is that the translation between perception and action can be studied and analyzed within the same framework used for study of these two general classes of processing. The assumption is that complex performance can be described in terms of its component parts and the relationship among the components.

Consider the task of naming a picture or naming a written word describing the same object. Identification would involve making contact with some knowledge about the object on the basis of its pictorial or written representation. Response selection might be said to involve a selection of the appropriate action such as programming the articulators to utter the appropriate word. These two stages of processing do not capture an important process between identification and response selection referred to as "perception-action translation" in the present development (see also Virzi & Egeth, 1985). In this case translation involves a transformation of information about the object made available by identification to information about some code that is meaningful to the response-selection stage. This perception-action translation process would differ significantly for the picture - and word-naming tasks. The translation between the identified word and naming is more direct than the translation between the identified picture and naming because the former carries information about what is to be articulated. (Even Chinese ideographs have a phonetic component that provides information about the sound of the word.) The word specifies for the skilled reader its pronunciation (to some extent), whereas the picture does not. Unfortunately, previous studies had overestimated the ease of processing pictures relative to words because of comparing large pictures to small words. Theios and Amrhein (1989), using an ingenious control for the visual properties of picture and words, found that the translation process was 134 ms longer for naming pictures than for naming words.

Hick's Law

An independent variable receiving much attention and presenting a challenge to interpretation within the information-processing framework is the number of stimulus-response alternatives in the task. Shannon's (1948) information measure was tested for psychological validity by assessing whether RT would be linear with information transmitted when measured in bits. As formulated in Hick's law, doubling of the number of alternatives should add a constant to RT. After some initial support (Hick, 1952), several studies reported a number of limiting conditions. These limiting conditions have been reviewed and interpreted by Teichner and Krebs (1974). Stimulus-response compatibility and practice were discovered to be important variables. Hick's law failed for cases of high stimulus-response compatibility, such as pressing a key with the finger stimulated by vibration (Leonard, 1959). In the prototypical task with test digits and key presses, practice effects can also drive the RT function to a zero slope as in the heroic performance for 39 000 trials by the single subject in Mowbray and Rhoades (1959). A lifetime of practice can be observed in the laboratory by having subjects simply pronounce written numbers or letters of the alphabet (Fraisse, 1967; Hellyer, 1963; Morin, Konick, Troxell, & McPherson, 1965; Mowbray, 1960).

Broadbent and Gregory (1962) even eliminated the robust difference between the Donders B and C conditions with highly practiced or compatible responses. The B condition requires differential responding to different stimuli, whereas the C condition requires a response to only one of the stimuli. With practiced or compatible stimulus-response (verbal naming and tactile key pressing) pairings, RT in the B condition was as short as it was in the C condition. This result was not found for unpracticed pairings (saying a word other than the verbal stimulus or incompatible responses - (hitting a key with a different finger than the one stimulated).

Another important variable noted in the early work (Morin et al., 1965) and studied intensively in the last decade is stimulus-response consistency. Morin et al. (1965) found that RTs increased with increases in the numbers of stimulus-response alternatives when drawings of animals were used as test stimuli that were pronounced in a naming task. On the other hand, we would expect little, if any, increase in RT with increases in the number of alternatives if the alternatives are presented as written or spoken words (Theios, 1973). Given the same names in both conditions, response familiarity is an inadequate explanation of the differences between the picture and word conditions. The rather mundane but all too real variable of association provides a handle on these observations. What seems critical is the relative associative strength between the stimulus and response. A picture of a dog might be associated with other responses such as collie, pooch or Lassie, whereas the word "dog" is associated first and foremost with the name dog.

The notion of associative strength between stimulus and response is translated easily into stimulus-response consistency. Using a Sternberg (1975) memory search task, Schneider and Shiffrin (1977) asked subjects to indicate whether or not a test item is a member of a memory set of items. A test item identical to an item in the memory set is a target item and requires a positive response, whereas a test item not in the memory set is a nontarget and requires a negative response. The makeup and the number of items in the memory set are varied systematically. With consistent mapping, the memory items are always chosen from the same superset of items, and these items are never presented as nontargets. In the varied mapping condition, the memory items are not constrained in this way, and a target item on one trial can be a nontarget on another trial. The results showed that memory search with arbitrary stimulus-response pairings is influenced by the consistency of the pairings throughout the experiment (Kristofferson, 1975). If the pairings change randomly, RTs will increase with increases in memory set size. If the memory set items are never presented as nontargets, however, a relatively small number of trials (2100) is sufficient for the subjects to produce RTs independent of memory set size. In addition, reversal of the target and nontarget sets produced a huge negative transfer. Subjects were asked to search for items that were previously nontargets in displays with nontargets that were previously targets. This task was found to be more difficult and time consuming than the original task without practice.

Longstreth, El-Zahhar, and Alcorn (1985) illuminate the importance of the percept-act translation stage in perception-action tasks. Subjects were visually presented with any one of a number of digits and had to press a corresponding key, as in the typical task, or hold a single key down for a duration given by the value of the test digit. The digit 2, for example, would require the duration of the key press to be two time units (usually about 30 ms per time unit), and so on for other test digits. Unpracticed subjects were fairly accurate in this task, but their RTs did not increase much with the number of test alternatives. The large difference in the RT functions from the two tasks is impressive (see Fig. 3). In the traditional task with a unique key for each alternative, RTs increased by 150 ms with a doubling of alternative between 2 and 4. The increase was only about one-tenth of this value, however, in the novel duration task using a single key. A number of obvious explanations, such as much of the processing occurring after the response was initiated, were eliminated.

Differences in memory requirements appear to be sufficient to account for the flattening of the RT function in the novel task of Longstreth et al. (1985) relative to the prototypical task. As explained in formal detail by the authors, a single stimulus-response code can be used by the subjects in the duration task: press the key for the duration specified by the test digit. The recognition of the test digit provides the numerical code needed for the duration task but does not describe the key to be pushed in the traditional multiple-key task. The efficient responding in the duration task was eliminated by changing the test items to letters and having the subject treat the letters ordinally. In this case, the letter had to be recognized and then transformed into a numerical code before response selection could occur. With few alternatives, subjects could hold these translations in memory and access them fairly quickly. Increasing the number of alternatives will make this process more difficult and, accordingly, lead to an increase in RT.

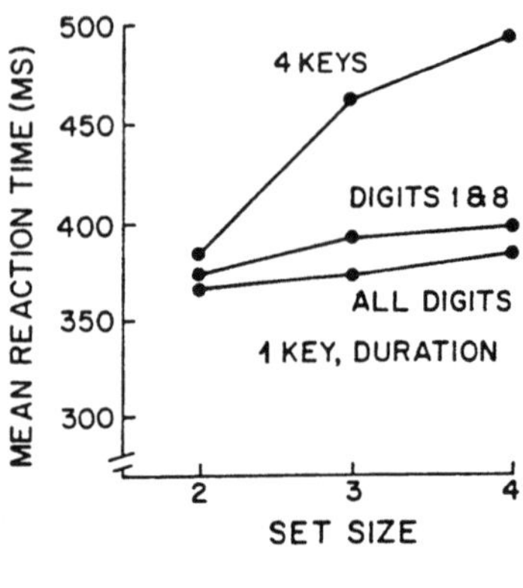

Fig. 3. Reaction time as a function of the number of alternatives in the typical task with a response key for each unique alternative (*4 KEYS*) and in the novel task of holding a single key down for the duration specified by the value of test digit (*1 KEY, DURATION*). (Longstreth et al., 1985)

A stage analysis illuminates the variables of stimulus-response compatibility, number of alternatives, practice, stimulus-response consistency, and memory requirements in the prototypical information-processing task. The number of stimulus-response alternatives interacts with stimulus-response compatibility, practice, consistency, and memory requirements. Stimulus-response compatibility refers to the ease with which a given response can be made to a given stimulus, everything else being constant. Young soccer players find it easier ti approach the ball rather than to approach the goal to receive a pass. We find it easier look to the location of a sound or light, rather than to look away from these events. Orienting responses to stimuli might offer some a priori specification of stimulus-response compatibility. Responses similar to the natural orienting response to a stimulus might well predict stimulus-response compatibility. Practice and stimulus-response consistency represent the number of times and the consistency with which the stimulus is responded to in a given manner, and they might be interpreted as the relative association strength between some stimulus and some response. Neisser (1964) described how efficiently employees in news-clipping services could spot their clients' names in newspaper articles. These agents have both a consistent history of clients and practice on their side. Memory requirements are related to the complexity of the translation between the stimulus and the response. To the extent that each additional alternative requires a unique translation into a response, we can expect some increase in RT.

No single variable or process will account for the range of performance that is observed. Analogous to the observations made by Ryan (1983), no single factor can explain whether controlled or automatic processing will emerge in search tasks. Performance is influenced by multiple factors as revealed, for example, by the range of variables influencing trial-to-trial effects in choice RT tasks (Soetens, Deboeck, & Hueting, 1984). A similar conclusion can be reached about the multiple contributions to context effects (Taylor, 1977). An information processing framework, although usually utilized to isolate cognitive processes (analysis), also provides a framework for the integrative functioning of the processes (synthesis).

Spelling-to-Sound Translation

Brooks (1977) translated real English words into an artificial alphabet and asked subjects to learn the original pronunciation of the words presented in the artificial alphabet. The same subjects were also asked to learn stimuli that had the stimuli and responses re-paired so that the new alphabet was no longer a useful guide to pronunciation. Figure 4 lists an example of the alphabets, and the stimuli and responses used in the experiment. In the orthographic condition, each of the artificial letters corresponds to an English letter. In the paired-associate condition, the stimuli are re-paired with the responses so the regularity between spelling and sound is lost.

Subjects were asked to read aloud lists of six items as fast as possible without error. Although the paired-associate list was initially read faster than the ortho-

Stimuli	Responses	
	Paired Associate	Orthographic
∩∪∧∞	TANS	PENT
∩∪∨∞	SANE	PEAT
I∨∧∪	PEAT	SANE
∞∨∩∪	SAPS	TAPE
∞∨∧I	PENT	TANS
I∨∩I	TAPE	SAPS

Fig. 4. Illustration of an experiment to test the importance of an alphabet on fluency. For each subject, one list has a functional alphabet and one does not. Each subject is given 400 trials of practice naming the words in each of the two lists. Which list is orthographic (has a functional alphabet) is counterbalanced across even- and odd-numbered subjects. (After Brooks, 1977)

graphic list, the asymptotic reading times of highly practiced subjects were significantly faster for the orthographic than the paired-associate condition. In the second experiment, the component letters were concatenated to form glyphic patterns, making it difficult to recognize the component letters. Even though the glyphic calligraphy was read faster than the words made up of discrete letters, the orthographic patterns were still read faster than the paired-associate patterns at asymptote. In the Brooks study (1977), words that looked similar were pronounced similarly in the orthographic condition but not in the paired-associate condition. Therefore, the advantage of the orthographic condition over the paired-associate condition could reflect the correspondence between spelling and pronunciation in the former but not the latter condition. This facilitation with the orthographic patterns would be located in the percept-act translation stage in our information-processing model.

Singer (1980) replicated the findings in a lexical decision task, even when the subject did not learn the pronunciations of the test words. Not surprisingly, the constraints built into the test words influence other stages in addition to percept-act translation. In this case, the orthographic redundancy is effective in facilitating perceptual recognition of the test words (Massaro, 1979). Letter and word recognition is easier to the extent that the letters conform to the spelling constraints in the language. Partial information about a string of letters, when combined with knowledge about spelling, can lead to correct recognition - even though each of these is insufficient taken alone (Massaro, 1975b).

Sound-to-Movement Translation

Simon (1969) presented a test tone to the right or left side of subjects, and their task was a move control handle toward or away from the side of the test tome presentation. Subjects were significantly faster in responding toward the test tone than away from the test tone. This result was described in terms of a natural tendency to react toward the source of stimulation (see also Simon, Small, Ziglar, & Craft, 1970).

Simon et al. (1970) extended the Simon (1969) study in the following manner. Two test tones differing in frequency were used. The test tone frequencies were 200 and 500 Hz, respectively. Half of the subjects were instructed to press a right key to the low tone and to press a left key to the high tone. The other half of the subjects were given the opposite tone-key rule. The tones were presented to either the left or right side of the subject. The RTs between the tone presentation and the onset of a key press were significantly longer when the ear of the test tone presentation conflicted with the meaning of the tone command. That is, longer RTs were observed when subjects had to hit a key on the opposite side of the tone presentation.

Logically, we might interpret the results of Simon et al. (1970) in terms of the perception-action stage of processing. Given recognition of the pitch of the test tone, subjects must select the appropriate response given the tone-key press rule. The rule states which key should be pressed to each of the two possible pitches of the test tone. Evidently, it takes subjects longer to select the response to the appropriate key when the location of the test tone conflicts with the location of the key to be pressed. Hearing the test tone at a particular location might produce a natural tendency to respond to that location. This would interfere with selecting the key in the opposite location and, therefore, would be longer in this condition.

However, the single-factor experiment of Simon et al. (1970) does not unambiguously locate the stage of processing responsible for the results. One might also argue that recognition time is responsible for the observed differences. Subjects might have more difficulty recognizing the pitch of the test tone when its location of presentation conflicts with the key-press location required by that test tone. Consider the situation in which the subject is instructed to respond with the right key when the high tone is presented. Given this rule, the subject might be biased to hear any tone on the right side as high. This bias would tend to increase the recognition time when the location of the test tone conflicts with the location of the response.

As can be seen by the above analysis, the stage of processing responsible for the effect of stimulus-response compatibility is not known. The additive-factor method can be used to locate the stage or stages of processing responsible for a particular experimental finding. The key feature of this method is to manipulate two or more variables in the experiment, and to observe the pattern of results. To disambiguate the issue, it is possible to manipulate a variable that is expected to influence the recognition stage of processing. The frequency difference between the two test tones can be assumed to influence recognition time. To the extent that there is a large difference between the tones, we can expect recognition time to be short. Decreasing the frequency difference can be expected to increase recognition time. The critical result of interest is whether this variable adds with or interacts with the stimulus-compatibility variable. If the latter variable influences only perception-action translation, then the effects of the two variables should be additive. If stimulus-response compatibility also influences recognition time, then the two variables should interact.

In an unpublished study, we manipulated the frequency difference between the two test tones. In the easy recognition condition, the test tones were 600 and 1400 Hz. In the difficult recognition condition, the test tones were 800 and 1200 Hz. Depending on the location of the test tone, the response will be either compatible or incompatible. If the tone and key are on the same side, the response is compatible. When the tone and key are on opposite sides, the response is incompatible. The results shown in Fig. 5 revealed the main effects of stimulus discriminability and stimulus-response compatibility but no interaction between the two factors.

We have interpreted the Simon effect terms of stimulus-response compatibility. Additional support for this interpretation has come from recent experiments by Inhoff et al. (1984) and Nicoletti and Umiltá (1984). Inhoff et al. showed additivity between the Simon stimulus-response compatibility manipulation and number of discrete responses in the judgment (see Fig. 2). Subjects were required to make one, or three responses to compatible or incompatible locations specified by the stimulus cue. Figure 2 shows that RTs of the first response were longer in the incompatible relative to the compatible condition and increased systematically with increases in the number of required responses. The joint effects of these variables was additive. Nicoletti and Umiltá showed that the Simon effect was equivalent to a spatial compatibility effect in which a subject is required to encode relative positions of the stimuli.

Dual-Task Performance

Results from dual-task experiments also support the important role of stimulus-response compatibility in human performance. McLeod and Posner (1984) had subjects perform the Posner Letter Match Task at every trial, producing a same-different response to two visually and sequentially presented letters. In half the trials, the subjects received one of two possible auditory probes. In one condition, the probes could be the word "high" requiring the subjects to say the word "high,

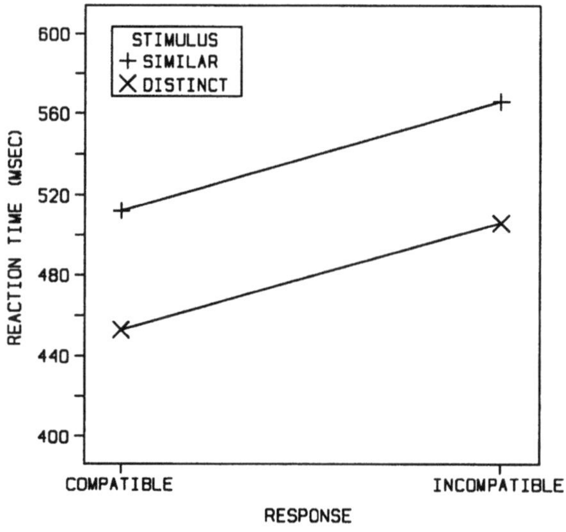

Fig. 5. Reaction times as a function of stimulus discriminability and stimulus-response compatibility (Massaro, unpublished results)

or a 400-Hz square wave tone requiring the subjects to say "low." RT to the probe tone varied systematically with the temporal location of the probe during the letter-matching task. On the other hand, the RT to the word "high" was independent of its temporal occurrence during the letter-matching task. The letter-match same-different RT was influenced by both types of probes so that the independence was not symmetrical.

McLeod and Posner (1984) propose that auditory-vocal (actually speech-vocal) tasks function as a privileged loop which is a special class of input-output transformation that are separate from the general information-processing network. This property allows these privileged loop operations to be performed with relatively little interference from other cognitive activities. However, Paap and Ogden's (1981) point about the appropriate control for the probe RT may be important here. The RT to naming the word "high" was significantly longer in the dual-task than in the single-task condition, and therefore we cannot say that the privileged loop task occurred independently of the same-different letter-matching task. In addition, we might expect that visual information with the same stimulus-response compatibility would be equally easy. For the primary task in this situation, we could present the auditory letters *a* and *b* or *a* and *a* and have the subject move a lever left and right to indicate whether the letters match or mismatch. The probe task would be either the reading of a word "up" and the reading of a nonsense symbol presented in a particular spatial location as "down". My proposal is that the probe RT task with reading the written word "up" will show very little interference from the same-different matching task analogous to the auditory word "high" in the probe task during the visual matching task. If this result obtains, there is not necessarily an auditory-vocal privileged loop. Both results would simply reflect high stimulus-response compatibility conditions in which information about the response is specified in the stimulus (see Greenwald, 1972; Greenwald & Shulman, 1973).

Perception-Action Compatibility

"Stimulus-response compatibility" is a misnomer in that perception-action compatibility is the predictive variable. Usually, stimulus-response compatibility maintains its predictive value because of its close correspondence with perception-action compatibility. A recent study pulled apart these two descriptions of compatibility in a serendipitous way (ten Hoopen, Akerboom, & Raaymakers, 1982). The authors failed to replicate Leonard's (1959) classic study showing that high stimulus-response compatibility can lead to a violation of Hick's law. Leonard stimulated the fingertips with a 50-Hz vibration, and the task of the subject was to press the key of the finger that had been stimulated. Ten Hoopen et al. (1982) observed an increase in RT between four and eight choices, and they realized that the vibrations in their experiment were relatively weak.

To assess whether the amplitude and frequency of the vibrations might be important, they systematically varied the frequency and amplitude of the vibrations along with the number of alternatives in the task. They found a systematic effect of both frequency and amplitude of the vibration. With high frequencies and high amplitude, a relatively flat RT function like that observed by Leonard (1959) was obtained. On the other hand, with a low-frequency, low-amplitude vibration, RTs increased systematically with increases in the number of choices. An analysis of the error rates revealed a systematic increase in errors with increases in the number of choices for the low-frequency, low-amplitude condition. The ten Hoopen et al. results replicate previous experiments showing that stimulus discriminability interacts with number of choices (Sternberg, 1969; Sanders, 1980). Given highly confusable percepts, the translation process going from perception to action becomes more difficult, and thus one observes an increase in RT. With highly discriminable stimuli, as in the high-frequency, high-amplitude vibrations, the high percept-act compatibility produces very little increase in RT with increase in the number of alternatives.

Ten Hoopen et al. (1982) used the concept "ideomotor compatibility" to explain the results. Their assumption is that an ideomotor action gives feedback resembling the activity of the stimulus. The argument is that the low-frequency, low-amplitude vibration is not ideomotoric in that the weakly vibrating button activates only Meissner corpuscles, but that the pressing of the button activates both Meissner and Pacinian corpuscles. Thus, the feedback of the button pressing does not totally resemble that of the vibration. On the other hand, the high-frequency, high-amplitude vibration stimulates both types of corpuscles which would then also be simulated with the appropriate action. However, this explanation seems to require some kind of advanced processing, in that feedback of the response must further activate the response to speed it up. A test of this idea would be to measure the dynamics of the movement, so that the onset of the movement could be measured, as well as the movement time itself. My prediction would be that the onset of the movement would show the same results as the key press observed by ten Hoopen et al., making their version of the ideomotor compatibility explanation unreasonable. This conclusion follows because the onset of the movement should not be influenced by feedback from the touch receptors, and, therefore, no increase in RT with the number of alternatives should be observed. It is not necessary to postulate separate systems since percept-act compatibility is sufficient to explain the results.

Percept-act compatibility will, in many cases, be an empirical question. Consider an impressive finding by Rosenbaum, Gordon, Stillings, and Feinstein (1987). Subjects were presented with a tone that was high or low in pitch and had to respond with a spoken syllable /gi/ or /gu/. The vowel /i/ is high in subjective pitch, whereas the vowel /u/ is low in subjective pitch. An intriguing possibility is that responding /gi/ to the high tone and /gu/ to the low tone would have more percept-act compatibility than the opposite pairing. Indeed, the reaction times for the onset of the spoken response in the compatible condition averaged 88 ms faster than the incompatible condition. In addition, this compatibility effect was additive with a motor-programming effect involving the number of syllables in the spoken response "Spelling-to-Sound Translation," (see Fig. 2). Given a percept-act compatibility effect with a pure tone and a spoken response, we might expect that there are a multitude of similar effects yet to be discovered. What seems problematic is the development of theoretical framework that would predict these effects a priori.

Communication Between Perception and Action

Miller (1982) has accumulated important relationships between perception and action. Although his work is formalized within testing the distinction between discrete and continuous stage models, it is relevant to the relationship between perception and action. If perception and action are considered as two stages of processing, an important question is whether these stages of processing can overlap in time. In Miller's terms, the question is whether response preparation can begin before stimulus identification is complete, or whether the stimulus must be completely identified prior to any response preparation or activation. Miller's novel technique is to control the information available at various times during stimulus identification by making some relevant stimulus characteristics easy to discriminate and some difficult to discriminate. The question is whether the stimulus characteristics which are easy to discriminate could be used for response preparation before the stimulus was completely identified, i.e., before the stimulus characteristics which are difficult to discriminate were also resolved.

Consider an experiment with four letters created by independently varying the size and identity of the letters. Two stimulus sets were used; there were large and small S's and T's in one condition, and capital and lower-case I's and T's in the other condition. Pilot results indicated that letter discrimination could be made about 85 ms faster than the size discrimination in the first stimulus set (sStT), whereas the size discrimination could be made about 72 ms faster than the letter discrimination in the second stimulus set (iItT).

The utilization of partial information about the stimulus could be observed by the response requirements in the task. The usefulness of partial information was varied by manipulating the assignments of particular stimuli to responses. On each trial, subjects responded to one of the four stimuli with one of four fingers, two on each hand. Two responses made by the same hand can be prepared together more efficiently than two responses made by different hands. The preparation of two response fingers on the same hand has been shown to be more effective than preparation of two response fingers on different hands (Rosenbaum, 1980).

Subjects responded more quickly if the large and small versions of a given letter were assigned to the same hand for both stimulus sets. Subjects also responded more quickly if the two large letters were assigned to one hand and the two small letters were assigned to the other hand, but only when the size discrimination was very easy relative to the letter discrimination, that is, in the letter set (iItT). Thus, separate or discrete attributes of the stimulus could be encoded before the stimulus was completely identified, and this information could be passed on and influence response preparation. The idea is that determining that the letter is a large letter even though the letter has not yet been identified can tell the subject which hand is relevant and thus decrease RT relative to the case in which the two letters of the same size are not assigned to the same hand. Analogously, and much more intuitively, if the letter name is assigned to a given hand so that both the large and small versions of a given letter are responded to on the same hand, then determination of the letter name can aid response selection even though the size of the letter has not been completely determined to identify uniquely the alternative as one of four alternatives.

However, now consider the stimulus set with the capital letters M, N, U, and V. Information obtained early in processing can constrain the stimulus to be one of the two letters M or N versus one of two letters U or V. Thus, this information derived before the letter was completely identified could inform the subject about which pair of letters was present in the display. If subjects are able to utilize this information for response preparation, then the subjects should be faster in responding when the letters M and N are assigned to one hand and U and V are assigned to the other relative to the case in which these letters are assigned to different hands. This was not the case, however, in that no facilitation was observed. The explanation is that the information could not be encoded in terms of a discrete code to pass on to response preparation. In these simple information-processing tasks, information appears to be transmitted discretely from perception to action in terms of stimulus codes. Information that is not encoded discretely does not constrain response preparation.

Meyer, Yantis, Osman, and Smith (1984) found similar results in a priming task. Subjects performed a choice RT task to a test stimulus preceded by a priming stimulus. The priming stimulus completely predicted the test stimulus and, therefore, the response that would be required. The test stimulus followed the prime after a short, medium, or long delay interval. A discrete stage model predicts, according to the authors, that the RT to the test will be determined by either the prime or the test. That is, there must be no interaction between these two horses in the race. A cascade model, on the other hand, predicts an interaction between the horses in the race. In fact, no interaction was found in a two-choice task with compatible stimulus-response mappings. Although these results can be taken as evidence for a discrete stage model, the opposite results would not necessarily disprove a discrete stage model. It is possible that there are successive evaluation and integration process leading to a response. Although the evaluation process might function in a manner that treats the two stimuli as independent horses in a race, the information from the prime and test stimuli could be integrated by the integration process (Massaro, 1987). This integration could provide more information than provided by either one of the two stimuli at the evaluation stage. The integration would lead to response facilitation not predicted by two independent horses in a race. In principle, a discrete stage could still be consistent with the results. Some results with four alternative responses and incompatible stimulus-response mappings, in fact, were inconsistent with their version of the discrete model (Meyer, Yantis, Osman, & Smith, 1985). However, it remains a possibility that the discrete stage model is valid, but that the information is integrated and the response is selected on the basis of the integrated information (Massaro, 1987).

The role of categorization in perception and action is apparent in studies of the relationship between speech perception and production (Pickett, 1985; Repp & Williams, 1985). A classic study of speech perception and naming was carried out by Chistovich, Fant, Serpa-Leitao, and Tjernlund (1966). A speech continuum between adjacent vowel categories was synthesized, and subjects were required to echo each vowel as it was presented in random order. There was no one-to-one relationship between the formants of the stimulus and the formants of the response; subjects achieved a good imitation for only the three canonical vowels. Repp and Williams (1985) found similar results for synthetic vowels between /i/ and /u/ when the authors imitated the vowels as much as possible. Given that it is well known that the vowels along the continuum could be differentiated, the observations reject a direct link between perception and action. Categorization intervenes between the two, and action is controlled by this categorization rather than the perceptual processing that led to the categorization. This mediation is not surprising for the discrete messages of language. Morton (1979) found it necessary to distinguish between input and output logogens in his model of language processing, for example, and it should be worthwhile to evaluate the relationship in other domains that are less discrete.

Categorization of Natural Objects

Attention to perception and action relationships illuminates research findings in the categorization literature. It is now well known that typicality predicts the facility with which subjects categorize instances as belonging to different categories (Rosch, 1978). Asked to categorize instances as members of categories such as bird, dog, car, and boat, subjects responded much more quickly to typical than atypical instances. The explanation of these results could be in terms of the "entry point" concept (Jolicoeur, Gluck, & Kosslyn, 1984) or the level of abstraction first identified by the observer. The time for categorization can be considered to be a function of the degree of association between the entry point of an object and the category name. Viewing a picture of a sparrow, the entry point level is bird, and categorization in the category bird is very easy. Viewing a picture of a duck, however, the entry point level is duck, and more time is needed to access the appropriate superordinate category. Supporting evidence also comes from neurological patients who succeed in reading aloud many more object names (banana and chair) than superordinate nouns (fruit and furniture) (Funnell & Allport, 1987). This explanation is at the level of percept-act translation in our six-stage model. Objects differing in typicality differ from one another in terms of the time needed to translate between their perceptual code and superordinate category code. We expect that similar results would occur if words rather than pictures were presented. In fact, Jolicoeur et al. (1984) found a positive correlation between words and picture in this task.

Spatial Linguistic Processing

Brooks (1968) offered a dramatic demonstration of the importance of the relationship between perception and action (i.e., between stimulus-processing and response-processing requirements in the task). Subjects performed some task with two different types of stimulus material and three different types of response. In the visual-spatial task, subjects imagined blocked letters such as the F shown in Fig. 6. They were required to scan the image from a particular starting position and in a given direction and to indicate at each corner "yes" when the point was at the extreme top or the extreme bottom of the letter, and "no" otherwise. For the example in Fig. 6, the answers would be "yes, yes, yes, no, no, no, no, no, no, yes". In the linguistic task, subjects were given a sentence to remember such as, "A bird in the hand is worth two in the bush." Their task was to indicate "yes" or "no" whether each successive word was a noun. Subjects made their responses for both tasks in one of three ways, by saying "yes" or "no", by tapping with their left and right hands, and pointing to Y's and N's arranged randomly on a sheet.

Table 2 gives the mean performance times to complete the six different tasks. Given that the two tasks differ in difficulty, the absolute times are not at issue but rather the interactions between stimulus material and response output. For the pointing and tapping responses, the visual spatial task is three and two times more difficult than the sentence task. For the vocal response, on the other hand, the sentence task is 20% more difficult than the visual task. Contrary to what might be expected from stimulus-response compatibility, subjects find it more difficult to act spatially to spatial perceptual processing and to act linguistically to linguistic input. These results occurred because of the relatively arbitrary relationship between the perception and action tasks. In this case, both perception and action required relatively unique operations, and it was more optimal for the systems to divide perception and action across spatial and linguistic processing rather than to utilize just one of these for both the perception and action tasks. In this situation, perception and action are two tasks, and there is more conflict between the two tasks when they are both carried out spatially than when the two tasks are performed spatially and linguistically, respectively.

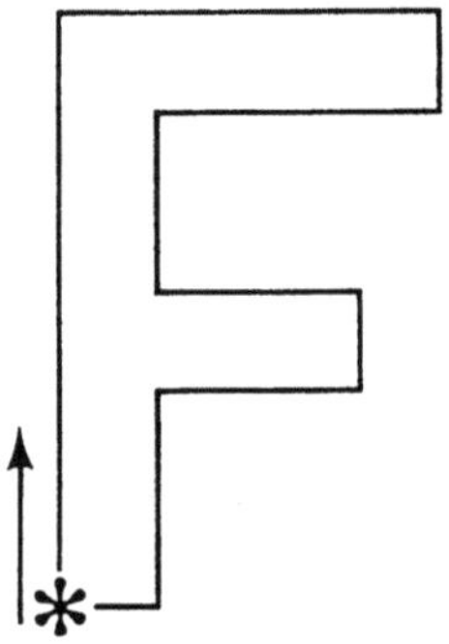

Fig. 6. An example of a simple block diagram used by Brooks (1968) to study the scanning of mental images. The *asterisk* and *arrow* showed the subject the starting point and the direction for scanning the image.

Table 2. Mean performance times in seconds for the visual-spatial and the memory-linguistic task

	pointing	tapping	vocal
visual-spatial	28.2	14.1	11.3
memory-linguistic	9.8	7.8	13.8

Specificational Versus Indicational Information

Pattee (1974) has developed the distinction between *specificational* and *indicational/injunctional* types of information, and this distinction has been important in recent neo-Gibsonian theories (Kelso & Kay, 1987; Turvey & Kugler, 1984). Specificational information implies that the environment constrains or specifies the appropriate action, whereas indicational information only indicates what actions are appropriate. The former supposedly specifies how the action is to be performed, whereas the latter indicates only that a given action should be performed. This distinction seems to illuminate to some extent the nature of the perception-action interactions that are observed. Consider the results of our extension of the Simon et al. (1970) study, however (Fig. 5). Can we claim that the tone in the compatible response situation specified the response, whereas it only indicated the response on the incompatible condition? No, because the influence of stimulus-response compatibility is a continuous variable and cannot be captured by a qualitative distinction. That is, it would be easy to create a situation between the two levels of compatibility such as presenting the tone in the middle of the subject's head. The results would resemble neither of the previous two conditions, but would fall somewhere between them. The qualitative distinction would fail to explain the three different results.

The specificational mode of information description can, on the other hand, illuminate some other perception-action relationships, however. Expansion of the retinal image of a surface can provide reliable information about the time to contact that particular surface (Lee & Reddish, 1981). How does this information specify some action, however? A gannet, a large seabird, diving for its prey, must retract its wings before hitting the water's surface. Although time to contact can not specify wing reaction to the gannet, It can provide continuous information about rate of contraction. Time to contact indicates wing retraction and specifies perhaps the time course of retracting its wings. Is the gannet behaving qualitatively differently from a teenager approaching a stop sign with her mother's new car? The stop sign indicates one must stop (or at least slow down) at the impending intersection, while, supposedly, the rate of expansion of the retinal image specifies how to slow down. Admittedly, the latter information provides the detail about the visual world needed to stop the car at the appropriate place and time, whereas this detail is not provided by the word *STOP*. The rate of expansion does not specify the action, however. The action must still be selected from a repertoire of actions; the driver could release the gas pedal, shift into a lower gear, apply the brakes, begin looking both ways to run the stop sign if nobody is coming, or engage in some subset or all of these actions. Gibson's concept of affordance notwithstanding, I remain to be convinced that the rules of perception will disambiguate rules of action and the relationship between the two. We will need rules of perception, action, and of the translation between the two.

Similarly, reference to different modes of description of information (Pattee, 1974) - the discrete, symbolic rate-independent mode and the continuous, dynamic, rate-dependent mode - will not predict perception-action interactions. The hope for the neo-Gibsonians might be that, although the former is not sufficient to control action, the latter is. We might expect that much less translation between perception and action would be needed for dynamic relative to symbolic information. A counter example to this expectation might be the vocal articulation of the word *bird* to either the object or to the written word. Reaction time as faster to the word than to a picture of the object (Lupker, 1979; Lupker & Katz, 1982; Theios & Amrhein, 1989), and yet the written word fits the symbolic mode to a greater degree than does the object. Although this counterexample should not be taken to preclude further study of these two different modes of description, it indicates to this observer that perception-action interactions will not be totally explained by the mode of information description.

Stimulus-Response Compatibility Versus Separate Cognitive Systems

We have used the notion of stimulus-response compatibility to clarify perception-action relations. An alternative view is to postulate separate cognitive systems such that performance is efficient if perception and action occur within a system and difficult if they occur between systems. [The Brooks (1968) study discussed in the section "Categorization of Natural Objects" above, would be a counterexample.] We address these contrasting interpretations in an old and venerable task which is as challenging today as it was at the time of its introduction (Stroop, 1935).

Stroop Color-Word Findings

The role of stimulus-response compatibility has been used to explain Stroop color-word interference first reported by Stroop in 1935. Subjects are asked to name as rapidly as possible the color of various stimuli, such as the print of a color name. A color word that specifies a different name than the color of its print can produce large interference and increase the naming time of the color. McClain (1983) modified the Stroop color-word situation to demonstrate the contribution of stimulus-response compatibility. An auditory analog of the Stroop task was developed with the words "high" and "low" sung in high and low pitches. Each stimulus had two dimensions, word and pitch, and the two dimensions could be either congruent or incongruent.

Subjects were asked to respond to either the pitch or the word dimension with either a high or low verbal response or a high or low humming response. When the word was the relevant dimension, there was no difference between incongruent and congruent test stimuli. That is, subjects could verbally name the appropriate word regardless of the pitch that the word was sung in. On the other hand, when subjects were required to hum the word, the incongruent stimulus gave significantly slower responses than the congruent stimulus. The analogous results occurred when the relevant dimension was pitch. No difference between the congruent and incongruent test stimuli was observed when subjects were required to hum the pitch dimension. On the other hand, large differences were observed between the congruent and incongruent test stimuli when subjects were required to name verbally the pitch dimension. Thus, subjects were able to ignore the irrelevant dimension when the response that was required was compatible with the relevant dimension. When the response was incompatible with the relevant dimension, and therefore compatible with the irrelevant dimension, significant interference effects were observed. Although these results might not account for all versions of the Stroop interference and facilitation, they point to the important role of stimulus-response compatibility in the task.

Virzi and Egeth (1985) replicated and extended these findings in both the color-word situation and a new situation. Subjects are presented with the word "left" or "right" presented in a spatial location to the left or right of fixation. The subjects respond to either the spatial location of the stimulus or its meaning. The response is either hitting a left or right button or responding with the words "left" or "right". This factorial design combines two modes of presentation with two modes of response. Half of the trials are congruent (e.g., word "left" presented on the left side) and half are incongruent (e.g., word "right" presented to the left of fixation), as in the Stroop Color-Word Task. Subjects make the appropriate decision, and reaction times are recorded. In the compatible situation, subjects are responding to spatial location with the button pushing, and, in the incompatible situation, subjects are responding to the word with a button push. If subjects are asked to respond to the meaning of the word, the vocal response RTs show no difference between the congruent and incongruent situations. However, there is a large difference if subjects are asked to make a manual response to the meaning of the word. Analogously, if subjects are asked to respond manually to the location of the word, there is no difference between the congruent and the incongruent situation but a large difference if subjects have to name or use linguistic output with respect to the location of the word in the display.

According to Virzi and Egeth (1985), these results cannot be explained by simple stimulus-response compatibility, but must be explained in terms of having two cognitive systems. The idea is that each cognitive system codes and processes information in a way that is specific to that system. Some systems have a response stage perfectly suited to the information that that system processes. This model is reminiscent of McLeod, McLaughlin, and Nimmo-Smith's (1985) idea of information encapsulation. A particular perceptual system can resolve and transmit information very quickly to a particular motor system. In their study, information about time to contact of an object approaching could be transmitted very quickly to a motor response of hitting the approaching object with a bat but could not be translated into a judgement about when the object would arrive at a given location in space. The idea of different systems is also reminiscent of the concept of modularity as articulated by Fodor (1983).

With respect to the Stroop task, translation between the two systems is required when the stimulus and response come from two different systems. For example, responding with the word "left" to the word "left" can be carried out within a single linguistic system. Responding with a button push, however, would require a translation between two systems. Although information may be translated from one system into the code of another system, the translation process takes time. Rather than postulating separate systems, however, it seems that the concept of stimulus-response compatibility might be capable of handling the results. Some translation is always required from a perceptual code into an action code . The nature of the translation will influence the time for this stage of processing.

A third condition of the McClain (1983) study might distinguish between the translation model of Virzi and Egeth (1985) and the present interpretation in terms of stimulus-response compatibility. In this condition, subjects had to respond by hitting a button, with the two buttons labeled "high" and "low." In this case, no difference between the congruent and incongruent conditions occurred when the relevant dimension was the word, whereas significant differences occurred between these two conditions when the relevant dimension was the pitch. According to the translation model, it seems that a translation would be necessary to go from the word dimension to the button push, because these would not be capable of being emitted by the same system, yet no interference was observed. The results indicate instead that encoding pitches into verbal labels is not necessary, and therefore no difference between the incongruent and congruent stimulus conditions is observed. Thus, in addition to stimulus-response compatibility, we have the factor of whether or not a stimulus dimension must be encoded. Perhaps if the subjects were taught that the buttons correspond to high and low pitches or if the buttons were placed high and low on a response panel, then differences would be observed between the congruent and incongruent conditions when the relevant dimension was a word, but not when the relevant dimension was the pitch. Thus, it would not be whether or not the response followed within a system without any translation, but rather the stimulus-response compatibility between the perceptual code and the action code.

Virzi and Egeth (1985) state, "The main conclusion reached is that the translational model provides a useful model of dual attribute processing" (p. 316). This conclusion must be considered an overgeneralization because the authors in no way provide a complete model of dual-attribute processing. The authors' work, like much of the previous work on Stroop, is aimed at discovering a single explanatory principle of the Stroop phenomenon. The literature reveals multiple influences and determinants of Stroop interference (e.g., Regan, 1978) analogous to other domains such as context effects and priming (La Heij, van der Heijden, & Schreuder, 1985). In addition to the focus on a single explanatory principle, the research is carried out in terms of a confirmation bias, in that research is developed only to confirm rather than to falsify the explanation. The Stroop phenomenon is a complicated piece of performance and will not be explained in terms of a single variable or process. It engages most of the stages of processing developed in our prototypical information-processing model and will thus be affected by variables that affect each of these processing stages. An explanation of the Stroop phenomenon will involve the contribution of several stages in the information-processing model.

The translation mechanism proposed by Virzi and Egeth (1985) also reflects the bias of investigators to assume qualitative mechanisms to explain experimental results. Thus, a Stroop effect occurs when a code must be translated from one system to another, whereas no Stroop effect occurs when no translation is necessary. However, one can expect that such an all-or-nothing process does not exist. In all cases, a translation is necessary between a perception code and an action code, and the translation process will depend on stimulus-response compatibility. Thus, translation requires time, and the amount of time will be directly related to the incompatibility between the perception and action codes. Also, there is no reason to assume different cognitive systems to explain the results. It is only necessary to assume, as in fuzzy logical model of perception (Massaro, 1987), that attributes are processed in parallel and independently of one other at the feature evaluation stage of processing. These attributes are usually integrated to provide perceptual identification, and a pattern is classified. The unique feature of the Stroop task, as in other attention experiments (Treisman, 1969), is that subjects are required to report a given attribute and ignore some other attribute (see van der Heijden, this volume). Given this type of task, we should ask the following kinds of questions: (a) are the two dimensions processed independently of one other (see Ashby and Townsend, 1986); (b) is one dimension processed more quickly than the other; (c) to what extent can a response be selected on the basis of only the relevant dimensions? The similarity between the Stroop situation and other tasks is reinforced by the observations of La Heij et al. (1985). They uncovered common processes in the priming task in which an irrelevant stimulus facilitates performance and the Stroop task in which interference is observed. Semantic facilitation and response competition occur in both types of tasks, but to different degrees depending on the number of semantic domains/response alternatives in the task.

The idea of different cognitive systems also loses its explanatory force, as anticipated by Virzi and Egeth (1985), when empirical results require postulation of plethora of cognitive systems. The authors acknowledge that one will have to assume separate cognitive systems for colors, written words, positions, auditorily presented words, pitch, numerosity, and digit identity. Given the ease with which it is possible to construct analogs of the Stroop Color-Word Task, the number of cognitive systems will have to grow with the number of independent attributes in our perceptual world. Thus, we can imagine Stroop results in situations in which subjects have to process the size, shape, depth, brightness of a word independently of its meaning. Similarly, several attributes could be varied in the auditory domain and even the tactile domain, and so on. One critical distinction, however, might be that linguistic and nonlinguistic codes are responsible for the Stroop effect. The two attributes, linguistic and nonlinguistic, are not usually integrated in our perceptual world, whereas other attributes within each of these two dimensions are usually integrated for decision and action.

One critical component of the Stroop Color-Word Task seems to be that one of the two dimensions be represented in a way that the surface structure stimulus specifies a given response, as in a written word specifying the name of the word. Interference occurs when the name of the word conflicts with the required response to the relevant dimension. In McClain's (1983) experiment, subjects required to hum the meaning of the word showed interference when the word was sung in a high or low pitch. In this case, it is necessary to assume that singing the word in a high or low pitch specified humming high or low which then interfered with humming the relevant word dimension. Consistent with previous interpretations of the Stroop task, it appears that the Stroop phenomenon requires that the response mode be primed in the Stroop task which is then easily captured by the irrelevant dimension. In McClain's experiment, subjects were primed to hum high or low, and the singing of a high or low pitch captured the response mechanism and therefore interfered with humming the meaning of the word.

One test between the two viewpoints might involve manipulating the stimulus-response compatibility in the Stroop task by re-pairing the stimuli and responses within a "cognitive system." In one condition, the word would be the relevant dimension but it would be pronounced by another color name. The word red would be called blue, for example, and so on. If the linguistic system could function independently of the color system, then no difference should be found between this condition and (a) a control condition with the words printed in a neutral color; or (b) a facilitation condition with the words printed in the color appropriate for the response. Although a translation between the color and the verbal label might not be mandatory (Dunbar & MacLeod, 1984) we expect effects of stimulus-response compatibility to interact with Stroop interference even when separate cognitive systems are not a viable explanation. Analogously, colors could be paired with colored patches pressed manually and the stimulus-response within the "color system" varied by repairing stimulus colors and response presses. Once again, stimulus-response compatibility should interact with Stroop interference even though the stimuli and responses are within the same color system.

Two Visual Systems

There has been much interest in the idea of two visual systems in vision, one serving cognitive function and one serving motor function. These systems have also been referred to as serving the functions of object recognition and object location. Consider the evidence that is normally used to support the idea of two visual systems. Subjects are given some visual display and asked to make some judgment about the display. The judgment consists of a verbal report or a motor response. For example, a target might be moved in a visual field, and the subject is asked either to report verbally whether the target moved or to point at the new location of the target. Results have shown that subjects are able to point accurately at the location of the dot even though they fail to report verbally a displacement in the target. This result has been taken to mean that the motor system has access to information that is unavailable to the verbal cognitive system (Bridgeman, Lewis, Heit, & Nagle, 1979). Similarly, the target could be induced to appear to move by movement of a background even though it did not (Bridgeman, Kirch, & Sperling, 1981). The background position did not influence pointing to the target, however. This double dissociation between the verbal report and pointing responses does not necessitate two visual systems. Both judgments could be mediated by the same visual information, but with the function relating the judgment to the visual information differing for pointing and verbalization. Birnbaum (1981) has given an analogous interpretation of the differences observed between cognitive and affective reports of experience (Zajonc, 1980). In this case, it is not necessary to assume separate affective and cognitive systems. Similarly, Roediger (1984) points out that a dissociation might reflect different processes rather than different systems.

More parsimoniously, consider relative movement with respect to a frame and absolute movement with respect to oneself as two forms of information. The first might be given the most emphasis in the verbal judgment and the latter given the most emphasis in the pointing judgment. The functions relating these two sources of information to the verbal and pointing judgments might differ from one another, as illustrated in Fig. 7. Both judgments are influenced by both forms of information, and there is no need to postulate two visual systems. Another interpretation of these phenomenon is to postulate not two visual systems, but a translation stage

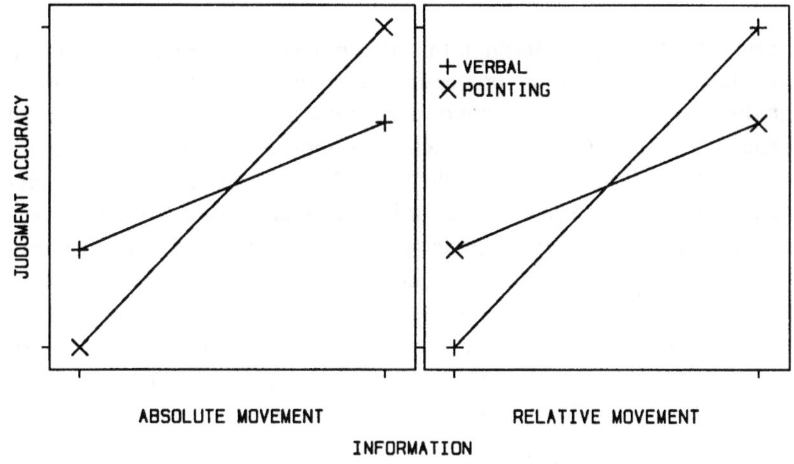

Fig. 7. Hypothetical functions relating judgment accuracy of the two types of judgment as a function of the two types of stimulus information

between perception and action. In this case, the same perceptual information would be available to verbal and motor processes, but the translation might work differently for the two modes of action. Stimulus-response compatibility might account for some of the differences that are found for these two modes of action. For example, pointing to target is more compatible than verbally reporting the location of the target changed, because the latter requires a more elaborate translation between a perceptual dimension and a verbal dimension.

The reluctance to propose two different systems is based on the eventual need for a plethora of cognitive systems. One might ask whether analogous phenomena in related domains demand separate systems. For example, does the visual information that support pointing come from different systems depending on the distance of the object from the observer? Do we have a different system handling information within our reach as opposed to outside our reach? Similarly, we can view the visual world in two different modes, the standard three-dimensional mode, and also in what might be considered to be a two-dimensional mode as viewing two-dimensional works of art. Thus, we can respond differently to the same display depending on the mode of responding. Different sources of information may be utilized within the two modes of responding, and, therefore, a double association could be established in this domain as in the cognitive-motor domain. There would also be cross-talk between the two domains, again analogous to more recent results in the cognitive-motor domain. One can see that the number of systems can grow considerably, and one must be cautious about contributing to such growth. If a cognitive report can be found to indicate better sensitivity to location information than a motor action, it would seem to weaken the case for two visual systems.

Conclusion

Survivors of a voyage through this article have experienced a somewhat involved treatment of information processing and how perception relates to action. One guiding principle has been that of parsimony. I have tried to hold onto the simplest explanation of relatively complex interactions. Stages of processing and information-processing principles provide a useful heuristic to impose some order on what might appear to be somewhat disorderly results. Most environment-behavior relationships are too complex to be predicted from simple information-processing principles. This failure does not reflect the limitations of the principles, as much as the complexity of the natural world and most experiments. There are no easy solutions; science of the mind and behavior is difficult work. Given the growing number of scientists resonating to the challenges of explaining our behavior, we can expect significant, although slow and gradual, progress in this enterprise.

Acknowledgements. The research reported in this paper and the writing of the paper were supported, in part, by grants from the National Science Foundation (BNS.83.15192), the Public Health Service (NINCDS Grant 20314), and the graduate division of the University of California, Santa Cruz. Ray

Gibbs, Lex van der Heijden, and other members of the PEP Program at UCSC and the group on Perception and Action assembled at Bielefeld made helpful comments on this enterprise. I am particularly grateful to Nancy McCarrell for her incisive comments on an earlier version of this chapter.

References

Allport, A. (1987). Selection for action: some behavioral and neurophysiological considerations of attention and action. In H.Heuer & A.F. Sanders (Eds.), *Perspectives on perception and action*, Hillsdale, NJ: Erlbaum.

Annett, J. (1983). Motor learning: A review. In H. Heuer, U. Kleinbeck, & K. Schmidt (Eds.), *Motor behavior: Programming, control, and acquisition*. Berlin, Heidelberg, New York: Springer.

Ashby, F.G., & Townsend, J.T. (1986). Varieties of perceptual independence. *Psychological Review, 93*, 154-179.

Birnbaum, M.H. (1981). Thinking and feeling: A skeptical review. *American Psychologist, 36*, 99-101.

Bridgeman, B., Lewis, S., Heit, G., & Nagle, M. (1979). Relation between cognitive and motor-oriented systems of visual position perception. *Journal of Experimental psychology: Human Perception and Performance, 5*, 692-700.Bridgeman, B., Kirch, M., & Sperling, A. (1981). Segregation of cognitive and motor aspects of visual function using induced motion. *Perception & Psychophysics, 29*, 336-342.

Broadbent, D.E., & Gregory, M. (1962). Donders' B- and C-reactions and S-R compatibility. *Journal of Experimental Psychology, 63*, 575-578.

Brooks, L. (1977). Visual pattern in fluent word identification. In A.S. Reber & D.L. Scarborough (Eds.), *Toward a psychology of reading. The Proceedings of the CUNY Conferences* (pp. 143-182). Hillsdale, NJ: Erlbaum.

Brooks, L.R. (1968). Spatial and verbal components of the act of recall. *Canadian Journal of Psychology, 22*, 349-368.

Chistovich, L.A., Fant, G., Serpa-Leitao, A., & Tjernlund, P. (1966). *Mimicking of synthetic vowels* (Speech Transmission Laboratory Reports, STL-QPSR, 2). Stockholm: Royal Institute of Technology.

Cutting, J.E. (1982). Two ecological perspectives: Gibson vs. Shaw and Turvey. *American Journal of Psychology, 95*, 199-222.

Dunbar, K., & MacLeod, C.M. (1984). A horse race of a different color: Stroop interference patterns with transformed words. *Journal of Experimental Psychology: Human Perception and Performance, 10*, 622-639.

Fodor, J.A. (1983). *Modularity of mind.* Cambridge, MA: Bradford.

Fraisse, P. (1967). Latency of different verbal responses to the same stimulus. *Quarterly Journal of Experimental Psychology, XIX*, 353-355.

Funnell, E., & Allport, A. (1987). Non-linguistic cognition and word meanings: neuropsychological exploration of common mechanisms. In D.A. Allprot, D.G. Mackay, W. Prinz, & E. Scheerer (Eds.), *Language perception and production: Shared mechanisms in listening, speaking, reading and writing*. Academic: London.

Gentner, D.R. (1985). *Skilled motor performance at variable rates: A composite view of motor control* (Chip 124). Center for Human Information Processing. San Diego, CA: University of California.

Gibson, J.J. (1979). *The ecological approach to visual perception*. Boston: Houghton Mifflin.

Greenwald, A.G. (1972). On doing two things at once: time sharing as a function of ideomotor compatibility. *Journal of Experimental Psychology, 94*, 52-57.

Greenwald, A.G., & Shulman, H.G. (1973). On doing two things at once. II. Elimination of the psychological refractory period effect. *Journal of Experimental Psychology, 101*, 70-76.

Hellyer, S. (1963). Stimulus-response coding and amount of information as determinants of reaction time. *Journal of Experimental psychology, 65*, 521-522.

Hick, W.E. (1952). On the rate of gain of information. *Quarterly Journal of Experimental Psychology, 4*, 11-26.

Inhoff, A.W., Rosenbaum, D.A., Gordon, A.M., & Campbell, J.A. (1984). Stimulus-response compatibility and motor programming of manual response sequences. *Journal of Experimental psychology: Human Perception and Performance, 10*, 724-733.

James, W. (1950). *The principles of psychology*. Mineola, NY: Dover. (Original publication 1890).

Jenkins, J.J. (1979). Four points to remember: A tetrahedral model of memory experiments. In L.S. Cermak & F.I.M. Crack (Eds.), *Levels of processing in human memory*. Hillsdale, NJ: Erlbaum.

Jolicoeur, P., Gluck, M.A., & Kosslyn, S.M. (1984). Pictures and names: Making the connection. *Cognitive Psychology, 16*, 243-275.

Keele, S.W. (1986). Motor control. In K.R. Boff, L. Kaufmann, & J.P. Thomas (Eds.), *Handbook of perception and human performance, Vol. II: Cognitive processes and performance* (pp. 1-60). New York: Wiley.

Kelso, J.A.S., & Kay, B.A. (1985). Information and control: A macroscopic of perception-action coupling. In H. Heuer & A.F. Sanders (Eds.), *Perceptives on perception and action*. Amsterdam: North-Holland.

Kelso, J.A.S., & Kay, B.A. (1987). Information and control: A macroscopic analysis of perception-action coupling. In H. Heuer & A.F. Sanders (Eds.), *Perspectives on perception and action*. Hillsdale, NJ: Erlbaum.

Kristofferson, M.W. (1975). On the interaction between memory scanning and response set. *Memory and Cognition, 3*, 102-106.

La Heij, W., van der Heijden, A.H.C., & Schreuder, R. (1985). Semantic priming and Stroop-like interference in word-naming tasks. *Journal of Experimental Psychology: Human Perception and Performance, 11*, 62-80.

Lee, D.N., & Reddish, P.E. (1981). Plummeting gannets: A paradigm of ecological optics. *Nature, 293*, 293-294.

Leonard, J.A. (1959). Tactual choice reactions I. *Quarterly Journal of Experimental Psychology, 11*, 76-83.

Longstreth, L.E., El-Zahhar, N., & Alcorn, M.B. (1985). Exceptions to Hick's law: Explorations with a response duration measure. *Journal of Experimental Psychology: General, 114*, 417-434.

Lupker, S.J. (1979). The semantic nature of response competition in the picture-word interference task. *Memory and Cognition, 7*, 485-495.

Lupker, S.J., & Katz, A.N. (1982). Can automatic picture processing influence word judgments? *Journal of Experimental Psychology: Learning, Memory, and Cognition, 8*, 418-434.

Lupp, U., Hauske, G., & Wolf, W. (1976). Perceptual latencies to sinusoidal gratings. *Vision Research, 16*, 969-972.

Manicas, P.T., & Secord, P.F. (1983). Implications for psychology of the new philosophy of science. *American Psychologist, 38*, 399-413.

Massaro, D.W. (1975a). *Experimental psychology and information processing*. Chicago: Rand-McNally.

Massaro, D.W. (Ed.). (1975b). *Understanding language: An information processing analysis of speech perception, reading and psycholinguistics*. New York: Academic.

Massaro, D.W. (1979). Reading and listening (Tutorial paper). In P.A. Kolers, M. Wrolstad, & H. Bouma (Eds.), *Processing of visible language* (Vol. 1, pp. 331-354). New York: Plenum.

Massaro, D.W. (1987). *Speech perception by ear and eye: A paradigm for psychological inquiry*. Hillsdale, NJ: Erlbaum.

Massaro, D.W., & Cohen, M.M. (1983). Evaluation and integration of visual and auditory information in speech perception. *Journal of Experimental Psychology: Human Perception and Performance, 9*, 753-771.

McClain, L. (1983). Stimulus-response compatibility affects auditory Stroop interference. *Perception & Psychophysics, 33*, 266-270.

McClelland, J.L. (1979). On the time relations of mental processes: An examination of systems of processes in cascade. *Psychological Review, 86*, 287-330.

McGill, W.J. (1961). Loudness and reaction time. *Acta Psychologica, 19*, 193-199.

McGill, W.J. (1963). Stochastic latency mechanisms. In R.D. Luce, R.R. Bush, & E. Galanter (Eds.), *Handbook of mathematical psychology, 1*. New York: Wiley.

McLeod, P., & Posner, M.I. (1984). Privileged loops from percept to act. In H. Bouma & D.G. Bouwhuis (Eds.), *Attention and performance X* (pp. 55-66). Hillsdale NJ: Erlbaum.

McLeod, P., McLaughlin, C., & Nimmo-Smith, I. (1985). Information encapsulation and automaticity: Evidence from the visual control of finely timed actions. In M.I. Posner & O.S.M. Marin (Eds.), *Attention and performance XI* (pp. 391-406). Hillsdale, NJ: Erlbaum.

Meyer, D.E., Yantis, S., Osman, A.M., & Smith, J.E.K. (1984). Discrete versus continuous models of response preparation: A reaction-time analysis. In S. Kornblum & J. Requin (Eds.), *Preparatory states and processes*. Hillsdale, NJ: Erlbaum.

Meyer, D.E., Yantis, S., Osman, A.M., & Smith, J.E.K. (1985). Temporal properties of human information processing: Tests of discrete versus continuous models. *Cognitive Psychology, 17*, 445-518.

Miller, J. (1982). Discrete versus continuous stage models of human information processing: In search of partial output. *Journal of Experimental Psychology: Human Perception and Performance, 8*, 273-296.

Morin, R.E., Konick, A., Troxell, N., & McPherson, S. (1965). Information and reaction time for "naming" responses. *Journal of Experimental Psychology, 70*, 309-314.

Morton, J. (1979). Facilitation in word recognition: Experiments causing change in the logogen model. In P.A. Kolers, M. Wrolstad, & H. Bouma (Eds.), *Processing of visible language (Vol. 1*, pp. 259-268). New York: Plenum.

Mowbray, G.H. (1960). Choice reaction times for skilled responses. *Quarterly Journal of Experimental Psychology, 12*, 193-202.

Mowbray, G.H., & Rhoades, M.U. (1959). On the reduction of choice reaction times with practice. *Quarterly Journal of Experimental Psychology, 11*, 16-23.

Neisser, U. (1964). Visual search. *Scientific American, 210*, 94-102.

Neisser, U. (1967). *Cognitive psychology*. New York: Appleton-Century-Crofts.

Nicoletti, R., & Umiltá, C. (1984). Right-left prevalence in spatial compatibility. *Perception & Psychophysics, 35*, 333-343.

Paap, K.R., & Ogden, W.C. (1981). Letter encoding is an obligatory but capacity-demanding operation. *Journal of Experimental Psychology: Human Perception and Performance, 7*, 518-527.

Pattee, H.H. (1974). Discrete and continuous processes in computers and brains. In M. Conrad, W. Guttinger, & M.D. Cin (Eds.), *Physics and mathematics of the nervous system*, (pp. 128-148). New York: Springer.

Pickett, J.M. (1985). Shadows, echos and auditory analysis of speech. *Speech Communication, 4*, 19-30.

Regan, J. (1978). Involuntary automatic processing in color-naming task. *Perception & Psychophysics, 24*, 130-136.

Repp, B.H. (1981). Perceptual equivalence of two kinds of ambiguous speech stimuli. *Bulletin of the Psychonomic Society, 18*, 12-14.

Repp, B.H., & Williams, D.R. (1985). Categorical trends in vowel imitation: Preliminary observations from a replication experiment. *Speech Communication, 4*, 105-120.

Roediger, H.L. (1984). Does current evidence from dissociation experiments favor the episodic/semantic distinction? Commentary of Tulving, E., Precis of elements of episodic memory. *The Behavioral and Brain Sciences, 7*, 252-254.

Rosch, E. (1978). Principles of categorization. In Rosch, E. & Lloyd, B.B. (Eds.), *Cognition and Categorization*. Hillsdale, NJ: Erlbaum.

Rosenbaum, D.A. (1980). Human movement initiation: Specification of arm, direction, and extent. *Journal of Experimental Psychology: General, 109*, 444-474.

Rosenbaum, D.A., Gordon, A.M., Stillings, N.A., & Feinstein, M.H. (1987). Stimulus-response compatibility in the programming of speech. *Memory and Cognition, 15*, 217-224.

Ryan, C. (1983). Reassessing the automaticity-control distinction: Item recognition as a paradigm case. *Psychological Review, 90*, 171-178.

Salthouse, T.A. (1981). Converging evidence for information-processing stages: A comparative-influence stage analysis method. *Acta Psychologica, 47*, 39-61.

Sanders, A.F. (1980). Stage analysis of reaction processes. In G.E. Stelmach & J. Requin (Eds.), *Tutorials in motor behavior* (pp. 331-354). Amsterdam: North-Holland.

Santee, J.L., & Egeth, H.E. (1982). Do reaction time and accuracy measure the same aspects of letter recognition? *Journal of Experimental Psychology: Human Perception and Performance, 8*, 489-501.

Schneider, W., & Shiffrin, R.M. (1977). Controlled and automatic information processing. I: Detection, search, and attention. *Psychological Review, 84*, 1-66.

Schweickert, R. (1978). A critical path generalization of the additive factor method: Analysis of a Stroop task. *Journal of Mathematical Psychology, 18*, 105-109.

Schweickert, R. (1983). Latent network theory: Scheduling of processes in sentence verification and the Stroop effect. *Journal of Experimental Psychology: Learning, Memory, and Cognition, 9*, 353-383.

Shannon, C.E. (1948). The mathematical theory of communication. *Bell System Technical Journal, 27*, 379-423, 623-658.

Shapiro, R.G. & Krueger, L.E. (1983). Effect of similarity of surround on target-letter processing. *Journal of Experimental Psychology: Human Perception and Performance, 9*, 547-559.

Simon, J.R. (1969). Reactions toward the source of stimulation. *Journal of Experimental Psychology, 81*, 174-176.

Simon, J.R., Small, A.M., Jr., Ziglar, R.A., & Craft, J.L. (1970). Response interference in an information processing task: sensory versus perceptual factors. *Journal of Experimental Psychology, 85*, 311-314.

Singer, M.H. (1980). The primacy of visual information in the analysis of letter strings. *Perception & Psychophysics, 27*, 153-162.

Soetens, E., Deboeck, M., & Hueting, J. (1984). Automatic aftereffects in two-choice reaction time: A mathematical representation of some concepts. *Journal of Experimental Psychology: Human Perception and Performance, 10*, 581-598.

Sternberg, S. (1969). The discovery of processing stages: Extensions of Donders' method. *Acta Psychologica, 30*, 276-315.

Sternberg, S. (1975). Memory scanning: new findings and current controversies. *Quarterly Journal of Experimental Psychology, 7*, 1-32.

Stroop, J.R. (1935). Studies of interference in serial verbal reactions. *Journal of Experimental Psychology, 18*, 643-662.

Taylor, D.A. (1977). Time course of context effects. *Journal of Experimental Psychology: General, 106*, 404-426.

Teichner, W.H., & Krebs, M.J. (1974). Laws of visual choice reaction time. *Psychological Review, 81*, 75-98.

Ten Hoopen, G., Akerboom, S., & Raaymakers, E. (1982). Vibrotactual choice reaction time, tactile receptor systems and ideomotor compatibility. *Acta Psychologica, 50*, 143-157.

Theios, J. (1973). Reaction time measurements in the study of memory processes: Theory and data. *The Psychology of Learning and Motivation, 7*, 43-85.

Theios, J. (1975). The components of response latency in simple human information processing tasks. In P.M.A. Rabbitt & S. Dornic (Eds.), *Attention and performance V*. London: Academic.

Theios, J., & Amrhein, P.C. (1989). Theoretical analysis of the cognitive processing of lexical and pictorial stimuli: Reading, naming and visual and conceptual comparisons. *Psychologica Review, 96*, 5-24.

Townsend, J.T., & Ashby, F.G. (1983). *Stochastic modeling of elementary psychological processes*. London: Cambridge University Press.

Treisman, A.M. (1969). Strategies and models of selective attention. *Psychological Review, 76*, 282-299.

Turvey, M.T., & Kugler, P.N. (1984). Information and the ecological approach to perception and action. In H.T.A. Whiting (Ed.), *Human motor actions: Bernstein reassessed*. Amsterdam: North-Holland.

Umilta, C., & Nicoletti, R. (1985). Attention and coding effects in S-R compatibility due to irrelevant spatial cues. In M.I. Posner & O.S.M. Marin (Eds.), *Attention and performance. XI: Mechanisms of attention* (pp. 457-471). Hillsdale, NJ: Erlbaum.

Virzi, R.A., & Egeth, H.E. (1985). Toward a translational model of Stroop interference. *Memory and Cognition, 13*, 304-319.

Zajonc, R.B. (1980). Feeling and thinking: Preferences need no inferences. *American Psychologist, 35*, 151-175.

A Common Coding Approach to Perception and Action

W. PRINZ

CONTENTS

Introduction

This chapter is concerned with some of the issues involved in understanding how perception contributes to the control of actions. Roughly speaking, the term of action refers to any meaningful segment of an organism's intercourse with its environment. Two important features of this preliminary definition can be brought out more clearly when "actions" are contrasted with "responses" and "movements". Unlike response-centered approaches to psychology, which consider the organism's activity more or less determined by the actual stimulus information, the action approach emphasizes intentional control as being simultaneous with (or even prior to) informational control of activity, assuming that intentional processes fix the rules for the selection and use of stimulus information (Heuer & Prinz, 1987; Neumann & Prinz, 1987). Unlike movement-centered approaches, which describe the organism's activity in terms of the dynamics of muscular contraction patterns and the kinematics of the resulting body movements, the action approach stresses the environmental consequences that go along with these bodily events, contending that meaningful interactions with the environment, rather than movements per se, should be considered the effective functional units of activity (Fowler & Turvey, 1982; Neisser, 1985).

The main goal of this chapter is to outline a coherent framework for conceptualizing the relationship between perception and action. This framework relies on the notion of common and commensurate codes for percepts and acts. If differs from more traditional approaches which hold (more or less explicitly) that

Relationships Between Perception and Action
Edited by O. Neumann and W. Prinz

perceptual codes and action codes are separate and incommensurate. Since the time of Descartes, the separate-coding view has been the standard model of the relationship between perception and action. In order to explain how perception contributes to the control of action, a framework like this requires a translation mechanism that allows the mapping of the two incommensurate classes of codings onto each other.

Recently, this standard model has been challenged in several places, assuming that there might be more continuity, compatibility, or commensurability between perception and action than the standard model allows for (e.g., Fowler & Turvey, 1982; Kelso & Wallace, 1978; MacKay, 1987a; MacKay, Allport, Prinz, & Scheerer, 1987; Mounoud, 1981; Neisser, 1985; Prinz, 1987; Turvey, 1977; von Hofsten, 1985). Still, the theoretical discussion is clearly dominated by what MacKay et al. (1987) have called the separate-and-unequal approach to perception and action. One reason for this may be that the linear stage theory of information processing has provided a powerful paradigm to elaborate this approach and to explore the nature of the operations involved in the assumed translation mechanisms (see Massaro, this volume). Another reason seems to be that comparable alternative views are lacking - comparable in the sense of being equally coherent and powerful, equally broad in scope, as well as equally oriented toward explanations in terms of processing models.

What follows is an attempt to outline an alternative view that meets these requirements. The first section gives a fuller account of the two basic frameworks to be contrasted, viz., separate vs. common coding of perception and action, and compares them at the theoretical level. The next section presents some basic observations from matching tasks (i.e., various kinds of tasks that require that the act matches the percept in one or the other respect) and relates them to the issue of separate vs. common coding. Finally, after the common coding framework has been shown to be a powerful alternative to the standard model, some further elaborations are suggested in the last section.[1]

Two Basic Frameworks

When studying performance in an information-processing task, it seems natural to regard the presentation of the stimulus as the starting point, both with respect to the description of the task and the explanation of performance in terms of hypothetical processing operations. Yet, on a closer look, it becomes apparent that this account fails to acknowledge the fact that the stimulus is usually not presented until the experimenter has made sure that the subject has attained an appropriate understanding of the instruction. The presentation of instructions must therefore be considered the real starting point of the task (Neumann & Prinz, 1987).

Though the requirement that the instruction must precede the information to be processed may be trivial at the level of task description, it is clearly less trivial in terms of a more functional account of the underlying operations. The instructions provide the subject with rules that specify what he/she is supposed to do under certain conditions. As subjects usually adopt these prescriptions, they must be capable of somehow establishing and maintaining the intention to act according to these rules. Thus we have to recognize that intentions precede information, and that the processing of the information cannot be fully understood without reference to the antecedent intentional conditions.[2]

If we assume that the instruction induces the implementation of rules in the processing machinery, the next step is to ask how these rules are implemented, and how they interact with the operations triggered by the stimulus. Basically, the instructions specify what to do under which conditions. The rules that take over these prescriptions must therefore refer to linkages between afferent processes (which provide information about the conditions) and efferent processes (which control what the subject does). How could these linkages be implemented in the functional architecture of the machinery that links perception to action?

Separate Coding

The separate-coding view of the perception-action interface is based on the conviction that afferent and efferent codes are incommensurate at all levels of coding. On the afferent side there may be various processing stages, each with its characteristic input codes, transformations, and output codes, and the same may be true of the efferent pathways. However, since it is assumed that afferent and efferent codes refer to patterns of stimulation and patterns of muscle activity, or to properties of environmental objects and properties of bodily movements (at low or high levels of coding, respectively), there is no way in which afferent and efferent codes could be commensurate, i.e., in which they could be formed in the same format and maintained in a common representational medium.

If one adopts this general picture of the system, it becomes apparent that the linkage machinery must be a device for coupling incommensurate codes. Such couplings can be conveniently described as *production rules*. These rules link tokens of afferent information to tokens of efferent information (condition vs. action components of production rules). In the present context, three features of such coupling rules are worth considering. First, the couplings are arbitrary in the sense that, at least in principle, any afferent code can get linked to any efferent code, irrespective of their contents. Second, they may differ in strength, depending on factors like frequency of usage, relative importance, etc. Third, they can be read from both sides, either specifying appropriate efferent codes under given afferent conditions, or appropriate afferent conditions for given efferent codes.[3]

As far as the nature of the operations in which these rules are used is concerned, two metaphors can be found in the literature: *translation* (from stimuli to responses) and *choice* (among responses on the basis of given stimuli). Though each of these two metaphors emphasizes a somewhat different aspect of the use of production rules, they both express very clearly the principle of separate coding.

The translation metaphor which was introduced by Welford (1960, 1968) is now widely used in the field (Massaro, this volume; Proctor & Reeve, 1985; Reeve & Proctor, 1984; Sternberg, 1969a,b; Virzi & Egeth, 1985). This metaphor seems to view the production rules from the input side: how can given afferent codes become translated into corresponding efferent codes? As both codes are expressed in different languages, a translation device is required to map one onto the other. In the most simple case, this device may take the form of a dictionary that links the incommensurate entities in the same way as a look-up table does for a transformation. This metaphor stresses the fact that the two codes belong to functionally different and, hence, separate coding systems, and that the problem is to bridge the gap between the two systems. At the same time, it is also confusing since it could be taken to imply that the two incommensurate codes refer to a common content or meaning - an implication obviously never intended by its proponents.

Such confusion is avoided in the *response choice* metaphor (e.g., Sanders, 1979, 1980; Smith, 1968; Welford, 1960, 1968) which seems to view the production rules from the output side: how can choice of one among several efferent codes be accomplished on the basis of afferent information? This metaphor, too, stresses the inequality of the entities to be linked: response-specifying codes get linked to stimulus-dependent codes in a way that leads to the selection of the appropriate code on the efferent side. Afferent conditions serve the function of evidence for decisions between efferent codes, and again these two classes of codes are completely incommensurate and belong to separate coding systems.[4]

One of the major problems with the separate-coding framework is that many of the processing models that have been conceived in its spirit and couched in its terms are not very explicit in specifying the nature of the codes that are supposedly involved. On the afferent side, it is far from clear which level of processing and/or degree of elaboration is required for an afferent code to be an efficient input for a production rule. Usually it is contended that some degree of identification of the stimulus must precede the translation-and-choice mechanism, but it is not clear how deep and complete it has to be. On the one hand, as a minimal requirement, one could think of a rather restricted sensory code that refers to some basic physical stimulus attributes, just sufficiently elaborated to distinguish the actual stimulus from the possible alternatives (Sanders, 1980). On the other hand, one could also think of a more fully elaborated perceptual code that refers to well-structured, meaningful distal objects (Massaro, this volume; Prinz, 1983, 1984). Likewise, on the efferent side, it is far from clear which efferent codes are supposedly involved. For instance, in considering the codes that might be involved in the control of the responses in a manual key-pressing task, one could think of codes at various levels: codes that specify the muscle groups or the limbs involved in the movements, codes that specify the form of the movements or the forces to be exerted, or even codes that specify the total patterns of the responses, including all of these components and feedback information as well. No translation-and-choice model has ever spelled out at which of these possible levels the supposed choice is accomplished.[5]

Common Coding

Let us leave the paradigm for a moment and take a brief look at the relationship between the perceptual and the intentional aspects of a person's conscious experience in daily life. Though we cannot expect that this is, in general, a promising method for attaining a better understanding of the functional architecture of the underlying system, it seems to make an interesting contribution to the present discussion, suggesting a different view of the relationship between perception and action.

The critical observation is that this person, in the analysis of his/her conscious experience, would not find any indication of a qualitative difference or even a gap between action-related and perception-related contents. There is no feeling of incommensurability, or incompatibility. Rather, there is a clear sense of acting in the same world as is perceived. Acts (i.e., actions one intends to perform) are completely commensurate and continuous with percepts (i.e., environmental events one perceives). Percepts and acts both refer to events with comparable attributes. Both are characterized by location (in space and time) and content (in terms of physical and nonphysical properties), the only difference being that percepts refer to ongoing, actor-independent events and acts to to-be-generated, actor-dependent events.

Consider, for illustration, a person who is instructed to press the left of two keys in response to stimulus *A* and the right in response to stimulus *B*. How is the intention to act according to this rule experienced? Though this is not particularly easy to tell, the best answer still seems to be the one suggested in James' analysis of voluntary action: when we perceive the stimulus event (say *A*) we only need to think of the response event supposed to follow it (left key press) and - "presto!" - it happens by itself (James, 1890, p. 527; cf., Prinz, 1987). In this account the act is entirely continuous with the percept. This is not only because the events are commensurate in terms of their intrinsic properties, but also because both share the status of being cognitive contents, i.e., contents that we perceive and of which we think. In essence, James' account suggests the reduction of intentions to cognitions: when one wants to do something, the only requirement is to think of the intended action in terms of its ultimate distal result, and one need not care about the intermediate proximal steps required to realize it.

As long as we believe that our conscious experiences are uniformative corollaries of the ongoing functional processing, we can ignore these observations. However, we could also assume that the phenomenal continuity between percepts and acts reflects some kind of structural continuity in the codes underlying them, implying that percept codes and act codes are formed in the same format and are stored or maintained in a common representational medium.

If so, there is no fundamental and deeply rooted difference between percept codes and act codes. The difference that remains is that percept codes stand for events-to-be-represented, whereas act codes stand for events-to-be-effected. It should be noted that this difference is not correlated with the nature of the coded events. Both codes can either refer to purely environmental events (e.g., right key moves down when stimulus *B* is presented) or to mixed environmental and bodily events (e.g., right index finger moves down when pricked by a needle). Thus, the distinction between percept and act codes is not anchored in the distinction between the body and its environment, but rather in the distinction between event representation and event effectuation.

If we take it for granted that such a medium for commensurate coding of percepts and acts exists, it is still an open question what function it serves. One possibility is that it is of no functional use for the perceptual control of actions. One might suspect, for example, that its function is to keep a (supposedly conscious) monitoring module informed about the action decisions of a (supposedly nonconscious) separate-coding-and-translation device. With this view, common codes would only serve to produce epiphenomenal copies of the ongoing processing.

Another possibility is that the common medium is functional in the sense that perception and action get linked together within that medium.[6] This implies the notion of an abstract, high-level representational medium in which perception-generated codes and action-generating codes take the form of percept codes and act codes, respectively, i.e., of codes that refer to purely distal events, without any reference to their proximal mediation.

With this view, a different kind of linkage machinery is required. Unlike the separate-coding framework that requires devices for coupling incommensurate entities, this view requires devices that permit the construction of intended events (acts) on the basis of given events (percepts) according to given rules. Unlike production rules that link incommensurate codes, these *construction rules* always link event codes to each other within a common representational medium.[7] Just as production rules can be traced back to S-R connectionism, construction rules can be related to S-S connectionism: like the rat in the maze that is supposed to learn what comes next (rather than what to do next), the subject in the choice-reaction time experiment is supposed to construct which event has to be effectuated next (rather than to decide which muscles to move).[8]

If percept codes and act codes share the same representational medium, they are commensurate in the sense that they can be evaluated with respect to their distance within that medium. An act code can be the same as (or similar to) a percept code with respect to some of the representational dimensions, and different or dissimilar with respect to some others. Therefore, the principle that all possible mappings are basically equivalent does not hold with this framework. Instead, for a given percept code there will always be some act codes located nearby and, hence, easy to construct, and many more remote ones that are more difficult to construct. As the same applies in reverse, an act code that is located in the vicinity of certain percept codes may become perceptually induced, or suggested when one of these codes gets activated. A linkage that should be particularly easy to construct is one where the task requires the subject to reproduce in his/her actions as precisely as possible a pattern of events he/she perceives. In this case, the act codes involved would be more or less completely specified by the respective percept codes.[9]

Evidence from Matching Tasks

In this section I will consider a special class of reaction tasks, where the instruction requires that actions match certain characteristics of environmental events, i.e., that the subject repeats or reproduces what he/she perceives. For instance, when the stimulus events are visual presentations of one out of three characters on each trial (say, *A*, *B*, *C*) and the response events are vocal utterances of the names of these characters, a *matching task* would be defined by a stimulus-response event assignment where each character must be responded to by its corresponding name. This preestablished correspondence is the critical difference between matching and *mapping tasks*, which term will be used to denote all kinds of arbitrary assignments of responses to stimulus events (e.g., $A \rightarrow$ [bi:]; $B \rightarrow$ [si:]; $C \rightarrow$ [bi:]). As the example shows, a given assignment may be a matching task for one person and at the same time a mapping task for another person (say, an illiterate), implying that the distinction between mapping and matching cannot only be derived from the intrinsic physical structure of the events involved but depends on their acquired meaning as well.

The distinction between mapping and matching plays a different functional role in the two theoretical frameworks. In the separate-coding scheme, where all possible assignments are equivalent, there should be no basic functional difference between matching and mapping (if we disregard a possible difference in the preexperimental frequency of usage of the links involved). However, in the common-coding scheme there should be a clear functional advantage of matching over mapping. In the matching assignment, the percept codes will specify the act codes more or less completely, yielding more simple construction rules (and, presumably, faster and more efficient processing) as compared to the mapping assignment where the percept codes will not by themselves specify the act codes.

Consider, furthermore, a matching task where the number of alternatives is left unspecified. For instance, a subject may be required to imitate patterns of body movements shown by a model under conditions where there is no prior knowledge of which patterns may occur. The separate-coding framework cannot readily account for performance in a task like this because there is no closed set of efferent codes on which the translation algorithms can operate. Thus, within this framework, some additional processing is required that somehow assembles the appropriate efferent code, rather than selecting it from a predetermined set of alternatives. The operations involved in such an open assembly of efferent codes should be more complex than those involved in the selection from among a closed set of alternatives. On the other hand, the common-coding framework can directly account for performance in an open task like this simply because it implies that, whenever a percept code is generated, the act code corresponding to it is thereby specified and, hence, immediately on hand.

Another way to phrase the major functional features of matching tasks within the common-coding framework is to say that percept codes can be directly converted into corresponding act codes so that code construction is reduced to code reproduction. This implies that matching should be easier and faster than mapping, that advance specification of alternative actions should be less important for matching than for mapping, and that, under certain conditions, percept codes should attain the power to suggest, induce, or elicit the activation of corresponding act codes, perhaps even in the absence of explicit instructions to match.

None of these implications would immediately follow from the separate-coding framework which considers matching and mapping equivalent, usually requires advance specification of alternative actions, and cannot easily account for act-suggestive effects of the ideomotor type. This is certainly not to say that theories derived from this framework could not be adapted to cover such phenomena: there have been various attempts to do so, more or less successful in themselves. Therefore, in the discussion to follow, the claim will not be that the common-coding approach can account for the observations from matching tasks, whereas the separate-coding approach cannot. The claim will rather be that the common-coding approach offers a simple and coherent framework to accommodate with these findings, whereas accounts in the spirit of the separate-coding approach would have to invoke various additional mechanisms in order to give satisfactory explanations of the facts.

The following discussion considers various types of matching tasks, i.e., tasks where the instruction requires the subject to reproduce in the actions (the actor-dependent events-to-be-effected) certain characteristics of environmental events (the actor-independent events-to-be-represented). These tasks may be classified along two major lines.

The first distinction refers to the *level of abstraction* at which environmental events and actions are considered. One rather obvious way is to relate these concepts to individual stimulus events and response events in the task. In this case, the instruction requires that the individual response mirrors, or corresponds to, certain characteristics of the individual stimulus. Another, perhaps less obvious way is to relate these concepts to relationships among individual events. In this case, the instruction requires that some relationships among the individual response events mirror, or be compatible with, certain relationships among the individual stimulus events - without requiring any correspondence in the individual events themselves (*correspondence vs. compatibility matching*).

The second distinction refers to the *nature of the aspects* of events or event relationships with respect to which a match is required. Matches can refer to physical aspects of the involved events (e.g., how they look, sound, or smell), but also to nonphysical, or symbolic aspects that concern their meaning (e.g., what they are and what they can be used for). These two aspects are not entirely independent since physical similarity tends to imply symbolic equivalence. But, since the reverse is not necessarily true, they must be considered separately (*physical vs. symbolic matching*).

Correspondence Matching

Physical Correspondence. When a person performs an action that, in its physical structure, corresponds with another person's action (or, with an inanimate environmental event) observed before, this person is said to imitate the observed action or event. Are there any functional privileges of imitative actions over other forms of stimulus-guided actions? We consider movement imitation first and imitation of (speech) sounds next.

Problems of *movement imitation* have long been an issue under debate in developmental and social psychology and in experimental psychology as well (see Scheerer & Schönpflug, 1984). Yet, the major portion of this literature deals with two basic issues that are beyond our present concern: first, which kinds of imitative behavior can be observed, under which conditions they occur, and which functions they serve in the social context (e.g., Bandura & Walters, 1963; Bandura, 1977); and, second, how and when the capacity to imitate arises in early development and how it later changes into adult performance (e.g., Piaget, 1946). For instance, as far as imitation in neonates is concerned, the last decade has seen both a fascinating growth in our knowledge (Meltzoff & Moore, 1977, 1983a,b, 1985) and a fascinating discussion of how to interpret it (e.g., Ewert, 1983; Jacobson, 1979; Neuberger, Merz, & Selg, 1983; Vinter, 1985).

For our present discussion, however, the main issue is not how imitation develops and under which conditions it occurs, but rather how it can be accomplished at all. Though there are some pertinent theoretical notions (mainly derived from observations on learning and development; e.g., Bandura, 1977; Bower, 1977; Gewirtz & Stingle, 1968; Piaget, 1946) and some attempts at a process-oriented analysis of imitation under experimental conditions as well (e.g., Williams, 1985, 1986; Whiting, 1987; Whiting, Bijlard, & den Brinker, 1987), a coherent tradition of theoretically guided experimentation is still lacking.

The theories in the field can readily be classified along the separate- vs. common-coding dimension. Separate-coding theories of imitation hold that afferent codes are translated into corresponding efferent codes, perhaps via the mediation of corresponding kinesthetic codes. Some of these theories have been formulated with a nativistic flavor, assuming that the correspondence rules are wired-in in the translation machinery (e.g., Bower, 1977). Some others are clearly empiristic, assuming that these rules must be learned (e.g., Gewirtz & Stingle, 1968). In contrast, common-coding theories of imitation hold that no translation is required because, in this case, the act codes resemble (and, hence, are more or less completely specified by) the percept codes involved. This implies the assumption that imitative actions are basically different from nonimitative actions to which such resemblance does not apply.

Historically it seems that Koffka (1925) was the first to suggest an interpretation like this.[10] According to Koffka, the only way to account for movement imitation is to assume that perceptual configurations (relating to the model events) are capable of directly producing the imitative movement configurations (Koffka, 1925, Chap. V, § 9; cf. Prinz, 1987). Later, Piaget (1946) outlined a theory of imitation which also relied on the idea of common coding, though at a more abstract level than that advanced by Koffka. This is so in Bandura's theory which invokes a common representational medium at the symbolic level (Bandura, 1977).

One source of evidence may be derived from observations on spontaneous occurrences of imitation. Everybody knows that imitation is easy and sometimes tends to occur involuntarily. As is shown in a nice demonstration by Eidelberg (1929), this does not only apply to babies, infants, and children, but to adults as well. Eidelberg created a Stroop-like situation in the following way: he instructed his subjects to point to their noses on the instruction "nose!" and to a lamp in the room on the instruction "lamp!" At the same time the experimenter would point either to his nose or to the lamp. As one would expect, his subjects were perfect as long as the verbal commands and the visual demonstrations were in correspondence. However, when these two sources were dissociated, it turned out that the subjects (who were always instructed to follow the verbal commands) were unable to avoid mistakes: they tended to copy what they saw rather than follow what they heard. Eidelberg concluded that under certain conditions we may be compelled to imitate - even if this goes against our voluntary intentions.

Related evidence comes from the pathology of action. It has long been known that various kinds of echo-reactions can be observed in aphasia, apraxia, low-rate mental deficiency, epilepsy, or catatonic states (Stengel, Vienna, & Edin, 1947). Though most of these observations refer to unintentional repetitions of words heard (echolalia; see p. 179), unintentional imitation of actions observed is also reported (echopraxia). For instance, Stengel et al. (1947, p. 602) described a severely brain-damaged patient who, when the examiner coughed, whistled, got up from his chair, sat down, sneezed, put his tongue out, stamped his foot three times, or tapped on the table four times, the patient, without being asked to do so, repeated those actions promptly and correctly. Interestingly, such spontaneous imitations did not occur with actions the patient could not possibly perform in the actual context. For example, when the examiner took his watch and rewound it, or when he took off his tie, the patient did not even attempt to repeat these actions.

When phrased in our functional language, such observations could be taken to suggest that under conditions of severe brain damage, where more complex forms of code construction are heavily impaired, percept codes may still be capable of specifying corresponding act codes, i.e., of inducing repetitions of the actions observed. This conclusion is also supported by an error analysis in various forms of verbal instruction and imitation tasks, showing repetitions to be the most prominent type of error in both aphasic and apraxic patients (Lehmkuhl, Poeck, & Willmes, 1983).

Another source of evidence comes from experimental studies on movement imitation. Unfortunately, there seem to be no studies in the field of movement imitation proper that allow one to compare matching and mapping conditions directly. Some related evidence is available, however, from tracking studies. In these tasks, too, there is a pattern of events that exhibits a particular physical structure (say, a target that moves in a two-dimensional plane), and a pattern of movements that exhibits another physical structure (say, the movement of a joystick control).

The result of the control movement is indicated by the position of a marker in the display, and the task requires keeping the marker on the target.

In a task like this, the relationship between the pattern of the display events (both target and marker) and the pattern of the control movements can exhibit various degrees of correspondence. At the one extreme the relationship may be "natural" in the sense that the movement is more or less to copy the target motion (for instance, when the subject is required to trace the target directly with a stylus or with proportionally transformed joystick movements). At the other extreme, it may be "arbitrary" in the sense of being determined by a mapping rule that is fixed by a convention (say, when two wheels have to be operated, one for horizontal and one for vertical shifts of the marker).

Within certain limits the matching-type conditions of such tasks can thus be regarded as requiring the imitation of inanimate events, and they should not substantially differ from tasks that involve imitation proper.[11] As one would expect, the results from studies where mapping and matching arrangements can be directly compared indicate that processing is privileged in matching conditions (for summaries see Poulton, 1974; Loveless, 1962). Unfortunately, the evidence from such tasks is scanty. This is because, with the exception of some early classic studies, tracking research has concentrated on tasks involving complex transformations rather than on seemingly trivial conditions where the movements copy the motions. However, the few pertinent studies (e.g., Craik & Vince, 1963; and further experiments by Vince & Mitchell, cited and discussed by Loveless, 1962) suggest some simple conclusions.

When it is required that the movements performed correspond to the motions observed with respect to direction and velocity, performance in terms of both response time and accuracy is usually superior to other, less imitative forms of mapping. Though subjects are capable of learning various forms of more complex transformations, even extensive practice will usually not raise them to the same performance level as is observed in imitative matching conditions from the beginning. Also, it has been shown several times that matching is much less susceptible than mapping to various conditions of increased task difficulty (e.g., in terms of low practice, high information rate, additional load due to distractors of secondary tasks; e.g., Mitchell & Vince, 1951; Simon & Fitts, 1952). These findings seem to suggest that, when the movements imitate the motions, the underlying processing is faster, more accurate, and less susceptible to stress and, hence, presumably more simple in its functional structure as compared to nonimitative movements.

In the area of *speech imitation*, there is a more coherent tradition of discussing theories with respect to data, and data with respect to theories (cf., MacKay et al., 1987). In this field, too, a considerable portion of the pertinent work comes from developmental studies (e.g., Studdert-Kennedy, 1987). In this work it has been pointed out that the productive and receptive side of speech and language are likely to be acquired together during ontogeny. Moreover, the same has been suggested for phylogeny (MacKay et al., 1987), since the ability to produce language is of no use when there is no one to listen, and the ability to understand language is of no use when there is no one to produce it.

The motor theory of speech perception which holds that speech perception is mediated by the motor codes involved in speech production has been one of the major approaches emphasizing common coding (Liberman, Cooper, Shankweiler, & Studdert-Kennedy, 1967; cf., Gordon & Meyer, 1984, for a summary; and MacNeilage, 1978, for a critical review of the evidence). In its classic form the approach has stressed the discrete and categorical nature of these codes. For instance, in one of its more recent versions, the phoneme is considered the basic discrete perceptuomotor unit in speech perception and production (Studdert-Kennedy, 1987). In other related forms of the approach, it has been suggested that the correspondence is not in terms of elementary units, but rather of higher-order properties of acoustic spectra and vocal tract shapes, or, their respective neural correlates on the sensory and the motor side (e.g., Porter, 1987).

One class of such higher-order properties refers to the sequencing and timing involved in the perception and production of speech. Common timing has been stressed in a model advanced by Meyer and Gordon (Meyer & Gordon, 1984; Gordon & Meyer, 1984) on the basis of reaction times in auditory-vocal matching tasks. Common sequencing has been stressed in a model of the functional organization of the human language cortex advanced by Ojemann and co-workers on the basis of findings obtained with electrical brain stimulation (Ojemann, 1983; Ojemann & Mateer, 1979). According to this model the language cortex has common sites for the identification of phonemes and the sequencing of orofacial gestures.[12] This could suggest a common representational medium for the involved afferent and efferent codes.

Some of the evidence in support of these approaches can be derived from matching tasks. One class of observations comes from speech repetition studies in normal subjects. There are four converging lines of evidence. First, it has been observed that reaction times for the repetition of auditorily presented speech are extremely rapid and accurate, as compared, for instance, to visual-verbal control tasks. This applies to individual phonemes where the response times for repetitions may be as short as about 130 ms in two- and three-choice situations (Kozhevnikova & Chistovitch, 1965; Porter & Lubker, 1980), and to syllables where repetition times range in the order of 180-220 ms for various degrees of choice (Davis, Moray, & Treisman, 1961; Pickett, 1985, for summary; Porter & Castellanos, 1980; Saslow, 1958). In a study by Marslen-Wilson (1973) it was shown that, even in the shadowing of connected discourse, the lag can become as short as 250 ms. These latencies are about half the time that is required for corresponding visual-verbal responses. This seems to rule out an explanation in terms of frequency of occurrence and/or usage since normal adult subjects will be much more practiced in reading what they see than repeating what they hear.

Second, the number of alternatives does not seem to play a major role (Davis et al., 1961; Saslow, 1958). In two studies by Porter and co-workers (Porter & Castellanos, 1980; Porter & Lubker, 1980), it was shown that repetetive vocal choice reactions can even be faster than nonrepetetive simple reactions, suggesting that mapping a single vocal utterance to one out of several noncorresponding auditory speech stimuli may be more difficult than matching each stimulus by repeating it.

The short lags at which auditory texts can be shadowed (Marslen-Wilson, 1973) also suggest that the number of alternatives cannot be a major determinant of performance.

Third, auditory-vocal matching is only weakly susceptible to interference from other tasks. This was demonstrated in a study by McLeod & Posner (1984) where an auditory-vocal word repetition task was combined with a visual-manual reaction task in a dual-task paradigm. As no interference was obtained, the authors concluded that the auditory-vocal link forms a special, privileged loop from perception to action. Another way to phrase the same result is to assume that matched act codes are automatically available by-products of the activation of their corresponding percept codes.

Fourth, some more detailed evidence was provided by a phonetic feature matching task by Meyer and Gordon (Meyer & Gordon, 1984; Gordon & Meyer, 1984). In this task the auditory-vocal stimulus-response pairs could be matched or mapped with respect to different features. The results indicated that matched voicing shortens reaction time, whereas matched place of articulation does not. As voiced and voiceless consonants differ in their voice onset times, it was suggested by the authors that a common timing mechanism might be an interesting candidate to account for the effect.[13]

Another line of evidence may be derived from the pathology of speech perception and production. A case in point is the symptomatology of transcortical aphasia. This term, which comes from Goldstein (1917, 1948), denotes a class of language disorders where either language comprehension or language production or both are severely degraded (in the sensory, motor, and mixed cases, respectively), whereas speech repetition is virtually unimpaired. Some of these patients (particularly mixed transcortical aphasics) exhibit all varieties of echolalic reactions. Though echolalic reactions can also occur in the context of various other pathologies (cf., Schuler, 1979), they are fairly typical of this particular form of disease. Goldstein argued that it must be assumed that the two speech centers of the brain - Wernicke's area and Broca's area - as well as the connections between them are still intact in these patients because otherwise the ability to echo heard utterances could not be explained. He speculated that the lesion should rather be such that the speech area was isolated from those areas of the brain where language is processed semantically and syntactically. This assumption was later supported by neuropathological studies (Geschwind, 1974, Chap. VIII; Geschwind, Quadfasel, & Segarra, 1968; Whitaker, 1976).

Another case in point is conduction aphasia where the symptomatology is partially reversed. Comprehension is near-normal, spontaneous speech is slightly disorganized (e.g., in terms of phoneme substitution, neologistic paraphasias), but speech repetition is severely disturbed. According to Wernicke (1874), this complex of symptoms could point to lesions that destroy the connection between Wernicke's and Broca's area while leaving the connections to the other language centers intact. Later this view was also confirmed by postmortem studies in a number of cases (Benson et al., 1973). A similar view was advanced by Strub and Gardner (1974) who held that in conduction aphasia the impairment resides in the structures responsible for the conversion of the product of the phonological analysis into corresponding target phonemes.

The findings from these two symptomatologies suggest a simple picture of the organization of language functions in the brain. Roughly speaking, we have to distinguish between two major systems, one for speech and another one for language. In the speech system language is processed at the phonological level, whereas semantic and syntactic processing is performed in the language system. Though there are two anatomically distinct subsystems in the speech system - one for analysis and another for synthesis - these subsystems will usually act as a unitary system, i.e., corresponding codes will be activated in the two subsystems at the same time. When the task requires or permits using this direct route, processing is fast and efficient as compared to conditions where the detour through the language system cannot be avoided.

Symbolic Correspondence. In the case of symbolic correspondence the action has the same meaning as the event which triggers it, but there is no correspondence in physical structure. This is the case in a variety of coding tasks where the subject is required to "translate" stimulus events into actions according to coding rules that are fixed by convention. Such tasks will involve one or the other form of linguistic coding or some other artificial coding and notation system, as in object naming, reading, writing, typewriting, or playing music. For instance, there is no correspondence in terms of physical structure between the visual pattern of an *E*, the auditory pattern of its name and the movement pattern required to press the corresponding key on a typewriter. If there is any correspondence at all, it is in terms of nonphysical, or semantic aspects of these various patterns.

Though it has become trivial to posit that there can be no perceptual coding that does not include a mandatory semantic component and no productional coding without reference to meaning as well, it is certainly less trivial to postulate that semantic communalities of percepts and acts may play an important functional role in the perceptual control of action decisions. Even if one is willing to accept the common coding principle with respect to structural correspondence, it still means a further theoretical step to accept it at the symbolic level, too. Unlike structural correspondence which derives from intrinsic properties of events and actions, symbolic correspondence refers to extrinsic properties that have been attached to them like arbitrary labels. Structural correspondence can be assessed and produced by the system via its in-built feedback loops, whereas symbolic correspondence can only be ascertained on the basis of acquired and stored correspondence rules.

However, as I have pointed out elsewhere (Prinz, 1983), the claim that semantic relationships between percepts and acts may be functional in action control loses much of its counterintuitive appeal when one questions two implicit assumptions that have been dominant in much of our theorizing about the relationship between structure and meaning: that structure is implemented in the hardware of the system and meaning in the software, and that - as a consequence - structure comes first and meaning second.

Posner's conception of psychological pathways is an example of an approach where these difficulties are avoided (e.g., Posner, 1978; Posner & Snyder, 1975). Though Posner's definition of the concept is still stimulus-oriented ("set of internal codes and their connections that are activated automatically by the presentation of the stimulus"; Posner, 1978, p. 90), he uses it in a way that also stresses its role in action control as well as the relative independence of the various codes in a pathway with respect to the perception-action mediation.

With this picture in mind, consider *naming* and *reading tasks* where the actions to be performed are verbal utterances that bear a symbolic relationship to the visual objects or words presented. Note that the degree of ambiguity of this relationship is different in the two tasks. When one reads a word, say, "canary," the response is completely determined by the stimulus, given the intention to follow the instruction. This does not apply when one has to name the picture of a canary, since "animal" or "bird" would be likewise appropriate responses. Unlike the instruction to read which unequivocally specifies the response (act code) corresponding to a given stimulus (percept code), the instruction to name may arouse more than a single code.

A nice illustration of the difference between reading and naming is provided by Fraisse, Lanati, Regnier, and Wahl (1965). In two conditions of their experiment where they had their subjects either name outline drawings of familiar objects (like a cat, a bottle, a rose, etc.) or read the corresponding words, they observed that reading times for words were much shorter than naming times for objects, though the verbal responses were identical. In this they replicated an old and well-known result (cf., Brown, 1915a,b; Brugmans & Heymans, 1916; Cattell, 1885, 1886). As the same result was obtained under conditions where stimulus discriminability was more carefully controlled (Fraisse, 1967, 1968, 1969), the effect suggests that reading is somehow easier than naming.

More interestingly, in two other conditions where the subject had to give categorical responses to both objects and words (e.g., "animal," "object," and "flower," respectively), the difference between words and objects was reversed. Though categorical responses were slower than specific responses in both stimulus conditions, they were now much faster for objects than for words, suggesting that the same categorical response is easier to address on the basis of an object as compared to a word. When a word is presented and its category name has to be produced, the required act code seems to be interfered with by the dominant act code for reading the word. When an object is presented, this competition will be much weaker. This is because the object will not address the unique and strongly dominant name code, but rather activate a variety of such codes simultaneously (e.g., "canary," "bird," and "animal"). As a consequence, the relationship between specific and categorical response times will then depend on factors like stimulus discriminability, word frequency, or degree of choice (Morin, Konick, Troxell, & McPherson, 1965; Oldfield, 1966; Oldfield & Wingfield, 1965).

In reading, these factors are less important or even irrelevant. In the reading of word lists it has sometimes been observed that the degree of choice, i.e., the size of the vocabulary from which the items are drawn, is irrelevant (Conrad, 1962; Pierce

& Karlin, 1957; Pollack, 1963), whereas word frequency may still be a critical factor (Pierce & Karlin, 1957). In letter and numeral naming tasks (where reading becomes equivalent to naming), it has also been shown that reading time is independent of vocabulary size. In an experiment by Mowbray (1960), the time for naming visually presented numerals did not depend on the size of the set from which they could be drawn in a given condition. The same was shown for letter naming by Morin et al. (1965). In this experiment, where the naming of objects (faces, animals, colors, symbols) could be directly compared to the reading of letters, letter reading was the sole condition that was not affected by the number of choices. This was also observed in an experiment by Brainard, Irby, Fitts, and Alluisi (1962). In this study there were four conditions, combined from two classes of stimulus events (lights vs. numerals) and two action modes (manual vs. vocal), and again the numeral/vocal condition, where the numerals had to be named was the sole task showing a zero slope in the plot relating reaction time to degree of choice.

Though the conversion from visible to audible speech is symbolic in the sense that there is no obvious structural correspondence between the two linguistic forms, it does not have to implicate mediation through semantic information proper as contained in the semantic lexicon. Though there is ample evidence in support of a direct route from a word's orthography to its lexical unit, and of lexical effects on its pronunciation as well (e.g., in irregularly pronounced words), the evidence is no less clear in support of sublexical routes for the conversion of graphemic codes into articulatory and/or auditory codes (Henderson, 1982, 1985). These two routes are illustrated by the symptomatologies of two forms of dyslexia. In surface dyslectics the ability to pronounce irregular words correctly is impaired, and at the same time the patients have problems to distinguish between real words and their pseudo-homophones (e.g., speech - speach). These symptoms seem to indicate that lexical control of the phonological assembly is absent, so that reading relies exclusively on spelling-to-sound conversion rules, and comprehension is based on the phonological codes thus generated (e.g., Patterson & Morton, 1985; Shallice & McCarthy, 1985). In phonemic or deep dyslexia the case seems to be reversed. Here, the patient is virtually incapable of deriving phonemic from graphemic codes, except via lexical reconstruction. It appears that in this case the sublexical machinery for spelling-to-sound conversion is defective, whereas the direct route from the orthographic information to the lexicon is more or less intact (cf., Allport, 1979; Marshall & Newcombe, 1973; Patterson & Marcel, 1977).

In a similar vein, findings from studies of word reading latencies in normal subjects also suggest that the phonological assembly of pronunciation is performed under both lexical and sublexical control (for a review, see Henderson, 1985). Studies using semantic priming techniques (e.g., Meyer, Schwaneveldt, & Ruddy, 1975; Monsell, 1987); have provided evidence in support of common coding at the lexical level - which appears to be a natural principle of organization at this level.

Less obvious is the feasibility of this principle at the sublexical level. However, at this level, too, the evidence at least suggests a common symbolic coding of graphemic and phonemic information. Particularly the results from studies demonstrating orthographic and phonological priming effects on reading latencies (Glushko, 1979, 1981; Henderson, 1985) make a strong case for the common symbolic coding of percept and act codes. Graphemic and phonemic information are commensurate via these symbolic codes - either in terms of their common content or of more abstract features that could refer, e.g., to the sequential and/or temporal structure of the events involved (cf., Keele, 1987; MacKay, 1987b; but see Monsell, 1987, for arguments in support of a separate coding view of sublexical sequence representation in input and output pathways).

To conclude, some pertinent observations on the difference between reading and naming can also be made with respect to our daily environment. The fact that words have strong unique verbal codes, whereas objects have several weaker ones makes it difficult, if not impossible, to avoid the reading of words that catch our attention. This does not apply to the naming of objects. Graphic artists have always tried to capitalize on this, for instance, in advertising or in poster design. Also, it can often be observed that object naming is impaired while reading is more or less intact, e.g., in the mild amnestic aphasic syndrome that can be seen in many elderly people. Reading words is unavoidable and suggests itself under appropriate attentional conditions, whereas naming objects is much less self-suggestive.[14]

Further examples of symbolic correspondence are provided by *transcription skills* like writing, typing, or playing music at sight (cf., Shaffer, 1976). Like in discrete reading and naming tasks, skilled performance in continuous transcription also exhibits a high degree of automaticity in what appears to be a direct conversion of percepts into acts. However, since most of the theoretical issues addressed by this work refer to problems that emerge from the serial nature of these skills, data pertinent to our present concern are scarce. For instance, in the typewriting literature, most studies are concerned with the timing and the chunking of stroke sequences in programming and execution (e.g., Ostry, 1980; Shaffer, 1973, 1976, 1984; Sternberg, Monsell, Knoll, & Wright, 1978). These data do not allow a detailed analysis of the time course of the perceptual guidance of action, particularly since the timing and chunking of the information intake is normally controlled by the subject and not by the experimenter.

Some evidence is available on dual-task performance in highly practiced subjects. Allport, Antonis, and Reynolds (1972) have shown that music students are capable of playing piano at sight while shadowing prose without interference. Shaffer (1975) has demonstrated that skilled typists can type from visual texts while shadowing. (It should be noted, however, that these findings can also be interpreted in the light of the privileged loop hypothesis advanced by McLeod & Posner, 1984. In this case they would be taken as another demonstration of a privileged auditory-vocal by-pass loop rather than a privileged form of visual-motor transcription.)

Coltheart and Funnell (1987) have recently suggested that writing (i.e., spelling to dictation) might be based on the same orthographic lexicon as reading aloud. Their study was based on a detailed analysis of reading and spelling errors in a patient who exhibited severe surface dyslectic and dysgraphic symptoms. One of the interesting observations was that there was one set of words that was both easy to read and easy to write, and another set of words that was difficult to read and to write for this patient. This observation might suggest that there is a common, single orthographic lexicon used for both reading and spelling.

Compatibility Matching

In the case of compatibility matching, we consider tasks that exhibit a nonarbitrary relationship between the relations among the percept codes involved and the relations among the act codes involved. These two sets of relations may again be compatible with each other in two basic ways, physical and symbolic. In the physical case a further distinction is in order between tasks involving spatial and temporal compatibility.[15]

Physical Compatibility: Spatial Relationships. The analysis of spatial stimulus-response compatibility effects has been one of the major concerns in reaction time studies since the early work of Hick (1954) and Hyman (1953) on the dependency of reaction time on degree of choice. Since this work has frequently been reviewed (e.g., Keele, 1986; Massaro, this volume; Smith, 1968, 1980; Welford, 1960, 1968, 1980), we can concentrate on an attempt to rephrase some of the major findings in terms of the present approach.

For this, one has to consider the ensemble of stimulus events and the ensemble of actions involved in a given task. The case of spatial compatibility may arise when both the possible events and the possible actions have different locations in space. In this case we can distinguish between an arbitrary mapping assignment and an apparently natural matching assignment of stimuli and responses. It has frequently been reported that, when a matching assignment is chosen, reaction times are, on the average, very fast and essentially independent of degree of choice. For instance, in a now classic experiment by Leonard (1959), the subject had to press the armatures of relays on which their fingers were resting whenever one of these armatures began to vibrate. Degree of choice was entirely ineffective within the range of two to eight choices (see also Smith, 1980). Though it later became apparent that this result can only be obtained within a certain bandwidth of vibration amplitude (ten Hoopen, Akerboom, & Raaymakers, 1982; cf., Massaro, this volume), it clearly demonstrates the power of spatially compatible relations between stimulus events and responding actions.

In Leonard's experiment compatibility was realized in the strict form of locational coincidence: the locus of the required action was to coincide with the locus of the stimulus event. This was also the case in a visual pointing task (pointing to lights) where the same result was obtained (cf., Fitts & Posner, 1967, p. 105). Other studies have shown that such coincidence is not a prerequisite for the effect to occur. Similar, though weaker effects where also obtained when events and actions (displays and controls) were separated by locations, provided that they both exhibited the same basic arrangements of elements in the sense of relational congruence. In the early studies by Fitts and his group (e.g., Fitts & Deininger, 1954;

Fitts & Seeger, 1953), it was shown that such compatible couplings yield much shorter reaction times and much higher accuracy scores than less compatible and more arbitrary couplings. Furthermore, similar effects have been observed when the task required responding to certain nonspatial aspects of the stimulus events under conditions where the spatial location of the stimuli varied as an irrelevant dimension. In this situation, too, congruent responses were much faster and much more accurate than incongruent ones (Simon, 1969; Simon & Craft, 1970). This seems to indicate that spatial relations among perceptual events can induce the unintentional reproduction of similar spatial relations among actions.

In compatible tasks of the congruence type, degree of choice is usually not entirely ineffective. For instance, in a task studied by Lamb and Kaufman (1965; cf., Kaufman & Levy, 1966; Griew, 1959, 1964), the apparatus consisted of a semicircular arrangement of two to nine stimulus lights and another, somewhat smaller semicircle of response buttons. When one of the lights went on, the subject had to press the corresponding button, starting from a home-key in the center of the semicircle. In these studies reaction time was still weakly dependent on the degree of choice, but it is unclear whether this effect indicates some residual processing for code translation, or whether it simply mirrors the increase in stimulus confusability that goes along with the increase in the number of alternatives.

From these results it may be concluded that relational congruence is a sufficient condition for a direct conversion of percept into act codes to occur. Still, the results can be interpreted in two different ways, allocentric and egocentric. The allocentric interpretation postulates that the spatial relations among the possible stimulus events have to be congruent with the spatial relations among the possible response events. Conversely, the egocentric interpretation postulates that it is the spatial relation between the agent and the intended action which has to be congruent with the spatial relation between the agent and the observed event.

The egocentric interpretation has been stated in two versions, anatomical and functional. In the anatomical version, brain anatomy has been considered critical. For instance, in a two-choice task, where stimuli can either be flashed to the right or the left of the fixation point, and a compatible response with the right or the left hand is required, the stimuli are always analyzed in the same brain hemisphere in which the movement is triggered. However, although this correspondence seems to be a critical determinant of simple reaction times (Anzola, Bertolini, Buchtel, & Rizzolatti, 1977; Berlucchi, Crea, di Stefano, & Tassinari, 1977), it cannot account for the effects of spatial compatibility in choice reaction times. This must be concluded from the fact that crossing the hands, which eliminates anatomical correspondence, does not eliminate the compatibility effect as long as relational congruence between stimulus locations and response locations is maintained (Anzola et al., 1977; Riggio, Gawryszewski, & Umiltà, 1986; Simon, Hinrichs, & Craft, 1970; Wallace, 1971, 1972). Thus, the critical factor seems to be not to which side of the body the responding limb belongs and by which hemisphere it is controlled, but on which side of the body the action is executed. As Wallace (1971) put it, it is the (functional) spatial code of the locations of the hands, and not the (anatomical) body code of their relation to the sides of the body, that has to match the spatial codes of the stimulus events. Riggio et al. (1986) later suggested the modification that the functional spatial code refers to the location of the resulting response event (i.e., the action itself) rather than that of the limbs involved in effecting it.[16]

Though the relative contributions of the egocentric and the allocentric reference system have still to be explored, it has become apparent that they both contribute to the processing benefits observed under conditions of relational congruence. As far as actions are concerned, it was shown in two studies by Nicoletti, Anzola, Luppino, Rizzolatti, and Umiltà (1982) and Nicoletti, Umiltà, and Ladavas (1984) that the effects of crossed vs. uncrossed hands are mainly due to the relative positions of the hands with respect to each other and not to the absolute position with respect to body side. This suggests that the locations of actions and/or limbs involved in them are mainly coded in terms of their relative positions, i.e., in the allocentric mode. As far as stimulus events are concerned, studies by Ladavas and Moscowitch (1984) and Umiltà and Liotti (1987) suggest that when they can be coded in terms of both absolute egocentric side or relative allocentric position, either of the two codes can become effective in action control, depending on task demands and instructions, and either will yield the same beneficial effects of relational congruence.

The theoretical notions offered by Wallace (1971, 1972) and elaborated by Riggio et al. (1986) can be considered a task-specific formulation of a common-coding approach to spatial compatibility. It holds that the benefits of relational congruence will arise when the spatial codes of represented events and of to-be-generated actions resemble each other. The spatial codes are always relational since there is no other way of specifying spatial location than by reference to other events in space or to the acting agent.

Physical Compatibility: Temporal Relationships. To the degree that the temporal relationships among events are regular and predictable, we are capable of reproducing or accompanying them in our actions. In some cases, some people may even find it difficult to avoid event-synchronized movements like in dancing or marching. One way to think of movement synchronization - both intentional and unintentional - is in terms of a temporal coincidence of events: an action sequence is organized such that individual actions (say, claps of the hand) coincide with individual environmental events (say, beats of a metronome). The sequence of actions thus exhibits the same pattern of temporal relationships as the sequence of events. The fact that the two sequences coincide temporally seems to indicate that the timing of actions is somehow related to the perceived or anticipated temporal pattern inherent in the event sequence.[17]

In the dancer's view, the rhythm captures the movement, and there is often no way to escape from it. How does synchronization come about, and what is synchronized with what? In a simple task like clapping one's hands or pressing a key simultaneously with a metronome beat, synchrony is usually attained after the third beat. Likewise, with periodically structured rhythmic patterns, synchronization is usually attained by the third period (Fraisse, 1956, 1980).[18] Various experiments have demonstrated the crucial role of feedback in synchronization. Karlovich and Graham (1966) studied the synchronization of finger movements with simple visual patterns. When auditory feedback was provided, the movements were much better synchronized with the pattern, as compared to a control condition where no auditory feedback was provided. When auditory feedback was delayed, performance became rather poor. Conversely, the task of synchronizing an auditory pattern turned out to be much less susceptible to both support and interference due to visual feedback (Karlovich & Graham, 1968). Thus, there seems to be an asymmetry between vision and audition with respect to a power to guide rhythmic body movements, and of the power to impair or support such guidance as well. In studies where an attempt was made to separate the relative contributions of kinesthetic, tactile, and acoustic information to feedback, it was shown that, though all of these sources are contributory, the availability of tactile information is a prerequisite for obtaining precise synchrony (Fraisse, Paillard, & Oléron, 1958; Oléron, 1961).

These findings suggest an interactive type of model where an internal source for action timing interacts with an external source that provides feedback. During the first beats or periods of the rhythmic pattern, the temporal structure is acquired and used to generate an action pattern exhibiting the same temporal structure. This raw pattern is then continuously recalibrated and thereby kept in phase with the event pattern on the basis of feedback. With slow rhythms untrained subjects will find it rather difficult to keep these two components separate (i.e., when the task requires tracing the time structure of the rhythmic event with a phase shift; Fraisse & Ehrlich, 1955). However, in rapid event-synchronized tapping it may occur that, while the same frequency is used in the two sequences, their phases get shifted relative to each other (Klemmer, 1967), suggesting that the processing of feedback could become impaired with rapid sequences.

These observations support the notion of a common mechanism for the timing of perceptual events and actions. Whether or not this view is sound depends, among other things, on the nature of the codes that become synchronized. The general logic of the common-coding approach leads us to expect that the synchrony must apply to observed events and intended actions. Note that this is not equivalent to positing a synchrony between the underlying sensory and motor patterns in the brain. Rather, when an intended action (e.g., a key stroke) is to coincide with a perceptual event (e.g., a metronome beat), the corresponding motor pattern has to be initiated much prior to the onset of the beat. Moreover, if the point in time at which the key stroke is perceived is mainly dependent on tactile information, the key stroke itself would even have to precede the beat, since it takes longer for the tactile information to travel from the hand to the brain than it takes for acoustic information to travel from the ear to the brain.

This is exactly the pattern of findings usually obtained in auditory-manual synchronization. Synchrony is perceived when movement onset has a small temporal lead over event onset (Dunlap, 1910; Fraisse, 1966; Klemmer, 1967). As Fraisse (1980) has shown, the size of this lead depends on which limb is used for the synchronizing movement. When the feet were used instead of the hands, an increase of the lead was observed which roughly corresponded to the difference in nerve conduction time between feet and hands. These findings suggest that the timing of the movement by the motor system is such that movements can be perceived in synchrony with the (auditory) pacing events. Thus, if there is any shared timing at all, it can neither refer to the two patterns of afferent and efferent neural activity, nor to the two physical events producing them and produced by them, but only to the central codes that represent these events.

In terms of a process-oriented account which considers the planning of the temporal relationships among the actions, the resulting coincidence is seen as arising from the use of the time structure of environmental events (either ongoing or anticipated) for the generation of intended actions. The act code for the intended action pattern uses the temporal specifications of the (anticipatory) percept code of the rhythmic pattern. This is a straightforward way to guarantee that events and actions will perceptually coincide in terms of time - provided, of course, that there is a smart motor control module capable of computing patterns of proximal efferent commands appropriate for the realization of the intended distal actions.

In the recent literature on sequencing and timing, it has become customary to view timing as an achievement of clocks, timers, or similar special purpose devices (e.g., Keele, 1987; Shaffer, 1985; van Galen & Wing, 1984). Since comparable devices for spacing - i.e., assigning spatial locations - have not been suggested, this metaphor seems to imply the view that temporal coding and spatial coding are functionally different. The present approach uses a different theoretical metaphor. It considers the temporal properties of events to be intrinsic aspects of their coded contents rather than products of extrinsic timers. Keele's (1987) problem of whether the timing mechanism is common for perception and action must thus be translated into the question of the extent to which temporal patterns in percept codes can be converted into temporal patterns in act codes. One of the advantages of such a uniform approach is that it can be more easily extended to cover the simultaneous guidance of actions in both spatial and temporal respects.

Symbolic Compatibility. There are two basic ways in which relationships among environmental events could be symbolically compatible with relationships among actions. One way might arise when the events exhibit differences on a symbolic dimension that corresponds to a spatial or temporal attribute of actions. Consider, for example, a task where, on each trial, one of two possible digits is shown, and the subject has to respond by pressing the left or the right of two keys. Two basic assignments of digits to keys can be realized in this task: the numerically smaller and the larger of the two digits can either be assigned to the left and the right key, respectively, or vice versa. The first assignment can be considered symbolically compatible, but not the second. In both cases the digits differ with respect to numerical size, i.e., a dimension of difference symbolically attached to them. However, this abstract dimension can also be given a compatible interpretation in terms of a spatial continuum that is extended horizontally from left to right (probably due to the reading conventions in our language). Accordingly, the left-hand side and the right-hand side should be natural places to respond to numerals that are small (or come first in a serial order) and large (or come later), respectively, whereas the reverse assignment should be less natural. In an unpublished pilot study in our laboratory, we did indeed find a latency difference between these two assignments in a key-pressing task with visual digits.

A superficially related effect was reported by DeRosa and Beckwith (1971), DeRosa and Morin (1970), and Morin, DeRosa, and Shultz (1967). In a memory scanning task they had their subjects decide whether or not a probe digit belonged to a positive set of four digits defined immediately before. In some conditions the positive set was closed, i.e., it contained a sequence of four consecutive digits (e.g., 3, 4, 5, 6, or 8, 7, 6, 5). When correct negative responses were plotted as a function of the absolute distance between the probe and the least remote digit in the positive set, response times were monotonously related to remoteness: the more remote the probe was from the set, the faster the negative decision could be reached. This finding suggests that the positive set items and the probe were both coded in terms of the same symbolic continuum of numerical size, and that the ease of the negative response depended on the discriminability between the probe and the set with respect to this symbolic dimension.

As the probes were classified by binary verbal responses (yes/no), it may appear that no spatial interpretation of numerical value could be involved in this task. Note, however, that in a binary decision task the two responses, though arbitrarily mapped to stimulus events, exhibit a special logical relationship among each other in that the negative category is defined as residual, i.e., with reciprocal reference to the positive category. This logical relationship has a straightforward topological and, hence, spatial interpretation which can be illustrated by a simple Venn diagram. With this view, closed consecutive positive sets are symbolically compatible because the required partition into a positive and a residual set can be accomplished by means of an appropriate segmentation on the dimension of numerical size. On each trial the act codes for the two responding actions can be assembled by superimposing a trial-specific Venn diagram on the numerical size dimension of the percept codes involved. As a consequence, one should expect weaker performance with open (nonconsecutive) positive sets. This was in fact observed for reaction times and errors in both response categories (DeRosa & Morin, 1970; Morin et al., 1967).[19]

Another form of symbolic compatibility may arise when the stimulus events differ with respect to a physical dimension that is represented in the symbolic content of the corresponding actions. Consider, for example, a task where lights are arranged in a row and the instructions require the assignment of numerals to them in an ascending order from left to right. An assignment like this is symbolically compatible (say, as compared to a reversed or random assignment) for the same reason as in the digit-pressing task above: at the symbolic level the abstract dimension of numerical size has a natural interpretation in terms of a horizontal, left-hand oriented continuum. It should not go unnoticed, however, that Brainard et al. (1962) who included both a numeral-to-key pressing task and a light-to-numeral naming task in their study, observed rather pronounced effects of degree of choice on reaction time in both of these conditions. This result questions the generality of the supposed commonality of numerical size and spatial location codes.

On the other hand, it has been shown that colors can be symbolically assigned to actions in a rather effective way (Hedge & Marsh, 1975; LeMay & Simon, 1969; Simon, Sly, & Vilapakkam, 1981). In these studies, symbolic compatibility was introduced by assigning red and green keys to red and green lights. A marked benefit due to color correspondence was observed when performance was compared to control conditions where the assignment was reversed. Since color is not an intrinsic physical property of the action of pressing a key (in the sense that color cannot contribute to the specification of the spatial and temporal parameters required to guide the action), the correspondence of the color relationship among the stimuli and the responses (i.e., the percept and act codes involved) must be effective at the symbolic level. It is only at the level of the symbolic representation of an action's environmental consequences that the color of the response key can come into play. The fact that compatibility at this level is effective thus emphasizes the heuristic value of an approach that regards actions as a special class of distal, environmental events and, hence, distal act codes as continuous with distal percept codes.[20]

Conclusion

Though a good deal of further research is needed in all these areas in order to clarify in more detail the nature of the processing privileges of matching, we may conclude for the time being that the common-coding approach has proven to be a useful, if not powerful, framework for the analysis and integration of the available evidence. In drawing this conclusion, it must be emphasized that the approach, as it now stands, should be considered a theoretical framework, not a theory. The same applies to its counterpart, the separate-coding approach. These two approaches should be viewed as theoretical frameworks in the sense that they offer two broad frames of reference for the conceptualization of questions and the integration of answers to these questions. As they are quite unspecific, each of them is, in principle, also capable of incorporating much of the evidence in support of the other. Therefore, we cannot expect to be able to show one of them to be true and the other to be false, as would be the case with strictly alternative, testable theories. Rather, our decision to prefer one over the other should be based on a comparison of their power to integrate as many known facts as possible in a conceptual scheme as parsimonious as possible.

When this criterion is applied, the common-coding approach looks at least like a promising candidate to win the competition. This is because the separate-coding approach will often have to invoke additional special purpose processing modules in order to give a satisfactory account of ease of matching, whereas the common-coding approach offers a framework for the coherent interpretation of both mapping and matching performance.

Several further elaborations of the framework are required. One of them refers to the learning and the developmental processes involved in acquiring the supposed common representational medium for percept and act codes. As far as the developmental issue is concerned, the main problem is how such a common system for the coding of distal events (cognitive contents) can emerge. The answer to this question will be strongly dependent on the assumed starting point for that development. Should we believe that the neonate comes with a single, undifferentiated system where afferent and efferent patterns are jointly coded and only later become differentiated, or should we believe that he/she comes with two separate and functionally distinct systems from the beginning? With the first view, the main problem is how distal event coding could develop from proximal pattern coding. With the second view, the main problem is how a level for the common coding of incommensurate entities could be developed at all.

As far as the learning issue is concerned, the problem is how, in an already existing, representational medium, percept codes can acquire the capability of specifying corresponding act codes. This problem may not be too difficult to solve in cases where act codes are directly specified by physical properties of percept codes. However, it appears to be quite complex in cases where there is no obvious physical correspondence that permits specifying the act code through the percept code by virtue of the structure of the representational medium itself. In this case it must be assumed that the correspondence is created a posteriori, i. e., that it arises in the course of practice, possibly through the creation of congruent symbolic codes in addition to the noncongruent physical components of the involved percept and act codes. To the extent that this occurs, percept codes will attain the capability to specify act codes, at least partially, and mapping will develop into matching.[21]

Footnotes

[1] Most of the discussion in these sections will concentrate on action initiation (the *what* aspect of action control) and only incidentally touch upon action execution (the *how* aspect). This is owing to the fact that most of it refers in one way or another to the choice reaction time paradigm which is virtually a method for studying action initiation under conditions of trivial demands on action execution.

[2] Intentional aspects of information processing have long been neglected in modern theories of human performance (cf., Heuer & Prinz, 1987; Neumann & Prinz, 1987), and have only recently been rediscovered (e.g., Kelso & Wallace, 1978; Shaffer, 1976, 1984). This was different at the beginning of the century. For instance, in Ach's program of the experimental analysis of the will, it was natural to understand the operations involved in reaction times as realizations of preestablished intentional states (Ach, 1905, 1910). However, intention-oriented approaches like these were then pushed aside by the more stimulus- and information-oriented thinking that has since come to dominate the field.

[3] One way to look at the concept of production rules is to consider it a modern term for what used to be called an S-R connection (or s-r) in traditional S-R theory. It should be noted that this correspondence is by no means an incidental analogy. Traditional learning theory has always drawn sharp distinctions between afferent and efferent processes, and the fact that the information-processing approach also tends to conceptualize the linkage problem in terms of separate coding, must be regarded as part of its heritage from that tradition. For the early proponents of the new approach, it was natural to view expe-

rimental instructions and their assumed functional consequences as a human equivalent of instrumental conditioning procedures and their assumed functional effects in the rat, viz., as a means of establishing rules that link "stimuli" to "responses." This heritage is still alive today.

[4] Within the separate-coding framework, there seems to be no way out of the dilemma of incommensurability. In an earlier paper, in an attempt to get rid of this dilemma, I suggested modifying the separate-coding framework to encompass also some degree of common coding of transient, stimulus-derived codes and permanent, response-related codes (Prinz, 1984). However, in the two versions of this modification, the assumed common coding did not refer to afferent and efferent codes proper. Rather, it either referred to afferent stimulus codes and afferent corollaries to efferent codes (in the case of "sensory match") or to efferent codes and efferent derivatives of afferent codes (in the case of "motor match"). Therefore, in these modified schemes, too, afferent and efferent codes are always unequal and incommensurate and, hence, separate.

[5] To be sure, this lack of explicitness poses no problem as long as the discussion is confined to the choice-reaction time paradigm itself. Given this paradigm, the sole requirement for the codes involved is that they have to be distinguishable from the other experimental alternatives, and there is no need to ask at which level of coding and elaboration that degree of distinctiveness is achieved. However, if one wants to apply these models to other fields of study the splendid isolation of within-paradigm discussion must be abandoned.

[6] Note that this view does not imply that the operation of the machinery is always consciously represented. The claim is not that the linking is accomplished between percepts and acts (i.e., the conscious contents themselves), but rather between percept codes and act codes (i.e., functional entities, some of which may be represented in consciousness under certain conditions).

[7] A metaphor from the language domain may help to shape the difference between construction rules and production rules. Construction rules are, in a way, reminiscent of the linguistic rules inherent in a language for the construction of syntactically well-formed and semantically meaningful expressions in connected discourse. The application of construction rules thus requires a syntax and a lexicon for a single mental language (rather than a dictionary for translating one language into another as is required for production rules).

[8] The question then is, of course, how the muscles ever get moved (in both the subject and the rat). Given an act code that specifies an event-to-be-effected - how is the intended event put into real action? It seems that the price we have to pay for the assumed commensurability of high-level percept and act codes is a rather deep gap between high-level (intentional) act codes and low-level (executional) motor codes. This is the puzzle hidden behind James' "presto!" His solution to the puzzle was the ideomotor principle of voluntary action which states that inherent in every mental content is a tendency to induce actions that produce corresponding states in reality (James, 1890). More recently, some theorists have raised similar claims, e.g., Bernstein (1967), Gibson (1979, 1982), Turvey and Kugler (1984), and Wolff (1987). One possible solution to the problem is a two-level model of action control with an abstract representation of distally defined goals and intentions at the high level, and a machinery for the realization of these intentions through body movements at the low level, e.g., Kelso and Wallace (1978), MacNeilage (1970), Roy (1983), Shaffer (1976, 1984). The issue of how codes that refer to distal, to-be-effected events get transformed into efferent codes that control proximal movement patterns is comparable to the reverse problem of how proximal stimulus information gets transformed into percepts of distal objects and events. It is reasonable to assume that the efferent transformation problem can be solved along similar lines as its afferent counterpart, or perhaps even in conjunction with it (cf., Bridgeman, 1983, and this volume; Jeannerod, 1983).

[9] This view is reminiscent of Welford's suggestion that, under conditions of high compatibility between stimuli and responses (see below) the "work" of translating stimuli into responses might be substantially reduced (Welford, 1960, p. 199) or even replaced by in-built connections (Welford, 1968, p. 87). A still more pertinent idea was Sanders' suggestion that, under compatible conditions, elementary stimulus codes might coincide with response codes so that no translation is required at all (Sanders, 1967). Though these notions were still couched in the terminology of the translation metaphor, they also abandon this metaphor by assuming that translation can become unnecessary when more direct linkages between stimuli and responses become established. Later, however, when the function of translation came to be considered the defining operation of one of the unavoidable stages within the functional

architecture of a linear stage processing system, this line of thinking was no longer followed up. In the linear stage approach, the issue is how translation can be facilitated, but not whether it can be dispensed with at all.

[10] Similar ideas were also proposed by Storch (1902a,b). Storch reasoned that voluntary movements were organized within the same representational medium in which perceptual space is contructed. Hence, imitating a perceived spatial pattern was equivalent to using a perceptual form for the guidance of the form of a motor act (cf., Hildebrandt, 1985).

[11] One of the differences that should certainly be kept in mind relates to the fact that the "events" and the "actions" that are considered in tracking performance are usually reduced to motions and movements of single points in space - unlike natural imitation which involves patterns of simultaneous movements of various parts of the model's and one's own body.

[12] There are some parallel findings in birds where common brain sites for song perception and song production have been described (Williams & Nottebohm, 1985). This may be an interesting parallel to the observation in humans, particularly since it is known that the species-specific shaping of bird song is strongly dependent on imitation (cf., Konishi, 1985; Marler, 1984).

[13] This is in line with some findings from selective adaptation studies in which the influence of selective auditory adaptation on the production of certain phonetic features of speech was examined. For instance, when the adapting stimuli contained voiced consonants (i.e., with short voice onset times), the onset times of utterances produced after adaptation where shifted in the voiceless direction, i.e., they were shorter than under appropriate control conditions (Cooper, 1974, 1975, 1979; Cooper, Billings, & Cole, 1976; Cooper & Lauritsen, 1974; Cooper & Nager, 1975). Unfortunately, the results of these studies have not always been convincing (cf., e.g., Summerfield, Bailey, & Ericksen, 1980) so that this potentially powerful approach is still awaiting further elaboration.

[14] An obvious implication of this account is that in tasks that involve the presentation of objects and words - either simultaneously or in brief succession - the costs and the benefits of their mutual interaction should both be more pronounced when the required responses refer to the objects (and the words are the primes or the distractors) than in the reverse case. The asymmetry of the Stroop effect provides a clear demonstration of this (as well as of the unavoidability of reading): reading incongruous color words causes only slight interference, whereas naming the colors of incongruous words leads to a marked impairment of performance. Similar results have been obtained in a variety of other tasks (e.g., Carr, McCauley, Sperber, & Parmilee, 1982; Flores d'Arcais & Schreuder, 1987; Lupker & Katz, 1982; Vanderwart, 1984; see Neumann, 1984, for a review of some of the work).

[15] It should be noted that the dividing line that separates correspondenbe between percept codes and act codes and compatibility of relationships among percept codes with relationships among act codes must sometimes be drawn quite arbitrarily. As long as we do not have precisely defined criteria of how to individuate "percepts" and "acts" and their respective codes, there will be a zone of overlap where a given task can either be classified as an example of correspondence or compatibility matching.

[16] In the functional version of the egocentric interpretation, it is supposed that environmental events (perceptual events and actions) are localized with reference to the agent. In this interpretation the agent's position marks the origin of a reference system within which objects and events are localized. Note that this reference system is different from its anatomical counterpart. An object may be seen on the right-hand side of the body, and at the same time appear in the left half of the visual field, depending on the location of the fixation point. In the same way, the left hand may operate a key on the right-hand side of the body. On the other hand, the functional egocentric localization is also independent of allocentric localization. An object on the right-hand side of the body may be right or left of another object, and a key to the right of the body may be localized to the left of another key.

[17] Note again that the compatibility of temporal relationships could equally well be described in terms of the correspondence of temporal structures of complex event configurations and actions, depending on what counts as an event or an action (cf., [15]).

[18] Once obtained, synchrony can be easily maintained for some time, and the rhythmic movement pattern can also be continued when the pacing stimulus is no longer presented. In many studies,

performance has been only studied during this continuation period (e.g., Chase, Rapin, Gilden, Sutton, & Guilfoyle, 1961; Vorberg & Hambuch, 1978, 1984; Wing, 1980; Wing & Kristofferson, 1973), whereas some other studies have looked into the free production of rhythmic patterns (e.g., Brown, 1911; Ehrlich, 1958). These studies are not considered here because our concern is with the process of synchronization itself.

[19] Similar results were obtained by Fitts and Switzer (1962) in a naming task where it was shown that the latency for naming a given letter or a numeral was dependent on the consecutiveness of the subset from which it could be drawn. For instance, the naming latency for the numeral *2* and the letter *B* was shorter when they appeared in familiar, consecutive subsets like *1, 2* or *A, B, C, D* as compared to, for example, *2, 7* or *I, B, P*. This finding also suggests that an underlying symbolic continuum may be effective in action control.

[20] Recently Proctor and Reeve (1985) have suggested a still more abstract use of the concept of symbolic compatibility which we will not follow because it provides another example of the problem that a term is first introduced for the description of certain findings and then becomes gradually used as a (pseudo-)explanatory term. In a series of precuing experiments where two different properties of possible actions (i.e., which hand and which finger to use) could be controlled by two different properties of stimulus events (e.g., which consonant and which vowel was contained in a CV pattern), they were able to show differences in the efficiency with which certain stimulus dimensions could be used to cue certain response dimensions. The authors claimed that this finding questions the motor preparation account of earlier results with this paradigm as advanced by Miller (1982) and they suggested an interpretation in terms of compatibility instead. Though the account they offer may be entirely correct, it uses the term "compatibility" in a rather unusual way. In the usual understanding the term denotes the degree to which perception-related and action-related properties are commensurate and/or resemble each other so that actions and events can be coded together. In their approach, however, the term denotes the ease with which arbitrary perceptual dimensions can get control over arbitrary action dimensions. In this account the critical factor is dimensional associability rather than commensurability. Such dimensional associations can be created irrespective of the nature of the dimensions involved and irrespective of the sign or the correlations, whereas commensurability requires dimensional similarity or, at least, equivalence.

[21] The effect of practice on choice reaction times (e.g., Mowbray, 1960; Mowbray & Rhoades, 1959) may be viewed as an illustration of this type of explanation - and, at the same time, as another illustration of the supposedly continuous nature underlying binary contrasts of theoretical terms like construction vs. reproduction, or, mapping vs. matching tasks, etc.

References

Ach, N. (1905). *Über die Willenstätigkeit und das Denken*. Göttingen: Vandenhoeck & Ruprecht.

Ach, N. (1910). *Über den Willensakt und das Temperament*. Leipzig: Quelle & Meyer.

Allport, D.A. (1979). Word recognition in reading. In P.A. Kohlers, M.E. Wrollstad, & H. Bouma (Eds.), *Processing of visible language 1* (pp. 227-257). New York: Plenum.

Allport, D.A., Antonis, B., & Reynolds, P. (1972). On the division of attention: a disproof of the single-channel hypothesis. *Quarterly Journal of Experimental Psychology, 24*, 225-235.

Anzola, G.P., Bertolini, G. Buchtel, H.A., & Rizzolatti, G. (1977). Spatial compatibility and anatomical factors in simple and choice reaction time. *Neuropsychologica, 15*, 295-302.

Bandura, A. (1977). *Social learning theory*. Englewood Cliffs: Prentice-Hall.

Bandura, A., & Walters, R.H. (1963). *Social learning and personality development*. London: Holt, Rhinehart & Winston.

Benson, D.F., Sheremata, W.A., Bouchard, R., Segarra, J.M., Price, D., & Geschwind, N. (1973). Conduction aphasia. A clinicopathological study. *Archives of Neurology, 28*, 339-346.

Berlucchi, G., Crea, F., di Stefano, M., & Tassinari, G. (1977). Influence of spatial stimulus-response compatibility on reaction time of ipsilateral and contralateral hand to lateralized light stimuli. *Journal of Experimental Psychology: Human Perception and Performance, 3*, 505-517.

Bernstein, N. (1967). Some emergent problems in the regulation of motor acts. In N. Bernstein (Ed.), *The coordination and regulation of movements*. Oxford: Pergamon.

Bower, T.G.R. (1977). *A primer of infant development*. San Francisco: Freeman.

Brainard, R.W., Irby, T.S., Fitts, P.M., & Alluisi, E.A. (1962). Some variables influencing the rate of gain of information. *Journal of Experimental Psychology, 63*, 105-110.

Bridgeman, B. (1983) Mechanisms of space constancy. In A. Hein & M. Jeannerod (Eds.), *Spatially oriented behavior (pp. 263-280). New York: Springer.*

Brown, W. (1911). Temporal and accentual rhythm. *Psychological Review, 18*, 336-346.

Brown, W. (1915a). Practice in associating color-names with colors. *Psychological Review, 22*, 45-55.

Brown, W. (1915b). Practice in associating number-names with number-symbols. *Psychological Review, 22*, 77-80.

Brugmans, H.J.F.W., & Heymans, G. (1916). Versuche über Benennungs- und Lesezeiten. *Zeitschrift für Psychologie, 77*, 92-110.

Carr, T.H., McCauley, C., Sperber, R.D., & Parmilee, C.M. (1982). Words, pictures, and priming: On semantic activation, conscious identification, and the automaticity of information processing. *Journal of Experimental Psychology: Human Perception and Performance, 8*, 757-777.

Cattell, J. McK. (1885). Ueber die Zeit der Erkennung und Benennung von Schriftzeichen, Bildern und Farben. *Philosophische Studien, 2*, 635-650.

Cattell, J. McK. (1886). The time it takes to see and name objects. *Mind, 11*, 63-65.

Chase, R.A., Rapin, I., Gilden, L., Sutton, S., & Guilfoyle, G. (1961). Studies on sensory feedback. 2. Sensory feedback influences on keytapping motor tasks. *Quarterly Journal of Experimental Psychology, 13*, 153-167.

Coltheart, M., & Funnell, E. (1987). Reading and writing: One lexicon or two? In A. Allport, D. MacKay, W. Prinz, & E. Scheerer (Eds.), *Language perception and production*. London: Academic.

Conrad, R. (1962). Practice, familiarity and reading rate of words and nonsense syllables. *Quarterly Journal of Experimental Psychology, 14*, 71-76.

Cooper, W.E. (1974). Perceptuomotor adaptation to a speech feature. *Perception and Psychophysics, 16*, 229-234.

Cooper, W.E. (1975). Selective adaptation to speech. In F. Restle, R.M. Shiffrin, N.J. Castellan, H.R. Lindman, & D.B. Pisoni (Eds.), *Cognitive theory, Vol. 1* (pp. 23-54). Hillsdale, NJ: Erlbaum.

Cooper, W.E. (1979). *Speech perception and production. Studies in selective adaptation*. Norwood, NJ: Ablex.

Cooper, W.E., & Lauritsen, M.R. (1974). Feature processing in the perception and production of speech. *Nature, 252*, 121-123.

Cooper, W.E., & Nager, R.M. (1975). Perceptuo-motor adaptation to speech: an analysis of bisyllabic utterances and a neural model. *Journal of the Acoustical Society of America, 58*, 256-265.

Cooper, W.E., Billings, D., & Cole, R.A. (1976). Articulatory effects on speech perception: a second report. *Journal of Phonetics, 4*, 219-232.

Craik, K.J.W., & Vince, M.A. (1963). Psychological and physiological aspects of control mechanisms with special reference to tank gunnery. *Ergonomics, 6*, 1-33 (part 1); 419-440 (part 2).

Davis, R., Moray, N., & Treisman, A. (1961). Imitative responses and the rate of gain of information. *Quarterly Journal of Experimental Psychology, 13*, 75-89.

DeRosa, D.V., & Beckwith, M. (1971). Retrieval of information from organized memory sets. *Psychonomic Science, 23*, 177-179.

DeRosa, D.V., & Morin, R.E. (1970). Recognition reaction time for digits in consecutive and nonconsecutive memorized sets. *Journal of Experimental Psychology, 83*, 472-479.

Dunlap, K. (1910). Reaction to rhythmic stimuli, with attempt to synchronize. *Psychological Review, 17*, 399-416.

Ehrlich, S. (1958). Le méchanisme de la synchronisation sensori-motrice. Etude expérimentale. *L'Année Psychologique, 58*, 7-23.

Eidelberg, L. (1929). Experimenteller Beitrag zum Mechanismus der Imitationsbewegung. *Jahresbücher für Psychiatrie und Neurologie, 46*, 170-173.

Ewert, O.M. (1983). Eine historische Nachbemerkung zu Neuberger, Merz und Selg: Imitation bei Neugeborenen - Eine kontroverse Befundlage. *Zeitschrift für Entwicklungspsychologie und Pädagogische Psychologie, 15*, 277-279.

Fitts, P.M., & Deininger, R.L. (1954). S-R compatibility: correspondence among paired elements within stimulus and response codes. *Journal of Experimental Psychology, 48*, 483-492.

Fitts, P.M., & Posner, M.I. (1967). *Human performance*. Belmont, CA: Brooks & Cole.

Fitts, P.M., & Seeger, C.M. (1953). S-R compatibility: spatial characteristics of stimulus and response codes. *Journal of Experimental Psychology, 46*, 199-210.
Fitts, P.M., & Switzer, G. (1962). Cognitive aspects of information processing: 1. The familiarity of S-R sets and subsets. *Journal of Experimental Psychology, 63*, 321-329.
Flores d'Arcais, G.B., & Schreuder, R. (1987). Semantic activation during object naming. *Psychological Research, 49*, 153-159.
Fowler, C.A., & Turvey, M.T. (1982). Observational perspective and descriptive level in perceiving and acting. In W.B. Weimer & D.S. Palermo (Eds.), *Cognition and the symbolic processes* (Vol. 2, pp. 1-19). Hillsdale, NJ: Erlbaum.
Fraisse, P. (1956). *Les structures rhythmiques: étude psychologique*. Louvain: Publications Universitaires de Louvain.
Fraisse, P. (1966). L'anticipation de stimulus rhythmiques. Vitesse d'établissement et précision de la synchronisation. *L'Année Psychologique, 66*, 15-36.
Fraisse, P. (1967). Latency of different verbal responses to the same stimulus. *Quarterly Journal of Experimental Psychology, 19*, 353-355.
Fraisse, P. (1968). Motor and verbal reaction times to words and drawings. *Psychonomic Science, 12*, 235-236.
Fraisse, P. (1969). Why is naming longer than reading? *Acta Psychologica, 30*, 96-103.
Fraisse, P. (1980). Les synchronisations sensori-motrices aux rhythmes. In J. Requin (Ed.), *Anticipation et comportement* (pp. 233-257). Paris: CNRS.
Fraisse, P., & Ehrlich, S. (1955). Note sur la possibilité de syncoper en fonction du tempo d'une cadence. *L'Année Psychologique, 55*, 61-65.
Fraisse, P., Paillard, J., & Oléron, G. (1958). Sur les repères sensoriels qui permettent de controler les mouvements d'accompagnement de stimuli périodiques. *L'Année Psychologique, 58*, 321-338.
Fraisse, P., Lanati, L., Regnier, J., & Wahl, M. (1965). Le tempes de réaction verbale. 2. Résponses spécifiques et catégorielles. *L'Année Psychologique, 65*, 27-32.
Freeman, R.B., Jr. (1984). The apraxias, purposeful motor behavior, and left hemisphere function. In W. Prinz & A.F. Sanders (Eds.), *Cognition and motor processes* (pp. 29-50). Berlin, Heidelberg, New York: Springer.
Geschwind, N. (1974). *Selected papers on language and the brain*. Dordrecht: Reidel.
Geschwind, N., Quadfasel, F.A., & Segarra, J.M. (1968). Isolation of the speech area. *Neuropsychologica, 6*, 327-340.
Gewirtz, J.L., & Stingle, K.G. (1968). Learning of generalized imitation as the basis for identification. *Psychological Review*, 75, 374-397.
Gibson, J.J. (1979). *The ecological approach to visual perception*. Boston: Houghton Mifflin.
Gibson, J.J. (1982). Notes on action. In E. Reed & R. Jones (Eds.), *Reasons for realism: Selected essays of James J. Gibson*. Hillsdale, NJ: Erlbaum.
Glushko, R.J. (1979). The organization and activation of orthographic knowledge in reading aloud. *Journal of Experimental Psychology: Human Perception and Performance, 5*, 674-691.
Glushko, R.J. (1981). Principles for pronouncing print: the psychology of phonography. In A.M. Lesgold & C.A. Perfetti (Eds.), *Interactive processes in reading*. Hillsdale, NJ: Erlbaum.
Goldstein, K. (1917). *Die transkortikalen Aphasien*. Jena: Fischer.
Goldstein, K. (1948). *Language and language disturbances*. New York: Grune & Stratton.
Gordon, P.C., & Meyer, D.E. (1984). Perceptual-motor processing of phonetic features in speech. *Journal of Experimental Psychology: Human Perception and Performance, 10*, 153-178.
Griew, S. (1959). Complexity of response and time of initiating responses in relation to age. *American Journal of Psychology, 72*, 83-88.
Griew, S. (1964). Age, information transmission and the positional relationship between signals and responses in the performance of a choice task. *Ergonomics*, 7, 267-277.
Hedge, A., & Marsh, N.W.A. (1975). The effect of irrelevant spatial correspondences on two-choice response-time. *Acta Psychologica, 39*, 427-439.
Henderson, L. (1982). *Orthography and word recognition in reading*. London: Academic.
Henderson, L. (1985). Issues in the modelling of pronunciation assembly in normal reading. In K.E. Patterson, J.C. Marshall, & M. Coltheart (Eds.), *Surface dyslexia. Neuropsychological and cognitive studies of phonological reading* (pp. 459-508). London: Erlbaum.
Heuer, H., & Prinz, W. (1987). Initiierung und Steuerung von Handlungen und Bewegungen. In M. Amelang (Ed.), *Bericht über den 35. Kongreß der Deutschen Gesellschaft für Psychologie in Heidelberg, 1986*. Göttingen: Hogrefe.
Hick, E. (1954). On the rate of gain of information. *Quarterly Journal of Psychology, 4*, 11-26.

Hildebrandt, H. (1985). *Ideomotorik: Ein neues Paradigma für ein altes Problem?* (Report No. 65/1985 of the Research Group on Perception and Action). Bielefeld: ZiF.

Hyman, R. (1953). Stimulus information as a determinant of reaction time. *Journal of Experimental Psychology, 45*, 188-196.

Jacobson, S.W. (1979). Matching behavior in the young infant. *Child Development, 50*, 425-430.

James, W. (1890). *The principles of psychology* (Vol. 2). New York: Holt.

Jeannerod, M. (1983). How do we direct our actions in space? In A. Hein & M. Jeannerod (Eds.), *Spatially oriented behavior* (pp. 1-14). Berlin, Heidelberg, New York: Springer.

Kalovich, R.S., & Graham, J.T. (1966). Effects of pure tone synchronous and delayed auditory feedback on key tapping performance to a programmed visual stimulus. *Journal of Speech and Hearing Research, 9*, 596-603.

Karlovich, R.S., & Graham, J.T. (1968). Auditorily paced key tapping performance during synchronous, decreased, and delayed visual feedback. *Perceptual and Motor Skills, 26*, 731-743.

Kaufman, H., & Levy, R.M. (1966). A further test of Hick's law with unequally likely alternatives. *Perceptual and Motor Skills, 22*, 967-970.

Keele, S.W. (1986). Motor control. In K.R. Boff, L. Kaufman, & J.B. Thomas (Eds.), *Handbook of perception and human performance* (Vol. 2, pp. 301-306). New York: Wiley.

Keele, S.W. (1987). Sequencing and timing in skilled perception and action: an overview. In A. Allport, D. MacKay, W. Prinz, & E. Scheerer (Eds.), *Language perception and production*. London: Academic.

Kelso, J.A.S., & Wallace, S.A. (1978). Conscious mechanisms in movement. In G.E. Stelmach (Ed.), *Information processing in motor control and learning* (pp. 79-116). New York: Academic.

Klemmer, E.T. (1967). Sequences of responses to signals encoded in time only. *Acta Psychologica, 27*, 197-203.

Koffka, K. (1925). *Die Grundlagen der psychischen Entwicklung*. Osterwieck: Zickfeldt.

Konishi, M. (1985). Birdsong: From behavior to neuron. *Annual Review of Neurosciences, 8*, 125-170.

Kozhevnikova, V.A., & Chistovitch, L.A. (1965). *Speech, articulation and perception* (Joint Publications Research Service Trans.) Washington: JPRS 30543.

Ladavas, E., & Moscowitch, M. (1984). Must egocentric and environmental frames of reference be aligned to produce spatial S-R compatibility effects? *Journal of Experimental Psychology, 10*, 205-215.

Lamb, J., & Kaufman, H. (1965). Information transmission with unequally likely alternatives. *Perceptual and Motor Skills, 21*, 255-259.

Lehmkuhl, G., Poeck, K., & Willmes, K. (1983). Ideomotor apraxia and aphasia: an examination of types and manifestations of apraxic symptoms. *Neuropsychologica, 21*, 199-212.

LeMay, R.P., & Simon, J.R. (1969). Temporal and symbolic S-R compatibility in a sequential information-processing task. *Journal of Experimental Psychology, 80*, 558-560.

Leonard, J.A. (1959). Tactual choice reactions: 1. *Quarterly Journal of Experimental Psychology, 11*, 76-83.

Liberman, A.M., Cooper, F.S., Shankweiler, D.P., & Studdert-Kennedy, M. (1967). Perception of the speech code. *Psychological Review, 74*, 431-461.

Loveless, N.E. (1962). Direction-of-motion stereotypes: a review. *Ergonomics, 5*, 357-383.

Lupker, S.J., & Katz, A.N. (1982). Can automatic picture processing influence word judgements? *Journal of Experimental Psychology: Learning, Memory and Cognition, 8*, 418-434.

MacKay, D.G. (1987a). *The organization of perception and action: a theory for language and other cognitive skills*. New York: Springer.

MacKay, D.G. (1987b). Constraints on theories of sequencing and timing in the perception and production of speech and other complex skills. In A. Allport, D.G. MacKay, W. Prinz, & E. Scheerer (Eds.), *Language perception and production*. London: Academic.

MacKay, D.G., Allport, A., Prinz, W., & Scheerer, E. (1987). Relationships and modules within language perception-production: an introduction. In A. Allport, D.G. MacKay, W. Prinz, & E. Scheerer (Eds.), *Language perception and production*. London: Academic.

MacNeilage, P.F. (1970). Motor control of serial ordering of speech. *Psychological Review, 77*, 182-196.

MacNeilage, P.F. (1978). Is the speaker-hearer a special hearer? In H.R. Pick & E. Saltzman (Eds.), *Modes of perceiving and processing information* (pp. 53-66). Hillsdale, NJ: Erlbaum.

Marler, P. (1984). Song learning: innate species differences in the learning process. In P. Marler & H.S. Terrace (Eds.), *The biology of learning* (pp. 289-309). Berlin, Heidelberg, New York: Springer.

Marshall, J.C., & Newcombe, F. (1973). Patterns of paralexia: a psycholinguistic approach. *Journal of Psycholinguistic Research, 2*, 175-199.

Marslen-Wilson, W. (1973). Linguistic structure and speech shadowing at very short latencies. *Nature, 244*, 522-523.

McLeod, P., & Posner, M.I. (1984). Privileged loops from percept to act. In H. Bouma & D.G. Bouwhuis (Eds.), *Attention and performance X. Control of language processes* (pp. 55-66). London: Erlbaum.

Meltzoff, A.N. & Moore, M.K. (1977). Imitation of facial and manual gestures by human neonates. *Science, 198*, 75-78.

Meltzoff, A.N., & Moore, M.K. (1983a). Newborn infants imitate adult facial gestures. *Child Development, 54*, 702-709.

Meltzoff, A.N., & Moore, M.K. (1983b). The origins of imitation in infancy: paradigm, phenomena, and theories. *Advances in Infancy Research, 2*, 266-288.

Meltzoff, A.N., & Moore, M.K. (1985). Cognitive foundations and social functions of imitation and intermodal representation in infancy. In J. Mehler & R. Fox (Eds.), *Neonate cognition: beyond the blooming buzzing confusion* (pp. 139-156). Hillsdale, NJ: Erlbaum.

Meyer, D.E., & Gordon, P.C. (1984). Dependencies between rapid speech perception and production: Evidence for a shared sensorimotor voicing mechanism. In H. Bouma & D.G. Bouwhuis (Eds.), *Attention and performance X* (pp. 365-377). Hillsdale, NJ: Erlbaum.

Meyer, D.E., Schwaneveldt, R.W., & Ruddy, M.G. (1975). Loci of contextual effects on visual word recognition. In P.M.A. Rabbitt & S. Dornic (Eds.), *Attention and performance V* (pp. 98-118). London: Academic.

Miller, J. (1982). Discrete versus continuous stage models of human information processing: in search of partial output. *Journal of Experimental Psychology: Human Perception and Performance, 8*, 273-296.

Mitchell, M.J.H., & Vince, M.A. (1951). The direction of movement of machine controls. *Quarterly Journal of Experimental Psychology, 3*, 24-35.

Monsell, S. (1987). On the relation between lexical input and output pathways for speech. In A. Allport, D.G. MacKay, W.Prinz, & E. Scheerer (Eds.), *Language perception and production*. London: Academic.

Morin, R.E., Konick, A., Troxell, N., & McPherson, S. (1965). Information and reaction time for "naming" responses. *Journal of Experimental Psychology, 70*, 309-314.

Morin, R.E., DeRosa, D.V., & Shultz, V. (1967). Recognition memory and reaction time. *Acta Psychologica, 27*, 298-305.

Mounoud, P. (1981). Cognitive development: construction of new structures or construction of internal organizations. In I.E. Sigel, D.M. Brodzinsky, & R.M. Golinkoff (Eds.), *New directions in piagetian theory and practice* (pp. 99-114). Hillsdale, NJ: Erlbaum.

Mounoud, P. (1983). L'évolution des conduites de préhension comme illustration d'un modèle de développement. In S. de Schonen (Ed.), *Le développement dans la première année*. Paris: Presses Universitaires de France.

Mowbray, G.H. (1960). Choice reaction times for skilled responses. *Quarterly Journal of Experimental Psychology, 12*, 193-202.

Mowbray, G.H., & Rhoades, M.V. (1959). On the reduction of choice reaction times with practice. *Quarterly Journal of Experimental Psychology, 11*, 16-23.

Neisser, U. (1985). The role of invariant structures in the control of movement. In M. Frese & J. Sabini (Eds.), *Goal directed behavior: the concept of action in psychology* (pp. 97-109). Hillsdale, NJ: Erlbaum.

Neuberger, H., Merz, J., & Selg, H. (1983). Imitation bei Neugeborenen - Eine kontroverse Befundlage. *Zeitschrift für Entwicklungspsychologie und Pädagogische Psychologie, 15*, 267-276.

Neumann, O. (1984). Automatic processing: A review of recent findings and a plea for an old theory. In W. Prinz & A.F. Sanders (Eds.), *Cognition and motor processes* (pp. 255-294). Berlin, Heidelberg, New York: Springer.

Neumann, O., & Prinz, W. (1987). *Diesseits des Rubikon: Kognitive Antezedentien von Willkürhandlungen* (Report No. 99 of the Research Group on Perception and Action). Bielefeld: ZiF.

Nicoletti, R., Anzola, G.P., Luppino, G., Rizzolatti, G., & Umiltà, C. (1982). Spatial compatibility on the same side of the body midline. *Journal of Experimental Psychology: Human Perception and Performance, 8*, 664-673.

Nicoletti, R., Umiltà, C., & Ladavas, E. (1984). Compatibility due to the coding of the relative position of the effectors. *Acta Psychologica, 57*, 133-143.

Ojemann, G.A. (1983). Brain organization for language from the perspective of electrical stimulation mapping. *The Behavioral and Brain Sciences, 2*, 189-206.

Ojemann, G.A., & Mateer, C. (1979). Human language cortex: localization of memory, syntax, and sequential motor-phoneme identification systems. *Science, 205*, 1401-1403.

Oldfield, R.C. (1966). Things, words, and the brain. *Quarterly Journal of Experimental Psychology, 18*, 340-353.

Oldfield, R.C., & Wingfield, A. (1965). Response latencies in naming objects. *Quarterly Journal of Experimental Psychology, 17*, 273-281.

Oléron, G. (1961). Attitude et utilisation des repères dans la synchronisation de stimuli périodiques. *L'Année Psychologique, 61*, 59-78.

Ostry, D.J. (1980). Execution-time movement control. In G.E. Stelmach & J. Requin (Eds.), *Tutorials in motor behavior* (pp. 457-468). Amsterdam: North-Holland.

Patterson, K.E., & Marcel, A.J. (1977). Aphasia, dyslexia and the phonological coding of written words. *Quarterly Journal of Experimental Psychology, 29*, 307-318.

Patterson, K.E., & Morton, J. (1985). From orthography to phonology: an attempt at an old interpretation. In K.E. Patterson, J.C. Marshall, & M. Coltheart (Eds.), *Surfaced dyslexia. Neuropsychological and cognitive studies of phonological reading* (pp. 335-359). London: Erlbaum.

Piaget, J. (1946). *La formation du symbole chez l'enfant*. Paris: Delachaux & Niestlé.

Pickett, J.M. (1985). Shadows, echoes, and auditory analysis of speech. *Speech Communication, 4*, 19-30.

Pierce, J.R., & Karlin, J.E. (1957). Reading rates and the information rate of human channel. *Bell System Technical Journal, 36*, 497-526.

Pollack, I. (1963). Speed of classification of words into superordinate categories. *Journal of Verbal Learning and Verbal Behavior, 2*, 159-165.

Porter, J.R. (1987). What is the relation between speech production and speech perception? In A. Allport, D.G. MacKay, W.Prinz, & E. Scheerer (Eds.), *Language perception and production*. London: Academic.

Porter, R.J., & Castellanos, F.X. (1980). Speech-production measures of speech perception: Rapid shadowing of VCV syllables. *Journal of the Acoustical Society of America, 67*, 1349-1356.

Porter, R.J., & Lubker, J.F. (1980). Rapid reproduction of vowel-vowel sequences: Evidence for a fast and direct acoustic-motoric linkage in speech. *Journal of Speech and Hearing Research, 23*, 593-602.

Posner, M.I. (1978). *Chronometric explorations of mind*. Hillsdale, NJ: Erlbaum.

Posner, M.I., & Snyder, C.R.R. (1975). Attention and cognitive control. In R.L. Solso (Ed.), *Information processing and cognition: the Loyola Symposium* (pp. 55-85). Hillsdale, NJ: Erlbaum.

Poulton, E.C. (1974). *Tracking skill and manual control*. New York: Academic.

Prinz, W. (1983). *Wahrnehmung und Tätigkeitssteuerung*. Berlin, Heidelberg, New York: Springer.

Prinz, W. (1984). Modes of linkage between perception and action. In W. Prinz & A.F. Sanders (Eds.), *Cognition and motor processes* (pp. 185-194). Berlin, Heidelberg, New York: Springer.

Prinz, W. (1987). Ideo-motor action. In H. Heuer & A.F. Sanders (Eds.), *Perspectives on perception and action*. Hillsdale, NJ: Erlbaum.

Proctor, R.W., & Reeve, T.G. (1985). Compatibility effects in the assignment of symbolic stimuli to discrete finger responses. *Journal of Experimental Psychology: human Perception and Performance, 11*, 623-639.

Reeve, T.G., & Proctor, R.W. (1984). On the advance preparation of discrete finger responses. *Journal of Experimental Psychology: Human Perception and Performance, 10*, 541-553.

Riggio, L., Gawryszewski, L., & Umiltà, C. (1986). What is crossed in crossed-hand effects? *Acta Psychologica, 62*, 89-100.

Roy, E.A. (1983). Neuropsychological perspectives on apraxia and related action disorders. In R.A. Magill (Ed.), *Memory and control of action* (pp. 293-320). Amsterdam: North-Holland.

Sanders, A.F. (1967). Some aspects of reaction processes. *Acta Psychologica, 27*, 115-130.

Sanders, A.F. (1971). *Psychologie der Informationsverarbeitung*. Bern: Huber.

Sanders, A.F. (1979). Performance theory. In J.A. Michon, E.G.J. Eijkman, & L.F.W. de Klerk (Eds.), *Handbook of psychonomics* (Vol. 2, pp. 409-468). Amsterdam: North-Holland.

Sanders, A.F. (1980). Stage analysis of reaction processes. In G.E. Stelmach & J. Requin (Eds.), *Tutorials in motor behavior* (pp. 331-354). Amsterdam: North-Holland.

Saslow, M.G. (1958). Reaction time to consonant-vowel syllables in ensembles of various sizes. *MIT Quarterly Progress Report, 50*, 143-144.

Scheerer, E., & Schönpflug, U. (1984). Nachahmung. In J. Ritter & K. Gründer (Eds.), *Historisches Wörterbuch der Philosophie* (Vol. 6). Basel: Schwabe.

Schuler, A.L. (1979). Echolalia: Issues and clinical applications. *Journal of Speech and Hearing Disorders, 44*, 411-434.

Shaffer, L.H. (1973). Latency mechanisms in transcription. In S. Kornblum (Ed.), *Attention and performance IV* (pp. 435-446). New York: Academic.
Shaffer, L.H. (1975). Multiple attention in continuous verbal tasks. In P.M.A. Rabbitt & S. Dornic (Eds.), *Attention and performance V* (pp. 157-167). London: Academic.
Shaffer, L.H. (1976). Intention and performance. *Psychological Review, 83*, 375-393.
Shaffer, L.H. (1984). Motor programming in language production - a tutorial review. In H. Bouma & D.G. Bouwhuis (Eds.), *Attention and performance X* (pp. 17-41). London: Erlbaum.
Shaffer, L.H. (1985). Timing and action. In J.A. Michon & J.L. Jackson (Eds.), *Time, mind and behavior* (pp. 226-241). Berlin, Heidelberg, New York: Springer.
Shallice, T., & McCarthy, R. (1985). Phonological reading: From patterns of impairment to possible procedures. In K.E. Patterson, J.C. Marshall, & M. Coltheart (Eds.), *Surfaced dyslexia. Neuropsychological and cognitive studies of phonological reading* (pp. 361-397). London: Erlbaum.
Simon, C.W., & Fitts, P.M. (1952). The effect of age and experience on direction of movement stereotypes. *American Psychologist, 7*, 391-392.
Simon, J.R. (1969). Reactions towards the source of stimulation. *Journal of Experimental Psychology, 81*, 174-176.
Simon, J.R., & Craft, J.L. (1970). Effects of an irrelevant auditory stimulus on visual choice reaction time. *Journal of Experimental Psychology, 86*, 272-274.
Simon, J.R., Hinrichs, J.V., & Craft, J.L. (1970). Auditory S-R compatibility: reaction time as a function of ear-hand correspondence and ear-response-location correspondence. *Journal of Experimental Psychology, 86*, 97-102.
Simon, J.R., Sly, P.E., & Vilapakkam, S. (1981). Effect of compatibility of S-R mapping on reaction toward the stimulus source. *Acta Psychologica, 47*, 63-81.
Smith, E.E. (1968). Choice reaction time: an analysis of the major theoretical positions. *Psychological Bulletin, 69*, 77-110.
Smith, G.A. (1980). Models of choice reaction time. In A.T. Welford (Ed.), *Reaction times* (pp. 173-214). London: Academic.
Stengel, E., Vienna, M.D., & Edin, L.R.C.P. (1947). A clinical and psychological study of echo-reactions. *Journal of Mental Science, 93*, 598-612.
Sternberg, S. (1969a). The discovery of processing stages: extensions of Donder's method. *Acta Psychologica, 30*, 276-315.
Sternberg, S. (1969b). Memory-scanning: Mental processes revealed by reaction time experiments. *American Scientist, 57*, 421-457.
Sternberg, S., Monsell, S., Knoll, R.L., & Wright, C.E. (1978). The latency and duration of rapid movement sequences: Comparison of speech and typewriting. In G.E. Stelmach (Ed.), *Information processing in motor control and learning* (pp. 118-152). New York: Academic.
Storch, E. (1902a). Der Mechanismus der Willkürbewegungen. *Centralblatt der Nervenheilkunde und Psychiatrie, 22*, 601-620.
Storch, E. (1902b). *Versuch einer psychophysiologischen Darstellung des Bewußtseins, zugleich ein Beitrag zur Lehre von der Function der Großhirnrinde*. Berlin: Karger.
Strub, R.L., & Gardner, H. (1974). The repetition defect in conduction aphasia: mnestic or linguistic? *Brain and Language, 1*, 241-255.
Studdert-Kennedy, M. (1987). The phoneme as a perceptuomotor structure. In A. Allport, D.G. MacKay, W. Prinz, & E. Scheerer (Eds.), *Language perception and production*. London: Academic.
Summerfield, Q., Bailey, P.J., & Ericksen, D. (1980). A note on perceptuomotor adaptation of speech. *Journal of Phonetics, 8*, 491-499.
Ten Hoopen, G., Akerboom, S., & Raaymakers, E. (1982). Vibro-textual choice reaction time, tactile receptor systems and ideomotor compatibility. *Acta Psychologica, 82*, 143-157.
Turvey, M.T. (1977). Preliminaries to a theory of action with reference to vision. In R. Shaw & J. Bransford (Eds.), *Perceiving, acting, and knowing. Toward an ecological psychology*. Hillsdale, NJ: Erlbaum.
Turvey, M.T., & Kugler, P.N. (1984). An ecological approach to perception and action. In H.T.A. Whiting (Ed.), *Human motor actions. Bernstein reassessed* (pp. 373-412). Amsterdam: North-Holland.
Umiltà, C., & Liotti, M. (1987). Egocentric and relative spatial codes in S-R compatibility. *Psychological Research, 49*, 81-90.
Vanderwart, M. (1984). Priming by pictures in lexical decisions. *Journal of Verbal Learning and Verbal Behavior, 23*, 67-83.
Van Galen, G., & Wing, A.M. (1984). The sequencing of movements. In M.M. Smith & A.M. Wing (Eds.), *The psychology of human movement* (pp. 153-181). London: Academic.

Vinter, A. (1985). La capacité d'imitation à la naissance: elle existe, mais que signifie-t-elle? *Canadian Journal of Psychology, 39*, 16-33.

Virzi, R.A., & Egeth, H.E. (1985). Toward a translational model of Stroop interference. *Memory and Cognition, 13*, 304-319.

Von Hofsten, C. (1985). Perception and action. In M. Frese & J. Sabini (Eds.), *Goal directed behavior: The concept of action in psychology* (pp. 80-96). Hillsdale, NJ: Erlbaum.

Vorberg, D., & Hambuch, R. (1978). On the temporal control of rhythmic performance. In J. Requin (Ed.), *Attention and performance VII* (pp. 535-555). Hillsdale, NJ: Erlbaum.

Vorberg, G., & Hambuch, R. (1984). Timing of two-handed rhythmic performance. *Annals of the New York Academy of Sciences, 4*, 390-406.

Wallace, R.J. (1971). S-R compatibility and the idea of a response code. *Journal of Experimental Psychology, 88*, 354-360.

Wallace, R.J. (1972). Spatial S-R compatibility effects involving kinesthetic cues. *Journal of Experimental Psychology, 93*, 163-168.

Welford, A.T. (1960). The measurement of sensory-motor performance: survey and reappraisal of twelve years' progress. *Ergonomics, 3*, 189-230.

Welford, A.T. (1968). *Fundamentals of skill*. London: Methuen.

Welford, A.T. (1980). *Reaction times*. London: Academic.

Wernicke, C. (1874). *Der aphasische Symptomkomplex*. Breslau: Cohn & Weigert.

Whitaker, H. (1976). A case of the isolation of the language function. In H. Whitaker & H.A. Whitaker (Eds.), *Studies in neurolinguistics* (Vol. 2, pp. 1-58). New York: Academic.

Whiting, H.T.A. (1987). Imitation and the learning of complex cyclic actions. In O.G. Meijer & K. Roth (Eds.), *Complex motor behavior. The motor system vs. action systems controversy*. Amsterdam: North-Holland.

Whiting, H.T.A., Bijlard, M.J., & den Brinker, B.P.L.M. (1987). The effect of the availability of a dynamic model on the acquisition of a complex cyclical action. *Quarterly Journal of Experimental Psychology, 39*, 43-59.

Williams, H., & Nottebohm, F. (1985). Auditory responses in avian vocal motor neurons: a motor theory for song perception in birds. *Science, 229*, 279-282.

Williams, J.G. (1985). *Movement imitation: some fundamental processes*. Doctoral Dissertation, University of London.

Williams, J.G. (1986). *Imitative processes and motor skill learning*. Paper presented at a workshop on problems of movement imitation, University of Bielefeld, December 5-7.

Wing, A.M. (1980) The long and the short of timing in response sequences. In G.E. Stelmach & J. Requin (Eds.), *Tutorials in motor behavior* (pp. 469-486). Amsterdam: North-Holland.

Wing, A.M., & Kristofferson, A.B. (1973). Response delays in the timing of discrete responses. *Perception and Psychophysics, 14*, 5-12.

Wolff, P. (1987). Perceptual learning by saccades: A cognitive approach. In H. Heuer & A.F. Sanders (Eds.), *Perspectives on perception and action* (pp. 249-271). Hillsdale, NJ: Erlbaum.

Visual Information Processing and Selection

A.H.C. VAN DER HEIJDEN

The light of the body is the eye: if therefore thine eye be single, thy whole body shall be full of light. But if thine eye be evil, thy whole body shall be full of darkness. If therefore the light that is in thee be darkness, how great is that darkness.

(Sermon on the Mount; Matthew, 6: 22-23)

CONTENTS

Introduction

The information-processing approach to perception and cognition tries to infer the internal structure and functioning of a behaving organism from the overt behaviour of that organism, so that it becomes possible to explain the organism's behaviour in terms of its internal structure and functioning. Such an endeavour to learn about what is not observed from what is observed is not unique for the information-processing approach but is a characteristic of many other sciences (see for example, Manicas & Secord, 1983). Einstein and Infeld (1938), for instance, compare a physicist's attempts to understand the physical reality with the activities of a man who tries to understand the internal mechanism of a closed watch; he sees the face and the moving hands and even hears the ticks, but he cannot open the watch. If he is inventive and smart, he can develop a model of a possible internal mechanism responsible for what he sees and hears, i.e., he can invent a set of interrelated concepts and principles that explain what is observed.

The concepts and principles forming the model of the mechanism of the watch or the model of the internal structure and functioning of the behaving organism are not only, and often not even mainly, determined by the behaviour that was observed. The observers themselves contribute substantially in developing the form and content of the model. Their theoretical likes and dislikes, their sense of what is elegant and their sense of parsimony and beauty function to shape and phrase the model. What they know about other mechanisms and other functioning structures forms, as a priori notions, the main content of the model.

Human information-processing research within psychology has been very strongly dominated by one a priori notion in the past 30 years: the notion that the human information processor has a limited capacity for processing information. Limited capacity was and still generally is seen as simply given; a basic fact of

Relationships Between Perception and Action
Edited by O. Neumann and W. Prinz

nature. What had to be explained is how, in spite of this handicap, people could still manage reasonably well. The following quotes contain this premise and suggest solutions. The first two stem from the information *transmission* period (up to about 1960), the latter two from the later information-*processing* era.

> The premise is that the human observer, placed before an amount of information which vastly exceeds his intake capacity, must, in one way or another, try to encode this information efficiently.... (van de Geer, 1957, p. 333, translated from the Dutch)

> The view that a basic function of psychoneural activity is the economical encoding of experience ... has extraordinary generality, applying to the most complex scientific thinking as well as to the basic processes of perception and memory. We know, of course, that organisms rarely if ever encode their experience with optimal economy, but an interesting way of studying their processes is by determining how, and to what degree, they fall short of this ideal. (Attneave, 1959, pp. 87-88)

> Large though the brain is, any conceivable mechanism which could cope simultaneously with all possible states of the eye, the ear, and other receptors would probably be even larger. The workings of the nervous system then are likely to incorporate a good many devices aimed at economizing on the mechanism necessary.... Broadbent (1958) ... held that the limited capacity portion of the nervous system was preceded and protected by a selective device or filter, which would pass only some of the incoming information. (Broadbent, 1971, p. 9)

> The human organism exists in an environment containing many different sources of information. It is patently impossible for the organism to process all these sources, since it has a limited information capacity, and the amount of information available for processing is always much greater than the limited capacity. Therefore the organism must process information selectively. (Garner, 1974, pp. 23-24)

It is interesting to note that the transmission view and the processing view propose different solutions to circumvent the capacity problems. The first view emphasizes efficiency of encoding information, the second stresses adequacy of selecting information (Broadbent, 1958, marks the transition between the two views). It is, furthermore, of importance to note that both solutions together nearly completely exhaust the field of research and theorizing for information-processing psychology. Encoding (or decoding) and selection are certainly the processes which are of most importance in this branch of psychology. And because limited capacity was taken as a given fact, this notion has dominated and directed nearly all research and theorizing in information-processing psychology. Let us briefly look at what exactly this notion of limited capacity was.

It has recently become clear, especially from the work of Neumann (1985, 1987) and Allport (1980a,b), that we must distinguish sharply between two notions of "limited capacity". On the one hand, we have limited capacity as an observable given. This notion is not problematic. People cannot do everything and certainly not everything at the same time. On the other hand, we have limited capacity as an explanatory theoretical principle, i.e. a principle reflecting a property of the information-processing system, which makes it clear that people cannot do everything simultaneously.

Theoretical work in information-processing psychology has concentrated almost exclusively on limited capacity as an explanatory theoretical principle, i.e., on the idea of limited capacity as a property of the information-processing system, in terms of which limited capacity, as an observable given, can be explained. The question was not so much how these limitations in the information-processing system arose. That was taken as being obvious by almost all researchers; they were simply there. In one way or another, evolution had designed the human information-processing system in such a way that something, essential for performance, was missing. The system lacked capacity. What had to be arrived at was a description of the information-processing system with a limited capacity specifying where and what the system's capacity limitations were.

It is remarkable that in all these descriptions peripheral limitations only played a very minor role. Limitations of the sense organs (the eye has obvious limitations) and of effectors (we only have a limited number of effectors) either were not seen or were considered to be of secondary importance. For instance, it was not known or not realized that there are only 2 million (peripheral) retinal ganglion cells, the cells which form the connections between eye and cortex, but that the number of (central) neurons in the cortex that further processes visual information is hundreds of times larger (see Barlow, 1981, 1985). Information-processing psychology was concerned with a description of *central* limitations, limitations in the machinery which have nothing to do directly with the sense organs and effectors.

Broadbent (1958, 1971) postulated a central information-processing channel with a limited capacity, the "limited capacity channel". Moray (1967) thought more in terms of a limited computation capacity such as in the central processing unit of a computer. According to Norman and Bobrow (1975), it is better to think in terms of insufficient transmission channels and stores, a lack of "resources". However, for Neisser (1967), the processing limitation was in "focal attention"; an "act of focal attention" cannot handle more than one "unit" at one time (whatever "unit" and "act of focal attention" may be). Kahneman (1973) thought in terms of a limited amount of general mental energy that had to be distributed over all the central processing components. And after this, other theoreticians postulated multiple resources to bring theories and data on one line. (See Neumann, 1985, 1987, for the origin, development and decay of the concept of limited capacity.)

Many questions can be asked about this line of research. Where did these limited capacity notions come from? (Information transmission channels and computers as analogies have certainly contributed to these ideas.) Why is limited capacity so attractive as a theoretical principle? (Each observable limitation can be "explained" by a suitable postulated internal limitation. See, for example, Neumann, 1985, 1987.) What insights in human information processing has this type of research ultimately produced? (Not much. For a critical evaluation of this type of theorizing, see Neisser, 1976; Allport, 1980a,b; Claxton, 1980; Neumann, 1985, 1987). What more do we know now than 30 years ago, as a result of this capacity-inspired research, about "efficient encoding" and "adequate selection", i.e. the central topics of information-processing psychology? [Again, not much. Information-processing psychologists were not really interested in what information actually is; it was something like "the informative aspects of the environment", or "that of which there is much too much". As we shall show in the second section, ("Encoding") information-processing psychology was not interested in information encoding either: "processing, more processing, still more processing" (Neisser, 1976) aptly characterizes what was said. There was no real interest in the selection of information (see "Selection"); filters were looked for (Broadbent, 1958, 1971), as were attenuators (Treisman, 1960) that had to protect a system with limited capacity from going under.]

The last and most important question remains to be asked: the question whether it is necessary and useful to take the limited capacity notion as a point of departure for information-processing research. Limited capacity is certainly not a necessary, and probably even an incorrect, starting point.

- There is almost no undebatable evidence in visual information-processing research for limited central capacity. From much research reporting central limitations, it could be easily shown that either the experimental design was incorrect or that limited central capacity did not necessarily follow from the data. Using improved experimental designs and more critical experiments, the evidence disappeared (see, for example, Hagenaar and van der Heijden, 1986; van der Heijden, Hagenaar, & Bloem, 1984).
- There is evidence suggesting a very large central capacity. In research directly concerning the question of central processing capacity, usually more processing capacity was found than could be expected on the basis of limited capacity theories (see Schneider & Shiffrin, 1977; Shiffrin & Schneider, 1977; van der Heijden, 1981; van der Heijden, La Heij, & Boer, 1983; van der Heijden, Schreuder, Maris, & Neerincx, 1984). Over the years, limited capacity theories repeatedly had to expand their central capacity (see, for instance, the series Broadbent 1958, 1971, 1982). In these theories the limited capacity approaches the unlimited capacity (see, for example, Kahneman & Treisman, 1984).

In general, it seems that whether evidence favouring central capacity limits or something like unlimited central processing capacity shows up completely depends upon the type of experiment used to interrogate the human visual information-processing system (see, for example, Allport, 1980b). In one way or another, central capacity is an elusive theoretical construct. Proponents of limited capacity theories always obtain evidence in favour of limited capacity in their experiments (see, for example, Broadbent, 1958, 1971, 1982). Proponents of unlimited capacity theories, however, always produce evidence in favour of unlimited capacity theories (see, for example, van der Heijden, 1981; van der Heijden et al., 1983; van der Heijden, Hagenaar, et al., 1984; van der Heijden, Schreuder, et al., 1984; Hagenaar & van der Heijden, 1986). In physics, a similar situation is encountered. Whether light shows up as waves or as particles completely depends upon the kind of experiment performed. Therefore, in physics neither "light consists of waves" nor "light consists of particles" is a correct theoretical construct. In one way or another, a theory of light has to encompass both properties. In the same way, neither "limited central processing capacity" nor "unlimited central processing capacity" is correct as a theoretical construct in a theory of human visual information processing. The experimental results force one to look for a theory that can handle both properties.

Limited capacity is also not a useful starting point. From limited capacity, efficient encoding and adequate selection follow as a last desperate resort, without any indication of further properties of these processes or how they could be productively investigated. Therefore, it is better to search for another starting point (one that has nothing to do with capacity) from which properties of encoding and selection (and possibly even limited capacity as an observable given) follow in a more natural way. A starting point that, contrary to limited capacity, imposes constraints on theories about the information-processing system and on concepts such as information, encoding and selection. What is a productive starting point?

To find such a starting point it is worthwhile to return briefly to Einstein and Infeld's man who is trying to understand the mechanism of the closed watch. From the slowly moving hands and the relatively large intervals between the ticks he might surmise that the watch has only a limited capacity for running (an a priori notion derived from his own running abilities). He might then try to use this notion to explain why the long hand moves faster than the short hand, why the two hands meet at 12.00, 13.05, 14.10, etc. But it is not very likely that this theorizing will greatly further his understanding of the mechanism of the watch. Things change considerably, however, when he knows what purpose the watch serves or what problems it solves. If he knows that (a) there is a human convention to distinguish within each day and night 2x12 equally spaced hours and, within each hour, 60 equally spaced minutes; and (b) the watch is designed in such a way that it has the position of its hands in a one-to-one relationship with these hours and minutes, a much better starting point is reached. The knowledge that the mechanism of the watch solves this mapping problem introduces powerful constraints on the modelling of the mechanism. (Furthermore, it is damaging to the limited capacity notion; in principle, the watch is capable of running much faster but then it is completely useless as a watch.)

As already briefly mentioned, the limited capacity notion is partly borrowed from technomorphic models: radios, televisions, and computers. These recent man-made artefacts provide solutions for minor problems and have all types of (central) limitations. The human brain was not made recently and in its history it has been confronted with a large range of problems. The solutions to the problems before which it was and is placed have been developed, tested and modified during millions of years of evolution. To be able to understand something of these solutions, we must have an idea of the problems with which it was confronted. It was the problems which forced the solutions and it is not very likely that "limited central capacity" has been a determining factor or an important constraint.

It seems, therefore, that we can find a productive starting point in the problems which face the human information processor. Because the world poses these problems, we shall have to find the most important constraints in the problems that have to be solved in the system's interactions with the world.

Encoding

It is possible that information-processing psychology gave up the research into encoding (visual) information rather early because it appeared that neurophysiology had better techniques at its disposal for studying this subject. There were indeed some good reasons for this. Psychological research only resulted in very global characterizations of encoding with concepts for initial encoding like visual information storage (Sperling, 1967) or iconic memory (Neisser, 1967) and mechanisms for later encoding like logogens (Morton, 1969) or even recognition units (Kahneman, 1973). Allport (1980a, p. 30) aptly characterized the results of this research: "... the only way that psychologists' theories ... have succeeded in doing justice to the richness and complexity of human cognition is simply by being so general and underspecified, in process terms, as to mean almost anything". But neurophysiologists made spectacular discoveries in their attempts to look into the closed watch.

It became more and more clear that the eye was not a light detector but a detector of changes in space and time; a detector of patterns corresponding to, for animals, relevant objects and events in the world. Kuffler (1953) found that neurons in the retina of the cat (the retinal ganglion cells) reacted to light contrasts and not to absolute light intensity. Lettvin, Maturana, McCulloch, and Pitts (1959) discovered that the neurons in the eye of the frog inform the brain about (moving) contrasts, about light changes and even about something like "flying flies" (the convex edge detectors). Barlow, Hill, and Levick (1964) discovered that the retinal ganglion cells of the rabbit reacted selectively to direction of movement, speed of movement, orientation of stimuli, etc., this being independent of, for example, absolute light intensity or contrast. In their now well-known research, Hubel and Wiesel (1962, 1968) demonstrated that higher-order neurons in the visual cortex of the cat and ape primarily reacted to (moving) light or dark lines or edges with a certain orientation. Later research has increased the list of spectacular discoveries considerably, for example, with "hand-detectors" (Gross, Rocha-Miranda, & Bender, 1972) and "face-detectors" (Perret, Rolls, & Caan, 1982).

Isolated spectacular discoveries are one thing, but a general theoretical framework by which these data can provide more insight into visual perception is a totally different thing. Hubel and Wiesel (1962, 1968) originally thought that the visual system could be regarded as a strictly serial information processor; visual perception was seen as a step-by-step and hierarchical process of abstraction with a gain in meaning through loss of detail (see also Barlow, 1972). But further research showed that these ideas could not be correct. Too many data pointed to massive parallel information processing. The picture that emerged was that of a very complicated system of parallel and serial activities with considerable dialogue between the channels. (See Stone & Dreher, 1982; van Essen & Maunsell, 1983, for overviews.) Within neurophysiology there remained a lack of a good and detailed theory (see Stent, 1980; but also Barlow, 1981). Neurophysiology collected much information about the activities of individual neurons in the brain, but knowledge about what the neural networks were actually doing remained very limited.

What neurophysiologists needed was a theory which would indicate what had to be searched for in individual neuron research. An information-processing psychology working on encoding (or decoding) - a good start had been made in the 1950s (see van de Geer, 1957) - could have produced and actually ought to have produced such a theory. But this psychology was much too busy with research related to the capacity issue. Thus it was not psychology but artificial intelligence (AI) that produced such a theory. And within AI it was mainly one man, David Marr of the Massachusetts Institute of Technology who, with many brilliant insights directly referring to neurophysiological and psychophysical findings, showed neurophysiologists and psychologists a new direction by developing a computational theory of human visual perception.

Marr's (1980, 1982) point of departure is that a complex system such as the human information processor cannot be understood by a simple extrapolation from the properties of elementary components. Such a system must be analyzed and described on various levels which are virtually independent. The highest level is the theoretical level (the "computational theory"): the definition of the information-processing problem that must be solved (computed) and the specification of the relevant properties of the physical world which, introduced as constraints, make the problem solvable (see "Introduction"!). The second level is that of the methods of computation ("algorithms") that can be used to solve the actual information-processing problem. The last level is that of realization ("implementation"). This concerns the question of how an algorithm in a system (the brain or a computer, for example) is physically instantiated.

Marr (1980, 1982) does not begin with the capacity question (in fact, he complains about the limited capacity of computers which makes "real-time implementation of human-like early vision" impossible for the present) but he starts with the question: what are the problems actually solved by the brain in vision? His answer is: the construction (the computation) of efficient symbolic descriptions of the world on the basis of light patterns projected on the retina (under "Selection" we briefly return to Marr's answer). Marr distinguishes three descriptions or representations of the visual world which are created in the process of perception.

1. The "primal sketch": a description of the spatial changes of light intensity on the retina. The underlying idea here is that information about contours, and thus about objects in the environment, is encoded by more or less abrupt intensity changes in light.
2. The "21/2-D sketch": a perceiver-oriented description of properties of surfaces such as depth, contour and orientation.
3. The "3-D model": a perceiver-independent object-oriented representation of three-dimensional objects which makes recognition possible.

In his work Marr attempts to extract information about the world from the retinal images in all possible ways, without appealing to higher-order specific knowledge (such as "chairs have legs" or "hammers have handles"). The 21/2-D sketch, for example, is built up from the primal sketch by means of processes which only have a general knowledge of the outside world (see below). Depth analysis is such a process. The starting point is two primal sketches, one for each eye. Due to the different positions of the two eyes, the sketches differ somewhat: the binocular disparity. Given the primal sketches and, as constraints, the general knowledge that (a) a physical object can only be in one place at a time (each surface element receives only one disparity value); and (b) physical objects are coherent (that is, adjacent surface elements have continuously varying disparity values except at those places where real surface boundaries are passed), it appeared possible to develop algorithms which computed the surface forms and the distance of the surfaces to the perceiver (Marr & Poggio, 1979). Movement analysis is another such process. The structure of a moving object can be computed from a sequence of primal sketches, given the general knowledge that the moving object is rigid (Ullman, 1979). Other early visual processes compute properties of the 21/2-D sketch using texture, shadow, etc. (see, for example, Brady, 1981).

Along these very briefly sketched lines, Marr and his coworkers began with the development of a general and explicit theory of human visual information processing. The AI approach forced them to formulate and solve problems that psychologists and neurophysiologists, who were also working at visual information processing, had not seen because seeing (and talking) comes so naturally. In many ways the theory complemented what neurophysiologists had discovered. But (the success of) the theory also contains important indications for theoretical work in information-processing research in psychology. For encoding visual information, the most important one is: use as much "bottom up" processes as possible (with only a general knowledge of the structure of the world) and as few "top down" processes as possible (aided by specific knowledge of objects and events in the world). (See also Brady, 1981, Koenderink, 1984, and Fodor, 1983, pp. 64-86, for the importance of this strategy.) Closely related to this point is another important indication which concerns the modularity of the (human visual) information-processing system.

As already mentioned, in Marr's theory the 21/2-D sketch is built up of data from the primal sketch by processes that "know" and apply the relevant general constraints in their computations. In the theory these processes work relatively independently. Each works on its own set of data (its input representation), applies the, for its job, unique constraints (i.e. is informationally encapsulated; Fodor, 1983) and contributes its own unique result (its output representation). In this manner, visual perception becomes the result of, or consists of, the simultaneous activity of a set of more or less isolated modules that each computes its own sums. Modularity is not unique in Marr's approach but is a main characteristic of contemporary computer vision. After not too successful attempts to construct entire visual systems which mobilized knowledge at all levels as well as information specific to some domain of application, this field of research turned its focus to more narrowly defined visual abilities, i.e. to single modules concerned with problems like stereopsis from disparity, surface layout from texture, grouping from motion, etc. (see Brady, 1981).

This modularity notion was not only prompted by the failures of earlier attempts or by the programming problems facing Marr and other workers in computer vision. There is abundant evidence that one successful visual information-processing system, the human visual information-processing system, is basically modular. (See also Fodor, 1983; Marshall, 1984.) For instance, neurophysiology and anatomy provide several indications that modularity was the (efficient) solution, chosen by evolution, for its programming problems. The primary visual cortex, containing an exact map of the visual field, distributes its information over a surprisingly large number of higher-order, topographically less precise, visual areas. These modules appear to be specialized in the analysis of limited aspects of the retinal image, often only one single parameter, such as (orientation of) contours, (direction of) movement, shape, colour, location, identity, etc. (See also Zeki, 1978; Cowey, 1979; 1981; van Essen, 1979; Barlow, 1981, 1985; Stone & Dreher, 1982; van Essen and Maunsell, 1983).

But human psychophysics and neuropsychology also offer clear evidence in favour of modularity. In psychophysical demonstrations, the trick is to isolate a particular source of information and to show that a specific function is still performed. Julesz's (1971) demonstrations of stereopsis without monocular cues and Land and McCann's (1971) demonstrations of the determination of lightness from an image are the prime examples (see Brady, 1981, p. 4). According to Marshall (1984) neuropsychological data - often showing specific and isolated breakdown patterns - have never seriously challenged the modularity hypothesis.

> Severe, yet isolated aphasias are found with well-preserved non-verbal "intelligence", visuo-spatial skills, and (non-verbal) learning and memory functions ...; severe visual agnosias are found in patients without general intellectual deterioration or sensory loss and who can furthermore make excellent copies of complex geometrical drawings or freehand sketches of the scene in front of them ...; severe impairment of spatial orientation and topology can be found in cases with preserved language skills, good verbal intelligence and essentially normal figural perception ...; isolated disorders of musical expression ... and perception ... are found in patients without aphasic involvement; arithmetic skills can be grossly impaired in cases where short-term memory and language functions are essentially normal.... (Marshall, 1984, pp. 225-226)

All this evidence seems to suggest that something like Marr's algorithms are implemented in relatively isolated networks of neurons. In these modules the constraints, the general knowledge of the world needed for the appropriate computations, is then implemented in a specific pattern of connections.

It is precisely on the basis of this type of idea that recent information-processing theories again sought and found a link with the rest of the "neurosciences": neurophysiology and also AI. Starting from pseudoneural networks (modules) and their internal and external connections, a language was developed - the "new connectionism" - by which biologically plausible complex information processing can be described and simulated on computers (see, for example, Hinton & Anderson, 1981; Dennett, 1984; Rumelhart & McClelland, 1986; McClelland & Rumelhart, 1986). It specifies the mechanisms that Allport (1980a) so badly missed in psychologists' theories. In this chapter we cannot really go into this excitation-inhibition language. Just for illustrative purposes, Fig. 1 gives some ideas of this type of model. What is shown is a small part of the (four letter) word recognition model of McClelland and Rumelhart (1981).

Three levels (or modules) can be distinguished in this model: a feature level with pseudoneurons ("nodes") standing for letter features, a letter level with elements representing letters and a word level with word elements. The lines indicate activating (arrows) and inhibiting (circles) connections. In a computer simulation the "nodes" serve as counters. An activating connection increases the count and a

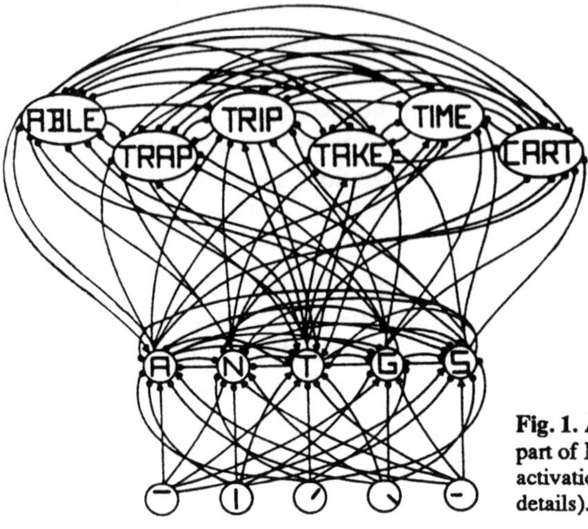

Fig. 1. An example of a connectionist model; part of McClelland and Rumelhart's interactive activation word recognition model (see text for details). (From McClelland and Rumelhart 1981)

inhibiting connection decreases it. In this model (parallel) activation at the feature level causes an activation-inhibition battle, until one of the word elements has a higher count than all the others. The word is then recognized. In this model, knowledge of words and word recognition is built in by means of a permanent wiring.

We shall briefly return to this approach under "selection". Here it suffices to say that this type of model can be best regarded as being stated at Marr's second level, i.e., the algorithmic and representational level. Theorists working with this type of model do not have the intention of saying something about how the functions modelled are really implemented in the brain, i.e. about the third, implementational, level (for a discussion of this issue see Broadbent, 1985; Rumelhart & McClelland, 1985).

Selection

By head and eye movements and adjustments of the lens, the eye has powerful means of selecting and analyzing different parts of the visual world. But apart from using external adjustments the (visual) system can also, by means of, not directly observable, internal adjustments, select relevant aspects from the surroundings. Aspects of this "central selection" (something like "looking out of the corner of the eyes") were already extensively described by James (1890), Wundt (1912), and von Helmholtz (1924). This selection possibility was rediscovered by information-processing psychology (for early overviews see Broadbent, 1958, 1971; Neisser, 1967). Inextricably interwoven with questions about the capacity of the information-processing system, central selection became one of the main topics of research and theorizing within this branch of psychology.

If we neglect the obscurities created by theories - we shall return to this later - we discover that, as far as central selection is concerned, one question was central: where, in a sequence of information representations, does the mechanism of central selection intervene? Is that relatively early, in a representation where the information is still encoded as more or less "concrete" physical characteristics (location, colour, brightness, size, shape, etc.) and where there is as yet no identified information (early selection)? Or is it relatively late, in an "abstract" representation which is derived or computed from earlier representations and where identity information (letters, numbers, which letter, vowel, etc) is present (late selection)? To use the terms mentioned under "Encoding": is selection from iconic memory or visual information storage (early selection) or from the logogen system or a system of recognition units (late selection)?

If we also neglect obscuring details of experiments, we see that the experimental literature provides a clear answer to this question. Almost all the evidence points to the fact that in typical visual information-processing tasks selection is early selection. The "partial report" task is such a task. From a briefly presented stimulus with many elements (letters and/or digits), subjects must name only those elements which meet a certain criterion. Subjects can do this very well when a concrete physical criterion is used, such as place (Sperling, 1960; Averbach & Coriell, 1961), colour (Clark, 1969; von Wright, 1968, 1972; Dick, 1969), brightness (von Wright, 1968), size (von Wright, 1968) and shape (von Wright, 1968; Turvey & Kravetz, 1970). But subjects have much more trouble with this task when a criterion derived or computed from the physical characteristics is used, such as letters (and not numbers; Sperling, 1960; von Wright, 1968, 1970), vowels (von Wright, 1970), or letters ending with an *e* (Coltheart, Lea, & Thompson, 1974). Early selection works well, and late selection is usually very difficult.

So the correct answer seems to be: early selection (see also Pashler, 1984; van der Heijden, 1986, 1987). Unfortunately, in the recent past this result was not seen as a good start for further research and theorizing, but more as the end of argumentation. The (wrong) conclusion was: the system does indeed have a limited central capacity. The grounds for this conclusion had already ripened in an early stage of theorizing. In Broadbent's (1958) influential filter model, limited central capacity and the protecting mechanism of early selection were inextricably connected. Limited central capacity required early selection. But from "limited capacity therefore early selection" it does not follow "early selection therefore limited capacity". And, early selection is what is observed; limited capacity is only an a priori notion. Even the few dissidents who rejected the limited capacity notion did not see that early selection and limited capacity are only inextricably connected in one direction. Their alternative was unlimited capacity and late selection (see Deutsch & Deutsch, 1963; Norman, 1968). Something like "unlimited" capacity (or capacity-irrelevant) and early selection could not be realized.

But why selection (and especially early selection), if the information-processing system has no capacity problems?

If there were really sufficient machinery available in the brain to perform such an analysis for every stimulus, and then to use the results to decide which should be selected, it is difficult to see why any selection at all should occur. The obvious utility of a selection system is to produce an economy in mechanism. If a complete analysis were performed even of the neglected messages, there seems no reason for selection at all. (Broadbent, 1971, p. 147)

This is the difficult question with which recent information-processing research is concerned. Treisman and her colleagues (see, for example, Treisman, 1977, 1982; Treisman, Sykes & Gelade, 1977; Treisman & Gelade, 1980; Treisman & Schmidt, 1982) and Crick (1984a, b) suggest an answer; in their views it is modularity that makes selection necessary. Modularity introduces problems:

> Most of the objects, people, and scenes we perceive produce complex, multidimensional, changing patterns of stimulation on the retina. At least some of their attributes appear to be registered by independent neural channels, specializing in different aspects, such as orientation, color, spatial frequency, brightness.... But this immediately raises the question of how the component properties are resynthesized into the correct compounds, so that we correctly see the shirt as blue and the trousers as gray, for example, rather than the reverse. (Treisman, 1977, p. 1)

> It is now well established that in the early visual system of primates there are at least ten distinct visual areas, arranged in a branching hierarchy. Different cortical areas specialise, to some extent, in different features, one responding mainly to motion, another to colour, etc. As one proceeds to areas higher in the hierarchy the mapping of the visual field onto the cortical surface tends to become more diffuse. This is not, however, how we see the world. Our inner visual picture of the external world has unity. How then does the brain put together all these different activities to produce a unified picture so that, for example, for any object the right colour is associated with the right shape? (Crick, 1984a)

The answer is that "focal attention" (Treisman & Gelade, 1980) or an "internal attentional searchlight" (Crick, 1984a, b) solves this problem.

> Stimulus locations are processed serially with focal attention. Any features which are present in the same central "fixation" of attention are combined to form a single object. Thus focal attention provides the "glue" which integrates the initially separable features in unitary objects. Once they have been correctly registered, the compound objects continue to be perceived and stored as such. (Treisman & Gelade, 1980, p. 98)

In this context "attention" refers to "selective or focused scanning of competing external stimuli, the type of attention involved in 'filtering' paradigms" (Treisman & Schmidt, 1982), i.e. attention as operating in the partial report tasks we briefly discussed. Crick (1984b) speculates on the neural machinery of focal attention.

Treisman's and Crick's theories are quite problematic, however. The basic issue is that it is far from clear whether there is indeed a *perceptual* problem, posed by modularity, to be solved by attention. What is the basis of claims like "Before any features can be consciously perceived, they must be recombined" (Treisman & Schmidt, 1982, p. 111), or "without focussed attention, features cannot be related to each other" (Treisman & Gelade, 1980, p. 98)? What leads to the suggestion that perceptual integration between different processing domains constitutes a necessary condition of perceptual awareness? (For similar ideas see also Allport, 1977, 1979; Marcel, 1983).

There is no justification for these claims. First, we do not know whether the different domains, distinguished by neurophysiologists and neuroanatomists, have anything to do with "see", "consciously perceive", "perceptual awareness" or, in Crick's terms, an "inner visual picture". Suggesting so goes far beyond what we know about the relation between brain and mind (in fact, we know nothing about this relation). Second, if these different areas have indeed to do with "see", "consciously perceive", etc., why then is the sheer presence of these domains (and possibly the assumption that the domains are in register) not a sufficient condition for a "unified picture"? In relation to a similar suggestion Cowey remarks:

> Once the visual image has been dissected and analyzed by cells in different retinotopically organized visual areas, there is no *necessity* that these cells should provide convergent input to cells in another visual area. Such an arrangement smacks of the homunculus inside the head "looking at" the combined activity of cells in a network.... (Cowey, 1981, p. 405)

Suggesting that more is needed, again goes far beyond what we know about brain and mind. Third, if the maps have indeed to do with "see", and if the sheer presence of these maps is not sufficient, how does "focal attention" then contribute to get from brain to "consciously perceive", i.e. how does focal attention solve the mind-brain problem? Why not something like "All neocortical outputs feed into action systems of the brain. The synthesis of the distributed cortical representations of the world is thus realized through the action elicited by their combinations"? (Creutzfeldt, 1979, p. 217). In short, the assumption that modularity creates problems for visual perception is based on a category error (Ryle, 1949). There is no reason to assume that modularity poses problems for perception. Evolution has solved problems, not created them.

But if there are no capacity problems and if we assume that modularity does not create problems for visual perception, why then selection and especially early selection? Something becomes clear when we begin with the correct questions: the question what the visual system is needed for and what problems it solves. This is not just "perceiving" or "seeing", as Marr more or less suggests, nor creating a consciously perceived unified picture as Crick and Treisman suggest. The most important function of the system lies in the providing of information for actions; in the steering and controlling of actions (walking, grasping, picking, etc.), in fitting the body to the world. The necessity of selection immediately follows from the difference in capacity of the central systems (no capacity problems) and the action systems or effectors (a very limited capacity). Because of the limited number of effectors and the fact that effectors generally can only perform one action at a time (see, for example, Gahery & Massion, 1981), only a limited (but for survival apparently sufficient) amount of information can and has to be used in behaviour (processing of the rest of the information is useful to be optimally prepared to react adequately and quickly to unexpected changes in the world). Thus, it is not perception but action, and not the capacity of the central information-processing system but the capacity of the body that makes selection necessary. Selection is selection for action (Allport, 1987). There is no reason to assume that the central system is limited but it must have mechanisms to limit itself in its expression in actions.

Some of these necessary limitations become clear if we take simple goal-orientated actions as the point of departure and examine these more closely. Picking an apple is a good example (see also Neumann, 1985, 1987; Allport, 1987). We can distinguish at least three necessary limitations and corresponding selection moments if we consider such a task:

- Action limitation and action selection: an orchard offers sufficient conditions for a large number of actions: mowing the lawn, chopping wood, lying down, etc. The limited capacity of the action apparatus - coordinated activity of almost all effectors is required for almost all actions - does not usually allow for more than one action at a time. Thus for picking, a limitation to pick is necessary; other actions must be temporarily excluded.
- Goal limitation and goal selection: an apple orchard offers many picking opportunities; leaves, branches and apples, for example. The limited capacity of the action apparatus - the hand can only be in one place at a time - makes restriction to one picking goal necessary and this restriction must be maintained until the action goal is reached. Therefore, for apple picking a restriction to one apple is necessary. Other goals must be temporarily excluded from influencing the action systems.
- Dimension limitation and dimension selection: apples have many properties that can control actions. The colour can be named, its location pointed at or its shape can be drawn. A limited number of these properties is important for picking. Location information must shape the distance and direction of the arm movement, form and size information must shape the precise hand movement. Adhering to these properties is necessary. Other properties, for example, the colour of the apple, must temporarily blocked from influencing the action systems (whether the apple is green or purple must not influence the performance of the action).

These restrictions and corresponding selection moments can be distinguished generally, and therefore also in typical experimental situations. In "name the colour of the square" (given a stimulus with a red circle and a blue square), (a) something must be named (and not added up, pointed at or sung); (b) it must be something about the square (and not of the circle, the table or the wall); and (c) only the colour of the square must be named (and not its position or size).

Using goal-directed behaviour as the point of departure (and the problem to be solved), a number of properties of the information-processing system and selection in a natural manner follow. We shall name a few:

- There are several selection problems to be solved. For each action, at least three selection problems (and appropriate selection mechanisms) must be distinguished: action selection, target selection and dimension selection. With one locus of selection (early or late selection) and one selection mechanism (attention?), no justice is done to the complexity and flexibility of behaviour.
- Central selection is necessary. While goal selection is (possibly) facilitated by peripheral adjustments (eye and head movements), selection of actions and dimensions will have to take place using central adjustments; the eye simply registers all the dimensions, irrespective of their relevance, and there are no peripheral means for action selection.
- Early selection is necessary. Especially in non-verbal behaviour, detailed and "concrete" visual information must shape the action or control the action. Picking cannot be guided by what an apple actually is or by an "abstract" represen-

tation of the apple. But not all "early" information is allowed to influence the action. Irrelevant information must, in one way or another, be prevented from influencing the action systems.

These properties, especially the last one, seem inevitably to lead to a picture of the information-processing system as a modular system. It is extremely difficult to see how an action can be guided by relevant properties only (size and distance, for instance) if this information were integrated, combined or "glued" together with irrelevant properties (such as brightness or colour) as Treisman, Crick and others have suggested. It is, however, easy to see how this can be done if relevant and irrelevant properties are physically separated as is supposed to be the case in a modular system.

The different visual fields can thus be regarded as potential sensori-motor links between the eye and visually guided behaviour (see also Creutzfeldt, 1985). And because multiple representation, i.e. modularity, is also found in "motor systems" - in the cortex of primates more than a dozen distinct fields are distinguished that appear to be involved in controlling different aspects of movements (Wise & Evarts, 1981) - *dimension* selection and *action* selection can be considered to consist of a temporary, task-specific, assembling and connecting of input and output modules. The term "skill" (Neumann, 1987) seems to be the appropriate one for such a task-specific perceptuo-motor system. In this view, modularity does not create problems that "focal attention" or a "searchlight" has to solve. Modularity seems one of the solutions nature has offered to solve a major selection problem; the problem that only a limited number of properties are allowed to influence an action.

Modularity, a mechanism for selective coupling (of modules that make task-relevant computations) and decoupling (of modules that do not) plus one or another intramodule selective enhancement mechanism for *goal* or *target* selection (see, for example, Bushnell, Goldberg, & Robinson, 1981; Wurtz, Goldberg, & Robinson, 1982; Harter & Aine, 1984; Allport, 1987) seem to be sufficient ingredients for a system to perform appropriate actions. This becomes apparent when we express the ideas outlined above in the language of the "new connectionism" which we introduced at the end of the section "Encoding".

Figure 2 shows a small part of a network that is used in a computer simulation for selection tasks of the type "name the colour of the square" (Phaf, 1986). The stimulus, a coloured square and a coloured circle, activates all corresponding "nodes" in the modules on the "feature-feature" level. These nodes activate and inhibit nodes in the modules on the "feature" and the "motor programme" level via fixed wiring. The instruction "name the colour" leads to extra activation (i.e. intramodule enhancement) in all nodes in the colour module and "of the square" does the same in all nodes corresponding to a square on the feature-feature level. The interaction of stimulus activation (world) and instruction activation (task) in this structured excitation-inhibition network selects the task-relevant modules and leads, after a number of iterations (the reaction time), to a high probability of reaching the correct answer. In this simulation, action and dimension (name and

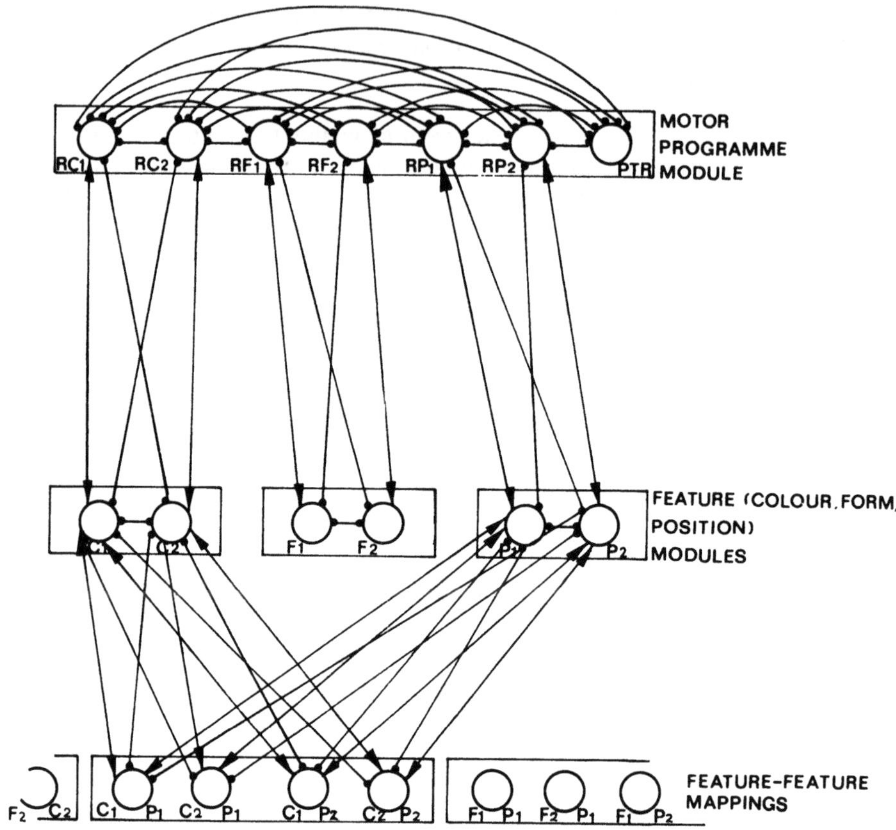

Fig. 2. An example of a connectionist model; part of Phaf's (1986) selective attention model (see text for details)

colour) are still combined in one selection moment (the network can only name). However, introduction of other action modules and the corresponding division of the selection effect over these action modules seems easy to realize.

Further Speculations

It is worthwhile to note that, as far as the encoding of visual information is concerned, neither workers in computer vision (see, for example, Marr, 1980, 1982; Brady, 1981), nor neurophysiologists (see, for example, Barlow, 1981, 1985) nor the new connectionists (see, for example, McClelland & Rumelhart, 1981) regard limited central processing capacity an important constraint that has to direct their theories.

...The massively parallel, apparently modular organization of processing in the nervous system does not suggest any compelling limitation on the current analysis of multiple visual objects, up to and including their categorization in terms of known forms and of learned actions they afford.... Of course, there may be such limitations; we simply observe that the available neurophysiology does not appear to require them. (Allport, Tipper, & Chmiel, 1985)

So it seems that, in principle, the human information-processing system has an unlimited capacity for processing information. The psychological literature, however, often strongly suggest that the system, as it functions in the world, shows severe central limitations. So, what we have to look for is (the beginning of) a theory that can handle both properties: unlimited central capacity as well as limited central capacity (i.e. for a "wave-as-well-as-particle" theory). One way to reconcile these apparently contradictory concepts starts from the basic assumptions that (a) the system has, in-principle, an unlimited capacity; but (b) has also central means or mechanisms that limit the expression of its capacity in order to deal effectively with the world. Starting from these assumptions, the remaining question to be discussed is why and how a system without capacity limitations can show up as a system with limited capacity in its interactions with the world. In this section we briefly discuss three possible answers.

The first answer concerns the consequences of action and dimension selection. Just as a watch, the human information-processing system serves functions. In functioning it has to fit the body to the world. Unlike the watch, however, the human information-processing system serves a diversity of functions. (If we term some of its functions or activities in terms of objects used, then we can say that it functions as a typewriter, bicycle, book, TV set, pencil, watch, car, lawn mower, etc.) Just because of the spatial structure of the world and the limited action capacities of the body, only one function can be performed at a time or only a limited number of functions can be performed simultaneously. To perform these functions the "general" human information-processing system has to assemble and programme itself into "special-purpose" processing systems (i.e. a car, a lawn mower, etc.) that allow that the goals that have to be attained can be attained without (much) interference. A modular system allows for such a flexible adaptation of the information-processing system and the body to the world. Modularity makes the inclusion and exclusion of specific-purpose processors in different perceptual-motor systems possible. Logan (1978) provides a more detailed description of how a set of "processing modules" (i.e. automatic special-purpose processors) that have to operate either simultaneously or successively in a task are temporarily assembled and organized (by attention, in Logan's view) to deal with the task(s) at hand. In his impressive theoretical work, Neumann (1985, 1987) confronts similar ideas with existing empirical data and convincingly shows how, from preventing behavioural chaos, all kinds of apparent limitations simply follow. For more details of this aspect of limited capacity, the reader is referred to Neumann's work (see also Allport, 1980b). Here it suffices to say that it is likely that it is not so much the capacity of the central information-processing system that limits people in processing the world and steering the body, but that it is properties of the world and the capacity of the body that limit psychologists in estimating the real capacity of the human information-processing system.

For a second possible answer to the question of why and how a system without capacity limitations can show up as a system with limited capacity we have to consider the perceptual modules in more detail. Cowey (1979, 1981) discusses the issue of why there are multiple representations of the retina (i.e. topographic maps) in the brain at all ("for it is possible to devise perceptual models which employ only random connectivity"; Cowey, 1981, p. 404). Cowey presents a simple answer.

> One of the best-known phenomena in the retina is lateral inhibition, whereby the physiological response to luminous contrast between adjacent regions is *relatively* enhanced, that is, signals about edges are less strongly inhibited than those about other aspects of the stimulus. If lateral inhibition also occurs at the cortex and is involved in highlighting not just sharp edges but also orientation, colour, disparity, spatial frequency, size, and movement, all this is most economically achieved by having a retinotopic representation so that the necessary interneurons for receptive field tuning lie close together. If there were no map and the cells concerned with a particular part of space were scattered throughout a visual area, the average length of their interconnecting processes would have to be much longer, and the problem of specifying them during development would be enormous. (Cowey, 1981, p. 404; see also Cowey, 1979, p. 10)

In short, Cowey's answer is that topographic maps are a desirable feature because then lateral inhibition between cells concerned with neighbouring regions of space and identical stimulus properties can be implemented in the most economical way. For Cowey, the short, intrinsic connections within the modules are concerned with the selective tuning of cells to some particular feature in a particular part of space. While he presents some evidence for this point of view (see Cowey, 1979, pp. 11-12; Cowey, 1981, pp. 404-405), he also admits that the reason why the brain initially needs to sharpen the tuning curves of individual neurons to attributes, such as contrast, spatial frequency, orientation, movement, retinal disparity, size, colour, etc., are *not* as obvious as often supposed (Cowey, 1979, p. 11). There are ample opportunities for sharpening the tuning curves in the retina, the lateral geniculate body and the cortex. Why is there then lateral inhibition within the modules?

Cornsweet (1970) has already suggested an answer. In his view, lateral inhibition is a mechanism that serves to suppress steady states.

> To any system that learns and remembers, information about steady states is redundant. Any given condition always began (or the organism first come into its presence) as a transient event, by definition. Once it has begun, if the organism has memory, it needs not be continuously reminded of the condition. If the condition stops, that too is a transient that will be transmitted. (Cornsweet, 1970, p. 438)

It will be clear that intramodule lateral inhibition, by producing steady-state suppression, can make a system with a large information-processing capacity appear as a limited capacity system. It is not clear, however, whether such a system really deserves the label of a limited capacity system. All information is processed, only priority is given to "news" and "old news" is suppressed. In previous work I have described a model showing the same properties: a fast initial build up of activation followed by an immediate autonomous decay even during stimulus presentation (see, for example, van der Heijden, 1981; van der Heijden, Hagenaar, et al., 1984). That model, however, reflects all essential features of an unlimited capacity processing system.

The third answer could be that it is also likely that "goal limitation" and the mechanism of what we called "intramodular selective enhancement" means that a system with an, in principle, unlimited capacity can show up as a limited capacity system. As described under "Selection", in typical visual information-processing tasks goal selection is early selection. This can be rephrased by stating that the mechanism of selective attention operates in the perceptual modules concerned with colour, size, position, etc. (In previous work I termed this selective mechanism "postcategorical filtering" to indicate that selection at this level does not imply that there are no higher-level, more abstract, representations in the system; see, for example, van der Heijden, 1981, 1984, 1987; Allport, 1987). After Cowey (1979, 1981) we have to assume that within these modules there are strong laterally inhibitory interactions. (In fact, in both McClelland & Rumelhart's, 1981, and Phaf's, 1986, connectionists' models, intramodular inhibition is essential for adequate task performance.) As briefly mentioned under "Selection", there are suggestions that the goal selection mechanism operates by means of selective enhancement of activity of cells (see, for example, Bushnell et al., 1981; Wurtz, Goldberg, & Robinson, 1980; Wurtz et al., 1982; see also Allport, 1987). (In my previous work I used the related assumption that the selection mechanism operates by maintaining the level of activation of a selected representation while the activation levels of non-selected representations simply decreased.)

Intramodular inhibition and selective enhancement together lead to an interesting conclusion. If the selective attention mechanism operates in the perceptual modules and if the mechanism of selective attention consists of enhancement of activity of a spatially isolated subgroup of cells, then selective attention results simultaneously in the decreased activity of at least some of the non-selected subsets (the exact effects depending on the kind of inhibitory connections postulated).

Figure 3 serves to illustrate this conclusion. In the left panel of the figure, the activation levels (in arbitrary units) of two units (*a* and *b*) are shown in a module without lateral inhibitory interactions. The second panel shows the activation pattern resulting from the same input activity in a module with lateral inhibition. (An inhibitory coefficient of 0.3 was arbitrarily chosen in Cornsweet's (1970, p. 299) lateral inhibition formula.) In other words, this panel gives the activation pattern in a module with rather strong steady-state suppression. The last panel shows what happens with "selective enhancement" of one unit in such a module. The input activity to group *b* was increased from 100 to 150; the input to group *a* was kept at 100. Panel 3 clearly shows, besides steady-state suppression and the enhanced activity of one unit, a simultaneously suppressed activity of the non-selected unit. In this way, the picture shows that selective attention may have two effects: enhancement of wanted information and simultaneous suppression of unwanted information. Keele and Neill (1978, p. 42) have already suggested such a mechanism "... when it (i.e., attention) selects some information, other conflicting information can be inhibited". Also Treisman (1964) can be interpreted as describing such a dual-effect mechanism; selection of one message and simultaneous attenuation of all others (see also James, 1890, pp. 409-410). On the one hand, simultaneous target facilitation (or maintenance against steady-state suppression) and distractor in-

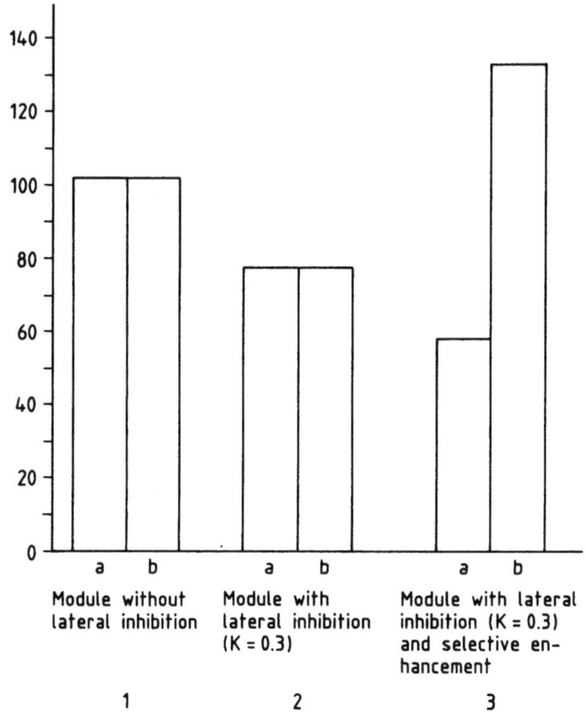

Fig. 3. Activity levels of two units, *a* and *b* (*panel 1*), steady-state suppression due to lateral inhibition (*panel 2*) and the effects of selective enhancement (*panel 3*) (see text for details)

hibition may explain the remarkable efficiency of selective attention. But on the other hand, the suppression part of this mechanism might again make a system without capacity problems show up as a system with limited capacity, just because selective attention efficiently suppresses unwanted information.

So far three possible answers to the question of why and how a system without capacity limitations can show up as a system with limited capacity in its interactions with the world (including laboratory settings). Space does not allow a real confrontation of the ideas developed in this paper with experimental data on important topics like object versus dimension selection; position versus form, colour, etc. selection, selection of a combined versus selection of a separate feature; sources of interference, etc.

One general conclusion seems clear, however. In my view it appears that, especially after Cornsweet's (1970) impressive lessons, visual information-processing research has neglected much too much the profound effects of (lateral) inhibition. It is not impossible to defend the position that it is inhibition for task selection (see Neumann, 1987) and dimension selection, and lateral inhibition for steady-state suppression and resulting from goal selection (this paper), that makes human beings appear as limited capacity information processors. Just as a watchmaker can remove a balance thereby making the watch run much faster, so can the great omnipotent designer remove inhibition, making much more information processing possible. But then neither the watch nor the human being performs its proper functions anymore. It seems that it is (lateral) inhibition that transforms an, in principle, unlimited capacity information-processing system into an, for practical purposes, severely limited capacity system that can perform its proper functions in the world. (But to arrive at this conclusion some glimpses in the watch were necessary.)

Acknowledgement. This paper was partly translated from the Dutch by A. Pleit-Kuiper.

References

Allport, D.A. (1977). On knowing the meaning of words we are unable to report: the effects of visual masking. In S. Dornic (Ed.), *Attention and performance VI*. Hillsdale, NJ: Erlbaum.

Allport, D.A. (1979). Conscious and unconscious cognition: a computational metaphor for the mechanism of attention and integration. In L.G. Nilsson (Ed.), *Perspectives on memory research*. Hillsdale, NJ: Erlbaum.

Allport, D.A. (1980a). Patterns and actions: cognitive mechanisms are content-specific. In G. Claxton (Ed.), *Cognitive psychology, new directions*. London: Routledge & Kegan Paul.

Allport, D.A. (1980b). Attention and performance. In G. Claxton (Ed.), *Cognitive psychology, new directions*. London: Routledge & Kegan Paul.

Allport, D.A. (1987). Selection for action: Some behavioral and neurophysiological considerations of attention and action. In H. Heuer & A.F. Sanders (Eds.), *Perspectives on perception and action*. Hillsdale, NJ: Erlbaum.

Allport, D.A., Tipper, S.P., & Chmiel, N.R.J. (1985). Perceptual integration and postcategorical filtering. In M.I. Posner & O.S.M. Martin (Eds.), *Mechanisms of Attention: Attention and performance XI*. Hillsdale, NJ: Erlbaum.

Attneave, F. (1959). *Applications of information theory to psychology*. New York: Holt, Rinehart & Winston.

Averbach, E., & Coriell, A.S. (1961). Short-term memory in vision. *Bell System Technical Journal, 40*, 309-328.

Barlow, H.B. (1972). Single units and sensation: A neuron doctrine for perceptual psychology? *Perception, 1*, 371-394.

Barlow, H.B. (1981). Critical limiting factors in the design of the eye and visual cortex. The Ferrier Lecture. *Proceedings of the Royal Society of London. B: Series Biological Sciences, 212*, 1-34.

Barlow, H.B. (1985). The twelfth Bartlett memorial lecture: The role of single neurons in the psychology of perception. *The Quarterly Journal of Experimental Psychology, 37A*, 121-145.

Barlow, H.B., Hill, R.M., & Levick, W.R. (1964). Retinal ganglion cells responding selectively to direction and speed of image motion in the rabbit. *Journal of Physiology, 173*, 377-407.

Brady, M. (1981). Preface - The changing shape of computer vision. *Artificial Intelligence, 17*, 1-15.

Broadbent, D.E. (1958). *Perception and communication*. London: Pergamon.

Broadbent, D.E. (1971). *Decision and stress*. London: Academic.

Broadbent, D.E. (1982). Task combination and selective intake of information. *Acta Psychologica, 50*, 253-290.

Broadbent, D.E. (1985). A question of levels: Comment on McClelland and Rumelhart. *Journal of Experimental Psychology: General, 114*, 189-192.

Bushnell, M.C., Goldberg, M.E., & Robinson, D.L. (1981). Behavioral enhancement of visual responses in monkey cerebral cortex: I. Modulation in posterior parietal cortex related to selective visual attention. *Journal of Neurophysiology, 46*, 755-772.

Clark, S.E. (1969). Retrieval of color information from preperceptual memory. *Journal of Experimental Psychology*, 82, 263-266.

Claxton, G. (1980). Cognitive psychology: a suitable case for what sort of treatment? In G. Claxton (Ed.), *Cognitive psychology, new directions*. London: Routledge & Kegan Paul.

Coltheart, M., Lea, C.D., & Thompson, K. (1974). In defense of iconic memory. *Quarterly Journal of Experimental Psychology, 26*, 633-641.

Cornsweet, T.N. (1970). *Visual perception*. New York: Academic.

Cowey, A. (1979). Cortical maps and visual perception; the Grindley memorial lecture. *Quarterly Journal of Experimental Psychology, 31*, 1-17.
Cowey, A. (1981). Why are there so many visual areas. In F.O. Schmitt, F.G. Worden, G. Adelman, & S.G. Dennis (Eds.), *The organisation of the cerebral cortex.* Cambridge, MA: MIT Press.
Creutzfeldt, O.D. (1979). Neurophysiological mechanisms and consciousness. *Brain and Mind, 69*, 217-253.
Creutzfeldt, O.D. (1985). Multiple visual areas: Multiple sensori-motor links. In D. Rose & V.G. Dobsen (Eds.), *Models of the visual cortex.* New York: Wiley.
Crick, F.H.C. (1984a). The machinery of focal attention. *Perception, 13*, 1, A2.
Crick, F.H.C. (1984b). The function of the thalamic reticular complex: The searchlight hypothesis. *Proceedings of the National Academy of Sciences of the USA, 81*, 4586-4590.
Dennet, D.C. (1984). Computer models and the mind - a view from the East Pole. *Psychology and Artificial Intelligence, 14*, 1453-1454.
Deutsch, J.A., & Deutsch, D. (1963). Attention: Some theoretical considerations. *Psychological Review, 70*, 80-90.
Dick, A.O. (1969). Relations between the sensory register and short-term storage in tachistoscopic recognition. *Journal of Experimental Psychology, 82*, 279-284.
Einstein, A., & Infeld, L. (1938). *The evolution of physics.* New York: Simon & Schuster.
Fodor, J.A. (1983). *The modularity of mind.* Cambridge, MA: MIT Press.
Gahery, Y., & Massion, J. (1981, August). Co-ordination between posture and movement. *Trends in Neurosciences*, pp. 199-202.
Garner, W.R. (1974). Attention: The processing of multiple sources of information. In E.C. Carterette & M.P. Friedman (Eds.), *Handbook of perception II.* New York: Academic.
Gross, C.G., Rocha-Miranda, C.E., & Bender, D.B. (1972). Visual properties of neurons in inferotemporal cortex of the Macaque. *Journal of Neurophysiology, 35*, 96-111.
Hagenaar, R., & van der Heijden, A.H.C. (1986). Target-noise separation in visual selective attention. *Acta Psychologica, 62*, 161-176.
Harter, M.R., & Aine, C.J. (1984). Brain mechanisms of visual selective attention. In R. Parasuraman & R. Davies (Eds.), *Varieties of attention.* New York: Academic.
Hinton, G.E., & Anderson, J.A. (1981). *Parallel models of associative memory*, Hillsdale, NJ: LEA.
Hubel, D.H., & Wiesel, T.N. (1962). Receptive fields, binocular interaction and functional architecture in the cat's visual cortex. *Journal of Physiology, 160*, 106-154.
Hubel, D.H., & Wiesel, T.N. (1968). Receptive fields and functional architecture of monkey striate cortex. *Journal of Physiology, 195*, 215-243.
James, W. (1890). *The principles of psychology* (Vol. 1). New York: Holt.
Julesz, B. (1971). *Foundations of cyclopean perception.* Chicago: The University of Chicago Press.
Kahneman, D. (1973). *Attention and effort.* Englewood Cliffs, NJ: Prentice-Hall.
Kahneman, D., & Treisman, A.M. (1984). Changing views of attention and automaticity. In R. Parasuraman & P.R. Davies (Eds.), *Varieties of attention.* New York: Academic.
Keele, S.W., & Neill, W.T. (1978). Mechanisms of attention. In E.C. Carterette & P. Friedman (Eds.), *Handbook of perception* (Vol. 8). New York: Academic.
Koenderink, J.J. (1984). What does the occluding contour tell us about solid shape? *Perception, 13*, 321-330.
Kuffler, S.W. (1953). Discharge patterns and functional organization of mammalian retina. *Journal of Neurophysiology, 16*, 37-68.
Land, E.H., & McCann, J.J. (1971). Lightness and retinex theory. *Journal of the Optical Society of America, 61*, 1-11.
Lettvin, J.Y., Maturana, H.R., McCulloch, W.S., & Pitts, W.H. (1959). What the frog's eye tells the frog's brain. *Proceedings of the IRE, 47*, 1940-1951.
Logan, G.D. (1978). Attention in character-classification tasks: Evidence for the automaticity of component stages. *Journal of Experimental Psychology: General, 107*, 32-63.
Manicas, P.T., & Secord, P.F. (1983). Implications for psychology of the new philosophy of science. *American Psychologist, 38*, 399-413.
Marcel, A.J. (1983). Conscious and unconscious perception: An approach to the relation between phenomenal experience and perceptual processes. *Cognitive Psychology, 15*, 238-300.
Marr, D. (1980). Visual information processing: The structure and creation of visual representations. *Philosophical Transactions of the Royal Society of London. Series B: Biological Sciences, 290*, 199-218.
Marr, D. (1982). *Vision.* San Francisco: Freeman.

Marr, D., & Poggio, T. (1979). A computational theory of human stereo vision. *Proceedings of the Royal Society of London. Series B: Biological Sciences, 204*, 301-328.

Marshall, J.C. (1984). Multiple perspectives on modularity. *Cognition, 17*, 209-242.

McClelland, J.L., & Rumelhart, D.E. (1981). An interactive activation model of context effects in letter perception: I. An account of basic findings. *Psychological Review, 88*, 375-407.

McClelland, J.L., & Rumelhart, D.E. (1986). *Parallel distributed processing. Explorations in the microstructure of cognition: Vol. 2. Psychological and biological models*. Cambridge, MA: MIT Press.

Moray, N. (1967). Where is capacity limited? A survey and a model. *Acta Psychologica, 27*, 84-92.

Morton, J. (1969). Interaction of information in word recognition. *Psychological Review, 76*, 165-178.

Neisser, U. (1967). *Cognitive psychology*. New York: Appleton-Century-Crofts.

Neisser, U. (1976). *Cognition and reality*. San Francisco: Freeman.

Neumann, O. (1985). Die Hypothese begrenzter Kapazität und die Funktionen der Aufmerksamkeit. In O. Neumann (Ed.), *Perspektiven der Kognitionspsychologie*. Berlin, Heidelberg, New York: Springer.

Neumann, O. (1987). Beyond capacity: a functional view of attention. In H. Heuer & A.F. Sanders (Eds.), *Perspectives on perception and action*. Hillsdale, NJ: Erlbaum.

Norman, D.A. (1968). Towards a theory of memory and attention. *Psychological Review, 75*, 522-536.

Norman, D.A., & Bobrow, D.S. (1975). On data-limited and resource-limited processes. *Cognitive Psychology, 7*, 44-64.

Pashler, H. (1984). Evidence against late selection: Stimulus quality effects in previewed displays. *Journal of Experimental Psychology: Human Perception and Performance, 10*, 429-448.

Perret, D.I., Rolls, E.T., & Caan, W. (1982). Visual neurons responsive to faces in the monkey temporal cortex. *Experimental Brain Research, 47*, 329-342.

Phaf, H. (1986). *A connectionist model for attention, restricting parallel processing through modularity*. Master's thesis, University of Leiden.

Rumelhart, D.E., & McClelland, J.L. (1985). Levels indeed! A response to Broadbent. *Journal of Experimental Psychology: General, 114*, 193-197.

Rumelhart, D.E., & McClelland, J.L. (1986). *Parallel distributed processing. Explorations in the microstructure of cognition: Vol. 1. Foundations*. Cambridge, MA: MIT Press.

Ryle, G. (1949). *The concept of mind*. London: Hutchinson.

Schneider, W., & Shiffrin, R.M. (1977). Controlled and automatic human information processing: I. Detection, search, and attention. *Psychological Review, 84*(1), 1-66.

Shiffrin, R.M., & Schneider, W. (1977). Controlled and automatic human information processing: II. Perceptual learning, automatic attending, and a general theory. *Psychological Review, 84*(2), 127-190.

Sperling, G. (1960). The information available in brief visual presentations. *Psychological Monograph, 74*(11, Whole No. 498).

Sperling, G. (1967). Successive approximations to a model of short-term memory. *Acta Psychologica, 27*, 285-292.

Stent, G.S. (1980, May/June). Thinking about seeing. *The Sciences*.

Stone, J., & Dreher, B. (1982). Parallel processing of information in the visual pathways: a general principle of sensory coding? *Trends in Neurosciences, 5*, 441-446.

Treisman, A.M. (1960). Contextual cues in selective listening. *Quarterly Journal of Experimental Psychology, 12*, 242-248.

Treisman, A.M. (1964). Verbal cues, language, and meaning in selective attention. *American Journal of Psychology, 77*, 206-219.

Treisman.. A.M. (1977). Focussed attention in the perception and retrieval of multidimensional stimuli. *Perception and Psychophysics, 22*, 1-11.

Treisman, A.M. (1982). Perceptual grouping and attention in visual search for features and for objects. *Journal of Experimental Psychology: Human Perception and Performance, 8*, 194-214.

Treisman, A.M., & Gelade, G. (1980). A feature integration theory of attention. *Cognitive Psychology, 12*, 97-136.

Treisman, A.M., & Schmidt, H. (1982). Illusory conjunctions in the perception of objects. *Cognitive Psychology, 14*, 107-141.

Treisman, A.M., Sykes, M., & Gelade, G. (1977). Selective attention and stimulus integration. In S. Dornic (Ed.), *Attention and performance VI*. Hillsdale, NJ: Erlbaum.

Turvey, M.T., & Kravetz, S. (1970). Retrieval from iconic memory with shape as the selection criterion. *Perception and Psychophysics, 8*, 171-172.

Ullman, S. (1979). *The interpretation of visual motion*. Cambridge, MA: MIT Press.

Van de Geer, J.P. (1957). Psychologische toepassingen van de informatie-theorie I, II. *Nederlands Tijdschrift voor de Psychologie, 12*, 295-357.
Van der Heijden, A.H.C. (1981). *Short-term visual information forgetting*. London: Routledge & Kegan Paul.
Van der Heijden, A.H.C. (1984). Postcategorical filtering in a bar-probe task. *Memory and Cognition, 12*, 446-457.
Van der Heijden, A.H.C. (1986). On selection in vision. *Psychological Research, 48*, 211-219.
Van der Heijden, A.H.C. (1987). Central selection in vision. In H. Heuer & A.F. Sanders (Eds.), *Perspectives on perception and action*. Hillsdale, NJ: Erlbaum.
Van der Heijden, A.H.C., La Heij, W., & Boer, J.P.A. (1983). Parallel processing of redundant targets in simple visual search tasks. *Psychological Research, 45*, 235-254.
Van der Heijden, A.H.C., Hagenaar, R., & Bloem, W. (1984). Two stages in postcategorical filtering and selection. *Memory and Cognition, 12*, 458-469.
Van der Heijden, A.H.C., Schreuder, R., Maris, L., & Neerincx, M. (1984). Some evidence for correlated separate activation in a simple letter-detection task. *Perception and Psychophysics, 36*, 577-585.
Van Essen, D.C. (1979). Visual areas of the mammalian cerebral cortex. *Annual Review of Neurosciences, 2*, 227-263.
Van Essen, D.C., & Maunsell, J.H.R. (1983). Hierarchical organization and functional streams in the visual cortex. *Trends in Neurosciences, 6*, 370-375.
Von Helmholtz, H. (1924). *Treatise on psychological optics*. Rochester, NY: Optical Society of America.
Von Wright, J.M. (1968). Selection in visual immediate memory. *Quarterly Journal of Experimental Psychology, 20*, 62-68.
Von Wright, J.M. (1970). On selection in visual immediate memory. *Acta Psychologica, 33*, 280-292.
Von Wright, J.M. (1972). On the problem of selection in iconic memory. *Scandinavian Journal of Psychology, 13*, 159-171.
Wise, S.P., & Evarts, E.V. (1981, December). The role of the cerebral cortex in movement. *Trends in Neurosciences*, pp. 297-300.
Wundt, W. (1912). *An introduction to psychology*. London: Allen & Unwin.
Wurtz, R.H. (1985). Stimulus selection and conditional response mechanisms in the basal ganglia of the monkey. In M.I. Posner & O.S.M. Marin (Eds.), *Attention and performance XI*. Hillsdale, NJ: Erlbaum.
Wurtz, R.H., Goldberg, M.E., & Robinson, D.L. (1980). Behavioral modulation of visual responses in the monkey: stimulus selection for attention and movement. *Progress in Psychobiological and Physiological Psychology, 9*, 43-83.
Wurtz, R.H., Goldberg, M.E., & Robinson, D.L. (1982). Brain mechanisms of visual attention. *Scientific American, 246*, 124-135.
Zeki, S. (1978). Functional specialization in the visual cortex of the rhesus monkey. *Nature, 274*, 423-428.

Visual Attention and Action

O. NEUMANN

CONTENTS

Introduction

Until recently, there has been relatively little interest in relationships between attention and action. When modern research into attention began in the 1950s, the dominating approach was to analyze attention within the information-processing framework, and information processing was conceptualized mainly in terms of what happens to stimulus information after it has entered the human processing system (e.g., Broadbent, 1958, 1971; Neisser, 1967). Attentional processes were viewed as modifying the flow of information by selecting part of the incoming information for processing. Most of the theoretical discussions since the 1960s have centered around the question of where in the processing sequence this selection takes place. This has usually been phrased as the problem of whether selective attention operates "early," i.e., prior to complete stimulus analysis, or "late," after the information in the stimulus has been fully identified (e.g. Deutsch & Deutsch, 1963; Treisman & Geffen, 1967). Common to both theoretical positions has been the conviction that not all incoming information can be dealt with simultaneously by the system because processing capacity is limited. Hence, selection was thought

Relationships Between Perception and Action
Edited by O. Neumann and W. Prinz

to be necessary. Attention was thus conceptualized as related to the analysis and internal representation of incoming information rather than to the control of action.

Although the early-vs-late-selection controversy is still flourishing (e.g. Broadbent, 1982; Dark, Johnston, Myles-Worsley, & Farah, 1985; Hoffman, 1986; van der Heijden, 1987), the assumptions underlying the classical information-processing approach to attention have recently been challenged. There have been several theoretical developments that, taken together, mark the emergence of a new, more action-related theoretical approach to selective attention. First, instead of 'limited capacity,' the functional incompatibility between processing operations ('crosstalk,' 'outcome conflict') has been suggested as the major source of attentional interference (e.g. Allport, 1980; Hirst & Kalmar, 1987; Navon & Miller, 1987; Neumann, 1978a). Second, attentional selectivity at the level of motor control has been introduced as a new field of investigation (e.g. Goldberg & Segraves, 1987; Kerr, 1983). Third, basic functional properties of visual attention have been linked to eye movement control (e.g. Rizzolatti, Riggio, Dascola, & Umiltá, 1987; Tassinari, Agliotti, Chelazzi, Marzi, & Berlucchi, 1987; Wolff, 1984). Finally, there have been several attempts to theoretically reconsider the whole range of empirical findings on attention, based on the possible functions of attentional mechanisms for action control (Allport, 1987; Keele & Neill, 1978; Neumann, 1978b, 1983; Shallice, 1972, 1978; van der Heijden, this volume).

The aim of this chapter is to apply one version of this new approach (Neumann, 1985, 1987a, 1987b) to the study of visual attention. One reason for choosing visual attention is that it presents a particular theoretical challenge to an action-oriented view of attention. Relating attention to action control seems intuitively plausible with respect to 'output' aspects of attention, e.g., conflicts between physically incompatible action tendencies (Goldberg & Seagraves, 1987) or conflicts between the outcome of processing operations (Navon & Miller, 1987). It is much less obvious why and how *sensory* attention should depend on action control. Common experience tells us that we can attend to parts and aspects of the visual world without acting toward them. Here the traditional approach seems, at first glance, much more natural than the view that these phenomena are related to action control.

The content of this chapter may be viewed from two sides. Under one perspective, it is a plea for an action-oriented view of attention, using visual attention as an example. Under a second perspective, it is an attempt to structure some of the the major findings on visual attention, based on the general view that attentional mechanisms have evolved in order to cope with problems of action control. The next section gives a general overview of the approach. The subsequent sections are concerned with different aspects of visual attention and their relationship to action control.

Theoretical Framework

General Approach

The basic assumption of the present approach to attention (Neumann, 1978a, 1978b, 1980, 1983, 1984, 1985, 1987a, 1987b) concerns the functional relationship between limited capacity and attentional selectivity. This relationship is assumed to be just the reverse of what has been suggested by traditional capacity theories: The basic characteristic of attention is its selectivity, not its limited capacity. It is selectivity that explains why capacity is limited, contrary to the traditional view that selection is needed because capacity is scarce. Attentional selectivity exists because of the physical limitations on overt action that require selecting between alternative actions, and hence demand selecting the information which is to control action. Competing actions and competing ways of carrying out an action need to be inhibited, and competing information must be prevented from gaining access to action control. Performing all the processing operations that the brain's hardware may permit would be quite dysfunctional.

Powerful 'software' solutions to these action-related selection problems must have been developed in the course of evolution. They produce performance limits, i.e., 'limited capacity' in a descriptive sense, which is a functionally useful achievement rather than a shortcoming of the processing system.

Different selection problems exist in the control of action, and evolution is likely to have developed different mechanisms to cope with them. It would therefore be ill-advised to look for a general 'attentional system' behind the different phenomena that we classify under the descriptive label 'attention'. For example, the mechanisms of sensory attention may be quite different from those that produce interference between simultaneously performed tasks. By the same token, there is no a priori reason why the mechanisms of visual attention should be the same as those in other sense modalities. Visual selection operates within a system that has its particular functional characteristics, different from those of, for example, the auditory system. The eyes are highly mobile; optical information is spatially structured and available from all parts of a lighted environment, while acoustical information has a basically temporal structure and is available only occasionally, and only from those parts of the environment where a sound-producing event occurs. These different conditions pose different selection problems, and hence different attentional mechanisms are likely to exist (see Neumann, van der Heijden, & Allport, 1986a). Visual attention has therefore to be studied in its own right, not as just another case of how 'attention' works. (I plead guilty to having neglected these differences in my earlier writing on attention; e.g., Neumann, 1983, 1984, 1987a. This has been rightly criticized by van der Heijden, 1987).

While the *mechanisms* of visual attention should be analyzed separately, the action-oriented approach suggests that its *functions* cannot be properly understood in isolation. If it is true that selection mechanisms have evolved to cope with problems of action control, then their functional properties are likely to be tailored to the requirements of action control. Before turning to visual attention, we have therefore to briefly consider the general selection problems that must be solved if action is to be controlled efficiently. The following is in part a summary of a previous paper (Neumann, 1987a), to which the reader is referred for more detail and supportive evidence.

Two Types of Selection Problems

A first, crude classification suggests that there are basically two types of selection problems in action control, one related to the selection between alternative actions, and the other related to selecting the manner in which an action is executed. The first may be termed the problem of effector recruitment. Doing many things at a time is sharply limited by the fact that we have only two hands, one mouth, and one respiratory system, that we can run in only one direction at a time, etc. Attempting to recruit effectors for physically incompatible actions would produce behavioral chaos. Hence, the system needs mechanisms of behavioral inhibition, legislating between competing action tendencies in such a way that only mutually compatible effector movements will be attempted. Biological evidence (e.g. Hinde, 1970; Shallice, 1978; see also Dean, this volume) indicates that the underlying mechanisms usually allow only one action to be carried out at a time ('dominant action system'; Shallice, 1978). As suggested by Neumann (1987a), this is the principal functional basis of general, unspecific interference. 'Dual tasks' can be performed, nevertheless, if they are controlled by a common action plan. However, initiating a new action will always interfere with the execution of an ongoing action, and a dual-task action plan will often be more complex than a single-task action plan, which may reduce performance.

While these are sources of unspecific interference, there is another type of selection problem that produces task-specific interference. It is the problem of parameter specification. The system has not only to decide what to do, but also how to do it. An intended action can be carried out in many different ways, but to execute it, exactly one way has to be selected. This requires mechanisms which guarantee that, in using a skill, the parameters to be specified are assigned unequivocal values. These mechanisms also produce specific interference between actions that demand usage of the same skill(s). The same inhibition that prevents a parameter from being specified in more than one way if a skill is used for one task also prevents that this skill is used for two tasks at the same time, if this requires specifying the same parameter differently. Interference of this kind can be partially overcome by reducing the skill overlap between the competing tasks, i.e., by acquiring more differentiated skills. This is an important aspect of automatization.

Sensory Information as a Source of Parameter Specification

These two selection problems and the ensuing types of interference exist independently of whether or not sensory information is used to select an action and to specify its parameters. Competing action tendencies have to be suppressed, whether they are caused by external stimuli or by internal states of the organism. An action's parameters have to be specified unequivocally, whether the information that provides the specification stems from the environment or is retrieved from memory. These are problems of attentional selection, not specific problems of sensory attention. Normally, however, at least part of the selection is done by using appropriate information from the environment. This poses additional selection problems, and to cope with them a special type of selection mechanism is required. If selecting one action rather than another depends on environmental stimuli, then it has to be decided *which* stimuli will have access to action control, i.e., which will determine the choice of the action to be executed. If the way in which an action is carried out is to be adjusted to environmental conditions, then it must be determined *which* part or aspect of the information in the environment will be used to specify each particular parameter of the action.

Hence, the sensory control of action has a double-sided effect with respect to solving the problems of effector recruitment and parameter specification: On the one hand, environmental information constrains the set of alternative actions competing for effector recruitment as well as the set of alternative ways to carry out a selected action. But on the other hand, there is usually more information available which could in principle be used for these purposes than can actually be used. Environmental information helps the organism decide what to do, but only after the organism has decided which environmental information to use. Accomplishing this selection is, I suggest, the task of the mechanisms of sensory attention.

Readers who are used to the traditional information-processing view of attention may initially find this suggestion difficult to accept. First, one could argue that it covers at best a small portion of contemporary empirical research on sensory attention, the majority of which has not been concerned with action control, but with purely central selection. Second, the reader may feel that this view obscures the distinction between attentional selectivity and other instances of input selection. From monosynaptic reflexes to complex motor control structures, there are many functional levels at which sensory information enters into the guidance of action (see Cruse, Dean, Heuer, & Schmidt, this volume; Dean, this volume). At each level there is selectivity in the sense that only stimuli with particular characteristics will have an impact on the action. Subsuming all these instances under the heading of 'sensory attention' would stretch the concept of attention to a point where it becomes meaningless.

A third possible objection is the mirror image of the second: Not only is there selective sensory control of action without attention; it also seems obvious that sensory attention can manifest itself without overt action. As experimental psychologists have known for more than a century, visual attention can be shifted without an accompanying eye movement (von Helmholtz, 1866; see e.g., Posner, 1980; Reeves & Sperling, 1986; van der Heijden, 1986). Auditory selection without head movements has been extensively since the 1950s (e.g. Broadbent, 1971, 1982). More recently, nonbehavioral measures such as event-related potentials (e.g. Hillyard, Münte, & Neville, 1985) and single-cell recordings (e.g. Fischer, 1986) have been used as dependent variables to study sensory attention independently of overt responding. As noted earlier, it is also a matter of common experience that attending need not be coupled to physical action. Clearly, then, paying attention cannot be equated with selection-for-action in the sense that the attended object or event is invariably used either to initiate an action or to specify action parameters.

A New Look at Some Experimental Paradigms

These three objections carry different weight. The second and third are of central importance. Much of the rest of this chapter will address them directly or indirectly. The first objection can be dealt with more easily. It is true that visual attention has usually not been investigated in a research context that has related it to action control. But there is a difference between the processes and factors that are *explicitly addressed* by experimental research and those that are *actually involved* in the experiments. While the large majority of experimenters in the area of sensory attention have been interested in central selectivity and central 'capacity limitations,' the experimental paradigms that they have typically used nevertheless involved action control. The simple reason is that, in behavioral research, one has to ask the subject to perform some action in order to get some performance measure. The theoretical analysis usually stopped at some point where 'information processing' was thought to be completed, but the experimental situation involved the whole process of the sensory control of action.

To illustrate, consider two classical paradigms that have been extensively used in attention research, the shadowing experiment (Cherry, 1953) and the Stroop experiment (Stroop, 1935). In the shadowing paradigm, subjects utter a sequence of words by repeating back one of two auditory messages presented simultaneously to each ear. Obviously, many sequences of words can be produced by the speech apparatus, and there are many sources of information in such an experimental environment that could determine the utterance. (For example, the subject could describe the recording apparatus.) Which of the many possible utterances is actually selected - i.e., which parameter values specify the act of speaking - is determined by one particular subset of the information from the environment, viz, the message that the experimenter has designed as relevant. The situation is similar in the Stroop experiment. Here the task is to name the ink color in which an incongruent color word is printed (e.g., respond "red" to the word BLUE, printed in red). Again, the proper execution of the task requires selecting part of the information available from the environment (the color of the word, but neither the meaning of the word nor, say, the color of the background screen on which the stimulus is projected) to specify the action of speaking.

In the Stroop paradigm, sensory information is used both to initiate an action and to specify its parameters. As is usual in choice reaction time experiments, different aspects of the stimulus contain the relevant information: The decision to act is contingent upon the apparition of the stimulus (the transition from 'background alone' to 'background-plus-stimulus'), whereas the decision which word to utter depends on its color (see Neumann, 1980, p. 143f; and van der Heijden, La Heij, Phaf, Buijs, & van Vliet, 1988, for theoretical discussions of this difference). In other paradigms that have been used in the study of sensory attention, information selection may be required for only one of these purposes. For example, in one version of the cuing paradigm introduced by Posner and associates (Posner, Nissen, & Ogden, 1978; Posner, 1980), the subject's task is always to press a button as soon as a visual stimulus is presented. Stimuli are presented either to the right or to the left of fixation. (The standard finding is that cuing one of the locations prior to target presentation shortens reaction time if the target appears in the cued box and lengthens reaction time if it appears in the opposite box.) In this paradigm, the action's parameters are completely specified by the instruction. Information has only to be selected to initiate the prespecified action.

These three paradigms - shadowing, Stroop, and cuing - are representative of current experimentation in attentional research. Although nominally directed at central factors and processes, these experiments actually require (and implicitly investigate) the sensory control of overt action. The first of the three above-mentioned objections can therefore be dismissed: Although an action-related approach to attention goes beyond traditional theorizing, it does not necessarily require new types of experimentation. The kinds of paradigms that it calls for are already available.

The other two objections require more detailed treatment. The action-oriented approach will be successful only

- If it can account for the cases where attention seems to be independent of overt action
- If it can specify what distinguishes attentional from nonattentional selectivity in the sensory control of action

To put it in terms of Allport's (1987; Allport, Tipper, & Chmiel, 1985) concept of 'selection-for-action': We have to consider not only attentional selection-for-action, but also both attentional selection-without-action and nonattentional selection-for - action.

The rest of the chapter is organized around these topics. There are two main sections. The first addresses the differences between nonattentional and attentional selection-for-action. In the second I discuss why attentional selection need not imply the control of overt action; this will lead to a consideration of the more general problem of how selection is related to representation. Although these questions can be asked for all sensory modalities, the discussion will be restricted to visual attention.

Nonattentional and Attentional Selection-for-Action

If attentional selection is conceived as selection-for-action, then it must clearly be a *special* kind of selection-for-action. The question of what makes it special can best be approached (for the visual modality) by first considering examples of the visual guidance of action where *no* attentional mechanisms are needed. We can then ask why this mode of control is not generally applicable, i.e., under which conditions attentional selection-for-action is required and which are its functional characteristics.

Property Selection in the Case of Automatic Parameter Specification

The simplest way of using visual information for action control is to employ a stimulus property for the automatic specification of an action's parameter(s). 'Automatic' does not imply that the specification has the character of 'direct perception' in Gibson's (e.g. 1979) usage of the term. Complex computations may be required on the way from the retinal information to the motor output pattern, but there is no need to categorize, store, or represent in a declarative format the information prior to using it for action control. The information consists in the presence of a certain stimulus property, e.g., the rate of change of a visual pattern on the retina, and the corresponding parameter specification can take place as soon as the sensory analysis of this stimulus property has been completed.

This is the phylogenetically oldest use of visual information to guide action. Simon (1969) has described the example of an ant's locomotion that may follow a very complicated trajectory, which is, however, not the product of intelligent planning, but of the ant's continuously adjusting to the surface of the terrain, e.g., the position of obstacles. The ant does not need to "know" anything about the environment and the objects of which it is composed; it has only to adjust its action to certain stimulus properties. Another example is a flying locust that corrects deviations from its flight attitude on the basis of optical information (see Dean, this volume, where more examples of this type of parameter specification can be found).

In man, property selection for automatic parameter specification has been investigated in various kinds of visually guided behavior, such as controlling body balance (Lee & Aaronson, 1974; Lee & Lishman, 1975), pointing to an object (e.g. Bridgeman, Kirch, & Sperling, 1981), car driving (e.g. Kramer & Rohr, 1982; Lee, 1976; Lee & Lishman, 1977), hitting a ball (Alderson, Sully, & Sully, 1974; Lee, Young, Reddish, Lough, & Clayton, 1983; McLeod, McLaughlin, & Nimmo-Smith, 1985), and jumping (Lee, Lishman, & Thomson, 1982).

The last two examples illustrate cases of interceptive actions that are extremely precise in their timing. For example, the subjects of Alderson et al. (1974), who had to catch a ball, initiated their grasps within a "window" of 14 ms before and after the optimal time. The subjects of McLeod et al. (1985), whose task was to hit a falling ball with a bat, performed with a standard deviation of about 5 ms. A similar precision (a standard deviation of about 10 ms) is attained by ski jumpers with respect to the time of take-off at the end of the run-in (Lee, Lishman, & Thomson, 1982). There is evidence that these actions are controlled by detecting a particular visual parameter, called tau, which is the inverse of the relative rate of dilation of the retinal projection (Lee, 1976; Lee & Young, 1986; see also Cruse et al., this volume).

Besides the extraordinary precision of the sensory control of action, these examples show several interesting common characteristics. First, there is some (possibly complex) stimulus property (e.g., an object's location; the optical parameter tau; an optical flow pattern in the low spatial-frequency range) that specifies the required parameter(s), but it is not the stimulus as an object, with its particular combination of properties (shape, size, color, etc.) that enters into action control. Second, it is unlikely that parameter specification is mediated by a conscious representation, and in some case there is direct evidence that it is not (e.g. Bridgeman et al., 1981; McLeod et al., 1985; for an overview see Neumann, 1989). Third, this type of parameter specification either is innate or needs to be acquired through extended practice.

Taken together, these characteristics suggest that this type of selection-for-action takes place within, and is controlled by, specific control structures, or skills. (See Neumann, 1983, 1984 where I suggested the term 'skill' for acquired, specific control structures in man; for a discussion of the more general term 'motor control structure' see Cruse et al., this volume.) Such control structures serve to guide specific types of actions (e.g., walking, throwing, speaking, or typing). They specify the action's parameters in part by internally generating the appropriate motor commands and in part by using information from the environment. In neural terms, one may think of such control structures as systems of interconnected neurons, extending from the analysis of sensory input to the innervation of motor control units; systems that, by means of specific neuronal connections, translate specific stimulus properties into specific action parameters. (For examples from biology see Dean, this volume. Models of this type have recently also made their entry into psychology within the connectionist approach; see e.g. Ballard, 1986; McClelland & Rumelhart, 1987; Rumelhart & McClelland, 1987; Schneider, 1987.)

Stimulus processing within such a control structure is, of course, highly selective. But this selectivity is not the work of a selection mechanism that exists *in addition* to the control structure. Rather, this structure itself provides the required selectivity through the way in which sensory neurons and motor neurons are connected. Just as the human visual system selects electromagnetic radiation roughly between 400 and 700 nm without the need for an extra mechanism to prevent the processing of lower or higher wavelengths, stimulus properties that are not linked to motor output through specific connections within a control structure simply cannot be processed by this control structure. No mechanisms are needed to prevent what cannot be done.

This point has been aptly made by Neisser (1976). Neisser was wrong, however, in regarding this as a general argument against the existence of attentional selection mechanisms. This will become apparent when we now consider more complex cases of selection.

So far we have made the simplifying assumption that property selection within a control structure is always sufficient to specify the required action parameters. In some of the examples that I have cited (e.g., ski jumping), this may be a realistic assumption. In other cases (e.g., hitting a ball) a different selection problem may arise, however. Suppose that *two* balls are being thrown toward a cricket batsman. Then he has not only to select the stimulus parameter(s) that are relevant for the control of his movement; he must also select between the two stimuli that both provide these parameters. This requires mechanisms in addition to those that subserve property selection.

A Simple Example of Attentional Selection

One of the classical investigations of mechanisms of property selection was the study of receptive fields in the frog's retina by Lettvin, Maturana, McCulloch, and Pitts (1961). Recording from individual fibers in the optical nerve of the frog, these authors discovered four types of receptive fields that could be related to particular classes of behavior. One of them was a 'bug detector' that was activated by a small dark form moving in the frog's visual field. This detector is probably part of a motor control structure subserving the action of snapping at bugs. It illustrates the kind of built-in selectivity that we have so far discussed. The selective response to bug-like stimuli is a functional property of the control structure itself and does not require any further, attentional mechanisms.

Consider now an experiment by Ingle (1975) in which *two* such bug-like stimuli moved simultaneously in a frog's visual field. Obviously, a frog cannot snap at two stimuli at the same time. What happens is that in most cases the frog will snap at one of the stimuli (although it may occasionally also snap at a position in between the 'bugs' or show no reaction at all; see also Arbib, 1981). Even more interestingly, when Ingle (1975) primed the position of one of the stimuli by a slight stimulus movement, itself insufficient to elicit a snap reaction and occurring up to a few seconds before the two 'bugs,' the frogs snapped toward the 'bug' at the primed position in 85% of the cases. Looking for a neural correlate of this type of selectivity, Ingle found cells in the frog's tectum whose spatiotemporal characteristics closely matched the behavioral data.

Ingle (1975) concluded from these data that the frog was exhibiting 'focal attention,' and termed the cells that apparently subserve this selectivity 'attention units.' Indeed, the mechanism that he describes illustrates a kind of selectivity that is basically different from property selection. Both competing bugs possess the properties (e.g., size, movement speed) required to specify the relevant parameters (e.g., timing and direction) of the snapping response. The intrinsic selectivity of the control structure is therefore unable to secure unequivocal parameter specification. An additional selection mechanism is required.

The mechanism described by Ingle (1975; for a formalized version see Lara & Arbib, 1982) shows two interesting characteristics. First, the selection is spatial. In Ingle's experiments, the first stimulus primed a certain location in the frog's visual field, presumably by preactivating the corresponding 'attention unit.' The selection between the subsequent bug-like stimuli was determined mainly by their spatial position relative to this preactivated attention unit. Second, the selection occurs at a sensory level, the optical tectum. The function of the selection process is quite obviously to secure an appropriate action, but the selection itself does not take place at a motor level, e.g., by inhibition between motoneurons.

It can easily be seen that, in combination, these two characteristics provide an extremely efficient way of coping with the parameter specification problem in the presence of competing visual stimuli. *Spatial* selection is efficient because of the structure of the physical world: The properties that compete for the specification of the same parameter within the same control structure usually belong to different objects at different spatial locations (Neumann, 1980; Wolff, 1977). Hence, selecting one location will, under ecological conditions, unequivocally specify the action in most cases. Selection at the *sensory* instead of the motor level has the decisive advantage that all parameters will be specified consistently (Neumann, 1983). If all competing stimuli were allowed to contact the motor control structure, selection between alternative parameter specifications would be required locally for each parameter (e.g., the timing, the direction, and the extent of the frog's snapping reaction). Without additional mechanisms, there would be no way of preventing the inconsistent outcome that different parameters are determined by different stimuli. The result would be that, for example, the frog snaps at an angle specified by bug X, but with a timing appropriate to catch bug Y. Sensory selection provides a most elegant and efficient solution to this consistency problem.

Given their high efficiency, it is not surprising that both functional characteristics have been preserved, and further developed, throughout the evolution of mechanisms of visual attention, up to the primate visual system. In the following I will first discuss the features that have been preserved: As in the frog, visual attentional selection in primates seems to be fundamentally spatial. As in the frog, it occurs at a sensory level although it serves the control of action. Based on this common functional design, specific characteristics of visual attention in higher vertebrates and man can be identified. These will be the topic of the final section.

The Sensory Character of Attentional Selection in Vision

Since the study by Ingle (1975), there has been much progress in the physiological investigation of visual attention. Most of it has been based on the study of single cell responses in mammals, mainly the cat and the monkey (for reviews see Allport, 1987; Bruce & Goldberg, 1984; Hyvärinen, 1982; Lynch, 1980; Mountcastle, 1978; Rizzolatti, 1983; Robinson & Petersen, 1986; Wurtz, Goldberg, & Robinson, 1980, 1982). The areas in the brain that have been investigated include the superior colliculus (e.g. Mohler & Wurtz, 1976), the pulvinar (e.g. Petersen, Robinson, & Morris, 1987), the prelunate area (e.g. Fischer, 1986), the frontal eye field (e.g. Bruce & Goldberg, 1984; Crowne, 1983), and the parietal cortex (area 7; e.g. Bushnell, Goldberg, & Robinson, 1981; Mountcastle, 1975).

There are two basic observations that justify relating these areas to visual attention. In some of them (especially the superior colliculus) an *inhibitory effect* has been observed. Cells that respond strongly to a visual stimulus S_1 become inhibited when a stimulus S_2 is presented at a different location in the visual field. There are several findings which suggest that this inhibition reflects attention being 'drawn away' from the first stimulus by the appearance of the second stimulus (Rizzolatti, Camarda, Grupp, & Pisa, 1974): First, unlike the usual lateral inhibition, this effect does not diminish as a function of the distance between S_1 and S_2. Second, the effect can be reduced and even abolished by presenting S_2 repetitively, similar to the behavioral finding that after repetition stimuli cease to attract attention. This makes it likely that the inhibition reflects the orienting response (passive sensory attention) to a novel stimulus, although other interpretations are not excluded (see Allport, 1987; Wurtz et al., 1980, for an alternative interpretation). Another type of suppression effect has recently been found in visual area V 4 (Moran & Desimone, 1985).

The second major observation is an *enhancement effect* that has been observed in several areas. For brevity, the following discussion will be restricted to findings from the posterior parietal cortex (area 7). The enhancement effect consists in an increase in a unit's activity if the corresponding stimulus has to be responded to or is relevant for the required response. For example, in one of the tasks used by Bushnell et al. (1981) - the saccade task - the monkey initially fixated a central spot. When the central spot went out and a small peripheral stimulus came on, the monkey had to make a saccade to the new stimulus. Neural activity was recorded in area 7. The appearance of the peripheral stimulus always elicited a burst of activity in the neural unit in whose receptive field it was presented. However, this response was much stronger when the stimulus was the saccade's target than in a control condition in which it was irrelevant.

Taken in isolation, this finding can be interpreted in different ways. The neural unit that shows this behavior could be sensory, but receive additional input from the eye movement control system. Or it could be a motor unit responsive to visual input. When the enhancement effect was first explored in the late 1970s, there was some debate about this. Mountcastle and his associates (e.g. Mountcastle, Lynch, Georgopoulos, Sakata & Acuna, 1975) favored the motor view and suggested that the posterior parietal cortex be regarded as a 'command system.' Others (e.g. Robinson, Goldberg, & Stanton, 1978) stressed the sensory nature of the neurons that show the enhancement effect. Meanwhile, it has become apparent that this dichotomy probably overstated the sensory-vs-motor issue. Indeed, it may be the very essence of the functions performed by the parietal cortex that they cannot be classified in this dichotomous manner. As Hyvärinen (1982) has put it, "In such a system neurones and neuronal ensembles are not merely sensory nor merely motor; their essential role is the integration of these aspects.... As a whole these neurones comprise part of a machinery that is able to interact between outside stimuli and motor control" (Hyvärinen, 1982, p. 167). A similar integrative view has been suggested by authors who originally favored either a 'motor' or a 'sensory' interpretation of the enhancement effect (e.g. Lynch, 1980, p. 495; Wurtz et al., 1980, p. 74).

Among the findings that have led to this integrative view are those that indicate most clearly the attentional character of the enhancement effect. For example, one of the tasks in the study by Robinson et al. (1978) was the 'peripheral-attention task.' The monkey was taught to fixate a central stimulus and to keep fixation when a second stimulus was presented in the visual periphery. The critical signal, in response to which the monkey had to release a lever, consisted in a dimming of either the central or the peripheral stimulus. In a control condition the peripheral stimulus was irrelevant to the task. In the experimental condition, a strongly enhanced response to the appearance of the peripheral stimulus was observed. Since this effect occurred even in trials where the peripheral stimulus did not dim and hence did not require a response, it is clear that the enhancement does not depend on the execution of an actual motor response. The effect is sensory in the sense that it is conditional upon the presentation of the stimulus, not up on the execution of an action. On the other hand, it is not *purely* sensory, since it depends on the stimulus-response mapping required by the task, which defines the stimulus as 'relevant.' In other words, the *locus* of the effect is sensory; *functionally*, it is related to action control.

Another important finding supports this conclusion. While initially units in the posterior parietal cortex were described that show task-specific enhancement effects ('saccade neurones,' 'hand projection neurones'; e.g. Robinson et al., 1978), there is now evidence that the large majority exhibit the enhancement effect independently of the type of task. For example, a cell that yields an enhanced response in a saccade task will in most cases show the same behavior in a hand-reaching task as well as in the peripheral-attention task described above (Bushnell et al., 1981; Goldberg & Bruce, 1985). Interestingly, however, there seems to be a functional specialization with respect to different regions within sensory space. Results from lesion studies (Rizzolatti, Gentilucci, & Matelli, 1985) suggest that some areas (e.g. 7b) represent the near surrounding (peripersonal space) while others (e.g. 7a) represent the far surrounding (extrapersonal space). The corresponding actions are reaching and eye movements. Again, the conclusion is that the enhancement effect reflects selectivity in the service of action control, but the selection itself takes place between stimulus representations, not between motor control structures.

If this conclusion is correct, then primates, despite their much more complex processing system, still use basically the same strategy by which the frog solves the selection problem in the presence of competing stimuli. The selection mechanisms may be driven by input from action systems into the cells of the posterior parietal cortex, as suggested by Wurtz et al. (1980). But the parietal system itself constitutes a representation of the sensory environment, not of actions. The selection strategy seems to be similar to what military strategists would call 'forward defense.' The function of the selection mechanism in the posterior parietal cortex is selection-*for*-action, but this mechanism is not localized *at* action control.

Rizolatti and Camarda (1987) have recently suggested an interesting extension of this account. They propose that the selection is not only sensory, but also spatial: "The motor plan elicited by the stimulus activates the representation of the space sector where the plan will be transformed into action" (Rizzolatti & Camarda, 1987, p. 307). As has been argued earlier, a combination of sensory selection with spatial selection constitutes a powerful solution to the parameter specification problem in the case of the frog. In the next section I discuss findings which indicate that the same basic solution has been preserved in primates, i.e., that attentional selection in vision is not only sensory, but also spatial.

The Spatial Character of Attentional Selection in Vision

In the frog, spatial selection is probably based on, or at least facilitated by, the somatotopic representation of visual information in the frog's tectum. In the primate visual system, somatotopic representation is found up to the striate cortex (area 17), were there are multiple maps for features such as color and orientation (see e.g. Barlow, 1986; Cowey, 1979; Kaas, 1987; van Essen & Maunsell, 1983). Beyond the striate cortex, the somatotopic organization is not generally preserved, and neither is responding of single cells always spatially selective. There are areas where there is little or no spatial selectivity; i.e., cells respond to a stimulus pattern in the central visual field largely irrespective of its localization. This seems to be the case in the inferotemporal cortex (areas 20 and 21), which is probably related to the identification of complex stimulus patterns (Desimone, Schein, Moran, & Ungerleider, 1985; Mishkin, Ungerleider, & Macko, 1983; Ungerleider & Mishkin, 1982).

By contrast, the enhancement effect in the posterior parietal cortex is spatially selective, although most of the receptive fields (which differ vastly in size) are larger than those in the striate cortex (Robinson et al., 1978). While position is important, the visual attributes of the stimulus (color, orientation, etc.) are not (Robinson et al., 1978). These are the characteristics that one would expect in a system whose function it is to select between competing stimuli on the basis of their spatial position. Indeed, it has been suggested that the posterior parietal cortex is part of a visual 'where' system as opposed to the geniculostriate-inferotemporal 'What' system (e.g. Mishkin et. al., 1983; Ungerleider & Mishkin, 1982; see also Harter & Aine, 1984, and the instructive review by Allport, 1987).

The evidence for the inherently spatial nature of visual attentional selection is not restricted to findings from single-cell recordings. There are at least three further, independent sources of corroborative evidence.

First, it has long been known that lesions in the posterior parietal cortex affect selective attention as well as spatial orientation. Patients with the so-called hemineglect (or unilateral neglect) syndrome (for reviews see, e.g., De Renzi, 1982; Heilman & Valenstein, 1985; Jeannerod, 1987; Lynch, 1980; Weinstein & Friedland, 1977) are typically impaired both in their ability to actively attend to the side contralateral to the lesion (e.g. Posner, Walker, Friedrich, & Rafal, 1984, 1987) and in their orientation in space. For example, they are often unable to make or use maps, to draw a diagram of a familiar room or to describe what a model would look like from a different perspective. This suggests a close functional connection between the control of visual attention and the internal representation of space.

Second, findings from event-related potential (ERP) studies indicate that space is functionally different from other dimensions with respect to visual selective attention. If subjects are asked to attend to a location in space, stimuli presented at that position produce an enhanced ERP that begins to differ from that at a nonattended position as early as 80-130 ms after stimulus onset (Harter & Aine, 1985; Harter, Aine, & Schroeder, 1982; Hillyard & Münte, 1984; Hillyard et al., 1985; Rugg, Milner, Lines, & Phalp, 1987). This difference is most pronounced at scalp locations that probably reflect mainly activity in the posterior parietal cortex (Harter & Aine, 1984; Rugg et al., 1987).

The picture is different if subjects attend to a dimension other than space. With color or orientation as the critical attribute, the ERP enhancement occurs later (starting at about 150-200 ms) and is most pronounced at fronto-central electrode sites. Particularly interesting is a result from Hillyard and Münte's (1984) study, in which both the color and the (easily discriminable) location of the stimulus could be either attended or unattended. When a stimulus was at an attended position, then the entire wave sequence from 80-90 ms to 300 ms was enhanced, regardless of whether the color was attended or unattended. Conversely, the enhancement of later components, indicative of color selection, was much larger at the attended than at the unattended location. This asymmetry strongly suggests that subjects could attend to location independently of color, but not to color independently of location.

The third group of relevant findings comes from behavioral studies. In the experimental psychology of attention there has recently been a shift of interest from the auditory to the visual modality (see, e.g., Neumann, van der Heijden, & Allport, 1986b; Bisiach & Rizzolatti, 1987; Umiltá, 1988). One particularly useful methodological tool in this rapidly expanding research area has been the cuing experiment described earlier in this chapter. In its presently most-often used form it was introduced by Posner and associates (Posner et al., 1978; Posner, Snyder, & Davidson, 1980), although similar paradigms were used by earlier investigators (e.g. Engel, 1971; Eriksen & Hoffman, 1973,1974; van der Heijden & Eerland, 1973).

The basic finding from this paradigm is that spatial expectancy (usually induced by a cue signal that informs the subject where the next response signal is likely to occur) affects the efficiency (speed and/or accuracy) with which the response signal is processed. There are two kinds of effects: First, signals that are presented at a nonexpected location are processed less efficiently than signals in a control condition with no specific spatial expectancy (the 'cost' effect). While this effect is observed very regularly, the opposite effect - 'benefit' from attending to the position where the response signal appears - has been found often, but not always (for recent summaries see, e.g., Eriksen & Yeh, 1985; Müller & Findlay, 1987; Stoffer, 1988; Tassinari et al., 1987; van der Heijden, Wolters, Groep, & Hagenaar, 1987; Zimba & Hughes, 1987).

Another often-reported finding has been that the amount and direction (cost vs benefit) of the cue's effect depend on its spatial distance from the target (e.g. Downing & Pinker, 1985; Hughes & Zimba, 1987; Posner et al., 1980; Rizzolatti et al.,

1987; Shulman, Sheehy, & Wilson, 1986; Tassinari et al., 1987; Zimba & Hughes, 1987). The usual interpretation of these effects has been in terms of an attentional 'beam,' or 'spotlight,' or 'zoom lens' of variable size that can be shifted across the visual display (e.g. Crick, 1984; Downing & Pinker, 1985; Eriksen & Yeh, 1985; Hurlbert & Poggio, 1985; Neumann, 1980; Posner, 1980; Shulman, Remington & McLean, 1979; Tsal, 1983). Though the usefulness of this metaphor has recently been questioned (La Berge & Brown, 1989; Müsseler & Neumann, submitted; Shepherd & Müller, 1989; Stoffer, 1988), the main notion that it has been intended to capture - that visual selective attention always involves the selection of a location in space - seems to be empirically well founded.

To summarize this section, there is converging evidence for the intrinsically spatial character of attentional selection in vision. Findings from at least four fields of investigation, obtained with different methodologies (single-cell recordings, clinical observations, ERP recordings, cuing experiments), all indicate that space has a special status as a dimension for selection. Like attributes such as color or form, spatial location can be used as a selection cue. But in addition, and more importantly, *all* attentional selection processes in vision seem to take place within a spatial representation of the visual environment, and they seem to always involve the choice of a location within this spatial representation.

We began with the question of what makes attentional selection-for-action special as compared with other instances of selection-for action. The findings that have been reviewed in this section suggest a fairly simple answer: There is a built-in selectivity in all motor control structures, based on their specific input-output connections. This selectivity - i.e., the fact that only certain kinds of inputs can be used for parameter specification within each control structure - is *not* attentional. It does not require any mechanisms over and above those contained in the control structure itself. However, these mechanisms cannot successfully cope with the parameter specification problem in situations where different stimuli compete for the specification of the same parameter.

This is where attentional selection comes into play. The presently available evidence suggests that, in the primate visual system, the selection mechanisms are primarily (but not exclusively) located in the posterior parietal cortex; that the selection is sensory (i.e., consists in the selection between stimuli in the service of action control, not in the direct selection between motor programs); and that it is intrinsically spatial, i.e., involves the selection of a position in space.

This basic strategy of solving the parameter specification problem in the presence of competing input information seems to have been invented early in the evolution of vertebrates. As we have seen, there is evidence that it is already present in the frog. I have so far stressed this evolutionary continuity, because it indicates that the general functional characteristics of visual attention are indeed related to problems of action control, problems such as effector recruitment and parameter specification that are not fundamentally different for frogs and men. On the other hand, of course, evolutionary continuity has its limits. The mechanisms by which these problems are solved are far more complex in primates than in amphibians. New problems of action control appeared as the behavioral repertoire of animals became richer and more flexible. New solutions had to be invented to cope with these problems. In the next section I look into some aspects and consequences of this development.

Selection and Representation

Evolutionary Changes

There seem to be at least four major differences between the mechanisms of visual attention in the frog and those in primates. They regard, first, their neuroanatomical locations; second, their degree of specificity; third, the outcome of the selection process with respect to the selected input; and, finally, the fate of the nonselected input. I first briefly describe these differences and then discuss their possible functional relationship and their implications for an action-oriented approach to visual attention.

First, the mechanisms described by Ingle (1975) suggest the optical tectum as the major *neuroanatomical site* of attentional selection in the frog. Although both single-cell recordings in monkeys (e.g. Mohler & Wurtz, 1976) and evidence from split-brain human patients (Gazzaniga, 1987; Holtzman, Volpe, & Gazzaniga, 1984; Holtzman, Sidtis, Volpe, Wilson, & Gazzaniga, 1981) indicate some subcortical control of visual attention in primates, the majority of the anatomical sites that have been found to be probably involved in the control of visual attention (the pulvinar, the prelunate cortex, the posterior parietal cortex, and the frontal eye field) belong to the thalamico-cortical circuitry. This is an example of the increase in the functional significance of the neocortex during the evolution from lower to higher vertebrates (see e.g. Altman, 1978).

Second, there is a change in *degree of specificity*. The 'attention units' in the frog's tectum described by Ingle (1975) respond primarily to bug-like stimuli. This indicates that the selection mechanism is essentially an addition to the motor control structure for snapping at a bug. By contrast, as we have seen, the neurons in the primate posterior parietal cortex that seem most closely related to visual attention are relatively unselective with respect to both the attributes of the stimuli to which they respond and the motor control systems which can modulate their response. This suggests a common attentional system for different kinds of visual stimuli and in the service of different types of actions, instead of specific attentional mechanisms for each type of sensory feature and motor control structure[1].

In human experimental psychology, this unspecific character of visual attention seems to be so obvious that it is usually taken for granted by experimenters. For example, in the cuing experiments described in the last section, experimenters have used various kinds of cues (e.g., the brief dimming of a box or the flashing of a light), presumably on the assumption that the visual features of the cue do not matter. Similarly, different kinds of responses (e.g., button pressing or letter naming) have been employed interchangeably, again implying that different kinds of motor output are controlled by the same attentional system.

The two further major differences between visual attention in the frog and in primates concern the results of the selection process with respect to (a) the selected stimulus and (b) the rejected stimuli. In the frog, the selected 'bug' is used to control the snapping response, and the rejected 'bug' is not used at all. In both respects the outcome is different in primates.

What happens to *ignored information*? In the frog's tectum, rejection is reflected in the strong suppression of activity in the neural unit that is stimulated by the nonselected stimulus. This 'early selection' clearly makes functional sense. The frog selecting a fly to snap at has to discard the information from other flies. The more quickly and the more completely the neural activity caused by all flies but one is suppressed, the better for the frog's survival (see Arbib, 1981, Lara & Arbib, 1982, and Arbib & House, 1987, for suggestions as to how such a mechanism might work). As we have seen, similar examples of selection by suppression have been found in higher vertebrates, e.g., in the cat's and the monkey's superior colliculus and in area V4 of the monkey. However, the more common selection mechanism seems to be based on enhancement; i.e., the attended stimulus causes additional neural activity, but the activity produced by unattended stimuli remains unaffected by the selection process.This suggests some functional advantage of not completely obliterating the nonselected information in the selection process.

The fourth evolutionary change that deserves consideration regards the further *fate of the selected information* subsequent to selection. In the frog, the selection of one from among the competing bugs serves the unambiguous specification of the snapping response's parameters. Once the bug has been selected, the further processing is presumably not different from that in the other cases of parameter specification for the immediate control of action, described earlier. To put it negatively, there is no reason to assume that, as a consequence of attentional selection, the selected information serves any purpose other than that of guiding the ongoing action. This is clearly different in higher vertebrates. Mammals, whose sensory systems are often equipped with their own motor apparatus (see Gibson, 1966; Trevarthen, 1978), have the capability to actively explore their environment. Actions like sniffing, touching and - most important in visual animals such as primates - visual exploration by means of eye and head movements, are similar to other actions in that their parameters (e.g., where to look, what to touch) are partially specified by sensory information. But obviously the sensory information does not serve *only* this purpose. In exploratory activity, not only is information used to guide action; more important, action is used to acquire information.

The information is usually not acquired for the guidance of other simultaneous actions. Exploratory activity competes for effector recruitment just as any other action does, and exploration normally engages the whole organism, leaving little room for current actions. The obvious main function of exploration is not to control ongoing action, but to gain knowledge, i.e., to store the acquired information for later use in the guidance of future action. A classical experimental example is the demonstration by Tolman and his co-workers that rats who are allowed to explore a maze will later learn a path through this maze faster than rats who had no opportunity to explore it (for a review see Thistlethwaite, 1951). This implies that the information about the spatial layout of the maze, acquired through exploration, is stored in a format unrelated to any particular movement sequence (as a 'cognitive map,' in Tolman's terminology).

To summarize, there is evidence for four major evolutionary changes from lower vertebrates to mammals, and especially primates, with respect to the control of visual attention: A corticalization of the attentional mechanisms, a shift from specific to unspecific mechanisms, a trend from suppression-type to enhancement-type selectivity, and the acquisition of exploratory behavior, i.e., selection of information from the environment for storage and later use in addition to selection for the guidance of ongoing action.

In the following I will argue that these changes can be viewed as different aspects of a general reorganization of the attentional system: In addition to selection-for-action, attentional mechanisms were developed for what may be called selection-for-exploration. I will first elaborate this hypothesis and then discuss some of its implications for attentional theory.

One way of elaborating the hypothesis is to link it to more general changes in the course of evolution from lower vertebrates to mammals. There is, indeed, a suggestion about the evolution of the brain and intelligence that provides a functional framework for our hypothesis. It was first put forward by Jerison (1973; see also Jerison, 1982, 1988).

The starting point of Jerison's (1973) theory is the finding that there is a difference in encephalization between lower vertebrates (amphibians and reptiles) and higher vertebrates (birds and mammals). Within each of these categories, we find a remarkable stability of the encephalization quotient (a measure for the ratio of brain weight to body weight). Large animals need a larger brain than smaller organisms do. But if one takes body weight into account, then there is, for example, no systematic difference between saurians and present-day reptiles. Similarly, the encephalization quotient is roughly the same for most mammals, ancient or recent, large or small (with a few exceptions, notably primates and dolphins). This suggests that mammals have developed their higher degree of encephalization not as a result of a general evolutionary premium on being more intelligent than others, but due to a particular ecological pressure.

Jerison's hypothesis is that this development began in the late Triassic period about 200 million years ago. At this time, small reptiles that were to become the ancestors of the mammalian branch occupied the ecological niche of night activity, based on their ability to keep temperature essentially constant despite changes in the temperature of the environment (homoiothermy). Night activity requires relying on other than visual information, in particular on acoustical information, for guiding action. Hence, new neural circuitry had to be added to the brain, capable of translating the temporally encoded acoustical signal into a spatial map of the environment. For anatomical reasons this could not be done peripherally, after the model of the neural apparatus of the retina. Instead, there was the appearance of cortical mechanisms for analyzing stimuli and transforming the information conveyed by them into a map of the environment. This, according to this hypothesis, was a first step toward representing information, not merely using it for the immediate control of action.

Jerison suggests that there were two further steps. First, the ability to acquire distance information from two different sensory systems led to mechanisms of intersensory integration, which implied representing objects rather than patterns of stimulation. A second, major breakthrough came at the end of the Mesozoic era, about 70 million years ago, when the saurians died out and mammals could invade the ecological niches open to diurnal land animals. This required 'returning' to visual stimuli as the main source of information, but this was not simply a falling back to the ancient, mainly retinal, mechanisms of the reptiles. Rather, the newly developing visual system was modeled after the auditory sense which had been evolving for 100 million years. Cortical rather than peripheral mechanisms were invented for visual processing. The result was the capacity to construct a visual representation of the world that contains objects and their relations - the basis for the development that finally produced human intelligence.

Though Jerison's theory is speculative in its details, his central postulate, that the development of mammalian intelligence was critically related to the acquisition of a representational system, seems now to be widely accepted (see e.g. Boden, 1988; Csanyi, 1988; Hoffman, 1988). In our present context, this hypothesis is attractive because the four evolutionary changes in visual attention described above can be directly related to the development of the capacity to internally represent the visual environment:

(a) Corticalization: Recall that Jerison's original observation was an evolutionary step in encephalization from lower vertebrates to mammals. Although encephalization cannot generally be equated with corticalization (see, e.g., Hodos, 1988), the difference in relative brain weight between higher and lower vertebrates is due mainly to the growth of the neocortex (cf. Altman, 1978). Indeed, Jerison's suggestion that the major new acquisition of mammals was their capability to internally represent their environment bears a striking similarity to current neurophysiological conceptions of the functions of the neocortex. As summarized by Barlow (1986), the cortex controls behavior "...that requires accurate knowledge and understanding of an individual's normal environment .... Thus ... the cortex forms and stores what Craik (1943) called a working model of the environment and what would nowadays probably be called a cognitive map of it" (Barlow, 1986, p. 88). If this view of the function of the cortex is correct, then the corticalization of the mechanisms of visual attention suggests that these mechanisms are at least in part related to internally representing the environment, and not merely to the selection of information for the immediate control of action.

(b) Unspecific selection: The spatial character of visual attention implies that selection takes place from a somatotopic map that includes the stimuli to be selected as well as those to be rejected. In the frog, with its restricted repertoire of relatively rigid actions, it is conceivable that each action has its corresponding map for this purpose; at least this seems to be the case for the snapping response. If there is, however, an internal representation of the visual environment not tied to any specific action, then this representation can be efficiently used for the guidance of action only if it is complemented by an equally action-unspecific selection mechanism. (An analogy is a data bank in a multi-user configuration. Although each user may employ it for a different purpose, all use the same program for accessing the stored data.) An additional advantage of such an unspecific attentional system is that its selections will constrain *all* actions, thereby reducing the chance that physically incompatible actions will be attempted. The development of a common attentional system may therefore have been a necessary concomitant to the evolutionary acquisition of the large repertoire of possible visually elicited and visually guided actions that characterizes higher vertebrates and especially primates.

(c) Fate of the selected information: As argued above, the updating of the internal representation is achieved mainly through exploratory behavior. In accordance with this suggestion, the high mobility of the sense organs, which is a prerequisite for active exploration (see Bridgeman, this volume), first appears in mammals (Trevarthen, 1978), i.e., it has developed together with the capacity for internally representing the environment. Active exploration as a means for updating the internal representation has several functional advantages over passive receptivity. First, it procures information that is not available to a passive receptor system (e.g., active vs passive touch; see Gibson, 1966, 1979). Second, the visual environment of an animal is continuously changing. Many of these changes are irrelevant to the control of further action. If the internal representation is updated through active exploration, then only those changes that elicit exploratory behavior will affect its content. Thus, the internal representation will reflect important changes in the environment, while it is relatively stable against minor fluctuations.

(d) Fate of the nonselected information: As noted earlier, the finding that, in primates, attentional selection in vision works mainly by enhancement rather than by inhibition suggests some functional advantage of not completely obliterating the 'rejected' information. According to Allport (1987; Allport et al., 1985), the biological advantage of selective enhancement as a method of priority assignment is that it requires neither the suppression of information that has already been encoded nor the restriction of processing to a small subset of the available information. This is useful because "if and organism's choice of action at any instant is to be efficient, that choice must be based on as rich as possible a description of its surroundings" (Allport et al., 1985, p. 109).

The 'description' of Allport et al. is similar to the internal representation, as the term is used here. However, it is important to note that the internal representation in the present sense is not identical to the result of processing all the information that reaches the senses at a given moment. On the one hand, it does not encompass *all* the presently available sensory information. As I have just argued, only selected information (i.e., information currently selected for exploration) enters into the updating of the internal representation. On the other hand, the internal representation does not consist *only* of contents that correspond to presently available sensory information; it is also based on information acquired during earlier exploration, and it is this property that makes it a rich description, despite the selectivity of updating at any given moment.

Thus, according to the present view, those characteristics of visual attention that are specific to mammals, and especially primates, are closely related to the ability of these higher vertebrates to internally represent their environment. In the rest of the chapter I will explore some theoretical consequences of this suggestion. I will begin by elaborating the concept of internal representation. This has to be done in some detail, because the subsequent argument will be that, just as the basic properties of visual attention that we share with lower vertebrates derive from the requirements of immediate action control, so the specific properties of more highly developed mechanisms of visual attention derive from the requirements of updating the internal representation.

Representation, Perception, and Memory

Basic to the following considerations is the notion that the internal representation is comprehensive (comprehends the whole visual environment) without being complete (without representing unselectively the complete input information at a given moment). The reason why it can be comprehensive without being complete is what might be called its cumulative nature: Its state at a given moment in time reflects not only sensory information available at that moment, but likewise information that was acquired earlier. The internal representation has thus both 'perceptual' and 'memory' aspects.

To illustrate these two characteristic of the internal representation - its comprehensiveness and its cumulative character - let us consider the following two examples. One is from neuropsychology, the other is from experimental psychology.

As mentioned earlier, lesions in the posterior parietal cortex produce the syndrome known as unilateral neglect, hemineglect, or hemi-inattention (e.g. De Renzi, 1982; Heilman & Valenstein, 1979; Weinstein & Friedland, 1977). The deficits are complex and probably involve perceptual as well as motor difficulties in orienting and responding to stimuli from the side contralateral to the affected hemisphere, including 'covert' shifts of attention (Posner et al., 1984, 1987). The syndrome would not normally be classified as a memory impairment. Yet it includes striking memory deficits. Bisiach and Luzzatti (1978) and Bisiach, Capitani, Luzzatti, and Perani (1981) asked right brain-damaged, hemi-inattentive patients to describe a place that they were familiar with, the Cathedral Square of Milan, as viewed from two specified observation points. The patients showed a strong tendency not to report details that would have been on their left, viewed from the respective points. Bisiach et al. (1981) also asked patients to draw the hours in a blank clock dial from memory and found that they left out the left half of the dial. Similar observations have been reported by Bisiach, Luzzatti, and Perani (1979), Bisiach et al., (1985), and Bisiach, Berti and Vallar (1985). Thus, the deficit in selecting information from the contralateral hemifield goes together with an impairment in retrieval from the corresponding memory representation, resulting in "the inability to conceive of space as extending to the neglected side" (De Renzi, 1982, p. 111).

Besides pointing to the interrelatedness of 'perceptual' and 'memory' aspects of the internal representation of the visual environment, these clinical observations highlight the comprehensiveness of the internal representation: Normally, without such an impairment, the internal representation of the visual environment is comprehensive in the sense that it represents the whole surroundings. It extends in all directions, and there are no 'holes' in it. Further, the representation of the visual environment is nested within more general representations that, together, form what has been termed 'generalized reality-orientation' by Shor (1969). Related concepts are Sokolov's (1963) 'neuronal model,' MacKay's (e.g. 1978, 1984) 'state of conditional readiness,' Csanyi's (1988) 'dynamic model of the environment,' and the 'model of the world' as discussed by Arbib (1981) and Yates (1985). With the possible exception of states such as hypnotic trance (e.g. Shor, 1969) or awakening from deep anesthesia, or states of pathological disorientation, attentional selection takes place within the functional framework of these internal representations; i.e., the internal representation guides selection, and selection serves to update the internal representation.

One field of investigation in which this interplay between selection and representation can be studied for visual perception is our second example, the exploration of visual scenes. Even in experiments where subjects are confronted with a new visual scene by tachistoscopically presenting a picture - i.e., where the picture appears in a 'ghostlike manner' (Neisser, 1976) - there is clear evidence that the representation of the scene is comprehensive from the beginning. For example, Antes, Penland, and Metzger (1981) presented line drawings for only 100 ms. Following each presentation, subjects were required to choose from a set of four objects the one that had been contained in the drawing. Performance was dramatically affected by whether or not the drawing represented a coherent scene, and whether or not the target was an object usually to be expected in the scene. Together with further aspects of the overall pattern of results, this suggested to the authors that, with such a short exposure, the stimulus "provides the subject with a general idea of the meaning (or gist) of the picture but little in the way of local information" (Antes et al., 1981, p. 287).

Since the representation is comprehensive from the beginning, further inspection fills in details rather than concatenating isolated elements. Which details are filled in seems to depend both on semantic factors and on the action context. When subjects are given time to explore a picture by eye movements, the first saccades are typically directed to the most informative locations, i.e., those locations that are judged by independent raters to contribute most to the whole information conveyed by the picture (e.g. Antes, 1974; Loftus, 1976; Mackworth & Morandi, 1967). The influence of the action context has been demonstrated, for example, by Yarbus (1967; the fixation pattern depends on the question asked before subjects examine a picture), Gould (1976; trained visual quality inspectors fixate on areas of a chip likely to contain a target), and Kundel and Nodine (1978; the fixations of radiologists examining a chest film depend on the clinical history of the patient).

One might argue that these examples demonstrate nothing more than well-known interactions between perception and cognition, or between "bottom-up" and "top down" processes. They certainly show such effects, but at the same time they illustrate a more general principle. Authors who have reflected on the relationships between memory and perception have frequently pointed out that perception is extended in time, and hence it may be difficult or impossible to draw a dividing line between perception and memory (e.g. Gibson, 1979; Neisser, 1976). Indeed, it seems obvious that the internal representation of the visual environment is *never* based solely on the momentary retinal stimulation. It is always the result of an integration over time.

A simple demonstration of this is the Tandem Effect (Müsseler, 1987; Müsseler & Neumann, submitted). It consists of the following visual illusion: When two vertical bars move through a horizontal window in succession (i.e., the first has left the window before the second enters), they are (under appropriate conditions) perceived as being *concurrently* in the window. A succession of physical events is transformed into a simultaneous representation, even though the interval between the physical events is far above the psychophysical successiveness threshold. Similar observations have been reported for more than a century (e.g. Parks, 1965; Rock, Halper, DiVita & Wheeler, 1987; Vierordt, 1868; Zöllner, 1862; for reviews see Klopfer, 1985; Müsseler & Neumann, submitted).

At a different level, a similar integration over time takes place when the information from several saccades is combined into a representation of a scene. Although there is a succession of fixations, the scene is perceived as stationary ('integration of snapshots' in Neissers [1967] formulation). This is an example of temporal constancy, analogous to the directional constancy and positional constancy during saccades. While these spatial constancies have been intensively investigated (see Wolff, 1984, for an excellent review and theoretical treatment), the temporal constancy during saccadic exploration seems to have received little experimental or theoretical interest.

At still another level, there is integration of the visual information acquired during head movements, positional changes, and locomotion. As pointed out, e.g., by Gibson (1979), the effect of these overt bodily movements is to reveal hitherto invisible parts of the visual environment. Others disappear from sight but are not erased from the internal representation. Phenomenally, they are experienced as part of the real environment, not as something imagined (see e.g. Metzger, 1954). Functionally, they can be used to guide action.

Indeed, it can easily be seen that only a representation that integrates information across bodily movements is suitable for the control of action. This is because locomotion and manipulatory activity need to be based on an allocentric representation of the environment, i.e., a representation that depicts the surroundings in environmental coordinates rather than retinotopic or head-oriented coordinates (see e.g. Feldman, 1985; Saltzman, 1979). To illustrate, take a soccer player who has approached the goal and wants to head in the ball. The target location that he selects has to be specified in environmental coordinates, since it is likely that he will move after having selected it and before hitting the ball. Hence, the information that he acquires successively, while his body is in motion, has to be integrated into one simultaneous representation.

There is a simple principle behind all these examples: The internal representation has both 'perceptual' and 'memory' aspects, because it is, at several different levels, the result of scanning-like operations. A scanning device (e.g., a television camera) transforms a spatial array into a temporal sequence. To reconstruct the original spatial array from the scanning process, time has to be discarded below the order of one scanning cycle; i.e., all the information that is taken up during a scanning cycle must be treated as simultaneous (see e.g., Stroud, 1949). The cathode ray tube (and/or the eye of the human observer that watches it) 'remembers' the information transmitted by the ray for at least one scanning cycle (due to sufficiently slow decay). This is the 'memory' aspect. But the result is a stationary representation. This is the 'perceptual' aspect. Similarly, the internal representation is the result of exploration. Exploration is extended in time, but the representational format does not reflect this temporal structure. The succession of exploratory steps, each involving only a small portion of the environment, is transformed into a simultaneous representation of the whole environment. Exploration as a means of acquiring information and the existence of an internal representation that integrates information over time are two aspects of the same functional principle.

Consequences for Attentional Selection

This chapter began with two questions: First, if attentional selection is in the service of action control, how can we then account for nonattentional selection-for-action? Second, if attentional selection is in the services of action control, how can we then account for attentional selection that is not selection for the control of ongoing overt action? The first of these questions has been treated in the first part of the chapter, where functional differences between attentional and nonattentional selection of input information were discussed. We are now in a position to answer the second question.

Basically, the answer that emerges from the preceding considerations is simple. Higher vertebrates have developed a particular category of action - exploration - that serves to establish and update an internal representation of the environment. With respect to this category of action, attentional selection takes place in principally the same way as with respect to other categories of action. An organism cannot explore everything in the environment at the same time, just as it is impossible to grasp all apples on a tree. By the same token, exploration competes with alternative actions. In short, there are the same problems of effector recruitment and parameter specification that we have discussed with respect to these other categories of action. This is why selective attention is "capacity limited" in exploration, just as it is in the control of other categories of action.

However, in exploration overt action is not an end in itself. It serves the acquisition of information. This makes it sometimes possible to reduce the necessary effector movements to a minimum. All actions include, of course, central components. At least in part, these can be activated without actually performing the overt action (among the terms for different aspects of this are motor preparation, set, action planning, motor programming, and mental training). However, the intended goal can normally be attained only if the physical action finally takes place. By contrast, exploration can sometimes be performed - albeit in a reduced and usually less efficient manner - without actually carrying out the peripheral movements. For example, in our case of visual exploration, attention can be decoupled from eye movements. However, this requires an active effort by the observer, and there can be no doubt that it is a highly artificial situation. (As was pointed out by Ribot in 1906, young children are unable to decouple covert attention from overt action).

What makes selection-for-exploration different from the selection of sensory information for the guidance of immediate action is less the basic functional architecture of the selection process than its result, which, according to the present suggestion, is the updating of the internal representation of the visual environment. I will now discuss some functional consequences of this shift in the task of attentional mechanisms.

If our assumption is correct, that selection-for exploration leads to an updating of the internal representation, then it is in fact a double selection process: It is, first, the selection of the section of the internal representation which is to be updated; second, it is the selection of the input information that will produce the updating. By the same token, the result of the updating can be viewed from two sides: The selected input has received processing that the rest of the input information was not subjected to, and the internal representation has been partially changed, based on this input information.

If the internal representation were identical to a representation of the *presently* selected information, then these two aspects would coincide. We would then arrive at the dichotomy assumed by most information-processing theories of attention (see next section): Part of the information from the environment is selected and hence will be represented at some central level (full processing, conscious representation, "recovery," etc.), while the rest of the information receives only "preattentive" processing.

The selective-updating assumption, however, suggests a different picture. As we have seen, there is reason to assume that the internal representation is comprehensive. It encompasses the presently selected information, as well as information about those parts and aspects of the visual environment that are not presently selected for updating. The two kinds of information differ with respect to how closely they represent the actual stimulus situation, but not with respect to their representational format. Updating adjusts the internal representation to the actual stimulus situation, but it does not create a different kind of representation that was not available prior to updating.

Hence, it is necessary to sharply distinguish between two questions: First, is a given portion of the input information selected for exploration and subsequent updating at a given moment in time? Second, is there a section of the internal representation that represents this portion of the input information? As to the second question, our foregoing considerations suggest that *all* the input information is represented in the internal representation. In this sense, the present view is a "late selection" view. However, regarding the first question, only a small proportion of it will be the subject of exploration and updating. In this sense, the present view is an "early selection" view.

This position will become clearer when we reconsider an argument that was cited earlier in a slightly different context. According to Allport et al., (1985), enhancement instead of inhibition as a mechanisms of selection makes functional sense, because it has the two advantages of (a) not suppressing already encoded information and (b) not restricting processing to a subset of the available information.

The first of these arguments is in agreement with the present position. The result of a previous updating of the internal representation should certainly not be obliterated by a subsequent updating of a different section of the internal representation. Once processed, i.e., included in the internal representation, information should remain available even after a shift of attention to another location in space. Otherwise, the internal representation would be restricted to the currently attended input; i.e., it would not be a representation at all. Selection by enhancement is indeed the appropriate mode for updating an internal representation without destroying the information that it already contains, just as selection by inhibition is the appropriate selection mode for the control of ongoing action.

As to the second argument, the present view suggests that it is true in one sense and wrong in another sense. There is one sense in which processing should not be restricted to a subset of the available information. It relates to the need for updating the internal representation if there is a substantial change in the environment. There must be a way of deciding when a change in the environment should elicit an updating of the internal representation (and, possibly, other ensuing actions such as fleeing) and when it should be ignored. The mechanism that subserves this 'interrupt' function is usually termed the orienting reflex or orienting response (Sokolov, 1963, 1975; for overviews see Lynn, 1966; Kimmel, van Olst & Orlebeke, 1979; Rohrbaugh, 1984; Sokolov & Vinogradova, 1975).

In part, the orienting response is probably mediated by subcortical mechanisms, but in mammals, and particularly in primates, it involves mainly cortical mechanisms, up to and including the frontal cortex (see e.g. Butter, 1987; Pribram, 1979). This corticalization reflects the growing complexity of the criteria that define 'novelty' and 'significance.' For example, in human subjects, the orienting response can be influenced via a verbal instruction that defines which stimuli are potentially significant for the task (e.g. Bernstein & Taylor, 1979; van Olst, Heemstra, & ten Kortenaar, 1979). Since the function of the orienting response is to react to stimuli that are not already attended, the mechanism that decides whether a stimulus is novel and/or significant according to these criteria must be responsive to nonattended stimuli. Complete suppression of this stimulus would render this impossible. In this sense, processing should indeed not be restricted to a subset of the available information. All stimulus information should have *access* to the internal representation, i.e., should be processed to a level that allows evaluating whether it should elicit exploration and a subsequent updating of the internal representation. Possibly, this includes some "semantic" analysis.

However, this does not mean, of course, that all the input information actually *affects* the internal representation. In this sense there is no "complete processing" according to the present view. If updating is mediated via exploration, and if exploration is subject to the physical and functional constraints that it shares with all actions, then complete processing in this sense is simply impossible. Even if it were possible, there would be no functional sense in adjusting the internal representation even to minor and accidental fluctuations in input information. As mentioned earlier, it is the selectivity of updating that gives the internal representation the necessary stability.

The Nature of Attentional Selection in Vision

The view of visual attention that has been developed in this chapter differs from most current theories of attentional selection that have been inspired by the information processing approach. In conclusion, I will contrast the two standpoints and present some arguments in favor of the present approach. The juxtaposition of the two views will also help to further clarify what has been suggested so far.

The mainstream, information-processing approach to attention, whose general theoretical framework - scarce capacity, selection in the service of coping with this limited capacity, etc. - was described in the introduction, includes a general point of view about selection. Let me call it the modal model. It encompasses several interrelated ideas, most of which rest on two basic assumptions: the two-stage assumption and the awareness assumption. In the following I will discuss only these two assumptions.

The *two-stage assumption* states that there are two distinct, functionally different stages of processing, one preattentive and the other attentional. This assumption seems to have originated from work on auditory and bisensory attention in the 1950s that led to Broadbent's (1958, 1971) filter theory[2]. While early work based on the filter theory was mostly in the auditory modality, a parallel distinction was introduced for visual information processing after the discovery of short-term visual information storage (Sperling, 1960; Averbach & Coriell, 1961; see van der Heijden, 1987, for a lucid summary of their work as well as of later related research). In this visual analogue of the filter model the first stage was iconic storage, as it was called by Neisser (1967); the second stage was readout from iconic memory. Neisser also introduced the term "preattentive processing," which he described as a crude and global stimulus analysis whose function it is to parse the stimulus pattern into primitive units which can then be selected for attentional processing. More recently, visual backward making has been used to distinguish between preattentive and attentional processing. According to Marcel (1983a,b), all stimuli are completely analyzed by processes with unlimited capacity, even if they do not reach a conscious representation, e.g., because of backward masking. In order to be represented in consciousness, the result of this first stage of processing has to be subjected to a second, capacity-limited process which Marcel calls recovery. Another variant of the two-stage assumption is Treisman and Gelade's (1980) proposal that only features are represented preattentively, while combining features requires attention.

These theories belong in part to the "early selection" and in part to the "late selection" camp and are usually regarded as being vastly different. However, they share the assumption that there is exactly one locus of selectivity. According to the present approach, this assumption is unwarranted. As discussed in the first part of the chapter, attentional selection is likely to be mediated by different mechanisms, some of them subcortical. For example, attentional effects have been found in the superior colliculus of the monkey (e.g. Mohler & Wurtz, 1976). Human ERP data suggests that part of the enhancement effect for stimuli at an attended position may be generated at the pulvinar (Rugg, Milner, Lines, & Phalp, 1987). It seems that some selectivity exists even further down toward the periphery. As reported by Eason, Oakley, and Flowers (1983) and confirmed by Eason (1984), the amplitudes of some components of the electro-retinogram (b-wave and afterpotential) are different for stimuli at attended and unattended positions. Thus, there seems to be a wide range of sites at which attentional selectivity operates, and some of them are probably located very early in processing. This is hard to reconcile with the two-stage view.

There are, to the best of my knowledge, no behavioral data that contradict this conclusion, although there is also no direct support for it from behavioral measurements. The ambiguity of most behavioral findings in this respect has a methodological reason: It is extremely difficult to assess the 'fate' of unattended stimuli by means of behavioral measurement. Only attended stimuli can normally be reported and responded to by subjects. The processing of unattended stimuli can be assessed only indirectly, based on uncertain generalizations.

Perhaps the most common of these generalizations are the following two: If it can be shown that *several attended* stimuli *can* be processed up to a given level (e.g., identification) without capacity limitation, then this indicates that *all unattended* stimuli are *always* processed up to this level[3]. Second, if *particular* stimuli (usually appearing at an attended position, and often semantically related to other components of the task, etc.) that the subject is unaware of can be shown to be semantically processed, then *all* stimuli are *always completely* semantically processed[4].

These generalization derive from the capacity model as discussed in the introduction. If one does not accept this model, there is no basis for them. If it can be shown that processing at some given stage is not capacity limited, this does not necessarily imply that it is unselective; there are other functional reasons for selective processing than limited capacity. By the same token, experiments such as backward masking studies tell us something about what kind of processing stimuli may undergo that, by some criterion - e.g., awareness - are processed "preattentively." But there is no reason to assume that these stimuli are processed unselectively. We cannot generalize from the processing of these particular stimuli to the processing of all stimuli. If, however, these generalizations are invalid, then there is, as far as I can judge, no empirical support for the two-stage assumption.

According to the present approach, the functional basis of "visual attention" combines phylogenetically old mechanisms that select information for the immediate control of action with more recent mechanisms that subserve the updating of the internal representation. Perhaps the internal representation may be called a processing stage. But access to it is not the only locus of attentional selectivity, and the internal representation comprises more than the presently selected information. Thus, if we regard the internal representation and the rest of processing as two separate stages, then these stages cannot be equated with the unselective/selective dichotomy.

Further, independent of whether there are two or more loci of selectivity, it is by no means certain that attentional selection has at all the character of a selective transition. The selective transition idea has been so dominant that theoretical alternatives seem to have been substantially overlooked. This is somewhat surprising, since the neurophysiological data on visual attention argue against the selective transition assumption. So far, neither single-cell recordings nor the study of ERPs seem to have produced evidence that attended visual stimuli are processed in subsystems which nonattended stimuli do not attain. The findings discussed in the first part of the chapter show as a rule a change in output strength (enhancement of a cell's response, increase in the amplitude of ERP components) as a result of attentional manipulations, not the complete absence of a response to unattended stimuli. To put it in neuroanatomical terms, there are several areas where attended and nonattended stimuli produce different effects, but so far, none seem to have been found where nonattended stimuli produce no effect whatsoever. This suggests that the selection performed in visual attention is not a *transition to* a particular representational system, but a *modulation within* systems to which both attended and unattended stimuli have access.

Of course, this is not yet conclusive evidence. It can be argued that functional subsystems need not be tied to brain areas; that a dichotomy at the functional level may be implemented as a difference in degree at the neuronal level; that the neuroanatomical correlate of the selective transition may not yet have been found. However, though not conclusive, the neurophysiological evidence is certainly suggestive. Possibly, many effects of attentional selection can be explained as a change in mode or degree of processing rather than as a transition to a different stage.

Although there is not yet sufficient evidence to settle this question, the present approach suggests that selective modulation and selective transition may be mutually complementary mechanisms rather than theoretical alternatives. If inhibition (probably mainly in selection for the immediate control of action) and enhancement (probably mainly in selection for exploratory updating of the internal representation) are the basic mechanisms of selection, then the selective process will at least *start* as a modulation of neuronal activity subsystems. This is likely to be true for all levels at which there is attentional selectivity, perhaps beginning as early as the retina.

However, if attentional selection is selection for the control action, then there is a need to *finally* transform the selective modulation into a selective transition. Decisions about whether or not to carry out an action, as well as decisions about how to specify an action's parameters, have an all-or-none character. This is also true for exploratory action and hence, according to the present suggestion, for the updating of the internal representation. A change in the environment will either trigger exploration or go unnoticed, leaving the internal representation in its present state. (But recall that exploratory updating will not change the relevant part of the internal representation from "precategorial" to "categorial," or from "literal" to "semantic." The internal representation is comprehensive; updating just makes it more complete.)

A second major assumption of the modal model is the *awareness assumption* which relates conscious awareness to attentional selection. Often, this is an implicit assumption. In its weakest from it says that awareness is an indicator of attentional selection; i.e., if a subject is aware of a stimulus, then this is evidence that the stimulus has been attended to. Stronger versions imply that attended stimuli are always represented in conscious awareness, nonattended stimuli never. In one or the other form, the awareness assumption seems to have been taken for granted by most proponents of the modal model. For example, it is the basis for the just-mentioned line of argument that uses data from backward masking experiments for theorizing about preattentive processing (e.g. Marcel, 1983a,b; for a review see Holender, 1986).

From the present point of view, there is no reason to tie attention to awareness. Attentional selection for the immediate control of action need not be, and probably often is not, accompanied by conscious awareness. As Neisser (1967) has noted, actions such as walking through a room or driving a car can be performed without awareness. Unfortunately, Neisser (1967) called this "preattentive control." Obviously, such actions require spatial selection as much as the action of picking an apple does; hence, as the term attention is used here, they are under attentional control.

The matter is more complicated when we turn to the internal representation of the visual environment. The updating of the internal representation may indeed be closely related to conscious awareness. Elsewhere (Neumann, 1987c) I have discussed the speculation that consciousness has evolved from the orienting response. So perhaps it is correct to say that, in exploration, attentional selection usually produces awareness of the selected stimuli. However, what about the internal representation if there is no updating, or those portions of the internal representation that are not being updated? Are they conscious or unconscious?

This is not an easy question. Since we are talking about a subjective phenomenon, a bit of introspection may help to clarify it. Let me reflect on visual attention from a phenomenological viewpoint for a moment. As I raise my eyes from the monitor screen that displays this text and look through the window, I see a huge wing of the university building facing me. Now I start shifting my attention, focusing first on the floor that is at about the same level as where I am working. Next I concentrate on one of the windows on this floor, and finally I focus attention on the piece of sky that is mirrored in the window. Thus at the outset of this attentional process the whole university wing was selected for attention, while at the end of it the object of my attention was the sky mirrored in one window. What has become of the university wing at this point in time? I am no longer attending to it, and yet it would be incorrect to say that I am not conscious of it. Metaphorically speaking, it has receded into the background of my consciousness. Its status has become that of my office room, the computer keyboard, my own body, and many other things that I am, in a sense, conscious of, although I do not pay attention to them.

It seems, then, that words such as "consciousness" and "awareness" do not refer to a unitary kind of experience (see Allport, 1988, for a strong argument on this point).

This phenomenological fact was, not surprisingly, well known to introspective psychology. Wilhelm Wundt, for example (e.g. Wundt, 1903), distinguished between the *field* (*Blickfeld*) of consciousness and the *focus* (*Blickpunkt*) of consciousness, i.e., the place where attention is directed. One may call the field of consciousness "preattentive," but is should be clear that this meaning of the term is completely different from the concept of preattentive processes in the sense of early steps in processing that have no conscious correlate whatsoever. To avoid this confusion, it would be better not to use the term "preattentive" at all.

In terms of the present theoretical notion of internal representation, the appropriate distinction is between a focal and a nonfocal component of the internal representation. Attentional selection during exploration makes previously nonfocal parts of the representation focal, while previously focal parts recede into the nonfocal component. To call the nonfocal component preattentive is misleading, and to call it nonconscious does not seem to be correct - at least if we trust phenomenology, which may be inevitable when talking about conscious awareness.

Conclusion

In a way, the main intent of this chapter has been to justify its title. Visual attention has usually been treated as a topic far away from action control. I have tried to show that it is possible to view it as closely related to action, and that such a view is profitable. It is, I believe, profitable at least in the sense that it provides a new look at some aspects of visual attention, integrates otherwise separate findings, and suggests new hypotheses.

Three decades ago, at the time when modern attention research began, very little was known about the physiology and neuropsychology of visual attention. This may have been one reason why the early psychological models took their inspiration more from communication technology than from the neurosciences. Meanwhile, the situation has radically changed. There have been methodological refinements in areas such as single-cell recording, EEG recording and analysis, and neuropsychological testing of patients, and these fields have added tremendously to our empirical knowledge. Psychological theories about visual attention that continue to ignore them simply disregard an important portion of what is empirically known about their subject matter.

This chapter's plea for an action-oriented view of visual attention has therefore also been a plea for a multidisciplinary approach. The present view is closer to the biological sciences than most other current psychological approaches are (for a notable exception see Allport, 1987). One aspect of this is that functional questions have played an important role in this chapter: What kinds of problems do organisms face in the control of their actions? How do mechanisms of visual attention contribute to solving these problems? How are the basic properties of visual attention - e.g., its 'limited capacity,' its sensory and its spatial character, mutual dependencies between selection and representation - related to the tasks of these mechanisms? Unlike biology, experimental psychology has often tried to circumvent functional questions of this kind. Sure enough, a functional analysis at such a general level is no substitute for more detailed analyses of more specific mechanisms. But it may be very difficult to understand how a mechanism works if we have no idea - or, worse, only implicit, unreflected ideas - about what it does.

Acknowledgements. The first draft of this chapter was written while I was a Fellow at the Netherlands Institute for Advanced Studies in the Humanities and Social Sciences (NIAS), Wassenaar, The Netherlands. I thank the NIAS for its generous hospitality and support. The chapter is partially based on an earlier, unpublished manuscript, on which I have received helpful comments from Donald G. MacKay, Dominic W. Massaro, Wolfgang Prinz, and A.H.C. van der Heijden.

Footnotes

[1] The suggestion that visual attentional mechanisms are not specific to particular motor control structures, but form a common system that can be used for the control of different types of actions, should not be confused with assumptions about the internal structure of this system. Rizzolatti and co-workers (Rizzolatti & Camarda, 1987; Rizzolatti, et al., 1985) have recently argued for a 'multicentric' view of visual attention instead of the notion of a 'single attentional center' in the posterior parietal lobe. This suggestion (which explains some findings from lesion studies, but needs further experimental support) is not necessarily incompatible with the notion of a common attentional system in the present sense. It states that there are different subsystems within this system, but not that each motor control structure has its own attentional subsystem.

[2] As Broadbent (1982) relates, one of the observations was that simultaneously presented stimuli (e.g., visual and auditory digits) could be recalled without impairment, indicating that there was memory storage not susceptible to capacity limitations. On the other hand, there was dual-task interference when events had to be noted and reacted to in both tasks, which pointed to the existence of a limited-capacity system located after the more peripheral memory stage. This led Broadbent (1958) to postulate two stages of processing, called the S system and the P system, with a filter between them whose function it was to protect the P system from being overloaded by the information that floats in from the S system.

[3] For example, in an otherwise excellent discussion of the problems involved in the 'early' vs 'late' selection debate, van der Heijden (1987) convincingly argues that in tachistoscopic experiments there is 'unlimited capacity processing' in the sense that, given certain conditions (such as sufficient spacing) targets can be detected in parallel without mutual interference (e.g., if there are several targets, each has an independent opportunity of being detected). Hence, as van der Heijden points out, the evidence from earlier partial report and masking experiments for a selective letter-by-letter 'readout' (e.g. Averbach & Coriell, 1961; Sperling, 1963) should not be interpreted as indicating selectivity at stimulus identification, but should instead be attributed to some later stage such as localizing the identified stimulus. So far, the argument is faultless. However, van der Heijden (1987) takes an additional step and concludes that there is 'unlimited capacity processing' in the sense that "all information registered is processed automatically and in parallel up to the stage of memory activation" (van der Heijden, 1987, p.

421). This is a possible conclusion, but it goes far beyond the empirical evidence. The data on the detection of multiple targets indicate that *several*, *attended* stimuli can be identified in parallel without 'limited capacity.' This does not imply that *all* visual stimuli - even those that are completely *irrelevant* to the task - are fully identified.

[4] This argument has been based on empirical indications that masked visual stimuli can be processed up to a semantic level without reaching conscious awareness, i.e., although subjects do not report having perceived them, or when the lack of a conscious representation has been ascertained by some other means, e.g., threshold measurement (e.g. Allport, 1977; Balota, 1983; Dagenbach, Carr, & Wilhelmsen, 1989; Leclerc & Freibergs, 1971; Humphreys, Evett & Taylor, 1982; Marcel, 1983a). While critical discussions of these studies have focused on methodological issues (Cheesman & Merikle, 1986; Holender, 1986), the logical fallacies that surround their interpretation have not always been noticed. The empirical findings suggest that under certain experimental circumstances it happens that a word of which the subject is not consciously aware is analyzed with respect to some of its semantic features, leading to some indirect effect, e.g., priming of a subsequent word. This does not imply that *all stimuli* that reach the retina are *always* analyzed with respect to *all* the meaning that they may convey. Such a generalization is particularly inappropriate for experiments in which the masked stimuli appeared at an attended position, as has usually been the case in these experiments (see Neumann, 1984; cf. Hoffman & MacMillan, 1985; Johnston & Dark, 1985).

References

Aldersen, G.J.K., Sully, D.J., & Sully, H.G. (1974). An operational analysis of one-handed catching using high-speed photography. *Journal of Motor Behavior, 6*, 217-226.

Allport, D.A. (1977). On knowing the meaning of words we are unable to report: The effects of visual masking. In S. Dornic (Ed.), *Attention and performance 6* (pp. 505-533). Hillsdale, NJ: Erlbaum.

Allport, D.A. (1980). Attention and performance. In G. Claxton (Ed.), *Cognitive psychology - New directions* (pp. 112-153). London: Routledge and Kegan Paul.

Allport, D.A. (1987). Selection for action: Some behavioral and neurophysiological considerations of attention and action. In H. Heuer & A.F. Sanders (Eds.), *Perspectives on perception and action* (pp. 395-419). Hillsdale, NJ: Erlbaum.

Allport, D.A. (1988). What concept of consciousness?. In A.J. Marcel & E. Bisiach (Eds.), *Consciousness in contemporary science* (pp. 159-182). Oxford: Clarendon.

Allport, D.A., Tipper, S.P., & Chmiel, N.R.J. (1985). Perceptual integration and postcategorical filtering. In M.I. Posner & O.S. Marin (Eds.), *Attention and performance 11* (pp. 107-132). Hillsdale, NJ: Erlbaum.

Altman, J. (1978). Three levels of mentation and the hierarchic organization of the human brain. In G.A. Miller & E. Lenneberg (Eds.), *Psychology and biology of language and thought* (pp. 87-109). New York: Academic.

Antes, J.R. (1974). The time course of picture viewing. *Journal of Experimental Psychology, 103*, 62-70.

Antes, J.R., Penland, J.G., & Metzger, R.L. (1981). Processing global information in briefly presented pictures. *Psychological Research, 43*, 277-292.

Arbib, M.A. (1981). Perceptual structures and distributed motor control. In V.Brooks (Ed.), *Handbook of physiology: Sec. 1. The nervous system: Vol. 2. Motor Control. Part 2* (pp. 1449-1480). Bethesda, MA: American Physiological Association.

Arbib, M.A., & House, D.H. (1987). Depth and detours: An essay on visually guided behavior. In M.A. Arbib & A.R. Hanson (Eds.), *Vision, brain, and cooperative computation* (pp. 129-163). Cambridge, MA: MIT Press.

Averbach, E., & Coriell, A.S. (1961). Short-term memory in vision. *Bell Systems Technical Journal, 40*, 309-328.

Ballard, H. (1986). Cortical connections and parallel processing: Structure and function. *Behavioral and Brain Sciences, 9*, 67-120.

Balota, D. (1983). Automatic semantic activation and episodic memory encoding. *Journal of Verbal Learning and Verbal Behavior, 22*, 88-104.

Barlow, H.B. (1986). Why have multiple cortical areas? *Vision Research, 26*, 81-90.

Bernstein, A.S., & Taylor K.W. (1979). The interaction of stimulus information with potential stimulus significance in eliciting the skin conductance orienting response. In H.D. Kimmel, E.H. van Olst, & G.F. Orlebeke (Eds.), *The orienting reflex in humans* (pp. 499-519). Hillsdale, NJ: Erlbaum.

Bisiach, E. & Luzatti, C. (1978). Unilateral neglect of representational space. *Cortex, 14*, 129-133.

Bisiach, E., & Rizzolatti, G. (Eds.) (1987). Visual selective attention. *Neuropsychologia, 25*. (Special issue).

Bisiach, E., Luzzatti, C., & Perani, D. (1979). Unilateral neglect, representational schema and consciousness. *Brain, 102*, 609-618.

Bisiach, E., Berti, A., & Vallar, G. (1985). Analogical and logical disorders underlying unilateral neglect of space. In M.I. Posner & O.S. Marin (Eds.), *Attention and performance 11* (pp. 239-249). Hillsdale, NJ: Erlbaum.

Bisiach, E., Capitani, E., & Porta, E. (1985). Two properties of space representation in the brain: Evidence from unilateral neglect. *Journal of Neurology, Neurosurgery and Psychiatry, 48*, 141-144.

Bisiach, E., Capitani, E., Luzzatti, C., & Perani, D. (1981). Brain and conscious representation of outside reality. *Neuropsychologia, 19*, 543-551.

Boden, M.A. (1988). Artificial intelligence and biological intelligence. In H.J. Jerison & I. Jerison (Eds.), *Intelligence and evolutionary biology* (pp. 45-71). Berlin, Heidelberg, New York, Tokyo: Springer.

Bridgeman, B., Kirch, M., & Sperling, A. (1981). Segregation of cognitive and motor aspects of visual function using induced motion. *Perception & Psychophysics, 29*, 336-342.

Broadbent, D.E. (1958). *Perception and communication*. Oxford: Pergamon.

Broadbent, D.E. (1971). *Decision and stress*. New York: Academic.

Broadbent, D.E. (1982). Task combination and selective intake of information. *Acta Psychologica (Amsterdam), 50*, 253-290.

Bruce, C.J., & Goldberg, M.F. (1984). Physiology of the frontal eye fields. *Trends in Neuroscience, 7*, 436-441.

Bushnell, M.C., Goldberg, M.E., & Robinson, D.L. (1981). Behavioral enhancement of visual responses in monkey cerebral cortex: I. Modulation in posterior parietal cortex related to selective visual attention. *Journal of Neurophysiology, 46*, 755-772.

Butter, C.M. (1987). Varieties of attention and disturbances of attention: A neuropsychological analysis. In M. Jeannerod (Ed.), *Neurophysiological and neuropsychological aspects of spatial neglect* (pp. 1-23). Amsterdam: Elseviers North-Holland.

Cheesman, J., & Merikle, P.M. (1985). Word recognition and consciousness. In D. Besner, T.G. Waller, & G.E. MacKinnon (Eds.), *Reading research: Advances in theory and practice* (pp. 311-352). New York: Academic.

Cherry, C. (1953). Some experiments on the recognition of speech, with one and with two ears. *Journal of the Acoustical Society of America, 25*, 975-979.

Cowey, A. (1979). Cortical maps in visual perception. *Quarterly Journal of Experimental Psychology, 31*, 1-17.

Crick, F. (1984). The function of the thalamic reticular complex: The searchlight hypothesis. *Proceedings of the National Academy of Sciences of the United States of America, 81*, 4586-4590.

Crowne, D.P. (1983). The frontal eye field and attention. *Psychological Bulletin, 93*, 232-260.

Csanyi, V. (1988). Contribution of the genetic and neural memory to animal intelligence. In H.J. Jerison & I. Jerison (Eds.), *Intelligence and evolutionary biology* (pp. 299-318). Berlin, Heidelberg, New York, Tokyo: Springer.

Dagenbach, D., Carr, T.H., & Wilhelmsen, A. (1989). Task-induced strategies and near-threshold priming: Conscious influences on unconscious perception. *Journal of Memory and Language, 28*, 412-443.

Dark, V., Johnston, W.A., Myles-Worsley, M., & Farah, M.J. (1985). Levels of selection and capacity limits. *Journal of Experimental Psychology: General, 114*, 472-497.

De Renzi, E. (1982). *Disorders of space exploration and cognition*. Chichester: Wiley.

Desimone, R., Schein, S.J., Moran, J., & Ungerleider, L. (1985). Contour, color and shape analysis beyond the striate cortex. *Vision Research, 25*, 441-452.

Deutsch, J.A., & Deutsch, D. (1963). Attention: Some theoretical considerations. *Psychological Review, 70*, 80-90.

Downing, C.J. & Pinker, S. (1985). The spatial structure of visual attention. In M.I. Posner & O.S. Marin (Eds.), *Attention and performance 11* (pp. 171-187). Hillsdale, NJ: Erlbaum.

Eason, R.G. (1984). Selective attention effects on retinal and forebrain responses in humans: A replication and extension. *Bulletin of the Psychonomic Society, 22*, 341-344.

Eason, R.G., Oakley, M., & Flowers, L. (1983). Central neural influences on the human retina during selective attention. *Physiological Psychology, 11*, 18-28.
Engel, F.L. (1971). Visual conspicuity, directed attention and retinal locus. *Vision Research, 11*, 563-576.
Eriksen, C.W., & Hoffman, J.E. (1973). The extent of processing of noise elements during selective encoding from visual displays. *Perception and Psychophysics, 14*, 155-160.
Eriksen, C.W., & Hoffman, J.E. (1974). Selective attention: Noise suppression or signal enhancement? *Bulletin of the Psychonomic Society, 4*, 587-589.
Eriksen, C.W., & Yeh, Y.-Y. (1985). Allocation of attention in the visual field. *Journal of Experimental Psychology: Human Perception and Performance, 5*, 583-597.
Feldman, J.A. (1985). Four frames suffice: A provisional model of vision and space. *Behavioral and Brain Sciences, 8*, 265-313.
Fischer, B. (1986). The role of attention in the preparation of visually guided eye movements in monkey and man. *Psychological Research, 48*, 251-257.
Gazzaniga, M.S. (1987). Perceptual and attentional processes following callosal section in humans. *Neuropsychologica, 25*, 119-133.
Gibson, J.J. (1966). *The senses considered as perceptual systems*. Boston: Houghton Mifflin.
Gibson, J.J. (1979). *The ecological approach to visual perception*. Boston: Houghton Mifflin.
Goldberg, M.E., & Bruce, C.J. (1985). Cerebral cortical activity associated with the orientation of visual attention in the rhesus monkey. *Vision Research, 25*, 471-481.
Goldberg, M.E., & Seagraves, M.A. (1987). Visiospatial and motor attention in the monkey. *Neuropsychologica, 25*, 107-118.
Gould, J.D. (1976). Looking at pictures. In R.A. Monty & G.W. Senders (Eds.), *Eye movements and psychological processes* (pp. 323-345). Hillsdale, NJ: Erlbaum.
Harter, M.R., & Aine, C.J. (1984). Brain mechanisms of visual selective attention. In R. Parasuraman & D.R. Davies (Eds.), *Varieties of attention* (pp. 293-321). New York: Academic.
Harter, M.R., Aine, C., & Schroeder, C. (1982). Hemispheric differences in neural processing of stimulus location and type: Effects of selective attention on visual evoked potentials. *Neuropsychologia, 20*, 421-438.
Heilman, K.M., & Valenstein, E. (1979). Mechanisms underlying hemispheric neglect. *Archives of Neurology, 5*, 166-170.
Heilman, K.M., & Valenstein, E. (1985). The neglect syndrome. In J.A.M. Frederiks (Ed.), *Handbook of clinical neurology: Vol 1(45). Clinical neuropsychology*. New York: Elsevier.
Hillyard, S.A., & Münte, T.F. (1984). Selective attention to color and location: An analysis with event-related brain potentials. *Perception & Psychophysics, 36*, 185-198.
Hillyard, S.A., Münte, T.F., & Neville, H.J. (1985). Visual-spatial attention, orienting and brain physiology. In M.I. Posner & O.S. Marin (Eds.), *Attention and performance 11* (pp. 63-84). Hillsdale, NJ: Erlbaum.
Hinde, R.A. (1970). *Animal behavior: A synthesis of ethology and comparative psychology*. New York: McGraw-Hill.
Hirst, W., & Kalmar, D. (1987). Characterizing attentional resources. *Journal of Experimental Psychology: General, 116*, 68-81.
Hodos, W. (1988). Comparative neuroanatomy and the evolution of intelligence. In H.J. Jerison & I. Jerison (Eds.), *Intelligence and evolutionary biology* (pp. 299-318). Berlin, Heidelberg, New York, Tokyo: Springer.
Hoffman, J.E. (1986). Spatial attention in vision. Evidence for early selection. *Psychological Research, 48*, 221-229.
Hoffman, J.E., & MacMillan, F.W. (1985). Is semantic priming automatic?. In M.I. Posner & O.S. Marin (Eds.), *Attention and performance 11* (pp. 585-599). Hillsdale, NJ: Erlbaum.
Hofman, M.A. (1988). Brain, mind, and reality: An evolutionary approach to biological intelligence. In H.J. Jerison & I. Jerison (Eds.), *Intelligence and evolutionary biology* (pp. 299-318). Berlin, Heidelberg, New York, Tokyo: Springer.
Holender, D. (1986). Semantic activation without conscious identification in dichotic listening, parafoveal vision, and visual masking: A survey and appraisal. *Behavioral and Brain Sciences, 9*, 1-66.
Holtzman, J.D., Sidtis, J.J., Volpe, B.T., Wilson, D.H., & Gazzaniga, M.S. (1981). Dissociation of spatial information for stimulus localization and the control of attention. *Brain, 104*, 861-872.
Holtzman, J.D., Volpe, B.T., & Gazzaniga, M.S. (1984). Spatial orientation following commissural section. In R. Parasuraman & D.R. Davies (Eds.), *Varieties of attention* (pp. 375-394). New York: Academic.

Hughes, H.C., & Zimba, L.D. (1987). Natural boundaries for the spatial spread of directed visual attention. *Neuropsychologia, 25*, 5-18.

Humphreys, G.W., Evett, L.J., & Taylor, D.E. (1982). Automatic phonological priming in visual word recognition. *Memory & Cognition, 10*, 576-590.

Hurlbert, A., & Poggio, T. (1985). Spotlight on attention. *Trends in Neuroscience, 8*, 309-311.

Hyvärinen, J. (1982). *The parietal cortex of monkey and man*. Berlin, Heidelberg, New York, Tokyo: Springer.

Ingle, D. (1975). Focal attention in the frog: Behavioral and physiological correlates. *Science, 188*, 1033-1035.

Jeannerod, M. (Ed.) (1987). *Neurophysiological and neuropsychological aspects of spatial neglect*. Amsterdam: Elsevier North-Holland.

Jerison, H.J. (1973). *Evolution of the brain and intelligence*. New York: Academic.

Jerison, H.J. (1982). The evolution of biological intelligence. In R.J. Sternberg (Ed.), *Handbook of human intelligence* (pp. 723-791). New York: Cambridge University Press.

Jerison, H.J. (1988). Evolutionary biology of intelligence: The nature of the problem. In H.J. Jerison & I. Jerison (Eds.), *Intelligence and evolutionary biology* (pp. 1-11). Berlin, Heidelberg, New York, Tokyo: Springer.

Johnston, W.A., & Dark, V. (1985). Dissociable domains of selective processing. In M.I. Posner & O.S. Marin (Eds.), *Attention and performance 11* (pp. 567-583). Hillsdale, NJ: Erlbaum.

Kaas, J.H. (1987). The organization of neocortex in mammals: Implications for theories of brain function. *Annual Review of Psychology, 38*, 129-151.

Keele, S.W., & Neill. W.T. (1978). Mechanisms of attention. In E.C. Carterette & M.P. Friedman (Eds.), *Handbook of perception: Vol. 9. Perceptual processing* (pp. 3-47). New York: Academic.

Kerr, B. (1983). Memory, action and motor control. In R.A. Magill (Ed.), *Memory and control of action* (pp. 47-65). Amsterdam: North-Holland.

Kimmel, H.D., van Olst, E.H., & Orlebeke, J.F. (Eds.). (1979). *The orienting reflex in humans*. Hillsdale, NJ: Erlbaum.

Klopfer, D.S. (1985). Constructing mental representations of objects from successive views. *Journal of Experimental Psychology: Human Perception and Performance, 11*, 566-582.

Kramer, U., & Rohr, G. (1982). A model of driver behavior. *Ergonomics, 25*, 891-907.

Kundel, H.L., & Nodine, C.F. (1978). Studies of eye movements and visual search in radiology. In J.W. Senders, D.F. Fischer, & R.A. Monty (Eds.), *Eye movements and the higher psychological functions* (pp. 317-328). Hillsdale, NJ: Erlbaum.

LaBerge, D. & Brown, V. (1989). Theory of attentional operation in shape identification. *Psychological Review, 96*, 101-124.

Lara, R., & Arbib, M.A. (1982). A neural model of interaction between tectum and pretectum in prey selection. *Cognition and Brain Theory, 5*, 149-171.

Leclerc, C., & Freibergs, V. (1971). L'influence d'indices subliminaux perceptifs et symboliques sur la formation d'un concept. *Canadian Journal of Psychology, 25*, 292-301.

Lee, D.N. (1976). A theory of visual control of braking based on information about time-to-collision. *Perception, 5*, 437-459.

Lee, D.N. (1980). Visuo-motor coordination in space-time. In G.E. Stelmach & J. Requin (Eds.), *Tutorials in motor behavior* (pp. 281-295). Amsterdam: North-Holland.

Lee, D.N., & Aaronson, E. (1974). Visual proprioceptive control of standing in human infants. *Perception & Psychophysics, 15*, 529-532.

Lee, D.N., & Lishman, J.R. (1975). Visual proprioceptive control of stance. *Journal of Human Movement Studies, 1*, 87-95.

Lee, D.N., & Lishman, R. (1977). Visual control of locomotion. *Scandinavian Journal of Psychology, 18*, 224-230.

Lee, D.N., Lishman, R., & Thomson, J.A. (1982). Regulation of gait in long jumping. *Journal of Experimental Psychology: Human Perception and Performance, 8*, 448-459.

Lee, D.N., & Young, D.S. (1986). Gearing action to the environment. In H. Heuer & C. Fromm (Eds.), *Generation and modulation of action patterns* (pp. 217-230). Berlin, Heidelberg, New York, Tokyo: Springer.

Lee, D.N., & Young, D.S., Reddish, P.E., Lough, S., & Clayton, T.M.H. (1983). Visual timing in hitting an accelerating ball. *Quarterly Journal of Experimental Psychology. A, Human Experimental Psychology, 35*, 333-346.

Lennie, P. (1980). Parallel visual pathways: A review. *Vision Research, 20* 561-594.

Lettvin, J.Y., Maturana, H.R., McCulloch, W.S., & Pitts, W.H. (1961). What the frog's eye tells the frog's brain. In W.R. Rosenblith (Ed.), *Sensory communication* (pp. 757-776). Cambridge, MA: MIT Press.

Loftus, G.R. (1976). A framework for a theory of picture recognition. In Monty R.A., Sanders, J.W. (Eds.), *Eye movements and psychological processes* (pp. 499-513). Hillsdale, NJ: Erlbaum.

Lynch, J.C. (1980). The functional organization of the posterior parietal association cortex. *Behavioral and Brain Sciences, 3,* 485-499.

Lynn, R. (1966). *Attention, arousal and the orienting reaction.* New York: Pergamon.

MacKay, D.M. (1978). The dynamics of perception. In P.A. Buser & A. Rougeoul-Buser (Eds.), *Cerebral correlates of conscious experience* (pp. 53-68). Amsterdam: North-Holland.

MacKay, D.M. (1984). Evaluation - The missing link between perception and action. In W. Prinz & A.F. Sanders (Eds.), *Cognition and motor processes* (pp. 175-184). Berlin, Heidelberg, New York, Tokyo: Springer.

Mackworth, N.H., & Morandi, A.J. (1967). The gaze selects informative details within pictures. *Perception & Psychophysics, 2,* 547-550.

Marcel, A.J. (1983a). Conscious and unconscious perception: Experiments on visual masking and word recognition. *Cognitive Psychology, 15,* 197-237.

Marcel, A.J. (1983b) Conscious and unconscious perception: An approach to the relations between phenomenal experience and perceptual processes. *Cognitive Psychology, 15,* 238-302.

McClelland, J.L., & Rumelhart, D.E. (Eds.). (1987). *Parallel distributed processing: Explorations in the structure of cognition: Vol. 2. Psychological and biological models.* Cambridge, MA: MIT Press.

McLeod, P., McLaughlin, C., & Nimmo-Smith, J. (1985). Information encapsulation and automaticity: Evidence from the visual control of finely timed actions. In M.I. Posner & O.S. Marin (Eds.), *Attention and performance 11* (pp. 391-407). Hillsdale, NJ: Erlbaum.

Metzger, W. (1954). *Psychologie.* Darmstadt: Steinkopff.

Mishkin, G., Ungerleider, L.G., & Macko, K.A. (1983). Object vision and spatial vision: Two cortical pathways. *Trends in Neuroscience, 6,* 414-417.

Mohler, C.W., & Wurtz, R.H. (1976). Organization of monkey superior colliculus: Intermediate layer cells discharging before eye movements. *Journal of Neurophysiology, 39,* 722-744.

Moran, J., & Desimone, R. (1985). Selective attention gates visual processing in the extrastriate cortex. *Science, 229,* 782-784.

Mountcastle, V.B. (1975). The world around us: Neural command functions for selective attention. *Neuroscience Research Progress Bulletin (Suppl. 14),* 1-47.

Mountcastle, V.B. (1978). Brain mechanisms for directed attention. *Journal of the Royal Society of Medicine, 71,* 1-16.

Mountcastle, V.B., Lynch, J.C., Georgopoulos, A., Sakata, H., & Acuna, C. (1975). Posterior parietal association cortex of the monkey: Command functions for operations within extrapersonal space. *Journal of Neurophysiology, 38,* 871-908.

Müller, H.J., & Findlay, J. (1987). Sensitivity and criterion effects in the spatial cuing of visual attention. *Perception & Psychophysics, 42,* 383-399.

Müsseler, J. (1987). *Aufmerksamkeitsverlagerungen und Relativität. Ein experimenteller Beitrag zur Raum-Zeit-Wahrnehmung anhand eines Kontraktionsphänomens (Tandem Effekt)* [Attention shifts and relativity. An experimental contribution to the perception of space and time, using a contraction phenomenon (Tandem Effect)]. Munich: Minerva.

Müsseler, J., & Neumann, O. (submitted). Apparent distance reduction with moving stimuli (Tandem Effect): Evidence for an attentional interpretation.

Navon, D., & Miller, J. (1987). Role of outcome conflict in dual task interference. *Journal of Experimental Psychology: Human Perception and Performance, 13,* 435-448.

Neisser, U. (1967). *Cognitive psychology.* New York: Appleton-Century-Crofts.

Neisser, U. (1976). *Cognition and reality.* San Francisco: Freeman.

Neumann, O. (1978a). Zum Mechanismus der Interferenz beim dichotischen Hören [On the mechanism of interference in dichotic listening]. *Reports from the Cognitive Psychology Unit, Ruhr University Bochum, No. 5.*

Neumann, O. (1978b). Aufmerksamkeit als 'zentrale Verarbeitungskapazität'. Anmerkungen zu einer Metapher [Attention as 'central processing capacity': Remarks on a metaphor]. In M. Tücke & W. Deffner (Eds.), *Proceedings of the 2nd Osnabrück Psychology Workshop.* Osnabrück: University of Osnabrück.

Neumann, O. (1980). *Informationsselektion und Handlungssteuerung. Untersuchungen zur Funktionsgrundlage des Stroop-Interferenzphänomens* [Information selection and action control. Studies on the mechanisms underlying the Stroop interference phenomenon]. Unpublished doctoral thesis, Ruhr University Bochum.

Neumann, O. (1983). Über den Zusammenhang zwischen Enge und Selektivität der Aufmerksamkeit [On the relation between the limits and the selectivity of attention]. *Reports from the Cognitive Psychology Unit, Ruhr University Bochum, No. 19*.

Neumann, O. (1984). Automatic processing: A review of recent findings and a plea for an old theory. In W. Prinz & A.F. Sanders (Eds.), *Cognition and motor processes* (pp. 255-293). Berlin, Heidelberg, New York, Tokyo: Springer.

Neumann, O. (1985). Die Hypothese begrenzter Kapazität und die Funktionen der Aufmerksamkeit (The limited capacity hypothesis and the functions of attention). In O. Neumann (Ed.), *Perspektiven der Kognitionspsychologie* (pp. 185-229). Berlin, Heidelberg, New York, Tokyo: Springer.

Neumann, O. (1987a). Beyond capacity: A functional view of attention. In H. Heuer & A.F. Sanders (Eds.), *Perspectives on perception and action* (pp. 361-394). Hillsdale, NJ: Erlbaum.

Neumann, O. (1987b). Zur Funktion der selektiven Aufmerksamkeit für die Handlungssteuerung [On the functions of selective attention for action control]. *Sprache & Kognition, 6,* 107-125.

Neumann, O. (1987c). An evaluation of three concepts of consciousness. *Reports from the Research Group on 'Perception and Action' at the Center for Interdisciplinary Research (ZiF), University of Bielefeld, No. 150*.

Neumann, O. (1989). Kognitive Vermittlung und direkte Parameterspezifikation. Zum Problem mentaler Repräsentation in der Wahrnehmung [Cognitive mediation and direct parameter specification: On the problem of mental representation in perception]. *Sprache & Kognition, 8,* 32-49.

Neumann, O., van der Heijden, A.H.C., & Allport, D.A. (1986a). Visual selective attention: Introductory remarks. *Psychological Research, 48,* 185-188.

Neumann, O., van der Heijden, A.H.C., & Allport, D.A. (Eds.) (1986b). *Visual selective attention (Psychological Research special issue)*. Berlin, Heidelberg, New York, Tokyo: Springer.

Parks, T. (1965). Post-retinal visual storage. *American Journal of Psychology, 78,* 145-147.

Peterson, S.E., Robinson, D.L., & Morris, J.D. (1987). Contributions of the pulvinar to visual spatial attention. *Neuropsychologia, 25,* 97-105.

Posner, M.I. (1980). Orienting of attention. *Quarterly Journal of Experimental Psychology. A, Human Experimental Psychology, 32,* 3-25.

Posner, M.I., Nissen, M.J., & Ogden, W.C. (1978). Attended and unattended processing modes: The role of set for spatial location. In H.I. Pick & E. Saltzman (Eds.), *Modes of perceiving and processing information* (pp. 137-157). Hillsdale, NJ: Erlbaum.

Posner, M.I., Snyder, C.R.R., & Davidson, B.J. (1980). Attention and the detection of signals. *Journal of Experimental Psychology: General, 109,* 160-174.

Posner, M.I., Walker, J.A., Friedrich, F.J., & Rafal, R.D. (1984). Effects of parietal injury on covert orienting of attention. *Journal of Neuroscience, 4,* 1863-1874.

Posner, M.I., Walker, J.A., Friedrich, F.A. & Rafal, R.D. (1987). How do the parietal lobes direct covert attention? *Neuropsychologia, 25,* 135-145.

Pribram, K.H. (1979). The orienting reaction: Key to brain representational mechanisms? In H.D. Kimmel, E.H. van Olst, & G.F. Orlebeke (Eds.) *The orienting reflex in humans*. Hillsdale, NJ: Erlbaum.

Reeves, A., & Sperling, G.A. (1986). Attentional gating in short-term visual memory. *Psychological Review, 93,* 180-206.

Ribot, T. (1906). *Psychologie de l'attention* (Psychology of attention). Paris: Alcan.

Rizzolatti, G. (1983). Mechanisms of selective attention in mammals. In J.-P. Ewert, R.R. Capranica, & D.J. Ingle (Eds.), *Advances in vertebrate neuroethology* (pp. 261-297). London: Plenum.

Rizzolatti, G., & Camarda, R. (1987). Neural circuits for spatial attention and unilateral neglect. In M. Jeannerod (Ed.), *Neurophysiological and neuropsychological aspects of spatial neglect* (pp. 289-313). Amsterdam: Elsevier North-Holland.

Rizzolatti, G., Camarda, R., Grupp, L.A., & Pisa, M. (1974). Inhibitory effect of remote visual stimuli on the visual responses of the cat superior colliculus: Spatial and temporal factors. *Journal of Neurophysiology, 37,* 1262-1275.

Rizzolatti, G., Gentilucci, M., & Matelli, M. (1985). Selective spatial attention: One center, one circuit, or many circuits? In M.I. Posner & O.S. Marin (Eds.), *Attention and performance 11* (pp. 251-265). Hillsdale, NJ: Erlbaum.

Rizzolatti, G., Riggio, L., Dascola, I., & Umilta, C. (1987). Reorienting attention across the horizontal and vertical meridians: Evidence in favor of a premotor theory of attention. *Neuropsychologia, 25,* 31-40.

Robinson, D.L., & Petersen, S.E. (1986). The neurobiology of attention. In J. LeDoux & W. Hirst (Eds.), *Brain and mind* (pp. 142-171). Cambridge: Cambridge University Press.

Robinson, D.L., Goldberg, M.E., & Stanton, G.B. (1978). Parietal association cortex in the primate: Sensory mechanisms and behavioral modulations. *Journal of Neurophysiology, 41*, 910-932.

Rock, I., Halper, F., DiViat, J., & Wheeler, D. (1987). Eye movement as a cue to figure motion in anorthoscopic perception. *Journal of Experimental Psychology: Human Perception and Performance, 13*, 344-352.

Rohrbaugh, J.W. (1984). The orienting reflex: Performance and central system manifestations. In R. Parasuraman & D.R. Davies (Eds.), *Varieties of attention* (pp. 323-373). New York: Academic.

Rugg, M.D., Milner, A.D., Lines, C.R., & Phalp, R. (1987). Modulation of visual event-related potentials by spatial and non-spatial selective attention. *Neuropsychologia, 25*, 85-96.

Rumelhart, D.E., & McClelland, J.L. (Eds.). (1987). *Parallel distributed processing: Explorations in the structure of cognition: Vol. 1. Foundations.* Cambridge, MA: MIT Press.

Saltzman, E.L. (1979). Levels of sensorimotor representation. *Journal of Mathematical Psychology, 20*, 91-163.

Schneider, W. (1987). Connectionism: Is it a paradigm shift for psychology? *Behavioral Research Methods, Instruments, and Computers, 19*, 73-83.

Shallice, T. (1972). The functions of consciousness. *Psychological Review, 79*, 383-393.

Shallice, T. (1978). The dominant action system: An information-processing approach to consciousness. In K.S. Pope & J.L. Singer (Eds.), *The stream of consciousness* (pp. 117-157). New York: Plenum.

Shepherd, M., & Müller, H.H. (1989). Movement versus focusing of visual attention. *Perception and Psychophysics, 46*, 146-154.

Shor, R.E. (1969). Hypnosis and the concept of generalized reality-orientation. In C.E. Tart (Ed.), *Altered states of consciousness* (pp. 233-261). New York: Wiley.

Shulman, G.L., Remington, R.W., & McLean, J.P. (1979). Moving attention through visual space. *Journal of Experimental Psychology: Human Perception and Performance, 5*, 522-526.

Shulman, G.L., Sheehy, J.B., & Wilson, J. (1986). Gradients of spatial attention. *Acta Psychologica (Amsterdam), 61*, 167-181.

Simon, H. (1969). *The science of the artificial.* Cambridge, MA: MIT Press.

Sokolov, E.N. (1963). *Perception and the conditioned reflex.* New York: Pergamon.

Sokolov, E.N. (1975). The neuronal mechanisms of the orienting reflex. In E.N. Sokolov & O.S. Vinogradova (Eds.), *Neuronal mechanisms of the orienting reflex* (pp. 217-235). Hillsdale, Nj: Erlbaum.

Sokolov, E.N., & Vinogradova, O.S. (Eds.). (1975). *Neuronal mechanisms of the orienting reflex.* Hillsdale, NJ: Erlbaum.

Sperling, G. (1960). The information available in brief visual presentations. *Psychological Monographs, 74*, (Whole No. 498).

Sperling, G. (1963). A model for visual memory tasks. *Human Factors, 5*, 19-31.

Stoffer, T.H. (1988). *Dynamische Aspekte der visuellen Aufmerksamkeit* [Dynamic aspects of visual attention]. Unpublished Thesis, Bielefeld: University of Bielefeld.

Stroop, J.R. (1935). Studies of interference in serial verbal reactions. *Journal of Experimental Psychology, 18*, 643-662.

Stroud, J. (1949). The psychological moment in perception. In H. v. Foerster (Ed.), *Transactions of the sixth conference on cybernetics* (pp. 27-63). New York: Josiah Macy Jr. Foundation.

Tassinari, G., Aglioti, S., Chelazzi, L. Marzi, C.A., & Berlucchi, G. (1987). Distribution in the visual field of the costs of voluntarily allocated attention and of the inhibitory aftereffects of covert orienting. *Neuropsychologia, 25*, 55-71.

Thistlethwaite, D. (1951). A critical review of latent learning and related experiments. *Psychological Bulletin, 48*, 97-129.

Tolman, E.C. (1948). Cognitive maps in rats and men. *Psychological Review, 55*, 189-208.

Treisman, A., & Geffen, G. (1967). Selective attention: Perception or response? *Quarterly Journal of Experimental Psychology, 19*, 1-17.

Treisman, A., & Gelade, G. (1980). A feature-integration theory of attention. *Cognitive Psychology, 12*, 97-136.

Trevarthen, C.B. (1978). Modes of perceiving and modes of acting. In H.I. Pick & E. Saltzman (Eds.), *Modes of perceiving and processing information* (pp. 99-136). Hillsdale, NJ: Erlbaum.

Tsal, Y. (1983). Movements of attention across the visual field. *Journal of Experimental Psychology: Human Perception and Performance, 9*, 523-530.

Umiltà, C. (1988). Orienting of attention. In F. Boller & J. Grafman (Eds.) *Handbook of neuropsychology* (pp. 175-193). Amsterdam: Elsevier.

Ungerleider, L.G., & Mishkin, M. (1982). Two cortical visual systems. In D.J. Ingle, M.A. Goodale, & R.J.W. Mansfield (Eds.), *Analysis of visual behavior* (pp. 549-586). Cambridge, MA: MIT Press.
Van der Heijden, A.H.C. (1986). On selection in vision. *Psychological Research, 48,* 211-219.
Van der Heijden, A.H.C. (1987). Central selection in vision. In H. Heuer & A.F. Sanders (Eds.), *Perspectives on perception and action* (pp. 421-446). Berlin, Heidelberg, New York, Tokyo: Springer.
Van der Heijden, A.H.C., & Eerland, E. (1973). The effects of cueing in a visual signal detection task. *Quarterly Journal of Experimental Psychology, 25,* 496-503.
Van der Heijden, A.H.C., Wolters, G., Groep, J.C., & Hagenaar, R. (1987). Single-letter recognition accuracy benefits from advance cuing of location. *Perception and Psychophysics, 42,* 503-509.
Van der Heijden, A.H.C., LaHeij, W., Phaf, R.H., Buijs, D.A.C.H., & van Vliet, E.C. (1988). Response competition and condition competition in visual selective attention. *Acta Psychologica, 67,* 259-277.
Van Essen, D.C., & Maunsell, H.R. (1983). Hierarchical organization and functional streams in the visual cortex. *Trends in Neuroscience, 6,* 370-375.
Van Olst, E.H., Heemstra, M.L., & ten Kortenaar, T. (1979). Stimulus significance and the orienting reaction. In H.D. Kimmel, E.H. van Olst, & J.F. Orlebeke (Eds.), *The orienting reflex in humans* (pp. 521-547). Hillsdale, NJ: Erlbaum.
Vierordt, K. (1968). *Der Zeitsinn nach Versuchen* (The time sense according to experiments). Tübingen: Laupp.
von Helmholtz, H. (1866) Handbuch der Physiologischen Optik. (Handbook of Physiological Optics). Leipzig: Voss.
Weinstein, E.A., & Friedland, R. (Eds.). (1977). *Hemi-inattention and hemispheric specialization.* New York: Raven.
Wolff, P. (1977). *Entnahme der Identitäts- und Positionsinformation bei der Identifikation tachistoskopischer Buchstabenzeilen. Ein theoretischer und experimenteller Beitrag zur Grundlagenforschung des Lesens* [Extraction of identity and position information during the identification of tachistoscopically presented letter rows. A theoretical and experimental contribution to the investigation of basic processes in reading]. Unpublished doctoral dissertation, Ruhr University Bochum.
Wolff, P. (1984). Saccadic eye movements and visual stability: Preliminary considerations towards a cognitive approach. In W. Prinz & A.F. Sanders (Eds.), *Cognition and motor processes* (pp. 121-137). Berlin, Heidelberg, New York, Tokyo: Springer.
Wundt, W. (1903). *Grundzüge der physiologischen Psychologie (Fundamentals of Physiological Psychology) (5th Ed.).* Leipzig: Engelmann.
Wurtz, R.H., Goldberg, M.E., & Robinson, D.L. (1980). Behavioral modulation of visual responses in the monkey: Stimulus selection for attention and movement. *Progress in Psychobiology and Physiological Psychology, 9,* 43-83.
Wurtz, R.H., Goldberg, M.E. & Robinson, D.L. (1982). Brain mechanisms of visual attention. *Scientific American, 246 (6),* 124-135.
Yarbus, A.L. (1967). *Eye movements and vision.* New York: Plenum.
Yates, J. (1985). The content of awareness is a model of the world. *Psychological Review, 92,* 249-284.
Zimba, L.D., & Hughes, H.C. (1987). Distractor-target interactions during directed visual attention. *Spatial Vision, 2,* 117-149.
Zöllner, F. (1862). Über eine neue Art anorthoskopischer Zerrbilder [On a new kind of anorthoscopic distorted pictures]. *Poggendorfs Annalen der Physik, 117,* 477-484.

Perception, Action, and Awareness: A Three-Body Problem

D.G. MacKay

CONTENTS

Introduction

Most chapters of the present book deal with a two-body problem, relations between perception and action. This two-body problem is difficult enough: as Miller, Galanter, and Pribram (1960, p. 11) point out, the "theoretical vacuum between perception and action" has been the subject of prolonged - and frequently violent - debate. However, two-body problems are well known to be solvable, and I myself have proposed a detailed theoretical solution to the relation between perception and action in the case of language (MacKay, 1987a) and other cognitive skills (MacKay, 1985, 1987b, pp. 14-140). The present chapter takes on a three-body problem that has proven much more difficult to solve: relations between perception, action, and awareness.

Relationships Between Perception and Action
Edited by O. Neumann and W. Prinz

The relations between perception, action, and awareness raise three broad classes of theoretically fundamental issues. The vicissitudes of awareness are central to the three-body problem, part 1, which includes questions such as: Why is awareness usually but not always associated with higher-level processing; In producing sentences, for example, why are low-level phonetic units normally produced without awareness but suddenly enter awareness when speakers misproduce a speech sound (MacKay, 1988)?

Differences between conscious versus unconscious processing are central to the three-body problem, part 2, which includes questions such as: Why is conscious processing so much slower than unconscious processing in perception and action; Why, for example, do subjects take so long (about 7.73 s on average) to become *conscious* of the two meanings of a lexically ambiguous sentence (MacKay & Bever, 1967), when available evidence indicates that both meanings have been processed *unconsciously* 300 ms after arrival of the ambiguous word (Swinney, 1979)?

Relations between awareness, attention, and practice (repeated retrieval) are central to the three-body problem, part 3, which includes questions such as: (a) Why do we normally become conscious of what is new in perception and action, while what is old or frequently repeated drops out of awareness, as in habituation and automaticity (Shiffrin & Schneider, 1977); (b) Why is practice essential for carrying out several perceptual and motor activities simultaneously, so that subjects can speak and play the piano concurrently without mutual interference, for example, but only after extensive prior practice (see Allport, 1980); (c) Completing the circle, why does action have such profound effects on attention and awareness; When subjects respond to an input arriving via one channel, for example, why is awareness of a target arriving simultaneously on another channel greatly attenuated (Ostry, Moray, & Marks, 1976)?

Structure of the Chapter

This chapter develops a detailed and explicit theory of attention and awareness in perception and action and divides into five sections. The first section describes the stages of development that the theory has already undergone and provides a preview of subsequent sections. The second section describes the theoretical foundation available when I began this chapter, the node structure theory of relations between perception and action. Included in this section are the basic processes in the theory (e.g., priming), the mechanism for activating nodes, and a general principle (the principle of higher-level activation) that determines what nodes become activated during perception. Because I have outlined the empirical and theoretical underpinnings of these basic processes elsewhere, the second section will present theoretical assumptions as if they were established facts, and will introduce theoretical terms without going into their rationale. Please follow closely as I lay out these theoretical building blocks because we will need them later in our discussion of attention and awareness[1].

The third and fourth sections show how awareness and attention can be incorporated into the theory outlined in the second section, and the fifth section shows how the extended theory applies to the three-body problem, the basic phenomena that any theory of relations between perception, action and awareness must address.

The Node Structure Theory: History and Prospects

Three stages of theory development have preceded the present paper. Some curious differences between the rate of producing internal vs. overt speech (MacKay, 1981) provided the original impetus for stage 1 of the theory. Stage 2 (MacKay, 1982) extended the theory so as to address more general issues: errors in action, the mechanisms underlying sequencing and timing in behavior, effects of practice on behavior, and speed-accuracy trade-off in skilled behavior. Stage 3 of theory development (MacKay, 1987a,b, 1990) added perception to the theory, including the perception of ambiguous inputs, relations between errors in perception and action, asymmetries in the ability to perceive vs. produce skilled behavior, the role of feedback in the perceptual monitoring of skilled behavior, and the effects of delayed and amplified auditory feedback on the production of speech and other cognitive skills. The node structure theory, stage 3, therefore addresses a wide range of phenomena, including the full scope of knowledge about relations between perception and action, and, because of its scope, the theory requires a sizeable number of assumptions.

The present paper outlines the node structure theory stage 4, an account of attention and awareness, and the remainder of this section previews this extended theory and how it bears on the three-body problem. The main addition to the theory is "pertinent novelty," which refers to a novel conjunction of internal or external events that fall into familiar categories or domains. Pertinent novelty triggers orienting reactions, awareness, and commitment learning, the process whereby new connections are formed. However, newly formed connections decay over a relatively brief period, so that nonrepeated events, e.g., most sentences, receive only temporary representation. This process of connection formation and decay is shown to be consistent with available data on amnesia, including the pattern of sparing and deficit in "hippocampal patients."

Explaining selective attention will require only mechanisms that are essential for other purposes in the theory: the "most-primed-wins" activation principle, and the mechanism that engages the activating mechanisms for systems of nodes representing one source of input rather than another. Although this second mechanism

[1] Readers interested in empirical support for the theory or in comparison with other theories such as those of McClelland, Rumelhart, and the PDP Research Group (1986) should consult the more extensive discussion in D.G. MacKay (1985, 1987a,b).

often determines what particular contents enter awareness, e.g., enabling the activation of target inputs instead of distractor inputs in selective attention tasks, the mechanisms for attention and awareness are nevertheless conceptually distinct in the theory.

Divided attention will likewise require no new "attentional mechanisms" if prior learning has established separate domains of nodes for the concurrent activities. However, if concurrent activities share nodes in the same domain, the theory predicts varying degrees of interference depending on the type of shared node (content, sequence, and timing nodes), on their temporal pattern of activity, and on the nature of their interconnections.

Finally, the theory will provide new insights into the three-body problem and its subsidiary issues discussed above. By way of preview, part 1 of the three-body problem (vicissitudes of consciousness) reflects the nature of pertinent novelty, the factor that triggers awareness in the theory. For example, errors represent instances of pertinent novelty, and hence can give rise to awareness of units that would otherwise be processed unconsciously. Part 2 of the three-body problem (differences between conscious versus unconscious processing) reflects the nature of the mechanisms underlying conscious versus unconscious processing in the theory: unconscious processing involves priming, a parallel and therefore rapid process, whereas conscious processing involves not just activation, a sequential and therefore slow process, but prolonged activation, an even slower process. These same mechanisms will also account for part 3 of the three-body problem (relations between awareness, attention, and practice). For example, awareness of inputs to unattended channels is attenuated during response to a target on an attended channel because responses require most-primed-wins activation mechanisms that are either-or in nature.

Basics of the Node Structure Theory of Perception and Action

The basic components of the theory are nodes, which are hypothetical processing units (as in Wickelgren, 1979) that share a set of relatively simple structural characteristics and processing capabilities, and respond in the same way to variables such as practice (repeated activation). Unless otherwise specified, the nodes discussed in the present paper play a role in both perception and action (see also Prinz, this volume). These "mental nodes" represent neither sensory input nor patterns of muscle movement, but higher-level cognitive components common to both perception and production, e.g., segments and syllables at the phonological level, and words and phrases at the sentential level. Mental nodes become active when we perceive a word (or sentence) and when we produce it, either aloud, or within the imagination (internal speech). During perception, including perception of self-generated feedback, mental nodes receive "bottom-up" inputs from sensory analysis nodes that represent patterns of, say, auditory input arriving via the basilar membrane and associated auditory pathways. During production, these same per-

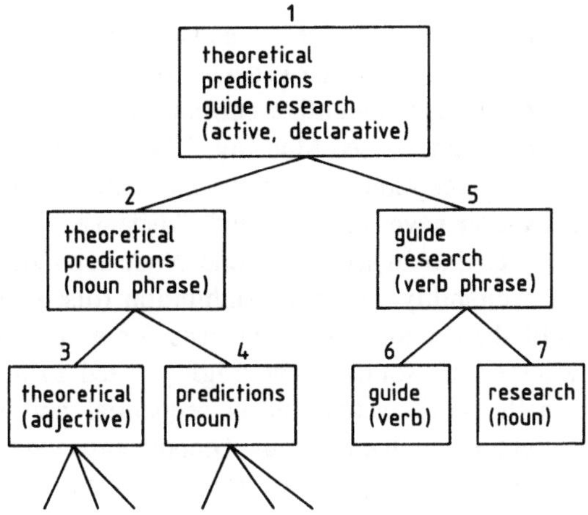

Fig. 1. The top-down hierarchy of nodes for producing the words in the sentence "Theoretical predictions guide research." The *numbers* represent the order in which the nodes become activated during production

ception-production units send "top-down" outputs to muscle movement nodes that represent patterns of movement for the speech muscles, producing contractions of respiratory, laryngeal, velar, and articulatory muscles.

Hierarchic Connections Between Nodes

In general, connections between nodes are "partially hierarchic" rather than "strictly hierarchic" (see MacKay, 1987b, pp. 17-22), but this distinction is not important for present purposes, and Fig. 1 illustrates the simpler case, a "strict hierarchy" of top-down connections between sentential nodes for producing the sentence, "Theoretical predictions guide research." Following MacKay (1982), I designate each node by a two-component label (see Fig. 1): the content that the node represents appears in italics, followed by its sequential domain (explained below) in brackets. For evidence supporting the particular units and connections illustrated in Fig. 1, the reader is referred to MacKay (1972, 1973b, 1978, 1979, and 1987b, pp. 14-38), and Treisman (1983). Omitted from Fig. 1 are the phonological nodes and the complex but otherwise similar hierarchy of nodes underlying the control of muscle movements.

Processing Characteristics of Nodes

Nodes exhibit three processing characteristics that are necessary for understanding awareness in the theory: activation, priming, self-inhibition, and linkage strength. Awareness itself arises from a fourth processing characteristic (prolonged activation) discussed later.

Node Activation

I use the term "activation" as short for "node activation" in the remainder of this chapter, and the reader is asked to keep in mind the following differences between node activation and other current uses of the term "activation" in the cognitive and neural sciences (see MacKay, 1987b, p. 9). Node activation is necessary for conscious perception and action and is all-or-nothing in nature: the intensity of node activation never changes with "distance," fatigue, or the number of other nodes connected to an activated node. Node activation is terminated by a period of reduced excitability called self-inhibition (discussed on p. 275) and is self-sustained until then: node activation lasts for a specifiable period of time, regardless of whether the sources that originally led to activation continue to provide input. Finally, node activation is sequential and nonautomatic in nature: a special activating mechanism must become engaged to determine when and in what order nodes within a system become activated (for an example of sequential activation during production, see the numbered nodes in Fig. 1).

Node Priming

"Node priming" refers to a transmission across a connection of subthreshold activity that spreads with decrement: an activated node primes its connected nodes most strongly (first-order priming), while an unactivated node receiving first-order priming primes its connected nodes less strongly (second-order priming), and so on up to nth-order (null) priming (the value of n being currently unknown: see MacKay, 1990). Priming prepares a node for possible activation, and all nodes must be primed in order to become activated.

Priming summates spatially (when two or more connections to the same node are simultaneously active), and temporally (during the time that any given connection remains active). However, summation of priming cannot by itself activate a node: a special activating mechanism is required for activation, and priming only accumulates to a subthreshold asymptotic level (see Fig. 2).

Unlike activation, priming does not self-sustain: it begins to decay as soon as input from its connected nodes stop. Also unlike activation, no period of self-inhibition and recovery follows priming, and no special triggering mechanism determines when and in what order nodes become primed. Indeed, priming is a parallel process: an active source simultaneously primes all connected nodes. Contrary to the usual assumption that priming in sentences is a left-to-right process (see Neumann, 1984), priming is nonsequential, and backward (right-to-left) priming can be as effective as forward (left-to-right) priming in the theory (see also Koriat, 1981).

Linkage Strength

Linkage strength is a relatively long-term characteristic of a connection that is determined by practice (the frequency with which a node has been primed and activated via a particular connection in the past). Linkage strength determines how

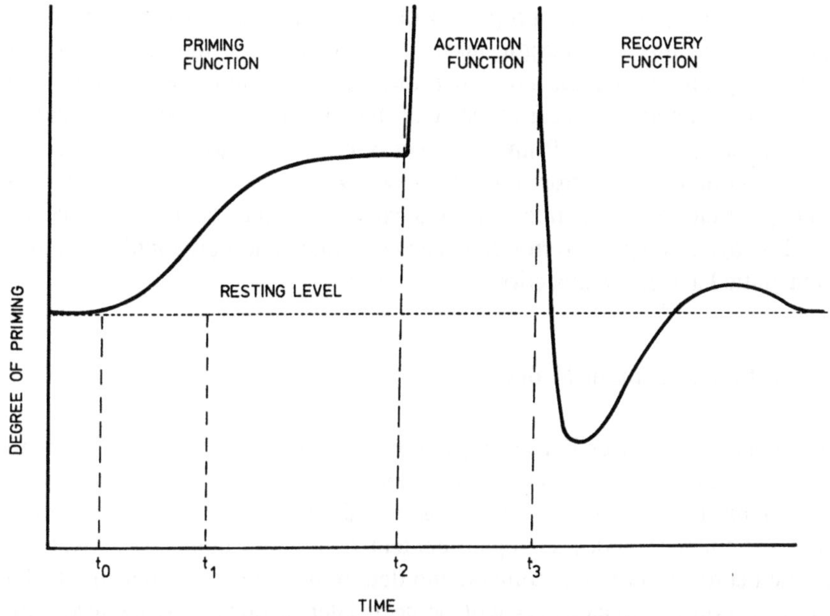

Fig. 2. The priming, activation, and recovery phases for a single node. The priming function shows how priming summates to asymptote following onset of priming at t_0. The activation function illustrates multiplication of priming and self-sustained activation until time t_3. The recovery cycle shows how priming first falls below resting level (self-inhibition) and then rebounds (the hyperexcitability phase)

much and how rapidly priming crosses a connection and is reflected in the initial slope and asymptote of a priming function: highly practiced connections transmit priming more rapidly (i.e., with a steeper slope) and up to a higher asymptotic level than do relatively unpracticed connections. These characteristics of linkage strength explain a wide range of practice effects in the psychological literature (see MacKay, 1982).

Self-Inhibition

After a node becomes activated, it undergoes a brief period of self-inhibition during which its level of priming falls below normal or resting level (see Fig. 2). The mechanism underlying self-inhibition is an inhibitory collateral or "satellite" that sends an inhibitory connection to and receives an excitatory connection from the "parent" node. After receiving sufficient first-order priming from its activated parent node, the satellite becomes activated and inhibits its parent node, which becomes deactivated, thereby deactivating its satellite and enabling recovery in the parent node to begin. Because linkage strength between the parent and satellite increases as function of repeated activation, practice determines when self-sustained activation ends and self-inhibition begins, and how long recovery lasts (unpracticed nodes require over 100 ms for recovery; see MacKay, 1987b, pp. 146-147).

In summary, dynamic properties of nodes such as activation, self-inhibition, priming, and linkage strength, taken individually as in the above discussion, are extremely simple, but taken together, these dynamic properties interact in complex ways that depend on the current state of the node and on its history of activity over the course of a lifetime. Priming is necessary for activation and is directly related to the probability of error (see MacKay, 1982). Activating a node increases the linkage strength of its connections and causes its connected nodes to become primed. Linkage strength in turn influences how much and how rapidly priming can be transmitted across a connection.

Activation of Content Nodes

The nodes I have been discussing so far are known as *content nodes*, because they represent the form or content of an action or perception, whether conscious or unconscious. I turn now to *sequence nodes*, the mechanisms for activating content nodes. Sequence nodes segregate content nodes into domains, activate the most primed content node in a domain, and determine the serial order in which content nodes become activated. As will be seen later, sequence nodes also call up the mechanism for prolonging activation and bringing the contents of perception and action into awareness.

Sequential Domains

Sequence nodes connect with a sequential domain, a set of content nodes representing units of behavior that all have the same sequential properties or privileges of occurrence in sequences involving other domains (see MacKay, 1987b, pp. 52-55). I use capital letters to denote sequence nodes and round brackets to denote a domain of content nodes. For example, the sequence node, COLOR ADJECTIVE[2] activates the domain (color adjective), the set of nodes representing color adjectives and sharing identical sequential properties or privileges of occurrence in English noun phrases.

Multiplication of Priming and the Most-Primed-Wins Principle

The Most-primed-Wins Principle is the basis for all node activation (see MacKay, 1987b, pp. 49-55) and follows directly from the way that sequence and content nodes connect with one another. Once a sequence node becomes activated, it repeatedly multiplies the priming of every node connect with it by some large factor within a relatively brief period of time. This multiplicative process has no effect on an unprimed node, but soon serves to activate (i.e., bring to threshold) the content

---

[2] MacKay (1990) notes that the rule (color adjective + noun) may be derived or "inherited" from the more general rule (adjective + noun) whereby all adjectives precede nouns in English.

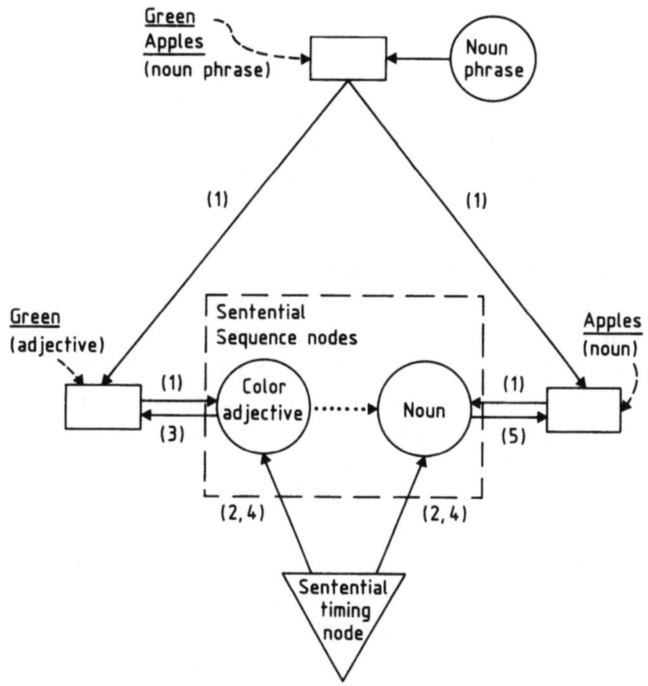

Fig. 3. The order (in *brackets*) of top-down processes underlying activation of content nodes (in *rectangles*), sequence nodes (in *circles*), and the sentential timing node (*triangle*) for producing the noun phrase *green apples*

node with the greatest degree of priming in its domain. For example, COLOR ADJECTIVE is connected to and, when activated, multiplies the priming of the dozens of content nodes in the domain (color adjective). Naturally, the node with more initial priming than all other nodes in its domain will reach threshold first, and this "most-primed" node will become activated.

During production, content nodes generally achieve their most-primed status via priming "from above." In producing the adjective *green*, for example, a superordinate node such as *green apples*(noun phrase) becomes activated, and strongly primes its connected nodes, including *green*(color adjective) (see Fig. 3). Being most primed when its activation mechanism is applied, the multiplied priming of this primed-from-above node reaches threshold sooner than the remaining "extraneous" nodes in its domain (i.e., nodes representing other color adjectives) and becomes activated.

During perception, content nodes achieve most-primed status mainly "from below." For example, visual inputs such as the color green or the printed word *green* will prime *green*(color adjective) from below. *Green*(color adjective) then passes second-order priming to its connected sequence node (see Fig. 3), enabling COLOR ADJECTIVE to become activated, and in turn to activate the most primed content node in its domain, *green*(color adjective) itself.

Quenching
Once a content node becomes activated, it quenches or inhibits, rather than further primes, its corresponding sequence node (see MacKay, 1987b, pp. 50-55), thereby ensuring that one and only one content node in a domain becomes activated at any one time. Quenching, together with multiplication of priming, therefore provides the functional basis for the most-primed-wins principle.

Activation of Sequence Nodes

Timing nodes activate sequence nodes and play a role in attention (discussed on pp. 290-293). They also control the rate of perception and action by determining how rapidly the sequence nodes become activated. Timing nodes connect with and activate sequence nodes in the same way that sequence nodes connect with and activate content nodes. However, timing nodes become activated according to an endogenous rhythm, and timing nodes for different systems of sequence nodes (e.g., the phonological system, the sentential system; see Fig. 1) have different endogenous rhythms. After each activation, timing nodes multiply the priming of sequence nodes connected to them, activating the most primed one on the basis of the most-primed-wins principle. This "most-primed sequence node" is, of course, usually the one that has just received second-order priming from a connected content node, e.g., *green*(color adjective) in the above example.

The Principle of Higher-Level Activation

Timing nodes for different systems, e.g., the phonological system and the speech muscle movement system, can be engaged or activated independently, and this provides the basis for selective attention (see pp. 290-292), for changing output mode (between, for example, *overt speech* where timing, sequence, and content nodes in all three systems become activated, versus *internal speech* where nodes in sentential and phonological systems activated but not those in muscle movement systems), and for a general perceptual principle called "higher-level activation" that plays a role in determining what units normally enter awareness. Under this principle, not all nodes in bottom-up hierarchies become activated during perception, the way they do during production in top-down hierarchies such as the one in Fig. 1: only nodes in higher level systems normally become activated and give rise to everyday perception. In particular, nodes in the sentential system normally become activated when perceiving conversational speech, whereas nodes in the phonological system do not (see MacKay, 1987b, pp. 74-84 for empirical, logical, and theoretical arguments for this principle).

Extension 1 of the Node Structure Theory: Awareness

Up to now I have been laying down a foundation on which to build a theory of awareness. Priming and activation are *necessary* for awareness, but not *sufficient*: awareness requires an additional mechanism. This section argues that the awareness mechanism is intimately related to novelty in perception and action, including the detection of self-produced errors, and I spell out an awareness mechanism for the node structure theory that has this property. I then argue that this awareness mechanism is necessary and sufficient for "commitment learning," the process whereby connections become functional in the node structure theory.

Novelty and Awareness in Perception and Action

The close relationship between consciousness and novelty has frequently been noted (e.g., Sokolov, 1963; Gregory, 1981). Consciousness seems necessary and sufficient for learning novel information: we normally become conscious of what is new, and what is old or frequently repeated drops out of awareness, as in habituation. When I say that I am aware of the familiar book before me on my desk, I am aware of the book in a novel context (e.g., a novel temporal context, or a novel spatial context relative to other objects on the desk). What is novel is not the book per se, but how its features conjoin in the present situation, and such novel feature conjunctions are a necessary condition for consciousness under the node structure theory. Similarly, when we become aware of a familiar word, it is not the meaning of the word per se that we become aware of, but word meaning in combination with its novel context of use.

The correlation between novelty and awareness has also been demonstrated experimentally. For example, MacKay (1973a) showed that novelty and awareness are correlated in language comprehension, so that awareness is necessary for forming novel (deep structure) integrations of the meanings of words in sentences. Specifically, MacKay showed that sentences undergo deep structure analyses during conscious but not during unconscious processing (via the unattended channel in a dichotic listening experiment). The highly familiar meanings and syntactic categories of words in the unattended channel received unconscious processing (priming), but failed to enter awareness. However, comprehending what was new - the particular conjunction or relation between words in the context in which they were spoken - required conscious processing. Treisman and Gelade (1980) demonstrated a similar correlation between consciousness and the integration of separately processed visual features (e.g., color and form) of objects. Without sufficient time for awareness, these separate features failed to become conjoined to enable object recognition.

Binding Nodes and Prolonged Activation: The Awareness Mechanism

The necessary and sufficient condition for conscious awareness in the node structure theory is prolonged activation: we become conscious of novel conjunctions of concepts during the time when the nodes representing those concepts are undergoing prolonged activation. This section examines the triggering mechanisms for prolonged activation, while subsequent sections examine the more complex process whereby binding nodes are called up and results in the formation of new connections.

How Binding Nodes Prolong Activation

Binding nodes prolong activation by inhibiting the self-inhibition mechanism of two or more sequential domains of content nodes. That is, binding nodes connect with two or more domains of content nodes, are inhibitory in nature, and connect not with the parent node but with its self-inhibitory satellite. An activated binding node therefore shuts down the self-inhibitory mechanisms of all content nodes to which it is connected. This causes currently activated nodes to remain activated for a prolonged period of time because they can no longer self-inhibit. Thus, although binding nodes connect with many content nodes, their effect is specific: they only alter the behavior of the small subset of connected nodes that happen to be activated at the time.

Comparison with Other Theories of Awareness

The node structure theory of awareness resembles other accounts in some respects but not others. Unlike other theories, the node structure theory relates consciousness to a strictly temporal factor: prolonged activation. Also, the same nodes represent a content either consciously or unconsciously, but not both simultaneously in the node structure theory, unlike the theories of, say, Baars (1983, 1988) and Thatcher and John (1977) where one system represents conscious contents and another (separate) system represents unconscious contents, with the same information or experience represented consciously in one place and unconsciously in the other. However, a conscious content is more than just a different state of a single, already-established representation in the node structure theory. Two different types of representation (committed versus uncommitted nodes) and a state change (increased linkage strength; see pp. 274-275) underly consciousness.

Pertinent Novelty and the Activation of Binding Nodes

I will use the term "pertinent novelty" to refer to the conditions that trigger binding nodes, causing prolonged activation and awareness. Pertinent novelty occurs whenever two or more committed nodes that have rarely or never been activated in simultaneous combination before become activated simultaneously or in temporal overlap and call up an existing higher-level sequence node via an uncommitted node. Understanding the distinction between committed versus uncommitted nodes is therefore necessary for understanding how binding nodes become activated.

Committed Versus Uncommitted Nodes

Up to now, all nodes discussed in the node structure theory have been committed nodes, defined in terms of the strength of their connections: connections *to* committed nodes are strong enough for the committed node to receive sufficient priming to enable activation when its activation mechanism (sequence node) is applied. Similarly, connections *from* committed nodes are strong enough for connected nodes to be able to become activated when their activating mechanism is applied. That is, in order to become activated, a node must achieve a minimal level of priming in addition to most-primed status in its domain. Without surpassing this minimal level, or commitment threshold, priming multiplied by an activating mechanism cannot reach the level required for activation.

Uncommitted nodes are likewise defined in terms of the strength of their connections: connections *to* uncommitted nodes are so weak that they cannot receive enough priming to reach commitment threshold or to achieve most-primed status in their domain and become activated when their activating mechanism is applied. Connections *from* uncommitted nodes are likewise so weak that they transmit too little priming for connected nodes to become activated when their activation mechanism is applied.

Most mental nodes begin with uncommitted status: they have weak or uncommitted connections. However, these uncommitted connections are prewired into hierarchically organized domains, such that nodes in two or more sequentially organized domains connect convergently with nodes in a single superordinate domain. That is, between any pair of uncommitted nodes in sequentially organized subordinate domains, there exists at least one uncommitted node in a superordinate domain that receives convergent connections from both of them. For example, every pair of uncommitted nodes in domains that come to represent (initial consonant group) and (vowel group) converges on at least one uncommitted node in the domain that comes to represent (syllable). This is not to say that the *particular content of a domain* comes prewired in the newborn. Only the organizational structure for sequential domains comes prewired: what particular content a prewired domain comes to represent is a matter of experiential factors such as order of acquisition. Nor is this to say that only a single pair of subordinate nodes contributes conjoint connections to any given uncommitted node, although this may in fact be true of nodes in the phonological system. However, uncommitted nodes in the sentential system will exhibit *multiple convergence*, or receive conjoint connections from many pairs of subordinate nodes.

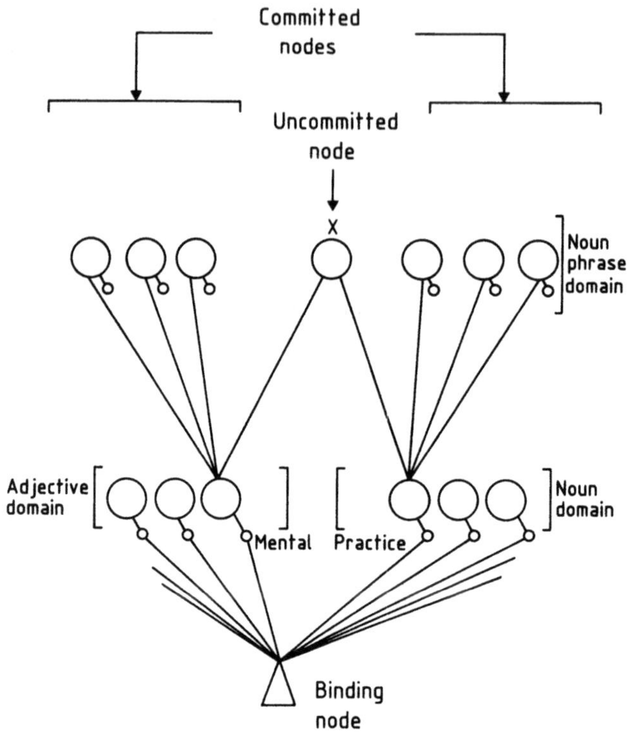

Fig. 4. The NOUN PHRASE binding node (*triangle*) prolongs activation of currently activated nodes, *mental*(adjective) and *practice*(noun), by inhibiting the inhibitory satellites (*small circles*) of committed parent nodes (*larger circles*) in the (adjective) and (noun) domains. NOUN PHRASE also commits uncommitted node X to represent the content *mental practice*(noun phrase)

How Pertinent Novelty Triggers Binding Nodes

The mechanism whereby pertinent novelty triggers binding nodes is as follows: sequence nodes are connected to a binding node with a very high threshold that cannot be reached when the sequence node is simply activated and then quenched. Now, when two or more committed nodes become activated simultaneously or in temporal overlap, they contribute conjoint first-order priming to an uncommitted node, and second-order priming to its sequence node via many nodes (see Fig. 4). The sequence node will therefore become activated but fail to activate the uncommitted node or any other nodes in its domain. The sequence node therefore "fails to quench" and remains activated for a prolonged period, and this prolonged activation triggers the binding node via temporal summation.

Pertinent Novelty, Orienting Reactions, and Errors

Convergent priming arriving at an uncommitted node can be said to signal pertinent novelty in the node structure theory; so does the "failure to quench" of its sequence node. For example, the expression "inhibitory satellite" constitutes an instance of pertinent novelty if the hearer "knows" that an adjective followed by a noun constitutes a noun phrase, but has never experienced this particular combination of adjective and noun before: no node has been committed to the content *inhibitory satellite*(noun phrase). Thus, when the sequence node NOUN PHRASE is activated, none of the nodes in its domain has accumulated enough priming to become activated, so that NOUN PHRASE fails to quench and triggers its binding node, causing prolonged activation, and awareness.

Orienting Reactions: Side-Effects of Pertinent Novelty

Besides activating binding nodes, pertinent novelty automatically triggers orienting reactions that include emotional components, e.g., surprise; autonomic components, e.g., increased skin conductance, cardiac deceleration, and pupil dilation; and behavioral components, e.g., inhibition of ongoing activity (Neumann, 1987). Thus, unlike other theories, such as Sokolov (1963) and Baars (1988), the node structure theory generates orienting responses without complicated mismatch mechanisms for comparing new models with prior models of the internal or external world.

Pertinent Novelty and Error Detection

Errors result in the activation or production of a sequence of units that is novel at some level. For example, *dump seat* misproduced as *sump deat* involves novel lexical units because *sump* and *deat* are nonwords in English. Similarly, *crawl space* misproduced as *crawl srace* involves a novel phonological unit because syllable-initial *sr* does not occur in English. *Fly the plane and buy the boat* misproduced as *Fly the boat and buy the plane* involves a novel propositional unit because boats do not fly. Similarly, *tool carts* misproduced as *cool tarts* in the intended sentence *They were moving tool carts down the assembly line* involves a novel propositional unit if the speaker lacks a committed node for *They were moving cool tarts down the assembly line*(proposition) (examples from Motley, Baars, & Camden, 1983).

Speech errors therefore introduce pertinent novelty that can trigger orienting reactions and signal occurrence of an error. However, different errors differ in the number of connections between the units produced in error and the uncommitted node that they prime, and this "distance" plays a role in error detection. Compare the effects of this distance for two phonological transposition errors (above): *crawl srace* instead of *crawl space*, and *cool tarts* instead of *tool carts* in the intended sentence *They were moving tool carts down the assembly line*. In *crawl srace*, no committed node represents *sr* (initial consonant group) for speakers of English, so that when *s*(initial stop) and *r*(initial liquid) are activated in error, first-order bottom-up convergent priming is transmitted immediately (distance 0) to an uncommitted phonological node, thereby triggering binding nodes, orienting reactions (causing output to terminate: see above), node commitment, and *awareness* (prolonged activation) that enables error detection. Indeed, this rapid detection sequence may explain why phonologically novel errors are so rare in overt speech (see Fromkin, 1971): these errors can be detected so rapidly as to prevent their occurrence before they appear in the surface output (see MacKay, 1990; Levelt, 1984).

However, error detection will be both less efficient and less likely when many intervening connections separate the uncommitted node from the phonological units produced in error. In the *cool tarts* error, for example, nodes higher in the hierarchy already exist for representing the segments *c*(initial consonant group) and *t*(initial consonant group), the syllables *cool*(stressed syllable) and *tarts* (stressed syllable), and the words *cool*(adjective) and *tarts*(noun). Even *cool tarts* (noun phrase) is likely to exist as an already committed node, precluding the possibility of orienting reactions and error detection at that level. However, the proposition node *They were moving cool tarts down the assembly line*(proposition) almost certainly does not exist as an already committed node, so that activating *cool tarts* (noun phrase) in the context *move down the assembly line* could potentially trigger orienting reactions and awareness of the error. However, *cool tarts*(noun phrase) is unlikely to achieve greatest priming in its domain and become activated while producing the remainder of this sentence because many connections separate cool tarts(noun phrase) from its source of first-order priming at the phonological level. In consequence, such an error is likely to pass undetected, and the node structure theory predicts that probability of error detection will vary with the proximity of units produced in error to the uncommitted node that they prime. Indeed, this proximity factor may contribute to the fact that speakers fail to detect or correct about 40% of the word substitution errors that occur. (See MacKay, 1990, for further discussion of error detection.)

Internally Generated Pertinent Novelty

As the above discussion suggests, the novel conjunctions providing the basis for awareness need not originate in the external world under the node structure theory. People can actively become aware of familiar objects and concepts by generating pertinent novelty internally: activating a novel combination of nodes in sequentially related domains will give rise to awareness regardless of whether the novel combination arises from internal or external sources. Indeed, novel conjunctions from internal sources not only provide conscious contents (images) without environmental help; they make possible the mental simulation of past and possible (future) events that is essential for planning adaptive actions.

Connection Formation: The Commitment Process

As discussed above, the brain comes equipped with an excess of uncommitted nodes with extremely weak but prewired or potential connections. The question is how one or more of the hundreds of uncommitted connections to an uncommitted node become committed or "functional", i.e., capable of transmitting enough priming to enable activation. This section argues that the awareness mechanism discussed above is necessary and sufficient for commitment learning. Uncommitted connections automatically become committed when inputs never previously experienced fall into familiar classes, and when outputs never previously produced are created, and give rise to awareness. Because most everyday sentences are new or never previously experienced, new connections to at least one new (proposition) node must become committed when comprehending or creating novel sentences, according to the theory. Conversely, if I remember novel information such as the place on a page where I read a particular fact, I must have been conscious of this page position when I initially learned the fact, according to the theory.

Binding Nodes and Prolonged Activation as Commitment Mechanisms

The uncommitted node receiving convergent connections from two or more committed nodes undergoing prolonged activation (see p. 282 and Fig. 4) achieves commitment levels of priming via temporal summation and becomes activated via its (unquenched) sequence node as the most-primed node in its domain. Activating the uncommitted node in turn causes a slight but relatively ling-term increase in linkage strength across its connections, thereby improving the asymptotic level and rate of priming via those connections. With further activations, linkage strength increases further, until bottom-up priming suffices to enable activation without the boost in bottom-up priming that results from prolonged activation and awareness. The uncommitted node has now been transformed into a committed node that consistently codes a particular cognitive content. Future activation proceeds automatically via the most-primed-wins principle and is followed by self-inhibition, so that consciousness can no longer occur: self-inhibition automatically shuts off activation after a set period, precluding the prolonged activation necessary for awareness.

Commitment of Bottom-up Connections

To illustrate how a binding node transforms an uncommitted node into a committed one in the theory, consider the child who knows the concepts *mental and practice* but has just encountered the expression *mental practice* for the first time. That is, there exist two parent nodes, represented *mental*(adjective) and *practice*(noun), each with an inhibitory satellite shown Fig. 4. The parent node labeled *mental* (adjective) is connected to several nodes in the (noun phrase) domain, including committed nodes such as, say, *mental arithmetic*(noun phrase), and uncommitted nodes such as the one labeled X in Fig. 4. The child also has a committed node *practice*(noun) connected to this same uncommitted node X as well as to perhaps several hundred other nodes, including, say, the committed node *piano practice* (noun phrase). Thus, despite its convergent (spatially and temporally summating) input, uncommitted node X may not receive greatest priming in its domain during the normal period that *mental*(adjective) and *practice*(noun) remain activated, and in any case, the convergent priming is too weak to enable activation of X when its triggering mechanism (NOUN PHRASE) becomes activated. NOUN PHRASE therefore fails to quench.

We now require a notational convention: bold capitals will denote binding nodes and square brackets will denote the corresponding domain of inhibitory satellites of the binding node. Thus, the domain for the binding node **NOUN PHRASE** (see Fig. 4) is [noun phrase], and includes the inhibitory satellites of content nodes representing immediate constituents of noun phrases, i.e., (adjective) and (noun), among others. When NOUN PHRASE in Fig. 4 fails to quench, **NOUN PHRASE** becomes activated, thereby inhibiting the self-inhibitory mechanisms of nodes in [noun phrase]. As a consequence, nodes in the domain [noun phrase] that are currently activated, specifically *mental*(adjective) and *practice* (noun), fail to self-inhibit, and so engage in prolonged activation. This provides uncommitted node X with the required boost in priming (via temporal and spatial summation), so that NOUN PHRASE can now activate X. Activation of X increases the linkage strength of its bottom-up connections, and with repeated activation enables X to code the content *mental practice*(noun phrase) without engaging **NOUN PHRASE** for introducing prolonged first-order bottom-up priming.

More generally, the bottom-up binding process goes as follows: two or more lower level committed nodes, A, B,..., become activated on the basis of priming from internal or external (environmental) sources and send temporally overlapping first-order priming to an uncommitted node, X. The problem is that, without outside help, the temporal and spatial summation of priming from A + B +... cannot reach commitment level, because A + B +... normally become self-inhibited soon after activation. The outside help comes indirectly from X's sequence node, which fails to quench, and thereby activates its binding node. The binding node inhibits the self-inhibitory satellites for A + B +..., thereby causing prolonged activation of A + B +.... This prolonged activation extends the duration of temporal summation at uncommitted node X, enabling X to reach commitment threshold and become activated.

Commitment of Top-down Connections

Once bottom-up connections to an uncommitted node become functional, top-down connections can become committed almost immediately, without further engagement or reactivation of the binding nodes. Specifically, once an uncommitted node becomes activated, it transmits first-order priming to all of its connected nodes, including the lower-level nodes that are still undergoing prolonged activation. Because a connection transmitting first-order priming to an activated node constitutes the basic condition for greatly increasing linkage strength, top-down connections become strengthened soon after their newly committed constituent node becomes activated. For example, during the time that X in Fig. 4 remains activated, the appropriate top-down connections can become strengthened almost immediately, because X now provides (two) activated nodes with first-order priming, the basic condition for a major increase in linkage strength. Of course, X also has uncommitted top-down connections to many other nodes, but only *mental* (adjective) and *practice*(noun) will meet this condition for increasing linkage strength to commitment levels. As a result, X now represents the specific content *mental practice*(noun phrase) for both input and output, albeit still relatively weakly.

Consciousness and Connection Formation In Vacuo

Because events and internal states are continuously changing and introducing pertinent novelties for humans in the waking state, conscious experience normally seems continuous (like a stream; James, 1890); nodes in one and often many input and output systems are undergoing prolonged activation, and we are continually aware of some novel aspect of our internal or external environment. However, without a changing environment or changing internal goals, the process of novel connection formation and consciousness seems to run off internally *in vacuo*: humans experience visual, auditory, and tactile hallucinations following prolonged periods of inactivity and sensory deprivation (Hebb, 1963). These hallucinations reflect internally generated pertinent novelty, or simultaneous activations of novel combinations of nodes, a necessary condition for connection formation and awareness. Dreams may represent another case where connection formation runs off internally in "unexpected and bizarre combinations" (Mandler, 1985, p. 80) due in part to reduced motoric activity and reduced bottom-up input from the environment.

Comparison with Retrieval Theories of Learning and Awareness

The node structure theory differs from retrieval theories of learning and awareness (see Baars, 1988). In retrieval theories, memory items enter consciousness (short-term memory) whenever they become activated, but, without rehearsal, no long-term connections whatsoever are established between the consciously represented items. The node structure theory goes beyond retrieval theories by providing a mechanism for explaining the effects of rehearsal (MacKay, 1981) and by providing a mechanism whereby new words, new phrases, and new experiences in general can be learned without rehearsal, simply by entering consciousness.

Decommitment, the Grandmother Cell Problem, and Amnesia

One or even several activations of an uncommitted node introduce only minor increases in linkage strength that can undergo atrophy or complete decay over a period of, say, a few days. As a result, unless weakly committed nodes undergo repeated activation, their connections can become decommitted or revert to uncommitted status. This means that a binding node may have to become engaged several times before the uncommitted node representing a particular instance of pertinent novelty becomes permanently committed and automatically activated. However, once a node's connections have become strongly committed, its remaining preexisting connections undergo atrophy so that it can no longer code any other content.

The Grandmother Cell Problem

The mechanisms of decommitment and multiple convergence provide an interesting solution to the "grandmother cell problem," the claim that theories incorporating "local" rather than "distributed" representations for relatively unique events such as sentences require an inordinate number of units. The node structure theory does not permanently code unique events, and the same uncommitted node can code many different contents, limited only by its degree of multiple convergence, and by the nonrecurrence of its contents: newly formed connections that remain unused undergo atrophy and become decommitted. Both of these preconditions are assumed to hold for nodes in sentential domains such as (active declarative proposition) where particular contents are repeated so rarely that permanent commitment is both unlikely and unnecessary. Only when the same sentence is perceived or produced repeatedly does its proposition node become permanently committed and incapable of coding alternate contents. Both of these preconditions hold in a similar way for visual cognition: permanent commitment is possible for a frequently experienced visual event such as one's grandmother, but not for an infrequently experienced visual event such as "my grandmother seen from behind walking past a desk in the parlor": if node commitment occurs at all for a nonrepeated event such as this, decommitment soon follows.

One interesting implication of the node structure solution to the grandmother cell problem is that uncommitted nodes require greater multiple convergence in higher level systems than in lower level systems. Uncommitted nodes in higher-level systems, e.g., the sentential system, where decommitment is common should exhibit an abundance of multiple conjoint connections, so that the same node can be reused repeatedly for representing novel contents. However, nodes with multiple conjoint connections should be less common in lower level systems, e.g., the phonological system, where decommitment is rare.

The node structure theory calls for a reanalysis of other assumptions implicit in formulations of the grandmother cell problem. One is that units exhibit only one type of processing (activation) and another is that units higher in a hierarchy exhibit the same behavior as units lower in a hierarchy (see D.M. MacKay, 1985). The fact that experimental evidence has not in general supported these assumptions (see, for example, D.M. MacKay, 1985) is consistent with the node structure distinction between types of processing (priming versus activation) with lower-level nodes behaving differently from higher-level nodes: under the principle of higher level activation, nodes higher in a hierarchy invariably become activated during perception, whereas nodes lower in a hierarchy can pass on priming without necessarily becoming activated. The node structure theory also calls for reanalysis of a third assumption implicit in formulations of the grandmother cell problem, namely that the filtering operations performed by feature detectors are directly responsible for "recognizing" the geometrical form of objects (see D.M. MacKay, 1985): conscious recognition requires more than simple classification in the node structure theory.

Binding Nodes and Amnesia

Binding nodes for speech perception-production are represented in the brain in structures such as the hippocampus and become engaged whenever an internally or externally generated verbal experience falls into a known category or domain, but is otherwise novel, so that a linguistic sequence node becomes activated but fails to activate any nodes in its domain. When this happens, the sequence node fails to quench and causes its connected high-threshold binding node to become activated.

The assumption that bilateral hippocampal and mediotemporal lesions destroy many of the binding nodes required for forming new long-term traces in higher-level language systems fits well with the pattern of sparing and deficit in amnesic patients (Squire, 1987; Shachter, 1985; Milner, 1968). Without input from the binding nodes for committing uncommitted connections, long-term learning of language inputs stops on the day of the operation. However, connections formed prior to the operation remain functional in hippocampal patients (see Squire, 1987, Milner, 1968) because bottom-up inputs still prime and enable activation of already committed nodes without help from the lesioned binding nodes: bilateral hippocampal damage cannot affect already established connections that are automatically primed and activated. As a result, densely amnesic patients show normal priming effects via already established connections, e.g., in a word completion task, even though they cannot form the new connections required to enable conscious verbal recall of having performed the task before (see, for example, Shachter, 1985).

However, higher-level language systems constitute a small sample of many systems that contain binding nodes for triggering awareness and connection formation. Other perception and action modules have their own (as yet undiscovered) systems of binding nodes, including those for classical conditioning of, say, the eyeblink reflex, and these other binding nodes are assumed to be intact in "language" amnesics. As a result, language amnesics are unaware of higher level language inputs,but not entirely unconscious or entirely incapable of learning: they can still learn nonlanguage behaviors such as the solution to a tactile maze, for example (see Corkin et al., 1985), even though they cannot verbally recall having seen or learned the maze before: nonverbal binding nodes suffice for learning a tactile maze, but verbal binding nodes are required for answering a verbal question about a tactile maze.

The fact that binding nodes are inhibitory in nature and only become engaged for novel inputs is consistent with recent observations on electrophysiological responses from the hippocampus: Smith (1986) and Smith, Stapleton, and Halgren (1986) recorded event-related potentials intracranially in humans performing recognition memory tasks involving repeated trials, and for novel inputs they observed temporally consistent long-latency (460 ms) potentials emanating from the hippocampus. Moreover, these hippocampal potentials appeared to be *inhibitory* in nature and disappeared after several trials, as if they reflected the inhibitory output from binding nodes for forming novel connections in the language system.

Finally, the fact that binding nodes become engaged automatically in response to pertinent novelty overcomes a problem with Wickelgren's (1979) proposal that the binding mechanism is excitatory in nature and selectively activates only uncommitted nodes by somehow forming a return inhibitory connection to the hippocampus once the uncommitted node has become committed. Wickelgren's otherwise excellent proposal requires connection formation for forming connections, a *reductio ad absurdum*.

Extension 2 of the Node Structure Theory: Attention

When we want someone to learn something new, we "call it to their attention," and it is often assumed that attention is necessary for learning and awareness (see Baars, 1988). Under the node structure theory, this common sense notion is only partially correct: commitment learning and awareness are coreferential in the theory, and although attention can help to activate particular contents, and activation is a precondition for awareness, attention is only necessary for learning and awareness under special circumstances. Moreover, the mechanisms underlying attention are essential for other purposes in the node structure theory, unlike theories such as Crick's (1984) where attention requires a special "searchlight" mechanism and a special type of (temporary) connection formation. The active direction of attention nevertheless remains a borderline phenomenon within the node structure theory: a complete theory of how attention is directed must provide an account of motivation, something the node structure theory does not do. What happens *after* a person chooses to attend to one source of input rather than another lies within the scope of the theory, but not the basis for choice per se. Within this limit, however, the node structure theory readily captures relations between awareness and attention, as I outline briefly below.

Selective Attention and Awareness

What are the theoretical mechanisms that enable people to selectively respond to one source of input rather than another? Two mechanisms already required for other purposes in the theory accomplish selective attention. One is the most-primed-wins principle that automatically activates target nodes whenever they receive more priming than distractor nodes in the same domain. For example, the most-primed-wins principle resolves ambiguities at various representational levels during perception by ensuring that only nodes representing the dominant (most frequent) or contextually most supported interpretation become activated and provide the basis for awareness (MacKay, 1987b, pp. 134-136). In short, the most-primed-wins principle automatically enables us to selectively ignore or fail to respond to the huge number of weaker inputs that are arriving simultaneously from the environment. Of course, the most-primed-wins principle also selectively activates target nodes whenever target and distractor inputs do not share domains of mental nodes, as when speech is presented to the left ear and a violin concerto to the right.

Here different-domain inputs are ignored because the timing and sequence nodes for activating nodes in these nonspeech domains are not applied.

The mechanism for engaging the activating mechanisms (timing nodes and sequence nodes) for systems of content nodes representing one source of input rather than another therefore provides another means of achieving selective attention (see MacKay, 1982). Consider shadowing, for example, a standard empirical paradigm for demonstrating selective attention: when different paragraphs are presented simultaneously to each ear, subjects can selectively shadow or reproduce with minimal lag one paragraph with little interference from the other paragraph. Under the node structure theory, shadowing tasks require activation of "target nodes" (that have been primed via one channel, say, the right ear) but not "distractor nodes" (that have been primed via other channels). This task becomes problematic when target and distractor inputs are similar and simultaneously prime nodes in the same domain: whatever node in this shared domain happens to be most primed will become activated under the most-primed-wins principle, regardless of its channel of origin, and when distractor rather than target nodes become activated, a "cross-talk" intrusion from the other channel will occur.

To prevent these cross-talk errors, activating mechanisms (timing nodes and sequence nodes) must become engaged at a lower than normal level for systems of content nodes that distinctively represent inputs from the target source rather than from the distractor source. The resultant node activations deliver a boost in priming to low-level nodes representing target inputs rather than distractor inputs, and this (first-order) boost in priming gets transmitted to target nodes in higher-level (shared) domains, enabling them to become activated under the most-primed-wins principle instead of distractor nodes (that lack this boost in priming). For example, when subjects are instructed to shadow speech inputs arriving at the right rather than the left ear, both sources of input prime nodes within shared domains in the phonological system (and above), but there exist lower-level domains of sensory analysis nodes (closer to the basilar membrane) that are unique to the right ear. Engaging the activating mechanisms for these unique lower level domains introduces first-order convergent priming that enables (connected) nodes for right ear (target) inputs to become most primed in (shared) phonological domains and become activated so as to determine the shadowing response[3].

One interesting implication of this view of selective attention is that comprehension will require more time under conditions requiring selective attention than under conditions not requiring selective attention. The reasoning goes as follows. Because lower-level nodes do not normally become activated during everyday comprehension (the principle of higher-level activation), but do become activated in selective attention tasks, more nodes than normal must become activated in selective attention tasks. Because activation proceeds sequentially and requires more time than priming (see MacKay, 1987b, pp. 77-78), comprehension will require more time when a task requires selective attention. For example, recognizing a

---

[3] This is not to say that attentional selectivity is *in general* based on distance from the sensory receptors: For example spatial selectivity is possible when both ears receive competing messages, and intensity or interaural time differences can be used to code spatial location.

semantic target (names of animals, say) will take more time under conditions requiring selective attention than under conditions not requiring selective attention. Similarly, as the rate of presenting a paragraph is speeded up via computer-compression techniques, comprehension should break down at slower rates when selective attention is required than when selective attention is not required.

Flexibility of Higher-Level Activation

As the above discussion suggests, higher-level activation is a relative rather than an absolute principle: it is not just possible but desirable to engage the timing nodes for activating lower-level systems of nodes whenever an input is especially degraded, or unfamiliar, or requires selective attention. Activating these lower-level systems incurs costs such as reduced rates of processing, but paying these costs is necessary in these situations to provide sufficient bottom-up priming to enable appropriate higher-level nodes to become activated.

Divided Attention and Awareness

"Dividing" attention between different output or perceptual-motor systems, as when we carry on a conversation and drive a car at the same time, requires the same mechanisms as selective attention in the theory (given the decision to carry on both activities simultaneously): the most-primed-wins principle, and the mechanisms for (simultaneously) engaging the activating mechanisms (timing nodes and sequence nodes) for the two or more systems of content nodes representing the concurrent activities. Divided attention is achieved within the theory by simultaneously activating nodes in different domains, on the input side (sensory analysis nodes), on the output side (muscle movement nodes), or both (mental nodes), and the theory predicts interference whenever the concurrent activities share nodes in the same domain.

However, nodes in shared domains represent only part of the reason why people tend to make errors when attempting to do two similar tasks simultaneously. The degree and nature of interference also depends on the type of shared node (content versus sequence versus timing nodes), on their temporal pattern of activity, and on the nature of their interconnections. Besides providing some interesting predictions, these mechanisms in the theory explain a wide range of already observed interference and noninterference effects. For example, the most-primed-wins principle allows error-free execution of two or more concurrent activities if the nodes that must be simultaneously activated inhabit different domains. This means that complex activities such as speaking, typing, or playing the piano can be executed concurrently and automatically without mutual interference, but only if prior learning and practice have established nodes for these activities in separate domains at every level of the system.

In conclusion, attention and consciousness are closely interrelated and interact with one another in the node structure theory: with conflicting sources of input, attention can play a role in determining what source of perceptual inputs become activated and enters awareness. However, awareness and attention reflect different processes under the theory. Awareness is a direct and ongoing consequence of commitment learning, whereas the learning that is required for divided and selective attention has already taken place. Moreover, awareness can occur automatically, without the aid of the motivational mechanism that is required to direct attention toward different types of perceptual or output activities.

Comparison with Other Theories of Attention

The mechanisms for attention in the node structure theory, e.g., the most-primed-wins principle, can be seen to exhibit general characteristics of two major, currently competing approaches to attention: the capacity limitation approach (Kahneman, 1973) and the distributed control approach (see Allport, 1980). However, the node structure theory achieves these characteristics using very different mechanisms from either of these approaches: capacity is limited because only one content node in a domain can become activated at a time under the theory, and processing is distributed because content nodes in different domains and systems can and often do become activated simultaneously.

New Insights into the Three-Body Problem

I now return to the empirical questions that began this chapter, starting with the three-body problem, part 1; vicissitudes of awareness in perception and action. My goal is to summarize observed relations within the vast literature on action, perception, and awareness that now make theoretical sense, although I must leave for a future publication the details of how the node structure theory explains some of these phenomena.

Vicissitudes: Level of Awareness Rules in Perception and Action

Because the node structure theory provides a detailed representation of what is above what in language perception and production, the concepts of "level of processing" and "level of awareness" can be well defined in the theory. The node structure theory also makes sense of the "level of awareness rule," the fact that we normally become conscious of higher- rather than lower-level units in perception and action (see MacKay, 1973a). The level of awareness rule is attributable to the inverse relation between level and linkage strength (practice) in the node structure theory. Phonological nodes generally receive more prior practice than lexical and phrase nodes (see MacKay, 1982), which in turn receive more prior practice than the proposition nodes that must become committed *de novo* when perceiving or producing sentences that have never previously been encountered. As a result, connection formation and awareness are usually limited to higher-level concepts because what is new (unhabitual and unhabituated) triggers consciousness, and what is usually new in everyday sentences is not phonemes but phrases and propositions.

Exceptions to the Level of Awareness Rule

The node structure theory also makes sense of the many exceptions to the level of awareness rule. For example, although awareness normally begins at the lexical level for adults producing everyday speech (MacKay, 1987b, p. 79), adults occasionally learn new words that only approach automaticity at the phonological and lexical levels after considerable practice. Moreover, if concepts at higher (suprasentential) levels receive sufficient practice, they too achieve automaticity or unconscious processing. Even everyday thought patterns involving discourse level units can receive so much practice as to become triggered automatically and unconsciously (see Freud, 1914; MacKay & Konishi, 1980).

Errors in speech and action illustrate another important exception to the level of awareness rule (discussed above). Speakers normally become acutely aware of a speech error, even a subphonemic error such as the slurring of a speech sound (see MacKay, 1990). The nature of the slur, what sound was slurred and perhaps also the higher-level implications of the slur for the speaker enter awareness. Similarly, when listening to an unaccustomed foreign accent, we become aware of *both* the low-level articulatory novelties *and* the functionally useful (pragmatic) message being conveyed. Finally, aspects of the environment often enter awareness automatically when an error occurs, as when seasoned drivers become aware of their previously unconscious driving behavior after experiencing a near miss, or an unexpected traffic light (see Mandler, 1985).

These exceptions show again that the level where conscious processing begins is not completely fixed. We adults *normally* only become aware of higher-level aspects of an input, i.e., the sentential and discourse levels in the case of language, and objects and ego space in the case of vision (Marcel, 1983), and we *normally* remain unaware of the sensory and lower level conceptual events, e.g., phonemes, whose priming contributes to determining that awareness. However, consciousness can begin at a lower than normal level when an input is attended to, novel, unexpected, degraded, or unfamiliar: either attention or pertinent novelty (as in the case of speech errors) can engage the mechanisms necessary for consciousness and connection formation at lower than normal levels, enabling phonological or even sensory and muscle movement information to enter awareness.

Vicissitudes in the Adaptive Value of Awareness

Consciousness is closely connected with the organization of action under the node structure theory: the prolonged and simultaneous activation of two or more nodes that results in consciousness forms new nodes for joining formerly disparate action components together in the service of a higher plan. The prolonged activation associated with consciousness also facilitates the rapid preparation of adaptive responses by priming or preparing for retrieval stored information relevant to the current situation. When a concept enters awareness, all of the mental nodes (information and actions) associated with it become very strongly primed or prepared for activation: the conscious organism is in a continual state of readiness to use a wide range of past experiences for responding to what's new in an ongoing situation. However, the benefits of prolonged activation and awareness vary with the structure of the overall network. If a (hypothetical) organism can respond only on the basis of preformed stimulus-response connections, responses that are unconscious, fast, and unmodifiable are preferable to responses that are conscious and modifiable, but slow. Prolonged activation and awareness are neither necessary nor necessarily adaptive for all perceptual-motor systems.

Language as an Imperfect Index of Awareness

Language is generally accepted as the main index of awareness because verbal systems constitute our most sophisticated and frequently used means of representing and expressing our awareness in everyday life: the large number of domains of nodes required for language facilitate the formation of new (conscious) representations, and once formed, language representations can be used immediately, e.g., in communicating with other people. However, contrary to Vygotsky (1962), our ability to verbally comment (either overtly or internally) on our percepts, thoughts. memories, or behaviors is not a necessary precondition for consciousness. When we become aware of something, we can *usually* identify it by means of a verbal response, but consciousness is neither synonymous with verbal awareness nor a direct product of our ability to speak. For example, one often becomes aware of a novel experience, say, an unfamiliar smell, without being able to find words to describe it appropriately. Words can even fail us for describing everyday objects and events of which we are undeniably aware, as during the tip-of-the-tongue state (James, 1890). Finally, most psychologists would hesitate to deny awareness to the aphasic who is incapable of speech but can compose a symphony or paint a portrait (Luria, 1980). In short, there exist other, nonverbal awareness systems as in the node structure theory.

Interactions Between Conscious and Unconscious Processes in Perception and Action

The three-body problem, part 2, consists of two questions. I consider the first of these questions here: how do unconscious processes give rise to perception and action without awareness on the one hand, and errors in perception and action on the other?

Perception Without Awareness

The vast literature on perception without awareness (see Dixon, 1981) illustrates effects of unconscious processes (priming) on conscious ones in ways that make systematic sense under the node structure theory. I touch briefly on several of the more salient examples. One is Corteen and Wood's (1972) study on conditioned galvanic skin responses (GSRs) to shock-paired words presented to the unattended ear in a dichotic listening-shadowing task. Corteen and Wood first paired the target words with shock and then presented these words interspersed among other, neutral words on the unattended channel. As commonly occurs during dichotic listening, the subjects were unaware of these unattended target words and unable to signal their occurrence by making a manual response or by stopping shadowing. However, both the shock-paired words and their semantic relatives elicited GSRs whether presented to the shadowed or to the unshadowed ear. These findings clearly illustrate semantic processing without awareness of the input to the unattended channel, and have now been successfully replicated in at least five published experiments (see Neumann, 1984).

MacKay's (1973a) experiments on selective listening also illustrate semantic processing without awareness and are consistent with the findings of Corteen and Wood and many other studies. When subjects in MacKay's study shadowed a lexically ambiguous sentence such as "They threw stones toward the bank yesterday", a semantically related word such as *river* presented simultaneously with *bank* on the unattended channel automatically influenced which meaning of the ambiguity they perceived, even though the subjects remained unaware of and unable to report what these unattended words were. Unconscious semantic processes (priming) clearly influenced what meaning entered awareness in this study.

Moreover, when two unattended words, e.g., *river* and *shore*, both related to the same meaning of the ambiguity were presented, the effect was greater than when either word was presented by itself. However, the effects of two unattended words denoting conflicting interpretations of the ambiguity, e.g., *money* and *river*, automatically cancelled each other out, so that the probability of perceiving the two interpretations remained unchanged, as if no words whatsoever had been presented. These findings provide evidence for the inflexible, automatic, and predictable nature of unconscious processing (priming).

Studies of "perceptual defence" (see Dixon, 1981, for extensive examples), where stimuli presented too briefly to enable conscious recognition nevertheless influence other behaviors, also illustrate perception without awareness, as do physiologically oriented studies such as Weiskrantz, Warrington, Sanders, and Marshall's (1974) demonstrations of "blindsight": following lesions to the visual cortex, patients are unable to report the presence of objects falling within large areas in the visual field (called scotomas). However, only verbal responses exhibit this deficit: blindsight patients can quickly and accurately point with their hand or move their eyes toward objects presented briefly within this "blind" region, even though verbally they continue to insist that their consistently accurate manual responses are only guesses. The lesion has dissociated the visual system representations for locating objects with the hand or eyes from the language system representations that enable conscious verbal description.

Action Without Awareness

As expected under the node structure theory, awareness neither causes nor consistently accompanies behavior in available data. For example, we can use unconscious processes (priming and activation) rather than conscious processes to maintain behavior, as when we orient ourselves in space and perform actions such as walking using visual cues that never enter our awareness (Lee & Lishman, 1974). Moreover, even when people *do* become conscious of the stimuli that trigger or guide their actions, *awareness of the stimuli is unnecessary for action*. As James (1890) pointed out, awareness cannot *cause* behavior because actions can precede rather than follow awareness: when pricked with a pin we withdraw the finger first and become aware of the pain later.

Perception-Production Errors: Deep Dyslexia

Because MacKay (1982, and 1987b, pp. 120-121) deals in detail with the effects of unconscious processes (priming) on everyday errors in perception and production. I will examine a somewhat different example here, deep dyslexia. Due to cerebral injury, deep dyslexics produce semantically similar word substitutions that are indistinguishable from the word substitutions of normal individuals except that the dyslexic cannot immediately correct the errors and makes them when reading printed words, misreading the word *table* as *chair*, or *uncle* as *aunt*, for example. The "near miss" nature of the target and error concepts indicates that the correct (visually presented) word must have received unconscious semantic processing (priming), but could not become activated to determine the response, perhaps because the lesion had selectively impaired the return connection from the sequence node to this particular content node. As a result, the target content node, say, *table*(noun), could not become activated, but nevertheless passed on priming to its semantic relative, *chair*(noun), via connections within propositions such as *Tables and chairs are furniture*, so that *chair*(noun) therefore became activated in error as the most primed node in the domain.

Differences Between Conscious Versus Unconscious Processes in Perception and Action

The second part of the three-body problem, part 2, is: what accounts for the differences between conscious versus unconscious processes? Conscious processes have been shown to differ from unconscious processes in six ways that make theoretical sense under the node structure theory; extensiveness, predictability, rate, flexibility, serial versus parallel character, and level of processing.

Extensiveness of Conscious Versus Unconscious Processing

Conscious processing is normally much more extensive than unconscious processing. Whereas unconscious processing is limited to the old and familiar, e.g., the meanings and syntactic categories of familiar words, conscious processing also extends to what is new, e.g., the particular conjunction or relation between words and the implications of this conjunction in the particular situation in which the word is spoken (MacKay, 1973a).

Predictability of Conscious Versus Unconscious Processing

With competing sources of input, priming is unconscious and predictable: familiar aspects of both input sources automatically become primed. However, activation and awareness are subject to effects of motivation which are unpredictable. By selectively directing attention (e.g., applying activating mechanisms at lower than normal levels), higher-level nodes representing one input source but not the other will become activated and introduce pertinent novelty, the precondition for awareness.

The Flexibility of Conscious Versus Unconscious Processing

Whereas unconscious priming spreads automatically up to some fixed level, the level at which conscious processing begins is flexible rather than fixed and automatic. We *normally* only become aware of higher-level aspects of an input, i.e., the sentential and discourse levels in the case of language, and objects and ego space in the case of vision (Marcel, 1983), and we *normally* remain unaware of the sensory and lower-level conceptual events that play a role in determining that awareness, e.g., phonemes. As we have seen, however, consciousness can begin at a lower than normal level when an input is attended to, novel, degraded, unexpected, or unfamiliar.

Serial Versus Parallel Nature of Conscious Versus Unconscious Processes

Whereas unconscious processes such as priming are fundamentally parallel in the node structure theory, awareness within a system is fundamentally sequential. The way that subjects become aware of the meanings of lexically ambiguous sentences clearly illustrates this sequential character of awareness: when searching for the two meanings of ambiguous sentences, subjects in MacKay and Bever (1967) perceived first one meaning, then the other, because only one node at a time can receive most priming and become activated in any given domain. But even though only one interpretation of an ambiguity becomes conscious at a time, both meanings receive *unconscious* processing (priming) simultaneously and in parallel (see Swinney, 1979; MacKay & Bever, 1967).

Experiments demonstrating effects of the unseen meanings of an ambiguous word on the interpretation of subsequent words further illustrate the parallel nature of priming. For example, Marcel (1983) tachistoscopically presented an ambiguous word such as *palm* followed by a patterned mask that prevented recognition or even better-then-chance guesses as to whether any stimulus whatsoever had preceded the masking pattern. The dependent variable was lexical decision time for a subsequent, consciously recognized word, either *maple* or *wrist*. The results showed that *palm* facilitated lexical decisions for both of these related words. The pattern mask clearly prevented conscious awareness (i.e., prolonged activation of lexical and phonological nodes representing) of *palm*, but did not prevent the priming that spreads unconsciously and in parallel from *palm* to words related to its two meanings (for a replication, see experiments 5 and 6 in Fowler, Wolford, Slade, & Tassinary, 1981).

The Rate of Conscious Versus Unconscious Processes

One implication of the fact that consciousness requires prolonged activation is that conscious processes are necessarily slower than corresponding unconscious processes such as priming and self-inhibited activation. This relative slowness of conscious processing explains why subjects searching for the two meanings of an ambiguous sentence take so long to become *aware* of both meanings (MacKay & Bever, 1967), even though both meanings are processed *unconsciously* (i.e., primed) shortly after reading the ambiguous word (Swinney, 1979).

Relations Between Awareness, Attention, and Practice in Perception and Action

I turn now to part 3 of the three-body problem; relations between awareness, attention, and practice in perception and action.

Relationship Between Consciousness, Learning, and Practice

Because the node structure theory postulates two fundamentally different types of learning, engrainment learning (see MacKay, 1982, 1990) and commitment learning, and only one type (commitment learning) determines awareness, the relationship between learning and awareness is relatively complex. Both types of learning are automatic, requiring neither intention nor volition. However, engrainment learning consists of unconscious increases in linkage strength that result from repeated first-order priming and activation of nodes via already existing connections (see MacKay, 1982, 1990) and is only very indirectly related to what enters or fails to enter awareness. Commitment learning, on the other hand, concerns the process of forming or "committing" new connections and gives rise to conscious awareness.

Distinguishing between these two types of learning in the theory explains two seemingly contradictory but persistent observations on the relation between practice, learning, and awareness: the fact (discussed on p. 279) that we learn and become conscious of mental contents *encountered for the first time* (commitment learning), and the fact that *repetition* (practice) often improves behavior *without awareness*. A recent example of the latter appears in Marcel (1983; see Hebb, 1963; and MacKay, 1981, for other examples). Words that are so effectively masked as to be unreportable facilitate lexical decisions for semantically similar words (see p. 299), and Marcel (1983) showed that repeating these unreportable words increased the facilitatory effect. Specifically, Marcel (1983) recorded lexical decision time for a word that followed 2-20 repetitions of a semantically similar but unreportable (masked) word, and found that lexical decision times improved systematically with repetition of the unreportable (masked) word up to an asymptote resembling that of the priming function in Fig. 2. Under the theory, the increased facilitation reflects engrainment learning: repeating the unreportable word improved its linkage strength and transmission of priming to connected nodes up to some asymptote, all in the absence of awareness.

Retrieval and Awareness

Nodes undergoing prolonged (conscious) activation prime and enable activation of nodes representing related concepts. This explains why it helps for people to consciously remind themselves of tasks that remain to be performed (Mandler, 1985): conscious reminders make remembering more likely by increasing linkage strength (engrainment learning), and by keeping relevant (connected) concepts highly primed and ready to be activated when conditions appropriate for retrieval or action appear.

Effects of Level in a Hierarchy

The inverse relationship between awareness and practice or repeated retrieval is one of the few generally agreed upon pretheoretical phenomena in the field: when our behavior becomes more skilled as a result of practice, we become progressively less conscious of how we execute it. Consciousness cannot occur when a perceptual or behavioral process has received so much practice as to become fully automatic (Sokolov, 1963; Shiffrin & Schneider, 1977). However, the usual explanation of this phenomenon, that consciousness constitutes a limited resource that must be reserved for unpracticed or nonautomatic processes, is at best circular and at worst untenable (see Allport, 1980).

Practice also plays a role in some of the well-known limitations of conscious processing, such as the fact that only the products of a mental activity enter consciousness, not the processes (Mandler, 1985). This limitation reflects the fact that only high-level content nodes normally undergo prolonged activation and give rise to consciousness under the node structure theory; awareness cannot arise when content, sequence, and timing nodes are highly practiced and automatically activated.

Effects of Action on Attention and Awareness

The node structure theory predicts profound effects of action on attention and awareness: when an action such as speech production must be based on inputs arriving via an attended channel, awareness of inputs via *unattended* channels should drop drastically under the theory. For example, compare two versions of a dichotic listening task where subjects respond as quickly as possible to a target word arriving at either ear: an *action* version where subjects produce (shadow) the input to one (attended) ear, and a *listening* version where subjects simply listen to that ear without shadowing its content. Under the node structure theory, detectability of the target word should be higher in the listening condition than in the action condition. Of course such a difference might be attributable to acoustic masking of unattended inputs by the shadowing output, but the node structure theory predicts this same difference without the possible masking artifact when shadowing is achieved by internal speech or mouthing (silent articulation) rather than overt speech. The reason is that under the principle of higher-level activation, perception (as in simple listening) does not engage the activating mechanism (sequence and timing nodes) for lower-level systems, and bottom-up input can automatically make the target node most primed in its domain, regardless of the target's channel of origin. Repeated application of the activating mechanism to the target domain will therefore ensure activation and awareness under the most-primed-wins principle. However, production (including silently articulated shadowing) must engage the activating mechanisms for lower- and higher-level systems representing the *attended* input, so that arrival of the target via the *unattended* channel will not guarantee most-primed status for the target node because of the boost in priming that nodes representing attended inputs will receive. As a result, shadowing will greatly interfere with detection of targets presented to the unattended ear.

Conclusion

The three-body problem is complex: an adequate account of relations between perception, action, and awareness calls for a general theory of mind with a scope ranging from errors, sequencing, and timing in behavior, to learning, memory, and amnesia. And if the present account is correct, awareness mechanisms in such a theory will be inseparable from mechanisms for representing novelty, and mechanisms for explaining selective and divided attention will be inseparable from mechanisms for perception and action in general.

Acknowledgements. The author thanks Drs. A. Allport, B. Baars, P. Bieri, D. Burke, H. Heuer, A. Meyer, H. Mittelstaedt, E. Scheerer, M. Smith, and especially O. Neumann for helpful comments on earlier versions of this chapter. Aspects of this chapter were presented to the conference on "Action, Attention, and Automaticity," held at the Center for Interdisciplinary Research, University of Bielefeld, November 26-30, 1984.

References

Allport, D.A. (1980). Attention and performance. In G. Claxton (Ed.), *Cognitive psychology: New directions* (pp. 112-153). London: Routledge & Kegan Paul.

Baars, B.J. (1983) Conscious contents provide the nervous system with coherent global information. In R.J. Davidson, G.E. Schwartz, & D. Shapiro (Eds.), *Consciousness and self-regulation* (Vol. 3). New York: Plenum.

Baars, B.J. (1988). *A cognitive theory of consciousness*. New York: Cambridge University Press.

Corkin, S., Cohen, N.J., Sullivan, R.A., Clegg, R.A., Rosen, T.J., & Ackerman, R.H. (1985). Analyses of global memory impairments of different etiologies. *Annals of the New York Academy of Sciences, 444*.

Corteen, R.S., & Wood, B. (1972). Automatic responses to shock associated words. *Journal of Experimental Psychology, 94*, 308-313.

Crick, F. (1984). Functions of the thalmic reticular complex: The searchlight hypothesis. *Proceedings of the National Academy of Sciences USA, 81*, 4586-4590.

Dixon, N.F. (1981). *Preconscious processing*. New York: Wiley.

Fowler, C.A., Wolford, G., Slade, R. & Tassinary, L. (1981). Lexical access with and without awareness. *Journal of Experimental Psychology: General, 110*, 341-362.

Freud, S. (1914). *Psychopathology of everyday life* (A.A. Brill Trans.). New York: Penguin.

Fromkin, V.A. (1971). The non-anomalous nature of anomalous utterances. *Language, 47*, 27-52.

Gregory, R.L. (1981). *Mind in science*. New York: Cambridge University Press.

Hebb, D.O. (1963). The semiautonomous process, its nature and nurture. *American Psychologist, 18*, 16-27.

James, W. (1890). *The principles of psychology*. New York: Holt.

Kahneman, D. (1973). *Attention and effort*. Englewood Cliffs, NJ: Prentice-Hall.

Koriat, A. (1981). Semantic facilitation in lexical decision as a function of prime-target association. *Memory and Cognition, 9*, 587-598.

Lee, D.N., & Lishman, J.R. (1974). Visual proprioceptive control of stance. *Journal of Human Movement Studies, 1*, 87-95.

Levelt, W.J.M. (1984). Spontaneous self-repairs in speech: processes and representations. In M.P.R. van den Broecke & A. Cohen (Eds.), *Proceedings of the Tenth International Congress of Phonetic Sciences* (pp. 105-111). Dordrecht: Foris.

Luria, A.R. (1980). *Higher cognitive functions in man*. New York: Basic Books.

MacKay, D.G. (1972). The structure of words and syllables: evidence from errors in speech. *Cognitive Psychology, 3*, 210-227.

MacKay, D.G. (1973a). Aspects of the theory of comprehension, memory and attention. *Quarterly Journal of Experimental Psychology, 25*, 22-40.

MacKay, D.G. (1973b). Complexity in output systems: Evidence from behavioral hybrids. *American Journal of Psychology, 86*, 785-806.

MacKay, D.G. (1978). Speech errors inside the syllable. In A. Bell & J.B. Hooper (Eds.), *Syllables and segments* (pp. 201-212). Amsterdam: North-Holland.

MacKay, D.G. (1979). Lexical insertion, inflection and derivation: creative processes in word production. *Journal of Psycholinguistic Research, 8*, 477-498.

MacKay, D.G. (1981). The problem of rehearsal or mental practice. *Journal of Motor Behavior, 13*, 274-285.

MacKay, D.G. (1982). The problems of flexibility, fluency, and speed-accuracy trade-off in skilled behavior. *Psychological Review, 89*, 483-506.

MacKay, D.G. (1985). A theory of the representation, organization, and timing of action with implications for sequencing disorders. In E.A. Roy (Ed.), *Neuropsychological studies of apraxia and related disorders* (pp. 267-308). Amsterdam: North-Holland.

MacKay, D.G. (1987a). The asymmetrical relationship between speech perception and production. In H. Heuer, & A. Sanders (Eds.), *Perspectives in perception and action* (pp. 301-334). Hillsdale, NJ: Erlbaum.

MacKay, D.G. (1987b). *The organization of perception and action: A Theory for language and other cognitive skills*. Berlin, Heidelberg, New York: Springer.

MacKay, D.G. (1990). Errors, ambiguity, and awareness in language perception and production. In B. Baars (Ed.), *The psychology of error: a window on the mind*. New York: Plenum.

MacKay, D.G., & Bever, T.G. (1967). In search of ambiguity. *Perception and Psychophysics, 2*, 193-200.

MacKay, D.G., & Konishi, T. (1980). Personification and the pronoun problem. In C. Kramarae (Ed.), *The voices and words of women and men*. London: Pergamon.

MacKay, D.M. (1985). The significance of 'feature sensitivity.' In D. Rose & V.G. Dobson, (Eds.), *Models of the visual cortex* (pp. 47-53). New York: Wiley.

Mandler, G. (1985). *Cognitive psychology: An essay in cognitive science*. Hillsdale, NJ: Erlbaum.

Marcel, A.J. (1983). Conscious and unconscious perception: Experiments on visual masking and word recognition. *Cognitive Psychology, 15*, 1197-1239.

McClelland, J.L., Rumelhart, D.E., & the PDP Research Group (1986). *Parallel distributed processing. Explorations in the microstructure of cognition: Vol. 2. Psychological and biological models*. Cambridge, MA: MIT Press.

Miller, G.A., Galanter, E., & Pribram, K.H. (1960). *Plans and the structure of behavior*. New York: Holt.

Milner, B. (1968). Visual recognition and recall after temporal lobe excisions in man. *Neuropsychologia, 6*, 191-209.

Motley, M.T., Baars, B.J., & Camden, C.T. (1983). Experimental verbal slip studies: a review and an editing model of language encoding. *Communication Monographs, 50*, 79-101.

Neumann, O. (1984). Automatic processing: A review of recent findings and a plea for an old theory. In W. Prinz & A.F. Sanders (Eds.), *Cognition and motor processes*. Berlin, Heidelberg, New York: Springer.

Neumann, O. (1987). Beyond capacity: A functional view of attention. In H. Heuer & A. Sanders (Eds.), *Perspectives on perception and action* (pp. 361-394). Hillsdale, NJ: Erlbaum.

Ostry, D., Moray, N., & Marks, G. (1976). Attention, practice and semantic targets. *Journal of Experimental Psychology: Human Perception and Performance, 2*, 326-336.

Shachter, D. (1985). Multiple forms of memory in humans and animals. In N.M. Weinberger, J.L. McGaugh, & G. Lynch (Eds.), *Memory systems of the brain*. New York: Guilford.

Shiffrin, R.M., & Schneider, W. (1977). Controlled and automatic human information processing. II. Perceptual learning, automatic attending, and a general theory. *Psychological Review, 84*, 127-190.

Smith, M.E. (1986). *Electrophysiology of human memory: scalp and intracranial event-related potentials recorded during recognition judgements and related tasks*. Unpublished Ph.D. dissertation, Psychology Department, University of California, Los Angeles.

Smith, M.E., Stapleton, J.E., & Halgren, E. (1986). Human medial temporal lobe potentials evoked in memory and language tasks. *Electroencephalography and Clinical Neurophysiology, 63*, 145-159.

Sokolov, Y.N. (1963). *Perception and the conditioned reflex*. New York: Macmillan.

Squire, L.R. (1987). *Memory and brain*. New York: Oxford.

Swinney, D.A. (1979). Lexical access during sentence comprehension: (re)consideration of context effects. *Journal of Verbal Learning and Verbal Behavior, 18*, 645-659.

Thatcher, R.W., & John, E.R. (1977). *Foundations of cognitive processes*. Hillsdale, NJ: Erlbaum.

Treiman, R. (1983). The structure of spoken syllables: Evidence from novel word games. *Cognition, 15*, 49-74.

Treisman, A., & Gelade, G. (1980). A feature-integration theory of attention. *Cognitive Psychology, 12*, 97-136.

Vygotsky, L.S. (1962). *Thought and language*. Cambridge, MA: MIT.

Weiskrantz, L., Warrington, E.K., Sanders, M.D., & Marshall, J.C. (1974). Visual capacity in the hemianopic field following a restricted occipital ablation. *Brain, 97*, 709-728.

Wickelgren, W. (1979). *Cognitive psychology*. Englewood Cliffs, NJ: Prentice-Hall.

Sensory and Perceptual Control of Action in Early Human Development

A. VINTER

CONTENTS

Introduction: About Perception and Action Relationships

Although the idea is not new, it is only recently that we are frequently able to read that action and perception are two inseparable poles of human and animal behavior[1]; it is a theory (or dogma?) which is so attractive that it unifies the adherents of apparently incompatible schools of thought, such as the cognitivists and the followers of "ecological psychology." Perception and action are inseparable in the sense that perception is constantly at the service of action, controlling and guiding action. Conversely, action may also be at the service of perception, by conditioning and determining it, although less research has been dedicated to the study of such a relationship (see Bridgeman, this volume, for a discussion of the role of eye movements in visual perception). Furthermore, it is through action that an object's properties are discovered. More specifically, most authors (Fowler & Turvey, 1982; Neisser, 1985; Mounoud, 1981) agree in the recognition that perception and action are intricately related in the sense that a unique language may be used to describe, and possibly explain, the functioning of the structures of both perception and action. For Neisser it is very likely that mental structures (schemata) that organize movements are represented internally in the same format as those responsible for perceptions.

But let us delude ourselves! The idyll between the adherents of these opposite schools of thought quickly evaporates as soon as the perceiver or the actor is im-

[1] It must be pointed out that the way the traditional approaches in psychology tackle the study of perception and action does not fit very well with this principle. For a long time, and it is still partially the case, perception and action were studied in a completely separate manner. Funnily enough, a similar gap between theoretical principles and effective methods of study can be found in the dichotomy between cognition and affectivity. It is very common to read that they cannot be separated, and it is equally as common to study them in a completely separate manner!

Relationships Between Perception and Action
Edited by O. Neumann and W. Prinz

mersed in his/her environment. For some, the environment is the prime factor in the subject-object relationship. It contains structured information which is directly pertinent to the guidance of action. Perception consists of a pick-up or extraction of invariant properties in the environment, one category of which is called affordances. Affordances are meaningful properties of objects in reference to the animal's action goals and "effectivities." Seeing a cup affords drinking; for a baby, a rattle affords reaching or grasping. A direct relationship exists between the object in its environment and the subject's characteristics, as well as between the subject's perceptions and his/her actions. "The information made available to the animal as perceiver should directly provide the information required by the animal as actor" (Fowler & Turvey, 1982). Affordances are more or less likely to be attended to, according to the animal's needs.

For others, neither the environment nor the object exists independently of the subject as perceiver or actor. For cognitivists, perception is equally a meaningful activity, as it is in ecological psychology, but the origin of the meaning is internal to the subject and not inherent to the object's structure. Consequently, the subject-object or "object-percept" relationship will be indirect, mediated by internal structures of thought, often called representations or schemas in cognitive psychology. The meaning of an object for a subject will change in relation to different internal characteristics: his/her intentions, knowledge, expectancies, beliefs, etc. The object is discovered through the subject's actions and by mediation of a mental construction. Representations organize the perceptual activities of the subject and allow the attribution of meaning to the products of these activities. Like perceptions, actions are centrally represented (for a discussion of the relationships between action and cognition see Zanone & Hauert, 1987; Hauert, Zanone, & Mounoud, this volume). The relationship between perception and action is therefore no more direct than that which links perceptions to the object's properties. Nevertheless, some authors such as Neisser and Mounoud support the theory of a direct compatibility between the languages in which the representations that mediate perceptions, on the one hand, and actions, on the other hand, are formulated - existence of an amodal language for Neisser, of a common code for Mounoud (see also Prinz, this volume, for a similar view).

As we can see, there are many ways of being inseparable, even for the perception-action couple! Consequently, there are many ways in which a sensory or perceptual control of action may be established, as we will see in this chapter in relation to a child's development in the first years of life. Action may be considered as being under a sensory or perceptual control in a double sense: in relation to its planning on the one hand, (i.e., to the eliciting dimensions), and in relation to its execution on the other. Within this perspective, two modes of action control are usually differentiated in the literature on infancy (Bower, Broughton, & Moore, 1970a; von Hofsten, 1979; White, Castle, & Held, 1964), as well as in classical neurophysiological studies (Jeannerod, 1982; Paillard & Beaubaton, 1976). In the so-called visually elicited or triggered control, movements are ballistic in the sense that their parameters are pre-programmed with no correction possible (this corresponds to the "advanced" processing mode of sensory information distinguished by

Cruse, Dean, Heuer, & Schmidt,this volume). By contrast, in the so-called visually guided control, some of these parameters at least are determined in the course of the action execution, on the basis of feedback which can be provided at different moments of the movement, and allow the subject to correct it (this corresponds either to the "continuous" or to the "intermittent" processing mode described by Cruse et al., this volume).

Although traditional in the literature on infancy, this distinction between a visually elicited and a visually guided control is not a happy one, (a) with respect to the term "elicitation," which suggests initiation of the action (i.e., motivation to act, for example), whereas this term refers essentially to the planning of the action (i.e., pre-programming of the movement's parameters before its execution); (b) with respect to the term "visually," which should be understood in the broad sense of "sensorily" or "perceptually" guided or elicited. Feedback used by the subject to perform his/her movement need not only be visual but also tactile or proprioceptive, or a combination of different sensory information. It has been proposed that movements are initially visually triggered until around 5-6 months of age, and then that a guided component is progressively integrated (Paillard, 1980; Jeannerod, 1982). Interestingly, the reverse passage, from a guided control to a triggered one has been described by other authors (White et al., 1964).

In this chapter, we will analyze which forms of action control exist in the first years of life and, in particular, if a simple dichotomy between an elicited and a guided control is sufficient to account for the different modes of sensory or perceptual control of action. It will progressively appear that this problem of action control is closely related to that of "segmentation" of behavior, i.e., to the degree of decomposability of action into elementary segments.

This problem will be tackled in relation to three kinds of sensorimotor behavior which have been selected according to the degree of integration between perception and action they require on the one hand, and to their ecological relevance on the other: imitation of facial and manual movements, reaching, and locomotion. Imitation requires a strong adaptation of the subject's actions to the perceived model, whereas such a constrained integration between perception and action is not required in reaching or locomotion. However, two components of the reaching act, the visual pursuit of the object (when necessary) and the shaping of the hand to grasp the object, also require a shaping of the body related to the properties of the perceived target. Integration between perception and action is therefore still relatively constrained in the reaching act. In contrast, this is not the case in the locomotor act. We will now review the most relevant data concerning early imitation, reaching and walking. Then a joint discussion of these data will permit an analysis of how perception and action are linked together at birth and how these linkages evolve during the first years of life.

Imitation of Facial and Manual Gestures

Mainly with reference to Piaget (1946), but in Europe also to Wallon (1942), it has long been assumed that very young infants are poor at imitating gestures, either manual or facial ones. Yet some authors (Preyer, 1887; Valentine, 1930) had previously noted some occurrences of early imitation in their diaries. Interestingly, the first experimental study of early imitation was carried out as early as 1928 by Guernsey, a student of Charlotte Bühler. Guernsey reported that 2-month-old infants were able to reproduce three different movements: the opening-closing of the mouth, the tongue protrusion, and a lateral rotation movement of the head. Furthermore, she observed that this imitation ability disappeared between 2 and 6 months of age. These imitation responses were conceived as reflex responses which must disappear for an active and conscious imitative ability to develop.

Much better known, the study by Maratos (1973, 1982) marked the beginning of a new field of research in infancy, which has seen an explosion of interest in the 1980s. Basically, the results obtained by Maratos confirmed those described above: at 1 month, infants imitate a tongue protrusion, an opening-closing of the mouth, and a lateral head movement. Moreover, these imitations disappear in the first months of life, at different ages depending on the model movement: between 2 and 3 months for the tongue protrusion, at around 3 months for the mouth movement, and between 3 and 4 months for the head movement. Maratos agrees with the reflex theory of early imitation but a reflex is conceived of in a Piagetian sense: a preestablished structure of action that can be modified by accommodation.

Meltzoff and Moore (1977) extended the repertoire of model movements that 2- and 3-week-old infants are able to reproduce (a lip protrusion movement, and a sequential finger movement) and demonstrated that newborns can also imitate facial movements (Meltzoff & Moore, 1983). According to the authors, the integration between perception and action, demonstrated by the imitation ability, cannot be learned, but are innate, existing from birth.

However, recently an important controversy has developed concerning the existence of this ability in very young infants. Three "parties" have clearly emerged: those in favor, those against, and those who are "undecided." The methodology, with its rich arsenal of extraneous variables, of bias, and of sources of internal validity, has become the touchstone of the debate on early imitation.

Among the studies in favor of its existence, Field, Woodson, Greenberg, and Cohen (1982) showed that newborns reproduce three different facial expressions. Dunkeld (1978) and Fontaine (1982, 1984) described cases of facial movement imitation, and Vinter (1985a,b) additionally described cases of manual movement imitation. Fontaine (1982, 1984) also mentioned rare but clear occurrences of eye blink and of check movement imitation at around 2 months, which seems impressive and must be confirmed by further studies. A recent study (Kugiumutzakis, 1985a,b), althought unfortunately methodologically very poor, also reports evidence for an innate imitation ability. In this experiment, neonates show themselves to be able to imitate an opening-closing movement of the eye and also the sound "a." Most of the studies of this group have examined development of early imitation and report a more or less progressive disappearances of these responses in the first months of life.

Studies which argue against the existence of early imitation are not numerous. Hayes and Watson (1981) concluded that Meltzoff and Moore's results are based on an experimental bias. MacKenzie and Over (1983) and Koepke, Ham, Legerstee, and Russell (1983) failed to replicate these results. Lewis and Wolan-Sullivan (1985) did not observe selective imitation of different facial and manual movements at 2 weeks, and at 3 and 6 months.

In the party of the undecided, Jacobson (1979) observed some imitations of the tongue protrusion movement at 6 weeks, and of the hand opening-closing at 14 weeks. But she found tongue protrusion to be elicited no more frequently by a person's protruding tongue than by a pen moving toward and away from the infant's mouth. Similarly, a dangling ring lifted up and down above the infant's hand was as effective as the adult model in eliciting hand opening-closing[2]. Early imitations are thus not selective - it would no longer be justified to speak in terms of imitation - and, according to Jacobson, they are possibly based on innate releaser mechanisms. Finally, Abravanel and Sigafoos (1984) described a very restricted imitation ability of the tongue protrusion movement: infants partially reproduce this movement when the time durations of model presentation are sufficiently long.

Since most studies are entirely devoted to a proof or disproof of the existence of early imitations, we rarely find those which include an analysis of how perception and action are linked together in the imitative act. With respect to the delineation of factors that may facilitate the elicitation of imitative responses, Fontaine (1984) showed that two factors related to the subject's state intervene: the level of visual attention and the frequency of general movements (of the arms, legs, etc.). The longer the infant looked at the experimenter and the less frequent his/her general movements, the higher the frequency of imitative responses. In this line of study, we may point out that many studies report important individual differences with respect to early imitation. A not negligable percentage of infants are consistently nonimitators. The neonate reported in a longitudinal study by Vinter (1983) was described as having important difficulties in establishing rich social relationships with adults (as attested by eye contact, visual pursuit of the face, interest in the voice). Thus a general ability to communicate with adults may also belong to the list of faciliting factors for early imitation.

From the perceptual point of view, Jacobson's study may suggest that, to some extent, the shape of the model and the movement are meaningful dimensions in eliciting imitations. Vinter (1986a) has more specifically investigated the role of movement. It turns out that neonates are unable to reproduce static models, either facial or manual, whereas they can imitate the corresponding dynamic models. She related this result to the distinction between the two visual systems, the peripheral and the foveal or central systems (Bronson, 1982; Paillard, 1980), by suggesting

[2] Let us just mention that in order to show that neonatal imitations are selective, most of the studies have used what Meltzoff and Moore (1983) call the "cross-model comparison." This method consists in the comparison of, for example, the tongue protrusion frequency obtained in response to the tongue protrusion model with that obtained when a different model is presented.

that imitation of static model movements requires an analysis of their intrinsic properties, such as shape, which is beyond the capacity of the neonate's visual system.

From the motor point of view, the neonate's imitative responses appear to be an integral part of a more complex behavioral sequence. The neonate does not imitate the movement exactly as it is performed by the experimenter, i.e., he/she is not able to reproduce a movement in isolation. Rather, Vinter (1985a) showed that the tongue protrusion movement, for instance, is reproduced in association with lateral head movements, whereas that hand opening-closing movement is performed together with arm movements, mostly extensions and flexions. To account for the neonatal imitative responses, this author suggested that they are rather primitive synergic movements. A synergy is conceived of as an incoercible sequence of movements that links proximal body segments. With development, these patterns undergo a dissociation: at 2 months, tongue protrusions are very likely to be performed independently of head movements, hand movements are dissociated from arm movements.

Unfortunately, little is known about the guidance of imitative responses. Maratos (1973) reported that infants' responses (at between 1 and 3 months) only gradually approximate and then equate to the model's movement, as if they were progressively correcting their movements on the basis of visual information. Meltzoff and Moore (1977) also observed a similar progressive approximation between the 2-3-week-old infant's and the experimenter's movements. By contrast, Vinter (1985b) described the neonatal imitations as more stereotyped responses in which an exteroceptive guidance of the movement was unlikely. Rather, the movement seems internally guided, as it is in the basic synergies present at birth, like the traction response or the flexion synergy described by Twitchell (1965). The discrepancy between these studies may be related to the difference in age of the observed infants, or more probably, to the difference in behavior coding.

How imitation ability later develops has been well established by Piaget (1946). Imitation of a facial movement - hence, invisible for the imitating agent - such as the tongue protrusion movement re-appears at around 12 months. The description Piaget gives of a facial movement imitation makes it clear how the infant, by this age, uses visual, tactile, and proprioceptive feedback information in order to correct his/her movements and to progressively match them with the experimenter's movements. This step, called by Piaget "homologation between the experimenter's face parts and one's own face parts," demonstrates that movements now come under differentiated perceptual control. The visual guidance of imitation of the hand opening-closing movement is also clear between 6 and 8 months. Such perceptual guidance of action then becomes unnecessary when the imitative skills are fully achieved, although it remains constantly available.

Reaching and Grasping Ability

Like imitation, reaching has been considered as nonexistent at the beginning of life, because of a lack of coordination between vision and the motor systems. Detailed descriptions of how a visually guided response develops between birth and 5 months can be found in Piaget (1936) and White et al. (1964). In 1970, however, Bower, Broughton, and Moore (1970b) revealed a remarkable ability of the neonate to correctly stretch out his/her arm in the direction of an object, opening and shaping the hand while so doing, and in some instances (30%) closing it around the object. This result has initiated a controversy which, at first, led to essentially negative replications (Dodwell, Muir, & DiFranco, 1976; Ruff & Halton, 1978). However, the same kind of early eye-hand coordination seems in fact to exist at birth (von Hofsten, 1982; Rader & Stern, 1980), although these recent studies reported that the infant never grasped the object and very rarely touched it. Von Hofsten showed that the neonate's arm movements, performed while he/she was fixating an object, were closer to that object than movements performed while the infant was looking elsewhere or had his/her eyes closed.

Which are the significant eliciting dimensions for the neonate's reaching or "pre-reaching" (Trevarthen, 1982) i.e., what kind of information is used for the planning of the movement? Movement again appears to be a determinant: pre-reaching is more easily elicited by a moving object than a static object. According to Bower et al. (1970a,b), the neonate's reaching takes into consideration different properties of the object. Reaching movements of 1-3-week-old infants are differentiated according to the size of the object and to its distance (distinction between within-reach and out-of-reach distances). Moreover, infants aged less than 1 month perform more reaching movements toward a solid tridimensional object than toward bidimensional representations of this object. They are also disturbed when reaching for a virtual object, as if they were sensitive to the solidity of the real object.

However, the list of failures to replicate or of objections to these findings is long. Size (Lockman & Ashmead, 1983), distance (Field, 1976), tridimensionality (Field, 1977; Rader & Stern, 1980) and solidity (Field, 1976; Yonas, 1979) do not seem to be significant properties for neonatal reaching, neither with respect to the planning of the movement's parameters nor with respect to its control. Calibration of the reaching movement with regard to some of these different object properties seems to appear later on, after 3 or 4 months of age.

What about the control of these pre-reaching movements? Von Hofsten (1982) demonstrated that the arm movements of the neonate may come under visual control in a particular sense. This does not mean that the neonate visually controls his/her own arm movements while approaching the target. Rather these movements are more precisely directed toward the object when it is fixated by the infant at the same time. This could imply that the spatial positioning of the arm is defined with reference to the spatial displacement of the eyes when they track the target, as if the visual map overlapped the motor map with regard to spatial orientation. Control of pre-reaching movements in the sense of correction is very rudimentary at birth, if not nonexistent. The accuracy of a movements is not a function of the previous movement's outcome. Each occurrence of movement seems to be planned in complete autonomy.

Bushnell (1985) reports that neonatal reaching differs from reaching behavior exhibited by infants aged around 4 months with respect to three dimensions. Neonatal reaching is less accurate, is "ballistic" (planned in advance), and is based on some kind of prewired visuo-proprioceptive coordination, while it is later accomplished by monitoring and reducing the "gap" between the seen target and the seen hand.

How does pre-reaching evolve? The best description we have can be found in von Hofsten (1984). The amount of reaching decreases between birth and 2 months and, furthermore, the hand starts to become fisted instead of open during the forward extension of the arm. Arm and hand movements are thus dissociated at 2 months, a finding we have also mentioned with regard to the imitation of the opening-closing movement of the hand. According to von Hofsten, this developmental pattern might be due to important changes in the organization of the nervous system. He proposed to relate them to the distinction introduced by Kuypers (1962) between a proximal (medial) and a distal (lateral) motor system. The asynchrony between the maturation of these systems may explain why arm and hand movements are momentarily uncoordinated: the proximal system matures earlier than the distal system. Clearly, the value of this explanation is limited, since arm and hand movements are nevertheless coordinated at birth, despite the supposed asynchrony of maturation of the two motor systems.

Each component of the reaching movement seems to be reorganized and to evolve independently of the others after the 1st month of life. Twitchell (1965, 1970) has analyzed the transformations of the grasping reflex, in particular how the opening and the closing of the hand progressively come under the control of tactile information, coming from different parts of the hand, independently of the synergy flexion.

As far as the approach phase is concerned, White et al. (1964) described the successive transformations of the reaching movement (swiping reaction, raising reaction), showing how visual feedback is used to calibrate the direction and amplitude of the forearm movement to the target. At around 4 months, a Piaget-type reach is described, in which the open hand is raised to the proximity of the object and then brought closer to it as the infant shifts his/her glance repeatedly from hand to object until the object is roughly grasped. Visual and tactual information are almost completely integrated. They are fully integrated around 2 weeks later ("top-level reach"); the opening of the hand is anticipated, and the infant reaches for and grasps the object in a direct and rapid movement without visual control.

Von Hofsten's (1979) observations are partly congruent with these findings. He reported that the number of movement elements in reaching decreases with age, which confirms the fact that the triggered component (i.e., the movement component of which parameters are planned in advance) again becomes predominant at around 5 months. Thus, a guided component in reaching appears between 2 and 4 months and then, from around 16-20 weeks, is substituted by a predominantly triggered reaching. But he did not observe alternating glances between hand and object: infants always fixated the object, never the hand.

Although the reaching of a 5-month-old baby has been qualified as top level, this ability undergoes new transformations later on. The most extensive studies we have at our disposal for an understanding of this further development come from Halverson (1931) and Castner (1932). Halverson reported the evolution with age, from 16 to 52 weeks, of different styles of reaching or approaches to a cube, in which the forearm cuts into the midline, aims at the cube or at the side of the cube. The first style (adductions to the midline) characterizes infants from 16 to 20-22 weeks and again illustrates that a planned-in-advance style of reaching is present by these ages. The second shows that reaching between 22 and 30-32 weeks can again be described as a visually and tactually guided response in which movements are discontinuous. Finally, from 30-32 to 52 weeks, infants achieve a new style of reaching, more similar to that of infants aged 16-20 weeks because of the smoothness of the trajectory, but at the same time different due to the integration of a correction component at least in the last part of the trajectory.

The McDonnell and Abraham (1979) study is also interesting with respect to this late development. In this experiment, infants had to reach for objects while wearing laterally displacing prism glasses. The authors found that the magnitude of perceptual adaptation decreased between 5 and 7 months and increased again by 9 months. This result could suggest that, at around 7 months, reaching is mainly guided, whereas the triggered component is predominant at around 9 months. The decline of visually guided reaching emphasized by Bushnell (1985), with support of her own studies (Bushnell, 1982), takes place toward the three.quarter mark of the 1 year, but a similar evolution has already been described between 3 and 6 months.

An object's properties, such as its distance and size, become progressively anticipated in the movement planning. Yonas, Cleaves, and Petterson (1978) and Yonas, Granrud, and Smith (1982) have extensively studied how reaching is adapted to the object's distance in relation to the kinds of information used by the infant to perceive depth. In short, depth perception is based on motion information at around 3 months, on binocular information at around 5 months, and on pictorial information or static monocular information at around 7-9 months. These results show that depth perception as required for reaching is first organized within the "space channel" (3 months) and then the "object channel" (7-9 months). As defined by Jeannerod and Biguer (1982) and Paillard (1971, 1980), the space channel is mainly concerned with the processing of inter-object relationships with respect to egocentric references (location, distance, displacement, direction, etc.), and is closely related to the processing of motion information. By contrast, the object channel is mainly concerned with intra-object relationships in relation to exocentric references (shape, texture, solidity, etc.), and relies essentially on static vision. It "requires the extraction of the invariant properties of object qualities at a perceptual level" (Paillard & Amblard, 1985, p. 321).

Calibration of the movement with regard to the object's orientation is particularly interesting. Five-month-olds do not orient their hand during the approach to a vertically or horizontally presented rod as 9-month-olds do, but do it just after having touched the object (Lockman, Ashmead, & Bushnell, 1984). By contrast, von Hofsten and Fazel-Sandy (1984) reported anticipation of the hand orientation in relation to the object's orientation, although in a less differentiated manner than in 30-34-week-olds. Morrongiello and Rocca (1986) obtained an interesting developmental progression with regard to the adjustment of hand orientation to the target's orientation. They compared trials in which the object's orientation was changed midway during a reach with no.change orientation trials. No adjustments of hand orientation were made by 5-month-olds. This parameter was pre-programmed on the basis of visual information initially available. Correction can be introduced only in the subsequent reaching trial. Seven-month-olds modified their hand orientation, but only when the target was in the new orientation. Finally, 9-month-olds modified the orientation of their hand during the change of the object's orientation.

If we try to summarize this evolution of reaching on the basis of the planned in advance/corrected or guided distinction (which corresponds to the more classical distinction between elicited and guided), we can show a complex development from birth to 1-2 years. Reaching at birth is completely pre-programmed or planned in advance, probably without any specific adaptation to the object's properties. No correction of any kind seems to be possible, and further studies are still needed for an understanding of the triggers of neonatal reaching behavior.

Reaching then becomes essentially guided, but this perceptual guidance refers independently to dissociated phases of the reaching act.

Planning of the movement is only local, i.e., it is related to small successive segments of the movement. Reaching is usually unsuccessful during this period. These different phases are simultaneously and progressively calibrated with regard to the properties of certain objects, which involves a progressive extension of the planned component of the movement (the so-called ballistic component).

At around 16-20 weeks, reaching can again be completely planned in advance, but with two essential differences in comparison with the neonatal level: as a result of the preceding evolution, an anticipated planning of the movement's parameters in relation to some of the object's properties is possible. Moreover, corrections can be introduced from one trial to the subsequent one, according to the outcome of the former, i.e., to the evaluation of the difference between the expected goal and the achieved goal. Rather than being replaced by the global planning, we may hypothesize that local guidance (or guidance in segments) is supplemented by the former. In some way, local guidance now takes the role of correcting the movement, but it cannot be carried out simultaneously with global planning.

Then, although the different phases of the movement are roughly coordinated between 20 and 32 weeks, prehension is again essentially perceptually guided, this perceptual control relying on specific "segments" of this movement, such as grasping. Put it in a more straightforward way, whether the approach or the grasp phase is concerned, a switch from a guided to an elicited mode of control, as well as the reverse switch, occurs at different moments in development. From 20 weeks onwards, intrinsic properties of the object, such as its shape or orientation, are progressively taken into consideration in the planning of the movement. There is still no smooth integration between the planned component and the guided component of the movement which, as a result, shows some discontinuities.

Finally, a combined and coordinated style of prehension, both planned in advance and guided, seems to characterize the final phase of this development. Prehension is then likely to be predominantly triggered or guided according to the degree of difficulty of the task. This evolution could be described in terms of programming (distinction between a local and a global programming) as Mounoud (1983) did, or in terms of representation of the object's properties (Mounoud & Vinter, 1981).

Locomotion

It has long been known that from birth and even before birth (Oppenheim, 1981) infants show a particular "locomotor" pattern called the "stepping reflex" which disappears between 2 and 5 months. However, it is only recently that detailed descriptions of this pattern have been available (Thelen, 1979, 1981; Forssberg, 1985). The stepping reflex period is then followed by what Forssberg calls an inactive period during which the feet either remain immobile or never touch the ground, raised by a powerful flexion of the legs. Then children start walking again with support and finally without support.

McGraw (1940) discussed the significance of the early stepping movements with regard to the development of adult gait. She emphasized the differences between them and claimed that different control levels must be involved. This hypothesis has been taken up by neurologists who have proposed that the stepping reflex must be cortically inhibited and must thus disappear for a voluntary skill to develop. The study by Zelazo, Zelazo, and Kolb (1972) was aimed at an understanding of the disappearance of the stepping reflex. They suggested that the stepping reflex and bipedal walking are directly linked, and that the first disappears owing to a lack of social reinforcement. They thus gave infants daily stepping exercises and found both a later disappearance of the reflex and an earlier onset of voluntary walking. Consequently, the neural network of stepping reflex does not disappear through cortical inhibition, but rather, the overt behavior is no longer performed because it is no longer reinforced.

The studies by Thelen and Fisher (1982, 1983) of stepping, kicking, and walking have opened up a new and original perspective on this locomotor ability. Basically, they showed that a spontaneous stereotyped pattern such as kicking shares the same muscle synergisms and the same neural network as a reflexive behavior such as stepping. Similarities between these patterns can be found at different levels: topographic (same left/right alternation), kinematic (similar temporal organization both for the absolute durations of the phase intervals and for the variability of these intervals), and electromyographic (same phasic flexor burst, followed by a passive extension phase). Moreover, some of these properties are equally present in the adult locomotor pattern. However, stepping movements disappear, whereas kicking remains and possibly increases in frequency during the first months of life.

Thelen and Fisher (1982) suggest a biomechanical explanation to account for this result, relying on the differential effects of gravity on muscle function in the two postures. Since muscle mass and muscle strength develop asynchronously in favor of muscle mass in the first weeks, the strength of the muscle contraction rapidly becomes insufficient to support the full body weight and to work against gravity when the infant is upright. In contrast, when supine, the body weight is supported and the movement is aided by gravity.

Thus, for these authors, the stepping reflex would disappear neither because of cortical inhibition, nor because of lack of social reinforcement, but as a result of internal biomechanical properties of the muscular system. Moreover, it is proposed that a reflex (stepping) and a spontaneous rhythmical stereotyped activity (kicking) are not qualitatively different. They share the same muscle synergies, but are distinct with respect to the effects of gravity and posture in terms of support or non-support. Stepping and kicking are also indistinguishable with regard to their eliciting factor. According to Thelen and Fisher, the arousal level (i.e., the behavioral state), and not the upright posture, is the appropriate releasing factor for stepping.

Thelen (1984) suggested an understanding of the antecedents of locomotion within the theoretical framework of the mass-spring model (Kelso, Holt, Kugler, & Turvey, 1980; Kugler, Kelso, & Turvey, 1982). In the mass-spring model, the trajectory is a function of the stiffness of the spring and of the load. In the infant's muscular system, the load corresponds to the mass of the legs, the stiffness is a function of the arousal level, i.e., of muscle tone. The five behavioral states defined by Prechtl (1974) may be differentiated according to their degree of muscle tone (very low in the first state, high in the fifth). During development, there are natural changes in both the load and the stiffness of the muscular system, which may be responsible for the qualitative changes observed in infants' movements. This hypothesis has been experimentally tested by first adding weights to a 4-week-old infant's legs in biologically appropriate amounts. As expected, the frequency of steppings as well as the amplitude of the steps decrease, as occurs naturally after 4 weeks. In contrast, reoccurrences of stepping can be obtained by immersing an infant's legs in water, i.e., by reducing the load of the muscular system.

Similarly, as predicted by the model with regard to the stiffness parameter, the more aroused the infant, i.e., the more tense his/her muscles, the more, the further, and the faster he/she kicks. But let us point out the danger of circularity in this reasoning. Awake behavioral states have been precisely differentiated on the basis of the presence or absence of gross movements (state 3 without gross movements, state 4 with gross movements). Thus, by definition, infants will be more likely to show stepping or kicking movements (i.e., gross movements) in a more aroused state. Evidence based on direct measurements of muscle tone would be much more convincing.

In summary, Thelen's work is based on the assumption of homology between infant stepping and kicking, and adult locomotion. The passage of one kind of locomotor pattern to another is not due to a central change in motor programming.

Biochemical constraints may account for the relationship between stepping and kicking, whereas a progressive transition from state-directed to task-directed behavior can account for the appearance of a voluntary locomotor pattern. This means that, instead of being under the control of the behavioral state, the parameters of the movement must be progressively tuned to specific task properties. This transformation involves both perceptual changes (extraction of affordances) and motor changes (assembly of new flexible muscular coalitions).

Forssberg's study (1985) has challenged Thelen's affirmation about the similarities between infant stepping and mature walking. Forssberg contrasted patterns of leg movement, of electromyographic activity, and of reaction forces in neonates (stepping), 6-12-month-olds (supported locomotion), and in 10-18-month-olds (independent locomotion). It turns out that these patterns are far from being similar. Essentially, a neonate's locomotor pattern is not plantigrade. In humans, the plantigrade gait is characterized by, for instance, a heel strike in front of the body. Neonates, on the other hand, place the forepart of their foot right under the body. The ankle extensors are activated prior to touchdown in stepping and not during the end of the stance phase as in the adult gait. According to Forssberg, neonate stepping is much closer to the digitigrade pattern of quadrupeds or bipedal monkeys than to the plantigrade pattern. The transformation of this digitigrade pattern to a plantigrade one is gradual and continuous to mature until the end of the 2nd year. One of the first major differences that can be observed during development is relative to the control of the locomotor pattern. Movement patterns in neonates seem to be "internally driven" in the sense that afferent inputs from the moving limbs themselves are needed to induce the locomotor activity. In contrast, later on the guidance of movements is under the child's active control.

On the basis of these findings, Forssberg favors the hypothesis of a hierarchical system of human locomotor control. Neonate stepping shows the existence of an innate control pattern generator which defines the basic rhythm. Other neural mechanisms develop later and progressively transform the original pattern in the typical human plantigrade gait. The theoretical differences between Thelen and Forssberg illustrate those we briefly developed in the introduction, i.e., between models which explain changes in action at a peripherical level (the ecological approach for instance) and those which suggest central changes (the cognitivist school for instance).

Conclusion: Emergence of Different Modes of Action Control

Early forms of imitation, reaching, and locomotion abilities present from birth clearly demonstrate that perception and action are coordinated from the beginning of life. The study of these abilities also shows that these perception-action linkages undergo a transformation in the very first months of life. Although there is disagreement between authors about the nature of this transformation, they all report that overt behavior disappears, i.e., is no longer elicitable under the same conditions as it used to be in the neonate. An appropriate understanding of neonatal perception-action linkages appears important in order to attain a deeper understanding of the development of sensorimotor coordinations.

With regard to the properties of objects that are relevant for the planning of an imitative or reaching response, just a few dimensions are revealed as significant. Motion is, without any doubt, one of them. Shape, size, and distance might also be significant but in a very nonspecific way. Stimulus properties that ensure optimal detection of the object by the neonate must be taken into consideration, not as triggers of a particular behavior, but as prerequisites for the elicitation of any behavior. For instance, the object to be reached for or the model to be imitated must be at a distance of around 20 cm from the neonate's eyes. Also, it is better if the object is small and highly contrasted rather than large and little contrasted. Such constraints on distance or size refer to the visual detection ability in the neonate and are thus not specific at all. Studies by Fantz, Fagan, and Miranda (1975), Atkinson (1983), and Banks and Salapatek (1983) contribute to the identification of such stimulus properties as far as the visual system is concerned.

As far as imitation is concerned, inferences relating to the significant properties of the models are more difficult to draw. We think that clear experimental pieces of evidence exist with regard to only three types of model: the tongue protrusion movement, the mouth opening-closing movement, and the lateral head movement[3]. It is remarkable that imitation of these facial gestures is differentiated from imitation of, say, manual movements in the sense that the neonate does not respond to the tongue protrusion model by an opening-closing movement of the hand. This ability makes clear the necessity to postulate an innate body schema representation. This representation may be more detailed or accurate for the face than for other body parts, but permits an appropriate matching between the perceived movements and one's own movements in terms of body parts. In fact, were perception amodal and direct, without mediation through a body representation, there would be no possibility to understand why neonates do not open and close their hands in response to an opening-closing of the mouth or vice versa. Among the three kinds of sensorimotor behaviors studied in this chapter, only imitation clearly claims the necessity to postulate internal representations from birth. The highly constrained integration between perception and action required by an imitative act is made possible by this internal body representation (but see Prinz, this volume, for a different point of view on imitation which stresses the notion of isomorphism between percept and act).

Imitation, reaching, and stepping show that a corrective or guided component is absent in a neonate's movements. Movements are predominantly stereotyped and very similar to synergies, which suggests that they are under an internal sensory control for which biomechanical and neurophysiological properties of the human organism are critical factors.

---

[3] I have shown that neonates are also able to imitate a manual movement, the opening-closing movement of the hand, by using a fine-grained video analysis (coding of the hand-finger position at 500-ms intervals). Most authors have failed to obtain manual imitation but none of them has used such a fine analysis.

The study of imitation, locomotion, and reaching has, above all, permitted the definition of five different modes of action control which we will briefly recall here by using the contrast between planned-in-advance and guided behavior. It will indeed appear clear that a simple dichotomy between these two modes is not sufficient, but that different kinds of combination of both are also necessary to be postulated.

The *basic mode* characterizes the neonatal period that has been more extensively discussed earlier, and in which behavior is completely planned without any guidance of movement being possible. A state of coordination between different sensory information sources and motor information predominates during these first weeks of life. Then a developmental sequence in four "steps"[4] can be postulated.

During the *first step*, behaviors become only locally or partially planned in advance. They are mostly under perceptually guided control, and they continuously need feedback information in order to be performed. But these feedbacks cannot be integrated during this step. Consequently, movements are discontinuous, made up of a series of elementary segments or units, and are not coordinated. It is likely that, at this step, the main kind of processing of sensory information (with respect to the skills in construction) is the "continuous" mode distinguished by Cruse et al. (this volume). During this period, the guided component cannot be conceived of as a correction component. Feedback is needed to perform the movement, whatever the difference between expected and achieved goals may be. Moreover, since the infant's capacity to anticipate is very limited, expected goals cannot be clearly differentiated.

In the *second step*, movements are again largely planned in advance and are now also adapted to some of the properties of the object. A disjointed, guided component can also be present at this step: a movement can be modified by feedback information and/or by knowing the outcome of previous movements. The analysis of locomotion development suggests that, at this step, some part of the neural network may be similar to the one present at birth, but that it is differently activated by hierarchically higher structures.

The *third step* is characterized by a partial integration of the pre-programmed and the guided components. We could say that they are in series, which permits movements to be corrected in the course of their execution, but with a resulting discontinuous character.

Then a full integration of both components takes place during the *fourth step*, in the sense that the guided component, which appears highly flexible and modulable, becomes embedded within the pre-programmed component. It is very likely that a feed-forward control mechanism exists at this level, which can be coupled with a feedback control if necessary. Movements are mainly continuous during this step which includes skillful behaviors. Linkages between perception and action may be

[4] The term "step" is preferred to "stage" because of its weaker meaning. Like stages, steps occur in a fixed sequence but, unlike stages, they are not supposed to be domain independent, i.e., to define general psychological functioning independently of content. By contrast, the steps may be content or task dependent. Furthermore, for us, the steps described in this paper characterize transition forms within a same stage, namely the perceptuomotor stage which takes place during the first 2 years of life.

considered as direct but they are only apparent owing to the high degree of prediction or anticipation of the subject's internal structures of behavior. A similar developmental sequence is described by Hauert, Zanone, and Mounoud (this volume) between 2 and 9-10 years.

Two important qualitative ruptures occur during this developmental sequence. The first takes place between the neonatal period and the first step in action control development. This rupture seems to be a rather radical change in the origin of behavior control. Thelen (1984) accounts for this change in terms of a transition from state-dependent to task-oriented behavior, which also means from an endogenous to an exogenous control. This transition involves, for her, the tuning of movement parameters to task space as well as the discovery of meaningful dimensions of the task space, i.e., of the affordances of situations. Mounoud (1981) suggests that during maturation the neonate acquires a new code, the perceptual code, i.e., a new system of information translation through which behavior will be progressively recoded during the first years of life. By contrast, the neonatal system of information translation is called the "sensory" code. Thus, this rupture is clearly related to a central change in Mounoud's theory, whereas in Thelen's theory it is not easy to understand what may motivate the critical parameters of movements, initially dependent on the infant's state, to change their determinants.

The second rupture intervenes when the second step occurs and strongly contrasts the first with the third or fourth steps. This rupture is related to what we may call the "*segmentation of behavior*". A graphic illustration may be of help in order to capture the opposition we would like to suggest between an "external" and an "internal" segmentation of behavior (Fig. 1).

Before the second step, behavior is segmented into independent modules without them being related to a common unit. For instance, the visual capture of the target, the approach phase, and the grasping define independent modules of reaching behavior at 2-3 months, and they are, still independent, progressively adapted to the properties of an object. This phase of external segmentation leads to the synthesis of new "wholes" or units (second step) through the chaining or concatenation of originally independent segments.

In contrast, after the second step, behavior is "internally" segmented, i.e., within a constituted unit. Relationships between the subunits and the global unit, as well as between the subunits themselves, are established. During this phase of "internal segmentation," a hierarchical structuring of wholes into new elements takes place, as well as a hierarchical structuring of wholes into subsystems[5].

Segments before and after the second step are of a radically different nature. It may be suggested that when segmentation is external, the size of the segments is more or less a result of the limits of the processing capacity of the system (maturation may be of prime importance during this period). On the other hand, when segmentation is internal, the segments may constitute "natural," in the sense

[5] If it were not too difficult or too confusing, we would suggest the distinction here between an "internal" hierarchical structuring of wholes, as represented in our graph - the wholes themselves are fully decomposable into subunits - from am "external" hierarchical structuring of wholes, where the whole itself may constitute a subunit within a flexible higher system.

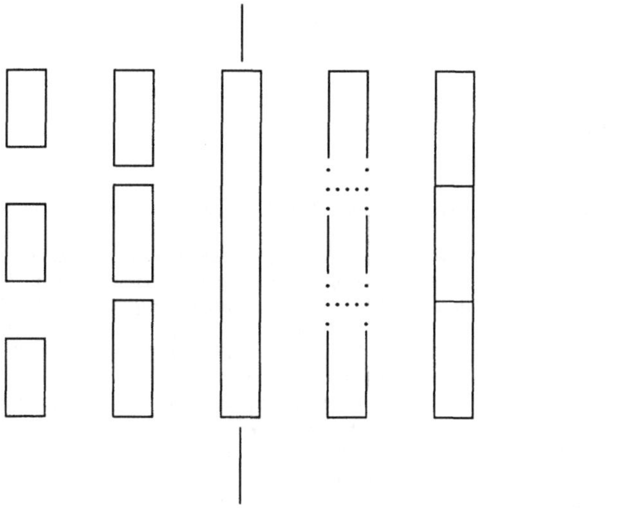

Fig. 1. Development of subunits (or segments) in behavior

of functional, subunits of the behavior (of the movement, considering the whole set of its aspects). We suppose the development of perception of an object, as well as that of speech perception, to be submitted to an identical and parallel process of segmentation (Vinter, 1988).

Again, we consider the origin of changes in segmentation to be internal to the subject. These changes reflect the evolution of changes in the meaning of objects for the subject; and meaning is provided by the internal representations the subject constructs of objects as well as of his/her body or him-/herself. In other words, how the subject perceives an object and acts on it is determined by what he/she knows about this object or how his/her knowledge about this object is represented.

Whether or not the developmental sequence we have described is age-dependent in a strict sense is not clear. As suggested by Mounoud (1981), development can be divided into stages which are defined by the maturational onset of a new code. However, we are not convinced that steps within a stage are also age related. Complexity of objects, situations, or events (what is often called a "task analysis") should be taken into consideration in order to relate this sequence to chronological ages. For instance, the fourth step of action control seems to be achieved at around 9 months when adaptation of hand orientation to the orientation of the object (a property directly perceivable) is considered. But it is only at around 16 months that adaptation of grasping to the weight of the object (a property to be inferred from size and/or texture) can be characterized by the same level.

Acknowledgement. The author wants to express her sincerest gratitude to C.A. Hauert, W. Prinz, P.G. Zanone, and to some of the members of the Perception and Action Research Group for very helpful comments on a preliminary draft of this chapter. Thanks are also expressed to P. Morse for his review of the English version of this chapter.

References

Abravanel, E., & Sigafoos, A.D. (1984). Exploring the presence of imitation during early infancy. *Child Development, 55*, 381-392.

Atkinson, J. (1983). *How does infant vision change in the first three months of life?* Paper presented at the first meeting of the Groningen Study Group on Developmental Neurology, Groningen.

Banks, M.S., & Salapatek, P. (1983). Infant visual perception. In P.H. Mussen (Ed.), *Handbook of child psychology* (Vol. 2). New York: Wiley.

Bower, T.G.R., Broughton, J.-M., & Moore, M.K. (1970a). The coordination of visual and tactual input in infants. *Perception and Psychophysics, 8*, 51-53.

Bower, T.G.R., Broughton, J.M., & Moore, M.K. (1970b). Demonstration of intention in the reaching behavior of neonate humans. *Nature, 5272*, 679-681.

Bronson, G. (1982). Structures, states and characteristics of the nervous system at birth. In P. Stratton (Ed.), *Psychobiology of the human newborn* (pp. 93-118). New York: Wiley.

Bushnell, E.W. (1982). Visual-tactual knowledge in 8-, 9 1/2-, and 11-month-old infants. *Infant Behavior and Development, 5*, 63-75.

Bushnell, E.W. (1985). The decline of visually guided reaching during infancy. *Infant Behavior and Development, 8*, 139-155.

Castner, B.M. (1932). The development of fine prehension in infancy. *Genetic Psychology Monographs, 12*, 105-193.

Dodwell, P.C., Muir, D.W., & DiFranco, D. (1976). Responses of infants to visually presented objects. *Science, 194*, 209-211.

Dunkeld, J. (1978). *The function of imitation in infancy*. Unpublished doctoral dissertation, University of Edinbourgh.

Fantz, R., Fagan, J.F., & Miranda, S.B. (1975). Early visual selectivity. In L.B. Cohen & P. Salapatek (Eds.), *Infant perception: from sensation to cognition* (Vol. 1). New York: Academic Press.

Field, J. (1976). Relation of young infants' reaching behavior to stimulus distance and solidity. *Developmental Psychology, 5*, 444-448.

Field, J. (1977). Coordination of vision and prehension in young infants. *Child Development, 48*, 97-103.

Field, T.M., Woodson, R., Greenberg, R., & Cohen, D. (1982). Discrimination and imitation of facial expressions by neonates. *Science, 218*, 179-181.

Fontaine, R. (1982). *Conditions d'évocation des conduites imitatives chez l'enfant de 0 à 6 mois*. Unpublished doctoral dissertation, University of Paris.

Fontaine, R. (1984). Imitative skill between birth and six months. *Infant Behavior and Development, 7*, 323-333.

Forssberg, H. (1985). Ontogeny of human locomotor control: I. Infant stepping, supported locomotion and transition to independent locomotion. *Experimental Brain Research, 57*, 480-493.

Fowler, C.A., & Turvey, M.-T. (1982). Observational perspective and descriptive level in perceiving and acting. In W.B. Weimer & D.S. Palermo (Eds.), *Cognition and the symbolic processes* (Vol. 2, pp. 1-19). Hillsdale, NJ: Erlbaum.

Guernsey, M. (1982). Eine genetische Studie über Nachahnung. *Zeitschrift für Psychologie, 107*, 105-178.

Halverson, H.M. (1931). An experimental study of prehension in infants by means of systematic cinema records. *Genetic Psychology Monographs, 12*, 107-285.

Hayes, L.A., & Watson, J.S. (1981). Neonatal imitation: fact or artifact? *Developmental Psychology, 17*, 655-660.

Jacobson, S.W. (1979). Matching behavior in the young infant. *Child Development, 50*, 425-430.

Jeannerod, M. (1982). A two-step model for visuo-motor development. In T.G. Bever (Ed.), *Regressions in mental development: basic phenomena and theories*. Hillsdale, NJ: Erlbaum.

Jeannerod, M., & Biguer, B. (1983). Visuo-motor mechanisms in reaching within extrapersonal space. In D.J. Ingle, R.J.W. Mansfield, & M.A. Goodale (Eds.), *Advances in the analysis of visual behavior*. Cambridge, MA: MIT Press.

Kelso, S.J.A., Holt, K.G., Kugler, P.N., & Turvey, M.T. (1980). On the concept of coordinative structures as dissipative structures: II. Empirical lines of convergence. In G.E. Stelmach & J. Requin (Eds.), *Tutorials in motor behavior*. Amsterdam: North-Holland.

Koepke, J.E., Hamm, M., Legerstee, M., & Russell, M. (1983). Neonatal imitation: two failures to replicate. *Infant Behavior and Development, 6*, 97-102.

Kugiumutzakis, J. (1985a). Imitation in newborns 10-45 minutes old. *Uppsala Psychological Reports, 376*, 1-16.

Kugiumutzakis, J. (1985b). Development of imitation during the first six months of life. *Uppsala Psychological Reports, 377*, 1-21.

Kugler, P.N., Kelso, J.A.S., & Turvey, M.-T. (1982). On the control and coordination of naturally developing systems. In J.A.S. Kelso & J.E. Clarke (Eds.), *The development of movement control and coordination*. Chichester: Wiley.

Kuypers, H.G.S. (1962). Corticospinal connection: post-natal development in rhesus monkey. *Science, 138*, 678-680.

Lewis, M., & Wolan-Sullivan, M. (1985). Imitation in the first six months of life. *Merrill Palmer Quarterly, 31*, 315-333.

Lockman, J.J., & Ashmead, D.H. (1983). Discontinuities in the development of manual behavior. In L.P. Lipsitt (Ed.), *Advances in infancy research* (Vol. 2). Norwood, NJ: Ablex.

Lockman, J.J., Ashmead, D.H., & Bushnell, E.W. (1984). The development of anticipatory hand orientation during infancy. *Journal of Experimental Child Psychology, 37*, 176-186.

MacKenzie, B., & Over, B. (1983). Young infants fail to imitate facial and manual gestures. *Infant Behavior and Development, 6*, 85-89.

Maratos, O. (1973). *The origin and the development of imitation during the first six months of life*. Unpublished doctoral dissertation, University of Geneva.

Maratos, O. (1982). Trends in the development of imitation in early infancy. In T.G. Bever (Ed.), *Regressions in mental development: basic phenomena and theories*. Hillsdale, NJ: Erlbaum.

McDonnell, P.M., & Abraham, W.C. (1979). Adaptation to displacing prisms in human infants. *Perception, 8*, 175-185.

McGraw, M.B. (1940). Neuromuscular development of the human infant as exemplified in the achievement of erect locomotion. *Journal of Pediatrics, 17*, 747-777.

Meltzoff, A.N., & Moore, K.M. (1977). Imitation of facial and manual gestures by human neonates. *Science, 198*, 75-78.

Meltzoff, A.N., & Moore, K.M. (1983). Newborn infants imitate adult facial gestures. *Child Development, 31*, 78-84.

Morrongiello, B., & Rocca, P. (1986). *Age-related changes in reaching behavior*. Paper presented at the International Conference on Infant Studies, Los Angeles.

Mounoud, P. (1981). Cognitive development: construction of new structures or construction of internal organizations. In I.E. Sigel, D.M. Brodzinsky, & R.M. Golinkoff (Eds.), *New directions in piagetian theory and practice*. Hillsdale, NJ: Erlbaum.

Mounoud, P. (1983). L'evolution des conduites de préhension comme illustration d'un modèle de développement. In S. de Schonen (Ed.), *Les débuts du développement*. Paris: PUF.

Mounoud, P., & Bower, T.G.R. (1974). Conservation of weight in infants. *Cognition, 3*, 29-40.

Mounoud, P., & Vinter, A. (1981). Representation and sensorimotor development. In G. Butterworth (Ed.), *Infancy and epistemology*. Brighton: Harvester.

Neisser, U. (1985). The role of invariant structures in the control of movement. In M. Frese & J. Sabini (Eds.), *Goal directed behavior: the concept of action in psychology* (pp. 97-108). Hillsdale, NJ: Erlbaum.

Oppenheim, R.W. (1981). Ontogenetic adaptations and retrogressive processes in the development of the nervous system and behavior. In H.F.R. Prechtl & K. Connolly (Eds.), *Maturation and development*. London: SIMP.

Paillard, J. (1971). Les déterminants moteurs de l'organisation de l'espace. *Cahiers de Psychologie, 4*, 261-316.

Paillard, J. (1980). The multichannelling of visual cues and the organization of a visual guided response. In G.E. Stelmach & J. Requin (Eds.), *Tutorials in motor behavior*. Amsterdam: North-Holland.

Paillard, J., & Amblard, B. (1985). Static versus kinetic visual cues for the processing of spatial relationships. In D.J. Ingle, M. Jeannerod, & D.N. Lee (Eds.), *Brain mechanisms and spatial vision* (pp. 299-300). Dordrecht: Nijhoff.

Paillard, J., & Beaubaton, D. (1976). Triggered and guided components of visual reaching. Their dissociation in split-brain studies. In M. Shabani (Ed.), *The motor system: neurophysiology and muscle mechanism* (pp. 333-347). Amsterdam: Elsevier.

Piaget, J. (1936). *La naissance de l'intelligence chez l'enfant*. Neuchâtel: Delachaux & Niestlé.

Piaget, J. (1946). *La formation du symbole chez l'enfant*. Neuchâtel: Delachaux & Niestlé.

Prechtl, H.F.R. (1974). The behavioural states of the newborn infant. *Brain Research, 76*, 1304-1311.

Preyer, W. (1887). *L'ame de l'enfant, observations sur le développement psychique des premières années*. Paris: Alcan.

Rader, N., & Stern, J.D. (1980). Visually elicited reaching in neonates. *Child Development, 53*, 1004-1007.

Ruff, H.A., & Halton, A. (1978). Is there directed reaching in the human neonate? *Developmental Psychology, 4*, 425-426.

Thelen, E. (1979). Rhythmical stereotypies in normal human infants. *Animal Behavior, 27*, 699-715.

Thelen, E. (1981). Rhythmical behavior in infancy: An ethological perspective. *Developmental Psychology, 17*, 237-257.

Thelen, E. (1984, March). *Toward an action-based theory of infant development*. Paper presented at the SRCD Study Group, New York.

Thelen, E., & Fisher, D.M. (1982). Newborn stepping: an explanation for a "disappearing" reflex. *Developmental Psychology, 5*, 760-775.

Thelen, E., & Fisher, D.M. (1983). The organization of spontaneous leg movements in newborn infants. *Journal of Motor Behavior, 4*, 353-377.

Trevarthen, C. (1982). Basic patterns of psychogenetic change in infancy. In T.G. Bever (Ed.), *Regressions in mental development: basic phenomena and theories*. Hillsdale, NJ: Erlbaum.

Twitchell, T.E. (1965). The automatic grasping responses of infants. *Neuropsychologia, 3*, 247-259.

Twitchell, T.E. (1970). Reflex mechanisms and the development of prehension. In K. Connolly (Ed.), *Mechanism of motor skill development*. London: Academic.

Valentine, C.W. (1930). The psychology of imitation with special reference to early childhood. *Journal of Psychology, 2*, 105-132.

Vinter, A. (1983). *Imitation, représentation et mouvement*. Doctoral dissertation, University of Geneva.

Vinter, A. (1985a). *L'imitation chez le nouveau-né*. Neuchâtel: Delachaux & Niestlé.

Vinter, A. (1985b). La capacité d'imitation à la naissance: elle existe, mais que signifie-t-elle? *Revue Canadienne de Psychologie, 39*, 16-33.

Vinter, A. (1986a). The role of movement in eliciting early imitations. *Child Development, 57*, 66-71.

Vinter, A. (1986b). A developmental perspective on behavioral determinants. *Acta Psychologica, 63*, 337-349.

Vinter, A. (1988). Perception of facial movements in early infancy: some reflections in relation to speech perception. In R. Campbell (Ed.), special issue of visible language, 23, 78-111.

Von Hofsten, C. (1979). Development of visually directed reaching: the approach phase. *Journal of Human Movement Studies, 5*, 160-178.

Von Hofsten, C. (1982). Eye-hand coordination in newborns. *Developmental Psychology, 18*, 450-461.

Von Hofsten, C. (1984). Developmental changes in the organization of pre-reaching movements. *Developmental Psychology*, 3, 378-388.

Von Hofsten, C., & Fazel-Sandy, S. (1984). Developmental of visually guided orientation in reaching. *Journal of Experimental Child Psychology, 38*, 208-219.

Wallon, H. (1942). *De l'acte à la pensée*. Paris: Flammarion.

White, B.L., Castle, P., & Held, R. (1964). Observations on the development of visually directed reaching. *Child Development, 35*, 349-364.

Yonas, A. (1979). Studies of spatial perception in infancy. In A.D. Pick (Ed.), *Perception and its development: A tribute to Eleonor J. Gibson*. Hillsdale, NJ: Erlbaum.

Yonas, A., Cleaves, W.T., & Petterson, L. (1978). Development of sensitivity to pictorial depth. *Science, 200*, 77-79.

Yonas, A., Granrud, C.E., & Smith, I.M. (1982). Infants perceive accretion/deletion information for depth. *Investigative Ophthalmology and Visual Science, 22*, 124.

Zanone, P.-G., & Hauert, C.-A. (1987). I am a poor lonesome psychologist, far away from neurosciences. *Cahiers de Psychologie Cognitive, 7*, 109-130.

Zelazo, P.R., Zelazo, N., & Kolb, S. (1972). "Walking" in the newborn. *Science, 276*, 314-315.

Development of Motor Control in the Child: Theoretical and Experimental Approaches

C.-A. HAUERT, P.-G. ZANONE, and P. MOUNOUD

CONTENTS

Introduction

This chapter is concerned with some general aspects of the ontogenetic development of motor planning and control in the child. According to classical theories on human development, psychologists describe the age of 2 years as a transition between two main steps in child development. However, this age can in no case be considered as an "endpoint" in the perceptuo-motor development, nor as a "startingpoint."

This preliminary remark is important if one considers the following apparent paradox. The perceptuo-motor coordinations the child exhibits at 2 years of age, as a result of his/her first development, are very numerous and fairly well adapted to many dimensions of the environment. As a matter of fact, compared with the neonate, the 2-year-old child is able to walk and run efficiently, or is able to grasp objects very accurately in a wide variety of situation with one hand or the coordinated activity of both hands, and so on (for a review, see Mounoud, Vinter & Hauert, 1985). But compared with a 9-year-old child, for example, he/she looks like a very incompetent, awkward, deficient "producer" and "controller" of perceptuo-motor behaviours. This remark holds true for a 9-year-old child compared with adult. In this chapter, we will try to provide some theoretical elements to discuss the way in which perceptuo-motor behaviours develop during childhood.

Relationships Between Perception and Action
Edited by O. Neumann and W. Prinz

As shown in many ontogenetic studies, children manifest dramatic changes in their actions from birth to adulthood. Some of these changes are qualitative, other are quantitative. If one accepts that the physical external world is in some way invariable, the question of the origin of these changes arises. With respect to this question, two theoretically opposed options can be distinguished. In the first one, these changes are assimilated to a maturational, physical, and neurobiological process allowing the subject to control his/her perceptuo-motor systems with increasing accuracy, and to coordinate them more and more adequately. We will try to argue for a second option that views maturation as a necessary but not sufficient condition to yield the changes occurring in perceptuo-motor development. Perceptuo-motor coordinations imply anticipatory and corrective adaptive mechanisms. In our opinion, such mechanisms depend on mediational-representational processes, enabling the subject to elaborate the relevant information involved before as well as during every motor task. Such an assumption has to be discussed at the theoretical level (for a general discussion about the relationship between cognitive and motor skills, see Mounoud 1986). On the other hand, this assumption has consequences at the methodological level: the most pertinent situations to assess the ontogenetic development of perceptuo-motor skills and address the issue of its nature have to present dimensions which can be clearly anticipated.

Theoretical Background

At the moment, literature devoted to human motor control in adults provides a consensual figure of the perceptuo-motor system as a hierarchical organization (Adams, 1976; Bernstein, 1967; Gentile, 1972; Keele, 1982; Newell, 1978; Paillard, 1980; Pew, 1974b; Schmidt, 1982; Shaffer, 1982). However, the definition of the very nature of the different organizational levels remains an open question. In Paillard's concept, for example - one of the most general models now available - the three lower levels of the perceptuo-motor hierarchy are conceived as follows: "servo-motor" control (first level, reflex control), "self-regulation" (second level, prewired programs of movements), "auto-adaptive" loops (third level, automatic adaptive process of prewired programs). In such a framework, if a low-level control mechanism cannot manage at a given movement - and in this case only - this function is run by the immediately superior level of control. Mediational processes do not arise in any of these three levels, which can certainly account for the major part of the perceptuo-motor competences of the subject. The fourth level, namely the "cognitive auto-organization," would only be involved in the conscious determination of the intended action. As a consequence, motor skills are implicitly considered as automatic, since their planning and control do not imply the so-called conscious "cognitive" level.

General Thesis

Our own thesis is based on experimental studies carried out from a developmental perspective. It also assumes different modes of motor control. However, only one of these consists, at some stages of the adaptive processes, in an automatized motor control. If the actual performance is considered, independent of its acquisition, such an automatization might suggest a complete lack of any cognitive mechanisms in the control of skills that would thereby exhibit an illusory automatic aspect. We will try to show that the different modes of action control are based on a general process that deals with the predictable aspects of the situations in which the actions have to be performed. Consequently, this process involves internal representations of the properties of the situations, whatever the general level of ontogenetic development or the specific level of acquisition of a given perceptuo-motor skill. Let us note here that the importance of an anticipatory process for action is accepted by many authors and is a basic postulate of Gibsonian theory (cf. Turvey & Kugler, 1984). From this point of view (Gibson, 1961), however, preparation for action should be limited to a nonmediational process that simply picks up the relevant environmental properties (affordances).

It is important to point out that the authors who regard perceptuo-motor coordinations as isolated functions, or who consider that they are not cognitively mediated (e.g. Adams, 1981; Kelso & Wallace, 1978; Paillard, 1980; or all the supporters of the "natural/dynamic approach", see Kugler, Kelso & Turvey, 1982) use the concept of cognition in a very restrictive sense. In their view, cognition consists in conscious and intentional operations that precede, accompany, or follow movement. In other words, all the mechanisms that are not linked to a clearly conscious experience, and whose contents cannot be "thematizable" or "expressible" (i.e., potentially a topic of discourse for the subject) are, by definition, of a necessarily non-cognitive and purely biological nature. Now, if carried to extremes, such a standpoint leads to absurd statements: any speech production, for example, would be considered as a purely biological activity! Actually, such misunderstandings are likely to originate in that the authors arguing for a biological concept of perceptuo-motor processes always assimilate automatized with automatic behaviours. MacKay (1984, p. 183) sums up the whole situation when he points out that, "1. Not all motor activities are conscious action; 2. not all sensory information-processing mediates conscious experience." Along the same lines, Newell and Barclay (1982, p. 205) state that, "Much of our knowledge about action is apparently tacit. By requesting subjects to be explicit on knowledge about action, an erroneous conceptualization could emerge."

From such a biological point of view, perceptuo-motor coordinations would be regulated by prewired programs, capable of an automatic adaptation to the changing conditions of their execution. Moreover, there would be several behaviours considered as basically automatic (walking, for example), that is, workable under the sole control of biologically determined levels. In our opinion, whereas motor behaviours under automatized control do exist in the spontaneous repertoire of the human subject, this repertoire does not include automatic behaviours. In human adults, Roll (1981) emphasizes the dramatic modifications that psychological factors can introduce in so-called automatic neurophysiological reactions (such as postural reflexes, segmental reflexes, the illusion of self-movement, etc.). He notes that, "It is classical that the occurrence of a 'functional stretch reflex' in a human subject depends on the instruction given to the subject not to interfere with, nor to resist the muscular stretching .... In the same way, in the postural control of an upright position in humans, the gain of rapid responses to stretch of the soleus muscle could depend only on the presence or absence in front of the subject of a support he could get hold of if he lost his balance" (p. 151, our translation). Thus, purely automatic behaviours are likely to occur only in artificial situations. To make ourselves clear, in no way can walking of a spinal cat (for example, Grillner, 1975) be compared to natural walking: the latter is an automatized activity since, in spite of its automatic appearance, it remains coercible and modulable. On the contrary, a spinal cat does not walk. If electrically or chemically stimulated, it may show coordinated patterns of body segments that would never allow it to catch a mouse.

Toward a Broad Concept of Cognition and Its Implication in Movement

Our thesis claims that cognition, as a conscious or unconscious process, in involved in planning, executing, and controlling every perceptuo-motor activity, even nociceptive reflexes, at least in adults (Cohen, Cranney & Hoffman, 1983). We conceive of cognition as ensuring the following functions: mediation of nervous signals by means of one or several internal codes (transformation of neural signals in information through a coding process), storage of the coded contents, generation of new contents by internal activity (anticipation, i.e., activity linking antecedents and consequents even in absence of specific external stimulus), activation or inhibition of such internal contents (choice, decision). The crucial point here is the first statement. As far as the concept of information is concerned, we reject the usual implicit assumption that nervous signals contain per se any relevant information for the perceptuo-motor system. Instead, information must be viewed as internal contents created by the system on the basis of incoming sensorial data with respect to the previous experiences. An argument for such a standpoint can be found in the changes of meanings (perception) of identical nervous events (sensation) occurring with age and individual features.

With regard to the topic of motor behaviour, this definition leads to a particular figure of the perceptuo-motor system as multilevel organization (Zanone & Hauert, 1987). Let us recall briefly the main aspects of our point of view. The highest level of this organization sets the nonmetrical aspects of movement: which body segments are involved; in what spatial direction is their trajectory going to develop, or in what sequence of directions; what is the final goal of the movement? Once these aspects are determined as a procedure, a general motor program (GMP) is selected. The notion of a GMP (Schmidt, 1975) designates a set of motor coordinations underlying a class of movements and is comparable to several classical concepts in the field: the motor engram (Bartlett, 1932; Pew, 1974a,b), the central program (Brooks, 1974), the motor scheme as discussed by Piaget (1936), or the motor control structure (cf. Cruse, Dean & Heuer, this volume). The GMP has to be conceived of as a rather abstract structure whose mutable parameters, from our point of view, are biomechanical (muscles and joints, i.e., the elements of a "coordinative structure" as defined by Kelso, Southard & Goodman, 1979), spatial (movement amplitude and trajectory), temporal (movement duration), kinematic (velocity and acceleration) and dynamic (intensity of active and passive forces). In this respect, let us recall Bernstein's famous example (1967) of the so-called motor equivalence in signature (see also Merton, 1972; Viviani & Terzuolo, 1982): the spatio-temporal characteristics of its components are invariant according to a homothetic principle across widely varying biomechanical, spatial, and temporal conditions of execution.

The actual matching of the movement with the spatio-temporal requirements of the task implies the anticipated instantiation of the mutable parameters of the GMP, allowing the initiation of the intended motor sequence. Then, as a function of the action outcome, an eventual updating of the GMP may occur that entails the generation of corrections during the ongoing movement. More specifically, procedural corrections lead to modifications in the nonmetrical aspects of movement, while instantiation corrections are related to changes in its spatio-temporal characteristics. Now, both instantiation and updating suppose the compilation of several sources of information - as defined above - pertaining to, on the one hand, the characteristics of the experimental situation and, on the other hand, to the biomechanical properties of the involved bodily segments.

GMP Instantiation

To discuss the process of GMP instantiation, the notion of a schema as defined by Schmidt (1975, 1976, 1982) is very powerful. Let us recall that, from Schmidt's point of view, a "recall schema" is supposed to be available to the subject. Such a schema is a kind of motor memory of the functional relationship (or rule, according to Shapiro & Schmidt, 1982) that has been progressively built during past experiences among: (a) the extero- and proprioceptive afferences ("the initial conditions;" IC); (b) the desired goals of action; (c) and the instantiation of the GMP. The recall schema is able, from this rule, to inter- or extrapolate a specific GMP instantiation for the actual action. In addition to the recall schema, the subject possesses another schema that is related to the sensory aspects of his/her actions: the "recognition schema," a sensory memory of the functional relationship between: (a) past IC; (b) the goals of actions; (c) and the past sensory consequences of actions. From this memory trace, the recognition schema generates the expected sensory consequences of the intended movement that provide a clear internal reference for the control of movement. Finally, a comparator is deemed to process the actual sensory consequences with respect to this reference, and to trigger an error signal in case of mismatch. It must be highlighted that a correction is then generated if, and only if, such a mismatch is detected with respect tot he *expected* sensory consequences of the ongoing movement.

From our point of view, these mechanisms have clearly cognitive dimensions. Indeed, the definition of the action procedure is a cognitive process. Then, the second source of information for both schemata, the IC, arises through an internal coding of the actual sensory consequences. Clearly enough, the schemata do not process directly the properties of the environment but internal translations of these. Thus, whatever the motor control mode, a movement never translates any intrinsic property of the recall or recognition schemata, but rather the very nature of the information on which they work.

According to these considerations, whatever movement an individual is asked to execute, preparation for action as well as its control are based on subjective internal representations of the goal and of the initial conditions of the intended movement.

GMP Updating

Let us recall that GMP updating, that is, changing some of its parameters during the execution of movement, may only result from the previous triggering of an error message by the comparator, following some mismatch between actual and expected sensory consequences. This point is particularly important insofar as it leads to a fundamental change in the way to conceive of the mechanisms responsible for motor control. The classical framework of engineering and cybernetics distinguishes several modes of movement control that have been nicely classified in three categories according to Cruse et al. (this volume). Their typology among "advance processing of sensory information," "intermittent processing of sensory information," and "continuous processing of sensory information" allows an understanding of when information is used by the system to control the to-be-executed or the current movement. Along with the argument about the concept of information we discussed above, the question remains of why some supplementary information is necessary during the execution of the movement. Two possibilities have to be envisaged. On the one hand, the actual performance did not follow the intended plan resulting from an advance processing of information because of some unexpected perturbations. A mismatch is then detected by the comparator that may entail a correction based on new token of information. On the other hand, the movement was consistent with its initial plan, but did not fulfill the intended goal. In the former case, a departure occurred with respect to the internal reference for the movement provided by the recognition schema, while, in the latter case, some gap was detected between the actual outcome of movement and the expected consequence on the environment. Whatever its origin, the crucial point is the occurrence of some mismatch between actual and expected sensory consequences. In other words, the issue is no longer at what rate sensory consequences are processed by the system, but at what time they become meaningful, namely when they are no longer consistent with the expected consequences of the action.

At a behavioral level, this distinction is somewhat confusing. Brooks, Cooke, and Thomas (1973) proposed a classification of movements into two categories according to their kinematics: continuous movements, characterized by a sequence of only one acceleration phase and one deceleration phase; and discontinuous movements, in which numerous sequences of this kind can be identified. Discontinuity of movement is attributed to the presence of one or several corrections and is then the behavioral clue of GMP updating. This means that some mismatch had been intermittently detected, but does not indicate at what rate the sensory consequences have been processed: as a matter of fact, continuous, as well as intermittent processing could have resulted in triggering a mismatch message; one may only discard pure advance processing. Conversely, continuous movements can be the consequence of any kind of processing. However, if no mismatch is detected, the resulting movement is ballistic (i.e., as traditionally defined, only one peak of velocity) and is due to pure advance processing, whereas it is continuous but nonballistic (i.e., modulations in both acceleration and deceleration phases) in cases of slight corrections following mismatch detections. Finally, larger corrections can result in discontinuous movements as well.

One must admit that continuous processing of sensory consequences can only be possible if the expected consequences of movement, with which the sensory consequences are to be compared, are defined for the entire course of the intended movement. On the other hand, nonballistic continuous movements are more probable when a continuous comparison between expected and actual sensory consequences prevents the occurrence of too large a mismatch.

Conversely, piecemeal, or incorrect expectations are likely to result in discontinuous movements because of the need for major corrections.

In terms of internal representations, the functional significance of such distinctions can be understood as follows (Hauert, 1980): discontinuous movement indicates a high level of uncertainty with respect to some or all dimensions of the situation, that is, a weak internal model of the action to be executed. The system needs to sample relevant information during the movement. Instead, a continuous movement indicates that the situation is sufficiently predictable, that is, the system has at its disposal a well-defined internal model. Slight corrections may nevertheless occur during a continuous movement, depending directly on such an internal model. Finally, a continuous ballistic movement witnesses a total certainty with respect to all the relevant dimensions of the situation. Obviously enough,such a certainty only translates the subject's point of view and may be, in reality, completely erroneous.

The Developmental Perspective

Following the above general assumptions, it becomes interesting to consider the perception-action relationships from the developmental perspective. Indeed, cognitive developmental psychology has clearly demonstrated that internal representations of reality are constructed by the subject him-/herself during the entire ontogeny (e.g., Piaget & Inhelder, 1941). At some steps of this construction, these internal representations are obviously complete and faithful with respect to the reality they mediate. At other steps, they are incomplete and distorted reflections of reality. But we have assumed that, in all cases, such representations are the inputs of the recall and recognition schemata. Now, the GMP responsible for a specific action is instantiated in a more or less complete and adequate way, to the same extent as these representations are complete and reliable according to developmental level. In parallel, the comparator is provided with a weak or strong internal model of the expected consequences of the movement by the recognition schema. As a consequence, some characteristics of the child's movements are expected to evolve qualitatively and not only quantitatively with age.

At the moment, literature about child development provides some experimental evidence for such developmental changes (Hay, 1979; Mounoud, 1983; Mounoud, Viviani, Hauert, & Guyon, 1985; Vinter, 1985; White, Castle, & Held, 1964) resulting from modifications with age in the mode of perceptuo-motor control. The available results suggest that a given perceptuo-motor behaviour evolves with age through a fixed temporal sequence: (a) movement control is based on an advance processing of sensory information; (b) it is assumed through discontinuous control; and (c) it becomes continuous. Moreover, such a sequence is likely to occur several times during ontogeny according to the different representational capacities that appear at different ages (for a discussion, see Mounoud, 1983).

The previous hypothetical considerations will be illustrated by two series of experimental studies that can be distinguished by the constraints they exert on the action. The first one concerns the development of visuo-manual pointing tasks, that is, situations mainly characterized by spatial constraints (orientation and location of various targets). The second series is interested in the development of visuo-manual tracking behaviour, that is, situations with spatio-temporal constraints (trajectory and kinematics of a moving target).

Methodologically, the common characteristic of these experimental paradigms is that the subject is presumably exposed to partly or totally predictable stimuli. Let us recall that all the data related to the development of perceptuo-motor skills in unpredictable situations exhibit progressive and monotonous increases in performance until adulthood. In the case of unpredictable visuo-manual tracking (Pew & Rupp, 1971), performance improves progressively with age, probably because such a task implies, by definition, the use of a discontinuous control of movement since no, or few, expectations may be available about the target motion. Thus, as far as ontogeny is concerned, the conclusions of this kind of study are very limited.

Visuo-manual Pointing Studies

Most developmental studies on pointing tasks are based on Fitts paradigm (i.e., a reciprocal tapping task under speed and precision constraints) (Connolly, Brown, & Bassett, 1968; Hay, 1981; Kerr, 1975; Salmoni, 1983; Sugden, 1980; Schellekens, Kalverboer, & Scholten, 1984). From a global survey, all these experiments converge to show a decrease in movement time with age, related to an increase in the mean velocity. In these studies, subjects are considered as information processors of limited capacity and, from this point of view, the decrease in movement time is interpreted as a progressive increase in the processing capacity with age.

Other studies on pointing tasks from a developmental point of view have attempted to assess experimentally the main theoretical postulates of schema theory (Schmidt, 1975, 1976), especially the effect of practice on schemata formation (Carson & Wiegand, 1979; Kelso & Norman, 1978; Kerr & Booth, 1977; cf. Shapiro & Schmidt, 1982, for a review). The question was asked whether variability in training favored performance in a new experimental situation - the so-called novelty problem - as assumed by Schmidt's prediction. As a matter of fact, the results of these experiments largely support this assumption.

Nevertheless, it is worth noticing that all the above results roughly sought to compare adult and child performance. Thus, the age scale is investigated using very large steps, if any. From our point of view, it may be suggested that such a gross observation along the age dimension could not lead to a real comprehension of the acquisition of pointing skills. Furthermore, this method is inappropriate to show any "U-shaped" evolutions that are reputed to occur within very narrow age intervals (Bever, 1982; Strauss, 1984).

In a study by Hay (1978), 4-11-year-old children and adults were asked to perform a visuo-manual directional pointing task without seeing their limb. Such a movement is usually defined as an open-loop task, implying visually triggered movement (White et al., 1964) or, according to the definition of Cruse et al. (this volume), relying on pure advance processing of visual information. In one experimental condition, subjects had to actively point their fingertips as accurately as possible in the direction of a light target using a horizontal swing of the arm. In a second condition, the arm was passively moved by the experimenter until the subject felt it just under the target and said "stop."

The results are very striking: in the active condition, children under 7 years showed a little undershoot - almost similar to adult performance. At the ages of 7 and 8 years, movement accuracy suddenly decreased and then progressively attained an almost adult level of performance until the age of 11. In the passive condition, accuracy showed a similar evolution across ages, but was lower than in the active condition, particularly in the older children.

This nonmonotonous trend in the acquisition of an open-loop pointing task is interpreted as the consequence of the appearance, at the age of 7 years of what the author calls "visual guidance" mechanisms in motor control, that is, a control based on an intermittent or continuous processing of visual information. Under this postulate, younger children - aged 4-6 years - produce mainly triggered or ballistic movements. Thus, they do not need any kind of information processing during the movement. On the contrary, the 7-year-old children are disturbed, because they do need the nonexistent visual information to monitor their arm position. The increased accuracy that is observed from this age onwards manifests the progressive use of proprioceptive cues to compensate for the lack of visual afferences. Such a process is assumed to require several years of perceptuo-motor experience. Once it is achieved, movement can be continuously controlled on the basis of a well-defined internal reference.

In a second experiment, Hay (1979) attempted to verify her hypothesis by defining more precisely the spatio-temporal characteristics of pointing movements in 5-11-year-old children. The procedure and the apparatus were the same as above. The results of this study confirmed the previous findings with respect to the evolution of movement accuracy in an open-loop pointing task, showing a less accurate performance at the age of 7. In a more detailed analysis, movements were classified into two gross categories according to their spatio-temporal characteristics. The first type included ballistic movements that showed only a sudden deceleration near the end of the movement. The second class included movements with one or several breaking activities either in the final part of the movement, leaving the initial ballistic phase undisturbed, or during the entire movement, reducing or even abolishing the ballistic phase. This dichotomy corresponds to our distinction between ballistic, on the one hand, and continuously or discontinuously controlled movement, on the other hand (see "GMP Updating").

From the developmental point of view, ballistic movements represent more than 60% of 5-year-olds' movements. This finding supports the postulate of a ballistic type of behaviour at this age, that is, based on advance processing of sensory information. Moreover, this type of movement disappears almost completely from the motor behaviour of older children. On the contrary, the rate of controlled movements increases steadily from the age of 7 years.

This classification, based on kinematic parameters, was confirmed by analyzing children's performance in a pointing task with the visual field rotated by wearing prismatic glasses (Hay, 1979). In this situation, the projected pointing movement had to be corrected to compensate for the apparent displacement of the target.

The moment of the onset of the trajectory correction in the ongoing movement varied as a function of age. At the age of 5 years, the correction occurred late, even after the pointing movement was completely achieved. This suggests that there is almost no visual guidance at this age. On the contrary, the 7-year-olds corrected their movement in half the time it took the 5-year-olds, whereas older children showed an intermediate moment of correction occurrence. These results provide some evidence that pointing movements are essentially ballistic at the age of 5, whereas they are mainly controlled at the age of 7. Nevertheless, highly efficient control does not occur before the age of 11 years.

From the two experiments by Hay, it could be argued that the observed evolution is paradigmatic of a general developmental trend. Thus, a general description of the evolution between the ages of 5 and 9 years can be attempted. First, there is a predominance of ballistic behaviours at the age of 5, that is, a predominance of an advance processing of sensory information. Then, a discontinuous control mode appears at the age of 7. Finally, from the age of 8, a continuous control mode gradually replaces the discontinuous one.

Visuo-manual Tracking Studies

Interestingly, a comparable developmental sequence can be found in a second experimental situation, that is, visuo-manual tracking of a simple predictable stimulus. Adult performance in tracking both predictable and unpredictable targets has been described in detail (cf. Ellson & Gray, 1948; Noble, Fitts & Warren, 1955; Pew, 1974a,b; Poulton, 1974; Stark, 1968). Several models, based on the concept of a servo-system (reduction of error on the basis of a continuous or intermittent processing of sensory information), have been devised to account for the experimental findings. Studies with children, however, are few. Let us recall the work of Pew and Rupp (1971) who investigated the performances of 10-, 13-, and 16-year-old children in tracking unpredictable targets. As mentioned above, the use of unpredictable targets necessarily constrains the subject to a discontinuous control mode and makes it difficult to explore the age-dependent evolution of the anticipations he/she can make about the target motion (Poulton, 1952) in order to instantiate the selected GMP.

Since we are interested in cognitive representations involved in perceptuo-motor coordinations, we have only considered predictable sinusoidal targets for which, unlike pseudorandom targets, an internal model could eventually be elaborated by the subjects. Indeed, Magdaleno, Jex, and Johnson (1970) showed that, while a feedback control mode may allow a successful pursuit of a target under a 0.5-Hz frequency, such a strategy cannot operate in tracking targets of frequency higher than 0.5-Hz. As a matter of fact, prediction and generation of a movement pattern are then required. Thus, the acquisition of tracking behaviour at two frequencies (0.2-Hz and 0.8-Hz) which are, respectively, lower and higher than this critical transition value has been studied (Mounoud, Viviani, et al., 1985).

Subjects were sitting in front of a screen on which it was possible to displace a red target spot horizontally (± 15 cm). The right forearm was fixed in a metal splint that could rotate in the same plane as the target. Forearm movements were recorded by an angular potentiometer mounted on the axis of rotation of the splint. A white light source at the end of the splint projected a circular marker spot on the screen. The task was to track the displacement of the target at 0.2 and 0.8 Hz with the white marker spot using forearm rotations during 35 full cycles of the target.

Adult subjects did not have difficulties in performing the task at either frequency. By contrast, some of the younger children were unable to accomplish the required task, especially at the higher frequency. By convention, a performance was defined as correct if, and only if, each stimulus cycle resulted in a response cycle. However, responses having the wrong amplitude or a phase difference with respect to the target, or showing distortions were tolerated. According to this criterion, the percentage of subjects who performed the task successfully was as shown in Table 1. Even the successful performances show a considerable variability in both amplitude and timing of the responses. In order to quantify this variability, the responses were analyzed cycle by cycle by measuring the gain (ration between the peak-to-peak amplitudes of the pursuit and target oscillations) and the phase lag with respect to the target. Phase lag indicates the temporal delay of fun-

Table 1. Percentage of subjects in each age group (n = 10) who performed the visuo-manual tracking task successfully at 0.2 Hz and 0.8 Hz, respectively

Age (years)		5	6	7	8	9
Frequency	0.2 Hz	70	100	100	100	100
	0.8 Hz	30	60	80	80	100

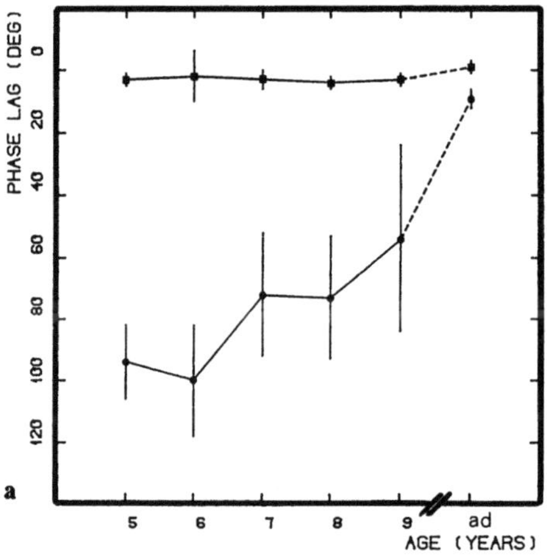

a

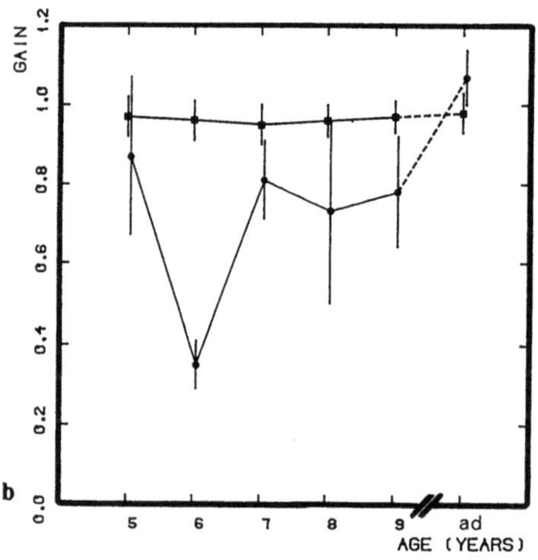

b

Fig. 1 a,b. *Gain and phase of the linear components.* **a** The phase differences between the target sinewave and the harmonic components of the pursuit with the same frequency of the target. Positive values indicate that the pursuit lags with respect to the target. **b** The ratio between the amplitude of the pursuit and the amplitude of the target. Each data point is the average of all subjects in the indicated age group who could execute the task successfully (*ad*, adult) of the within-trial mean values. Intersubject variability is indicated by *vertical bars* (± 1 standard deviation). *Squares*, 0.2 Hz; *circles*, 0.8 Hz. (From Mounoud, Viviani, et al., 1985)

damental harmonic of the response with respect to the target, expressed as a fraction of one complete cycle (360°).

Performances can be characterized by plotting the phase lag and gain of the response as a function of age. Figure 1 represents the dynamic properties of the tracking system at the two selected frequencies. At 0.2 Hz, the dynamics for those subjects who could perform the task seem to be independent of age. At 0.8 Hz, a 90° phase lag exists for the younger groups, which, with increasing age, monotonously approaches adult performance. The evolution of gain looks more complex. In particular, the 6-year-old children systematically (see the very low intersubject variability) produce movements which are much smaller than those of all other age groups. On the other hand, an evolution is no longer observed between 7 and 9 years.

However, successful performances show qualitative distortions with respect to a pure sinewave. The deviation from such a model - assumed to be related to fast corrections in the response - can be appreciated by spectral analysis, that is, by an analysis of its frequency components (Fourier analysis). In order to emphasize the distortions contained in the response, this analysis was carried out with respect to the instantaneous error between target and response velocities.

At least three types of distortions appear from the results of this analysis. The two first types are related to the components near the two target frequencies and indicate the difficulty of reproducing precisely the input frequency on each cycle of movements. The third type is represented by oscillations within the frequency range of 1.2-1.7 Hz. These components can be interpreted as the result of the effort to minimize the tracking error by a processing of visual information (Poulton, 1974). Under certain conditions (but not in our experiment), they also occur in adults (Pew, Duffendack, & Fensch, 1967) and are documented in ramp tracking too (Craik & Vince, 1963; Ellson, Hill & Gray, 1947).

These results can be summarized by grouping the total relative spectral components within three frequency intervals 0.2-0.9 Hz: low, 1.0-1.6 Hz: medium, 1.7-2.3 Hz: high. The age-related evolution of the total spectral power within each interval is shown in Fig. 2.

To summarize, Fig. 2 shows a clear decreasing trend for the low-frequency components, while high-frequency components increase with age.

The results demonstrate the following points: the tracking skill, as described in adults, is acquired progressively over a considerable number of years during childhood and cannot be adequately reduced to a monotonous maturational process. Typical adult performances are not yet attained at the age of 9 years. The evolution with age of the performance involves large changes in both qualitative and quantitative aspects of motor behaviour. Indeed, as we will try to speculate in the following discussion, at least one major change in control mode occurs at around 7 years of age that suggests a possible change in the strategy of perceptuo-motor coordination.

All types of tracking necessarily involve some kind of coordination between perceptual and motor systems. For the purpose of this discussion, we will begin by laying down a schematic description of what should be involved in this coordination from a logical standpoint.

One prerequisite for skillful tracking is the ability to represent centrally the trajectory of the target (Krendel & McRuer, 1960). For unpredictable targets, such a representation can only be piecemeal, while in the case of predictable targets, as

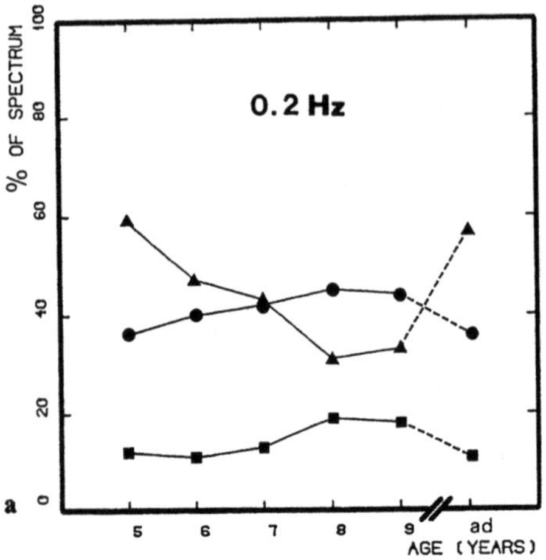

a

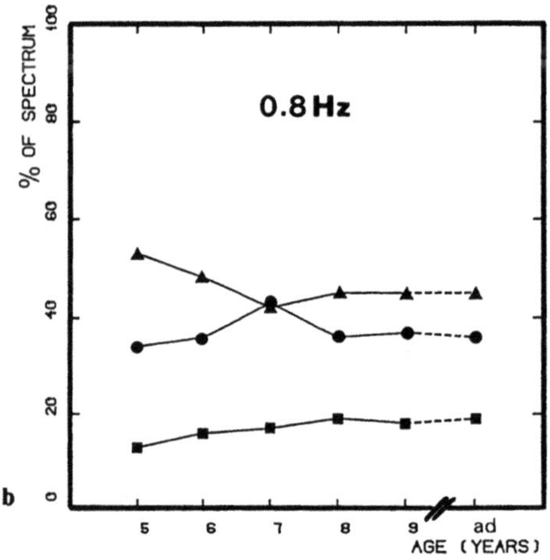

b

Fig. 2 a, b. *Frequency analysis of the nonlinear components*. Distortions in tracking a 0.2-Hz (**a**) and a 0.8-Hz (**b**) target with respect to age. Each *data point* represents the total relative amount of nonlinearities within three frequency intervals (*triangles*, low - 0.2-0.9 Hz; *circles*, medium - 1.0-1.6 Hz; *squares*, high - 1.7-2.3 Hz; *ad*, adult. (From Mounoud, Viviani, et al., 1985)

in our experiments, the entire trajectory can be apprehended as a whole. The second prerequisite is the subject's ability to build an appropriate procedure that reproduces the represented target trajectory. The above two conditions are tacitly assumed to be satisfied in normal adults. However, they are not trivial in children, for the relevant representations are only progressively established during childhood.

In addition, we assume that the motor system is able to select a GMP, to identify a set of parameters that completely specify the target motion, and to match these parameters into an appropriate setting for the mutable variables of the GMP (see Prinz, this volume, for a discussion of the concept of matching). We assume that the recall schema is responsible for these two functions.

Nevertheless, the actual pursuit will inevitably show some errors with respect to target motion. Therefore, we must postulate the possibility of updating the ongoing motor program on the basis of incoming information. Detecting these errors indirectly involves the recognition schema.

It must be stressed that classical models about tracking behaviour (cf. Young & Stark, 1963; Stark, 1968; Poulton, 1974) view the control problem mentioned above as a simple functional transformation from a sensory stimulus to a motor output. On the contrary, our concept implies a mediate matching between the perceptual and motor representations in planning and instantiating the motor command. Thus, the age-dependent evolution observed in tracking has to be considered as the expression of a concomitant change in the internal representation of the target motion. This leads to the conclusion that evolution toward adult behaviour is better described as a sequence of discrete steps, each one of which entails a (re)definition of the relevant representation, than as a smooth refinement process.

The input-output analysis of the pursuit linear component (Fig. 1) only discriminates adults and children at the highest frequency (0.8 Hz). Whereas the phase lag diagram indicates a monotonous reduction of the temporal delay between target and pursuit, which is compatible with the notion of a progressive improvement in the dynamics of the perceptuo-motor coordinations, the gain diagram is obviously discontinuous. This suggests the presence of at least one major modification in the strategy of perceptuo-motor coordination occurring between the ages of 6 and 7 years. With regard to the considerable improvement observed in 7-year-old children for both phase and gain, the sharp reduction in movement amplitude at the age of 6 cannot be interpreted as a regression from the level already attained at 5, but rather as the price to be paid when replacing one strategy with a better one that, however, only becomes effective at a later stage.

As a cue of this change, let us consider the distortions of performance at 0.2 Hz target frequency (cf. Fig. 2). Obviously, the low- and high-frequency nonlinearities show inverse tendencies with age. In particular, the 5-year-old boys stand out in this description, as they completely lack in high-frequency oscillations, while the low-frequency distortions are by far the most prominent with respect to all other ages. It is unlikely that these distortions are due to limitations in the motor execution abilities. On the basis of these considerations, we can tentatively assume that 5-year-old children master the basic motor skill to produce sinusoidal motions but, with few exceptions, cannot elaborate a global representation of the target motion in all its relevant aspects (frequency, amplitude, and phase). They resort to a sequential pursuit strategy in which successive, short segments of trajectory are independently planned and executed as independent ballistic movements. The large low-frequency distortions in their performances would then be a by-product of this piecemeal concatenation of responses separated by a reaction time. The high failure rate in the same age group at 0.8 Hz is probably due to the fact that the duration of the target period is too short for such a strategy to be operational (Magdaleno et al., 1970; Pew, 1974a,b). Instead, unsuccessful performances consist in true sinusoidal movements, but at a wrong frequency and/or with a wrong amplitude. However, from the point of view of the child, he/she feels that the task has been perfectly performed. This is a strong indication of a purely advance processing of sensory information at this age.

This mode of motor control subsides before the age of 7, as children can utilize the acquired ability to produce sinusoidal motions. Within this age span, the success rate at 0.8 Hz increases sharply and the amount of low-frequency distortions decreases. At the same time, there is moderate increase of the high-frequency harmonic components, which perhaps reflect the more numerous efforts to reproduce the temporal aspects of the target through a continuous processing of visual information. This strategy allows the sharp decrease in phase lag observed between 6 and 7 years. The pronounced reduction in gain at 0.8 Hz at the transitional age (6 years) could be the consequence of the processing load associated with the implementation of this frequency control through a temporarily discontinuous processing of visual information.

In conclusion, the results suggest that, although the basic motor ability involved in sinusoidal tracking is acquired very early in ontogeny, the construction of internal representations that would be appropriate to the fully developed skill requires the temporary neglect of this ability in favor of an altogether different strategy. It is then tempting to speculate that the transitory discontinuous control mode observed at the age of 6 is instrumental in the construction of these representations.

Conclusions

From a developmental point of view, the relationships between perception and action evolve in a very complex way. It seems clear that development cannot be seen as a progressive improvement in the planning of movement and the processing of sensory information. The nature of the changes occurring within the age range considered is not only determined by maturational factors that are intrinsic to the perceptuo-motor system. The results presented and discussed here strongly suggest that the construction of internal representations of the various properties of the physical world allows adapted executions of perceptuo-motor skills.

If one considers the developmental sequence described with respect to the skills discussed in this chapter, a striking similarity emerges. Each sequence is a succession of steps in which the child resorts to three different modes of perceptuo-motor control. As a matter of fact, the order of occurrence of these modes is rather clearly defined: the first step around 5-6 years, the child produces predominantly movements based on an advance processing of information that may or not attain the desired outcome. In the second step (about 6-7 years), action is temporarily controlled through a discontinuous mode of information processing. Finally (from 7-8 years), motor control becomes progressively carried out by a continuous mode of information processing. Let us emphasize the fact that this sequence is contingent on the actual prediction the child is able to make about the situation.

Finally, we must stress that this developmental sequence may be delayed for 1-2 years in the age scale, presumable because of some specific dimensions of reality that must be cognitively elaborated for a particular skill. From this point of view, development of the motor system per se seems a somewhat anecdotal factor if one expects to understand the acquisition of perceptuo-motor behaviours in children.

References

Adams, J.A. (1976). Issues for a closed-loop theory of motor learning. In G.E. Stelmach (Ed.), *Motor control: issues and trends*. New York: Academic.

Adams, J.A. (1981). Do cognitive factors in motor performance become nonfunctional with practice? *Journal of Motor Behaviour, 13*(4), 262-273.

Bartlett, F.C. (1932). *Remembering*. London: Cambridge University Press.

Bernstein, N. (1967). *The coordination and regulation of movement*. London: Pergamon.

Bever, T.G. (1982). *Regressions in mental development: Basic phenomena and theories*. Hillsdale, N.J.: Erlbaum.

Brooks, V.B. (1974). Some examples of programmed limb movements. *Brain Research, 71*, 299-308.

Brooks, V.B., Cooke, J.D., & Thomas, T.F. (1973). The continuity of movement. In R.B. Stein, K.G. Pearson, R.S. Smith, & J.B. Redford (Eds.), *Control of posture and locomotion*. New York: Plenum.

Carson, L.M., & Wiegand, R.L, (1979). Motor schema formation and retention in young children: A test of Schmidt's schema theory. *Journal of Motor Behaviour, 11*, 247-251.

Cohen, M.E., Cranney, J., & Hoffman, H. (1983). Motor and cognitive factors in the modification of a reflex. *Perception and Psychophysics, 34*(3), 214-220.

Connolly, K., Brown, K., & Bassett, E. (1968). Developmental changes in some components of a motor skill. *British Journal of Psychology, 59*(3), 305-314.

Craik, K.J.M., & Vince, M.A. (1963). Psychological and physio-logical aspects of control mechanisms with special reference to tank gunnery. *Ergonomics, 6*, 1-33.

Ellson, D.G. & Gray, F. (1948). *Frequency responses of human operators following a sine-wave input* (USAF Air Material Command, Memorandum Report MCREXD 694-2N). Dayton: Wright-Patterson Air-Force Base.

Ellson, D.G., Hill, H., & Gray, F. (1947). *Wave-length and ampli-tude characteristics of tracking error curves* (USAF Air Material Command, Memorandum Report TSEAA 694-2D). Dayton: Wright-Patterson Air-Force Base.

Gentile, A.M. (1972). A working model of skill acquisition with application to teaching. *Quest, 17*, 3-23.

Gibson, J.J. (1961). Ecological optics. *Vision Research, 1*, 253-262.

Grillner, S. (1975). Locomotion in vertebrate. Central mechanisms and reflex interaction. *Psychological Review, 55*, 247-304

Hauert, C.-A. (1980). Propriétés des objets et propriétés des actions chez l'enfant de 2 à 5 ans. *Archives de Psychologie, 48*, (185), 95-168.

Hay, L. (1978). Accuracy of children in an open-loop pointing task. *Perceptual and Motor Skills, 47*, 1079-1082.

Hay, L. (1979). Spatio-temporal analysis of movements in children: Motor programs versus feedback in the development of reaching. *Journal of Motor Behaviour, 11*(3), 189-200.

Hay, L. (1981). The effect of amplitude and accuracy requirements on movement time in children. *Journal of Motor Behaviour, 13*, 177-186.

Keele, S.W. (1982). Behavioral analysis of movement. In V.B. Brooks (Ed.), *Handbook of physiology*. Bethesda: American Physiological Society.

Kelso, J.A.S., & Norman, P.E. (1978). Motor schema formation in children. *Developmental Psychology, 14*(2), 153-156.

Kelso, J.A.S., & Wallace, S.A. (1978). Conscious mechanisms in movement. In G.E. Stelmach (Ed.), *Information processing in motor control and learning*. New York: Academic.

Kelso, J.A.S., Southard, D.L., & Goodman, D. (1979). On the co-ordination of two-handed movements. *Journal of Experimental Psychology: Human Perception and Performance, 5*, 229-238.

Kerr, M. (1975). Movement control and maturation in elementary grade children. *Perceptual and Motor Skills, 41*, 151-154.

Kerr, M., & Booth, B. (1977). Skill acquisition in elementary school children and schema theory. In D.M. Landers & R.W. Christina (Eds.), *Psychology of motor behaviour -II*. Champaign: Human Kinetics.

Krendel, E.S., & McRuer, P.T. (1960). A servomechanisms approach to skill development. *Journal of the Franklin Institute, 269*, 24-42.

Kugler, P.N., Kelso, J.A.S., & Turvey, M.T. (1982). On the control and coordination of naturally developing systems. In J.A.S. Kelso & J.E. Clarke (Eds.), *The development of movement control and coordination*. Chichester: Wiley.

Magdaleno, R.E., Jex, H.R., & Johnson, W.A. (1979). Tracking quasi-predictable displays: Subjective predictability gradations, pilot model for periodic and narrow band input. *Proceedings of the 5th Annual NASA-University Conference on Manual Control (NASA-SP215)*, (pp. 391-428).

MacKay, D.M. (1984). Evaluation: the missing link between cognition and action. In W. Prinz & A.F. Sanders (Eds.), *Cognition and motor processes*. Heidelberg, New York, Berlin: Springer.

Merton, P.A. (1972). How we control the contractions of our muscles. *Scientific American, 226*, 30-37.

Mounoud, P. (1983). L'évolution des conduites de préhension comme l'illustration d'un modèle du développement. In S. de Schonen (Ed.), *Le développement dans la première année*. Paris: PUF

Mounoud, P. (1986). Action and cognition. In M.G. Wade & H.T.A. Whiting (Eds.), *Motor development in children: aspects of coordination and control*. Dordrecht: Nijhoff.

Mounoud, P., Vinter, A., & Hauert, C.-A. (1985). Activités manuelles et développement. *Comportements, 3*, 103-124.

Mounoud, P., Viviani, P., Hauert, C.-A., & Huyon, J. (1985). Development of visuo-manual tracking in the 5 to 9 year-old boys. *Journal of Experimental Child Psychology, 40*, 115-132.

Newell, K. (1978). Some issues on action plans. In G.E. Stelmach (Ed.), *Information processing in motor control and learning*. New York: Academic.

Newell, K., & Barclay, C.R. (1982). Developing knowledge about action. In J.A.S. Kelso & J.E. Clarke (Eds.), *The development of motor control and coordination*. New York: Wiley.

Noble, M., Fitts, P.M., & Warren, C.E. (1955). The frequency response of skilled subjects in a pursuit tracking task. *Journal of Experimental Psychology, 49*, 249-256.

Paillard, J. (1980). Nouveau objectifs pour l'étude neurobiologique de la performance motrice intégrée: les niveaux de contrôle. In C.H. Nadeau, W.R. Halliwell, K.M. Newell, & G.C. Roberts (Eds.), *Psychology of motor behavior and sport - 1979*. Champaign: Human Kinetics.

Pew, R.W. (1974a). Human motor performances. In B.H. Kantowitz, (Ed.), *Human information processing: tutorials in performance and cognition*. Hillsdale, N.J.: Erlbaum.

Pew, R.W. (1974b). Levels of analysis in motor control. *Brain Research 71*, 393-400

Pew, R.W., & Rupp, G.L. (1971). Two quantitative measures of skill development. *Journal of Experimental Psychology, 90*(1), 1-7.

Pew, R.W., Duffendack, J.C., & Fensch, L.K. (1967). Sine-wave tracking revisited. *IEEE Transactions on Human Factors in Electronics, 8*, 130-134.

Piaget, J. (1936). *La naissance de l'intelligence chez l'enfant*. Neuchâtel: Delachaux & Niestlé.

Piaget, J., & Inhelder, B. (1941). *Le développement des quantités physiques chez l'enfant*. Neuchâtel: Delachaux & Niestlé.

Poulton, E.C. (1974). *Tracking skill and manual control*. New York: Academic.

Poulton, E.D. (1952). The basis of perceptual anticipation in tracking. *British Journal of Psychology, 43*, 295-302.

Roll, J.P. (1981). *Contribution de la proprioception musculaire à la perception et au contrôle du mouvement chez l'homme*. Doctoral dissertation, University of Aix-Marseille I.

Salmoni, A.W. (1983). A descriptive analysis of children performing Fitts reciprocal tapping task. *Journal of Human Movement Studies, 9*(2), 81-95.

Schellekens, J.M.H., Kalveboer, A.G., & Scholten, C.A. (1984). The microstructure of tapping movements in children. *Journal of Motor Behavior, 6*,(1), 20-39.

Schmidt, R.A. (1975). A schema theory of discrete motor skill learning. *Psychological Review, 86*, 225-260.

Schmidt, R.A. (1976). The schema as a solution of some persistent problems in motor learning theory. In G.E. Stelmach (Ed.), *Motor control: Issues and trends*. New York: Academic.

Schmidt, R.A. (1982). *Motor control and learning*. Champaign: Human Kinetics.

Shaffer, L.H. (1981). Performances of Chopin, Bach, and Bartok: studies in motor programming. *Cognitive Psychology, 13*, 326-376.

Shapiro, D., & Schmidt, R.A. (1982). The schema theory: recent evidence and developmental implication. In J.A.S. Kelso & J.E. Clark (Eds.), *The development of movement control and co-ordination*. Chichester: Wiley.

Stark, L. (1968). *Neurological control system: studies in bioengineering*. New York: Plenum.

Strauss, S. (1984). *U-shaped behavior and growth*. New York: Academic.

Sugden, D.A. (1980). Movement speed in children. *Journal of Motor Behaviour*, *12*(2), 125-132.

Turvey, M.T., & Kugler, P.N. (1984). An ecological approach to perception and action. In H.T.A. Whiting (Ed.), *Human motor action - Bernstein reassessed*. Amsterdam: North-Holland.

Vinter, A. (1985). *L'imitation chez le nouveau-né*. Neuchâtel: Delachaux & Niestlé.

Viviani, P., & Terzuolo, C. (1982). The organization of movement in handwriting and typing. In B. Butterworth (Ed.), *Language production: vol.II. The production of language in non-speech modalities*. New York: Academic.

White, B.L., Castle, P., & Held, R. (1964). Observations on the development of visually-directed reaching. *Child development*, *35*, 349-364.

Young, L.R., & Stark, L. (1963). Variable feedback experiments testing a sampled-data model for eye tracking movement. *IEEE Transactions on Human Factors in Electronics*, *4*, 38-51.

Zanone, P.-G., & Hauert, C.-A. (1987). For a cognitive conception of motor processes: a provocative standpoint. *Cahiers de Psychologie Cognitive/European Bulletin of Cognitive Psychology*, *7*, 109-129.

Informational Accounts of Perception and Action: Skeptical Reflections

P. BIERI

CONTENTS

Preliminary Remarks

For lack of detailed empirical knowledge, philosophers consider themselves specialists on the most general topics. Thus, my reflections will focus on some key terms and key ideas in this volume - on notions which are not explicitly discussed and justified because they form the backbone of many papers and define the theoretical framework with which they operate. There are two frames of mind you may adopt when working your way through a volume like this one. At first, you let yourself become immersed in the approach, the vocabulary, the leading assumptions, the type of questions, the style of reasoning. You get *used* to all of this, and in this sense you understand what you read. Later on, in a second and third round, you may take one step back from everything and ask yourself what you have really and truly understood. And in this second, sober frame of mind you may find that many things which at first seemed smooth and clear are confusing and deeply puzzling. My comments emerged from this sort of experience. The more I thought

Relationships Between Perception and Action
Edited by O. Neumann and W. Prinz

about the points I raise here, the more sceptical I grew. Finally, I decided that the best I could do was to describe the reasons for my confusion and puzzlement.

Let me prevent an almost classic misunderstanding which tends to occur when philosophers present critical conceptual comments on empirical work: I do not mean to legislate anything, to tell researchers a priori what to do. I fully realize that you have to start somewhere, that you need to operate with unfinished ideas, and that such ideas can be useful in terms of explanation and understanding even though they may be beset with difficulties. I am simply pointing out intriguing questions which everybody, I submit, would like to see answered.

I have no story of my own to tell about perception and action. Naturally, I have no empirical theory to propose. But I do not even have a coherent set of consequences to be drawn from my reflections. My commentary is best viewed as a sort of diary or memo where I probe into some of the things I find lacking or puzzling. I have purposely refrained from making substantial reference to the vast literature on the philosophy of mind. I wished to keep my comments free of specific philosophical doctrines. I did not want to impose any problems from the outside. The issues which keep worrying me emerge, I believe, entirely from within the texts assembled here, and I like to think that they will present themselves to anyone who is insistent enough in his questions. In order to make this point I am, to a certain extent, simulating naive curiosity.

I am not simulating anything, however, when I indicate that I feel very much as though I am struggling and stumbling in the fog. Presumably, I have misunderstood some of the things said in the texts. This is perhaps inevitable when different conceptual habits clash. Moreover, I had to resort to a certain amount of eclecticism in order to trace my worries through the texts. I apologize for any injustice done, and I hope that inadequacies of this sort do not prevent the text from being provocative in a productive way.

Intentional Mediation: Phenomena to Be Remembered

In commonsense psychology, the everyday psychological picture of man, there is a well-entrenched and important distinction between two ways in which perception may guide action, two ways in which action may be prompted by perception. First, an organism may receive a certain sensory (visual, auditory, etc.) input and may react to it right away by a certain motor output. Exposed to a sudden flash of light, we blink and shrink back from it. Standing in the bell tower, we wince and quickly shut our ears when the bells suddenly start to chime. Without much theory in mind we often call such sequences "reflex behavior," and we say about the transition from sensory input to motor output that it is "automatic." We do not mean thereby to suggest a mysterious sort of immediacy in the connection between perception and action. We take it for granted that even in such cases there are complex internal processes - causal or informational processes (see following section) - going on, which lead from the sensory stimulation to the motor response. What we consider typical about such cases, however, is that there are no *intentional* states or processes involved in the mediation between perception and action. Possible intentional mediators are simply bypassed. It is as if the organism decided to let the transition from sensory input to motor output run through a shortcut.

This observation becomes clearer when we look at the other, contrasting way in which perception may guide action. What we can observe here - once again according to commonsense psychology - is, first, that the process of perception gives rise to *thoughts* or *beliefs* about the perceived objects, events, or facts. We are referring here to states of a system which possess a semantic, usually propositional content to the effect *that* something is the case. In developing thoughts or beliefs a system achieves an identification of the items in its surroundings which exceeds the mere recognition of sensory patterns: the encountered items are identified in the strong sense of being subsumed under *concepts*. It is perception enriched by concepts and beliefs which constitutes the domain of perceptual knowledge that is the major concern of much traditional epistemology. For only when sensory discrimination is accompanied by concepts and beliefs do we have a logical space for operations such as gathering evidence, reasoning, and epistemic justification. Furthermore, this domain of perceptual beliefs is not causally idle with regard to action. On the contrary, whereas thoughts may be conceived of as semantic, propositional contents which can be merely entertained without resulting in action, it is an essential aspect of genuine beliefs that they make a difference in someone's behavior. That, after all, is how we detect their presence in others.

Beliefs are only one half of the intentional states which may intervene between perception and action. The other half consists in our *conative* nature: Apart from beliefs our actions are prompted by desires, wishes, volitions, plans, and intentions. In fact, we tend to reserve the term "action" for behavior with intentional antecedents of this kind. If a motor output or piece of behavior is supposed to be an action proper, we expect it to be the result of, among other things, intentional planning, however short-term. Our total intentional profile, then, comprises both beliefs and desires, volitions, or intentions in the narrow sense of the term. And these two strands of intentional states do not run disconnectedly. They are frequently interwoven in a process of practical reasoning: Noticing the goal of our desires, we mobilize our beliefs about the best means to reach it, and then, ideally, we act in accordance with these beliefs. When we follow this pattern, we behave as rational agents, and our rationality shows both in our capacity to use forms of practical reasoning and in the tendency to tune our actions to their conclusions. It remains to be added that the activation of a person's intentional profile need not be a conscious process. As a rule, we are aware of at least some of our beliefs and intentions which are momentarily relevant; we can report on them. But the notion of an intentional mediation between perception and action is in itself neutral with respect to the issue of consciousness or awareness.

From the point of view of commonsense psychology we have to distinguish, then, between, on the one hand, a transition from sensory input to motor output which manages without the intervention of beliefs and intentions and, on the other hand, a transition from perception to action which essentially involves intentional mediators and the activation of intentional profiles. Phenomenologically speaking, the first type of transition appears fast, definite, and rigid, whereas intentionally mediated transitions appear slower and more plastic. Now the first part of my commentary consists of reflections which arose when I asked myself: What happens to this and to related distinctions in the texts assembled in this volume?

There is no easy answer to this question, since the texts, even the predominantly psychological ones, are curiously silent about the things I have lumped together under the heading "intentional mediation." One has to make conjectures, and the following few paragraphs contain a discussion of some such conjectures.

A Radical Reaction

I start by mentioning a very radical reaction to the distinction I have sketched so far. I am not perfectly sure, but I do not think that this reaction is suggested by any one of the texts printed here. However, discussing it briefly serves our orientation by marking an extreme point of reference: One might declare the distinction between intentional and nonintentional mediation between perception and action to be wholly irrelevant in the context of scientific psychology. It might be claimed that the notion of perception-action sequences which pass through conceptual identification, beliefs, intentions, and practical reasoning is, although well entrenched in commonsense psychology, just a piece of bad theory. The answer to the question which preoccupies me would then be curt: What happens to intentional mediation in scientific psychology is that the whole notion is discarded.

This radical position would be hard to swallow. To all appearances, it makes a very real difference whether someone reacts to his sensory input automatically or whether his behavior is guided by intentional, thoughtful processes. For example, it makes all the difference in the world in terms of responsibility for one's actions. Furthermore, theoretically erasing the intentional profile of people would amount to cancelling the causal impact on action which we usually ascribe to perceptual knowledge as based on beliefs. Pointing out such things is not just begging the question. It is an argument in the sense that it makes us realize how far-reaching this radical move would be - how deeply our picture of responsible persons capable of knowledge would be affected. Once this is realized, the proposal becomes, I believe, overwhelmingly implausible.

Nor is this the only consideration which comes to mind. There is a reflection which results from looking at the practice of psychological research. It seems perfectly clear that the very idea of subjects fulfilling certain experimental tasks must rely on intentional mediation: People are given instructions which they need to *understand*; the experiment presupposes that they develop and sustain the firm *intention* to follow the instructions; the items encountered in the task must be *believed* to be such as to fall under a given rule of reaction. True, there is not much practical reasoning involved in the artificial situation of a laboratory task. The decision to participate in the experiment has been made (after much practical reasoning of the real-life sort), the intention has been stabilized, it is clear what to do. In a sense, the subject's mind or intentional profile has been put to rest. This fact may create the illusion either that there is nothing there in terms of intentionality, or that something is there but that it does not influence the perception-action sequence which is being investigated. But the structure of beliefs and intentions is still there, of course, and it is a structure of causal conditions, some of them standing. This will come to light quickly once the intentional mediators start to change. A change in the pertinent beliefs and intentions of a subject will strongly affect the reaction time or will even lead to a breakdown of the whole experiment. To sum up: The radical proposal simply to ignore the intentional mediation between perception and action would amount to destroying the essential difference between pure reflex behavior and the rule-governed, controlled behavior of a subject in an experiment.

Textual Evidence

The sweeping, negative answer to my question about intentional mediation is something you would have expected to find in the course of the behaviorist movement. You would not expect to find it in texts that were written long after the behaviorist interruption of cognitive psychology was terminated. What, then, do we find in these texts in terms of a more balanced, positive answer? Explicit references to the intentional mediators I have been talking about are very scarce. Beliefs and practical reasoning are not a topic at all. There are, however, some clues concerning intentions. Prinz acknowledges explicitly and with some emphasis the role of intentions, both in the intake of perceptual information and in the control of action. He speaks of "intentional control" (p. 167) and notes, quite generally, that "intentional aspects of information processing have long been neglected in modern theories of human performance" (footnote 2). Accordingly, he talks, e.g. about generation of intended actions and act codes for intended action patterns. Massaro's allusion to the present topic is more restrained. He recognizes a "continuum of perception-action relations," and the examples he gives for the two extremes are quite in the spirit of my opening distinction: "the automatic and fast striking out at the mosquito", and "the deliberate and slow rising from a warm bed" (p. 134). There is no explicit statement to the effect that a comprehensive theory about the relation between perception and action will have to include an account of their intentional mediation. But he urges that "it is necessary to measure this perception-action relation across the spectrum of human behavior to provide the constraints on potential theory" (p. 135). And since this spectrum includes "deliberate" actions, this amounts to demanding as a constraint on potential theory that intentional mediation be somehow accounted for. One more rather casual reference to this topic can be found in the paper by Cruse, Heuer, and Schmidt. They note that "it is rarely the case that an external stimulus is the only determinant of a movement" (p. 72) and that, in humans, "the assignment of output motor patterns to input signals is nearly arbitrary" (p. 74). The additional determining factor, they say, is an "intention" (in animals a "drive") or "intentional state" (p. 72).

Thus, we find some rare glimpses of the topic of intentionality, a topic so central to commonsense psychology. We find nothing resembling a systematic treatment. In a sense, this should not come as surprise. For unless the texts are outright physiological, they are committed to a theoretical framework (a "paradigm," as it is sometimes called, a little bombastically) which is built around a different idea: the idea of information processing. There has been a major shift of focus in theoretical psychology. Massaro is most explicit about this shift when he declares that William James' question about an intentional mediation between perception and action "is addressed in the general concern for how input processing relates to output processing" (p. 134). The general drift of this sort of account is that incoming perceptual information is coded in some representational medium, that this perceptual code is somehow transposed into an action code, and that this latter code triggers the action. Basically, this whole process takes place below the level of eventual intentional states like beliefs, intentions, or volitions, and below any level of consciousness or awareness. Later on (see following section) I will express doubts about key notions in this story and about the explanatory value of the story as a whole. For the moment, however, I want to leave everything untouched and want to ask: How is the intentional profile of a person as described in commonsense psychology related to the elements, parts, or stages of the information processing as described in this sort of scientific psychology? Given that we have dismissed the radical rejection of intentional notions, this question cannot be dodged, and it is pressing for anyone who seeks a comprehensive understanding of the human mind.

Concepts

The straightest answer would be to say that the phenomena of intentional mediation can be *redescribed* and thereby *explained* in terms of coding and information processing. I suppose the contributors to this volume would be inclined to give this answer. But what is the chance of its being true? Consider, first, concepts and conceptual identification as preconditions of beliefs. Neumann points out that frequently "there is no need to categorize, store, or represent in a declarative format the information prior to using it for the control of action" (p. 234). True; that is what happens in the "automatic," nonintentional transition from perception to action. What about the other case? Massaro describes an example (speech perception - speech production) where "the observations reject a direct link between perception and action. Categorization intervenes between the two, and action is controlled by this categorization rather than the perceptual processing that led to the categorization" (p. 153). Categorization, he says elsewhere, involves "pattern classification" or "the perceptual encoding of a stimulus event relative to various alternatives in memory" (p. 138). I wonder, how close this comes to concepts proper and to genuine conceptual control of actions. On the one hand, we are apparently dealing with recognition of sensory *types*, i.e. of something *general*, and this is the hallmark of conceptualization. On the other hand, something essential is lacking: the typical *inferential connections* between concepts, and the contribution that concepts make to *propositional* structures. Concepts proper owe their very identity to such things, and that means that they belong much more to the dimension of *semantics* than merely to the dimension of pattern recognition. In the text by Prinz there are passages which seem to do justice to this observation. He notes that "there can be no perceptual coding that does not include a mandatory semantic component" (p. 180). Furthermore, he notes about his "construction rules": "Construction rules are, in a way, reminiscent of the linguistic rules inherent in a language for the construction of syntactically well-formed and semantically meaningful expressions in connected discourse. The application of construction rules thus requires a syntax and a lexicon for a single mental language..." (footnote 7). Finally, we find the notion "that semantic communalities of percepts and acts may play an important functional role in the perceptual control of action decisions" (p. 180). So the information-processing approach seems to leave room for a dimension of semantic items and for the idea that semantic features may have an impact on action control. Is this the prelude to a satisfactory account of conceptual mediation in terms of information processing?

I see problems. First, I have trouble understanding the very notion of semantically relevant internal codes, i.e., of a "mental language." Prinz qualifies it as a "metaphor from the language domain" (footnote 7), thereby suggesting that it ought not to be taken literally. But what, then, are we talking about? What can it mean, for example, that the items in the "mental lexicon" have semantic contents which might serve as the scientific counterparts of concepts as conceived of in commonsense psychology and standard epistemology? This worry leads to a second point. Understanding conceptualization means understanding how certain *particular* states or items manage to carry or represent *general* information, namely information about types, classes, or properties as such. The difficulty in understanding this is easily underestimated. In fact, the only model we have for understanding it is natural language. Here, the trick is done by the complex process of assigning to linguistic expressions certain well-defined roles in language games.[1] On the level of internal codes, however, there is nothing in sight which might do an analogous job. For precisely this reason nominalists have always claimed that reference to conceptual processes does not make sense outside the domain of language proper. This suggests that to the extent to which intentional mediation between perception and action is conceptual mediation, it involves actual or potential use of language. It would seem to follow that in a comprehensive study of the perception-action relation the analysis of our language capacity should be given more weight than is the case in the present texts. Not that language is simply disregarded, but it is mainly treated as just one more case of a perception-action sequence. However, if the above nominalist reflection is not completely off the mark, we need a psychological analysis of language as the domain of concepts, propositional structures, and inferential patterns, in short: as the domain of *thinking*. And this is my third worry about an account of conceptual mediation in terms of information processing: Where in this account will we find

---

[1] See, for example, Sellars (1963).

the various modes of thinking which so often mediate between our perceptions and our actions?

Beliefs

This last point leads to the topic of beliefs. This topic is totally absent in the texts, at least under this heading. I wonder whether and how beliefs as conceived of in commonsense psychology (and, once again, standard epistemology) might be integrated into the picture of internal coding and information processing as it is drawn in this volume. Two observations make such an integration look difficult. First, the domain of beliefs is, typically, the domain of *reasoning* and *epistemic evaluation*. As believers we are capable of gathering evidence for or against our beliefs, of weighing the evidence, of questioning the inferential use we make of it, etc. This makes us *rational* beings, and the perception-action sequences which emerge in us are by no means untouched by that rationality. But how could these rational features be represented on a level of coding and information processing which is theoretically introduced as lying below the level of possible intentional mediators? Commonsense psychology has invented the notion of beliefs (and thoughts) in order to have something around which to build the description of rationality. How can the topic of rationality be saved in a framework in which reference to beliefs does not occur? Second, the transition from perception to action is sometimes mediated by very general and vague beliefs which may be reinterpreted simply in terms of certain behavioral patterns. Frequently, however, the mediating beliefs are very *specific*, and it is precisely their specificity which is decisive for the way in which perceptual input is exploited for action control. Now this specificity crucially depends on natural language as the medium of elaborating propositional contents.[2] And in this way an earlier theme recurs: How can propositional contents as fixed by sentences in a natural language be relocated on the level of information processing addressed by the theoretical approach we are discussing?

Intentions

I turn now to the conative aspect of the issue. How might intentions or volitions as entities introduced by everyday psychology be redescribed and explained in terms of coding and information processing? The material I have found in the texts is, once again, scarce. Massaro talks about "a stage of processing responsible for the mapping of perception into action" (p. 136), qualifies it as a stage of "percept/action translation", and implies that in this stage "the action is selected" (p. 138). Prinz speaks of "the perception-action interface" (p. 169) and of "devices that permit the construction of intended events (acts) on the basis of given events (percepts) according to given rules" (p. 172). We dispose of two sorts of codes: "...

[2] I have adopted this point from Dennett (1978).

percept codes stand for events-to-be-represented whereas act codes stand for events-to-be-effected" (p. 171). One crucial question concerns the details of how, on the basis of percept codes, we manage to "construct which event has to be effectuated next" (p. 172). But, Prinz says, there is another, equally crucial question: "Given an act code that specifies and event-to-be-effected - how is the intentional event put into real action?" (footnote 8).

Frankly, I am not sure to what extent these remarks are really related to my question. But I could not find anything closer, so I shall keep an eye on them. The crucial point is, of course, that the theoretical framework of coding and information processing is tailor-made for cognitive, but not for conative processes. The immediate consequence of this fact is that however detailed and sophisticated the descriptions of possible pertinent information-processing structures may be, you always feel like saying: "Fine, but where in all of this is the counterpart of intending, wishing, or willing?" Suppose, for example, that we knew everything worth knowing about Massaro's "percept/action translation," in particular about the crucial variables "percept-act compatibility" and "associative strength between stimulus and response." We could trace these variables "across the spectrum of human behavior," i.e., across the variety of direct and indirect ways in which perception helps in shaping action. Furthermore, we would know the various modes of programming various sorts of motor responses. The information processing story about the whole range of perception-action transitions would be complete. Would this yet amount to an understanding of the conative, intentional genesis of actions? Intuitively speaking, the answer would seem to be a clear "No!". For what is left out in the story is precisely my introductory distinction between perception-action sequences which are automatic and those in which our actions emerge from our perceptions because we *intend* or *want* to react to certain perceived situations by doing something specific. Basically, the same observation holds when we envisage an epistemically idealized version of the picture sketched by Prinz. Even if we knew the whole network of "construction rules" (p. 172) that lead from percept codes to act codes, so that we understood perfectly well how subjects "construct which event has to be effectuated next," we would still not know where and how to fit in intentions to act. At first sight, this point seems to be closely related to Prinz' worry about closing the gap between the presence of an act code and the realization of the action in terms of muscular activity. It may seem for a moment as if the lack of understanding were the same in both cases. This appearance is deceptive, however. For Prinz rephrases this problem as "the issue of how codes that refer to distal, to-be-effected-events get transformed into efferent codes that control proximal movement patterns" (footnote 8), and this shows that he is thinking of a problem which arises *within* the information-processing story, whereas the point I am trying to make concerns this story *as a whole*. To put it differently: Even if Prinz' problem were solved, we would *still* not know where to locate the very specific action control which is exercised by intentions or volitions. We would still feel that accounting for everything merely in terms of information processing takes our hands off the steering wheel to a greater extent than we are ready to accept.

The answer to my question seems to be, then, that intentions resist a redescription and explanation within the theoretical framework on which the texts in this volume are largely built. How might those who nevertheless consider this framework comprehensive react to my observation?

Saving the Approach

I want to mention, very briefly, three comments that might be made. First, someone may want to reconsider the radical, eliminative proposal that was discussed and rejected earlier on. He need not adopt the proposal in its sweeping form which extends it to our entire intentional profile. He might restrict it to such entities as intentions and volitions, and his reasoning might go something like this: Discarding these entities appeared scandalous just as long as we did not dispose of a detailed story about how actions originate from perceptual input. The intuitive resistance was due to the fact that, in this stage, dropping intentions would have left us with nothing at all in terms of action antecedents. However, now that we have found, in the information-processing approach, a rich source of detailed accounts to fill that lacuna, the intuitive insistence on the reality and power of intentions ought to dwindle. For while it is true that intentions are an important ingredient of commonsense psychology, it must be kept in mind that the everyday psychological picture of ourselves is not sacred. The notion of an intention was a placeholder for all the things we did not know about the origin of actions. Now that we have started to discover the workings behind the stage, it would be a mistake to cling to this notion and to turn it against the most promising project of scientific psychology. Strictly speaking - one would have to add - the references to intentions which I quoted earlier are misplaced relicts of prescientific days and ought to be removed from the texts.

A second comment might suggest a two-level model of action control, with an abstract representation of distally defined goals and intentions at the high level, and a machinery for the realization of these intentions through bodily movements at the low level. Searching for intentions or their counterparts in structures of information processing, it might be said, is a misguided enterprise. It ignores the fact that intentional action control and informational action control are phenomena on entirely different *levels*. It is true *both* that our actions are mediated by intentions *and* that they result from a chain of informational processes. The two sorts of antecedents ought not to be played off against each other. Doing this is the result of an inadequate, one-dimensional conception of the issue.

A third comment, finally, might claim that the misconception is even more fundamental. Not only are intentions on a different level from information processes; they are not even concrete, internal processes at all. Interpreting or elucidating actions in terms of intentions or the like is not explaining them in terms of inner processes. It is not analyzing action control as an internal mechanism. It is, rather, a sort of hermeneutic procedure which makes actions intelligible independently of knowledge about internal processes.[3] And for this reason it should not come as a surprise and should not strike us as a problem that the texts in this volume, devoted as they are to the study of the internal mechanisms of perception and action, are so silent about intentions.

The pivot in each of these comments is readily identified. In the first case it is the deep issue of what to say about the role and systematic value of commonsense psychology. The second comment hinges upon the difficult and problematic notion that phenomena can be distinguished with respect to levels. And the crucial point in the third comment is the internal coherence and intuitive adequacy of the claim that reference to intentions is not reference to internal processes. Evaluating these points would be a long and involved procedure. Neither do I have the space to do it, nor would I feel sure about its conclusion. And my purpose in sketching the different options was not to decide between them, but to point out that the topic of our intentional profile contains more urgent and at the same time opaque questions than you would think when reading the texts in this volume.

I expect to be told that you cannot do everything at once, that the information-processing approach is making rapid progress, and that, anyhow, "it's the best that's currently available" (Massaro, p. 134). I wouldn't want to dispute this. However, when I surfaced after immersing myself in the texts for some time, I found myself blinking, rubbing my eyes, and wondering: What happened to our intentional life? Where has it gone? I found this experience irritating, and the foregoing reflections are an attempt to conceptualize this irritation and the skeptical mood that goes with it.

Codes: Elusive Entities

I now turn to some key terms and key ideas which guide most of the explanatory attempts in this volume. This time I am not skeptical because I think that something essential has been left out. Rather, I have trouble understanding what is being said, and the more I try, the less I seem to understand.

The General Picture

The two most general notions in play are those of *coding* and *information processing*. The explanatory skeleton appears to be this: In perception a system undergoes the process of sensory stimulation. This process can be investigated and described in purely causal, physiological terms. But it can also be interpreted as the process of taking in or receiving *information* about the environment. In order to be available to the system, this information has to be put into some form or mode. This is the process of *coding* the incoming information. Now the system wants to

[3] Here I have in mind Daniel Dennett's account of intentional systems and intentional explanations, in Dennett (1981a; 1981b).

use the coded information for action control. The range of possible actions or motor responses is also available to it by way of a code. So it seems that the transition from perception to action can best be understood as a transition from elements of the perception code to elements of the action code. This process is an *information processing*, for the processed items - the elements of codes - are carrying information. Finally, the elements of the action code reached in the processing trigger the muscular activity. Given this overall idea, there are a lot of things to be investigated: the various sorts of codes; the principal relation between the perception code and the action code (common coding vs. separate coding); the number and nature of the stages in the information processing; the variables influencing the processing; the functional architecture required to manage the coding and processing, etc. It is agreed by everyone that the way from the sensory stimulation to the muscular activity could, in principle, be analyzed without invoking the notions of coding and information processing. It could be viewed in purely physiological or biochemical terms. The adherents of the information-processing approach claim, however, that we reach a better understanding of the perception-action sequence when we analyze it on the level of codes and informational processes. This level is usually qualified as "higher" and "more abstract" than the physiological level. Massaro calls it "the algorithmic mental level" (p. 137), and it is often said that it is the availability of this level of analysis and explanation which guarantees that psychology is an autonomous discipline.

Codes Proper

Why do I find this outlook difficult to understand? To begin with, let me focus on the notion of a code. Conceptual analysis reveals the following features of the original concept: First, a code presupposes a *language*. We engage in the operation of coding when we want to transmit a message which is expressed in sentences of a language. Designing a code or cipher system means establishing rules of substitution for letters, words, phrases, or sentences. Codes in the original sense are codes for propositional information, and the result of decoding is, accordingly, a text.[4] Second, the operation of coding is motivated by the intention to *conceal* a message from certain people. Third, the substitution rules which make up a code or cipher system are *conventional* rules, i.e., rules that are not prescribed by natural laws but are freely invented and are capable of being revised at any moment. It is precisely their nonnatural, arbitrary status which makes them suited to the goal of concealment. Fourth, the term "code" has a split reference: It refers both to an array of physical items (e.g., marks on paper, sounds, movements of a flag) and to the set of rules which assign these items to elements of the propositional material to be coded. Finally, a word about identity, difference, and similarity of codes. For two codes to be identical, they must be codes for the same elements of the material to be coded. They must, for example, both be codes for letters or codes for words.

---

[4] Regarding this point I am indebted to Hacker (1987).

When the same physical items are assigned to these elements, and when they are assigned according to the same rules, then we have, trivially, identical codes. But identity of the physical items need not be a necessary condition for the identity of codes. The same code, e.g., the Morse code, can be realized by different sorts of physical items, e.g., by written signals, acoustic signals, or flag signals. What makes certain patterns of such signals realizations of one and the same code is that these patterns are isomorphic. In other words, we count codes in terms of classes of isomorphic patterns of physical items which are assigned by the same substitution rules to the same elements of the material to be coded. Accordingly, we will speak of different codes in each one of three cases: when the pattern of the coding items is different, although the coded elements are the same, as when we compare, for example, the Morse code with a numerical code for letters; when the coded elements are different, although the coding items are the same (either in terms of physical identity or pattern isomorphism), as when, for example, numerals are used in one case for letters and in another one for whole words; finally and trivially, when both the coding items and the coded elements are different, as when we compare, for example, the Morse code with a numerical code for whole words. In all of these cases, there are two different codes. What might it mean to compare different codes in terms of similarity? We might have two points of reference in mind. On the one hand, we might say that two codes are similar if they encode the same sort of elements, e.g., letters. The Morse code would then count as similar to a numerical code for letters and dissimilar to a knocking or blinking code for words or whole sentences. On the other hand, we might consider two codes similar if they work on the same basis or with the same medium. The Morse code would then count as similar to a knocking or blinking code for words or sentences because all of these codes work with items where the relevant dimension is temporal extension. Accordingly, the Morse code and any numerical code would count as dissimilar because there is a contrast between sound or light signals as defined temporally and numerals as defined semantically, namely as referring to numbers.

Other Codes

How does the story about perception codes and action codes look in the light of these observations about the original idea of a code? It can be seen fairly quickly that the term "code" as used in this volume has little in common with the ordinary concept. First, the target of perception and action codes is not sentential material. Their point is not to encode propositional messages. True, codes are supposed to be involved also in perception and production of language. But the idea of such codes extends much further, and even in this special case the point is not to encipher propositional messages but to handle the perception and production of the linguistic material. The result of decoding perception and action codes is, accordingly, not a text. When we move from the codes to their targets, we arrive not at descriptions of sensory stimuli or actions, but at the stimuli and actions themselves. Second, the rules of these codes are not conventional, arbitrary rules, and they are not the result of the intention to conceal or of any other intention. They are fixed natural rules beyond or below any intentional influence. True, the texts occasionally mention "coding rules that are fixed by convention" (Prinz, p. 180), but such rules are restricted to the rather special cases of linguistic coding and to activities involving other artificial notations. The vast majority of perception and action codings are determined not by conventional facts but by natural facts. Third, there is a tendency to speak of "codes" neither in the sense of sets of rules nor in the sense of whole arrays or patterns of items, but in the sense of *individual* coding items. This can be seen clearly when Prinz, for example, says that "percept codes should attain the power to suggest, induce or elicit the activation of corresponding act codes" (p. 174), or that, under pathological conditions, "percept codes may be capable ... of inducing repetitions of the actions observed" (p. 176). Here, reference is made to individual internal states or processes and their causal properties. It is as if we called individual tokens of Morse signal "codes" and investigated their causal impact.

No essential connection to language, no concealing, no conventions, attention to individual tokens and their causal properties. All of this indicates that the word "code" in this volume expresses a concept which is very different from the original or ordinary one. Now this result by itself would not be particularly disturbing. Once the difference is clear, misunderstandings can be avoided (as in the case of "genetic code," for example). There is, however, a deeper worry. As my brief analysis has shown, the original concept of a code is a thoroughly *semantic* concept. The elements of a code form units which acquire reference and meaning by being conventionally linked to units of a natural language. Codes proper are themselves languages, although, functionally speaking, they are merely substitutional languages. They owe their character as languages to the fact that they are systematically related to natural languages and guided by a framework of intentions and conventions. By contrast, the new concept of a code as used in this volume is no longer a semantic concept, for the conventional and intentional elements and the essential reference to natural languages have dropped out. This means that the level of analysis and explanation in terms of perception and action codes cannot be a semantic level. The operations and processes on this level ought not to be described in semantic terms, for the elements of perception and action codes cannot be understood as bearers of genuine semantic properties.

The only appropriate stories to be told on this level - indeed the only possible stories - are, it seems, *causal* stories about the effects of sensory stimulation and the genesis of actions. And now we face, I believe, a very serious and fundamental problem concerning psychological accounts of the sort represented in this volume. Such accounts are supposed to be autonomous with respect to purely physiological accounts precisely because they are couched in terms of coding and of operations on codes. But reference to codes would constitute an autonomous, independent level of analysis and explanation only if this reference opened up a new logical space beyond causal explanation. The choice of the key term "code" seems to guarantee this, for it appears to open up the whole logical space of semantic relations and operations. However, if this appearance is deceptive, as I have argued, the autonomy of the psychological accounts would seem to vanish. If the framework of perception and action codes cannot, when scrutinized, claim to be more than a framework of causal explanations, it is no longer clear why explanations in terms of such codes are supposed to provide a better understanding of perception-action sequences than straightforward physiological explanations. And to some the situation may appear even worse: Whereas we have a relatively clear idea of some of the phenomena that would have to figure in physiological explanations, we are not given - at least not in the present texts - a clear picture of the intrinsic characteristics of perception and action codes. In fact, the examples of such codes to be found in the texts are amazingly scarce. There is no systematic exposition of these entities, and there is therefore really no way of knowing whether they possess causally relevant, nomic properties, and what these are. Playing the devil's advocate, someone might sum up the situation by saying: "These codes are indeed queer and shadowy entities. On the one hand, they are too far removed from ordinary codes to serve as the medium for a new sort of explanation in semantic terms. On the other hand, it is not even clear whether they can figure in old-fashioned causal explanations, for their causal, nomic properties remain in the dark. Is the very notion of such codes perhaps a chimera?"

Textual Evidence

Before turning to possible replies by advocates of the coding approach, I want to illustrate this skepticism with a few references to the texts. Massaro uses a group of concepts which, to my mind, are hybrids in the sense that they draw both on the semantic and on the causal domain without really belonging to either one. Take his and his formula of "the translation of perception into action" (p. 136). The proposed stage model to study these things has aspects which are clearly causal. For example, the initial and the final stage consist of purely physiological processes, and the whole study of variables can be understood as a study of causal influence. Why, then, an obviously semantic concept like "translation"? Massaro wants to be able to say things like, "objects differing in typicality differ from one another in terms of the time needed to translate between their perceptual code and superordinate category code" (p. 153), or, "A stroop effect occurs when a code must be translated from one system to another" (p. 159). He chooses the concept of translation because he has decided to use the concept of a code which appears to be a semantic concept. However, if this appearance is deceptive - as I have argued - transitions between codes can no longer be described as "translations," at least not without distorting this concept. Furthermore, codes are described not only from a semantic, but also from a causal perspective, when Massaro says, for example, that "quality of the perceptual code could influence the time needed to translate it into an action code and possibly even the time needed to program the action" (p. 141). But what would causal laws involving codes look like? Similar considerations could be applied to Massaro's talk about "percept-act compatibility" (p. 138), "stimulus-response consistency" (p. 143), and "associative strength between stimulus and response" (p. 143). On the one hand, these concepts have inevitable semantic overtones; on the other hand, they occur in the context of causal analysis, as when it is said, for example, that "translation requires time, and the amount of time will be directly related to the incompatibility between the perception and action codes" (p. 159). This tension between reaching out for explanations in semantic terms and falling back on causal accounts is never resolved. At least I was not able to find the resolution. Is "associative strength" a relation of semantic closeness or causal smoothness? The same sort of unresolved tension can also be found in Prinz. He ties the idea of codes very closely to the notion of a "mental language" when he says about the "separate coding approach" to perception and action that "both codes are expressed in different languages" (p. 170), and about the "common coding approach" that it requires "a syntax and a lexicon for a single mental language" (footnote 7). Although he qualifies this as a "metaphor," it is clear that codes are conceived of as bearers of semantic properties and as figuring in a type of explanation which exploits the idea of semantic relations among coding items. On the other hand, codes are also viewed in causal terms when Prinz wonders how act codes could generate efferent codes that eventually control the activity of the muscles, or when he says about a matching task that "matched act codes are automatically available by-products of the activation of their corresponding percept codes" (p. 179). To my mind, his theoretical notion of "construction rules" (p. 172), which control the transition from perception codes to action codes provides a perfect illustration of the problem I am trying to pin down. These rules could be understood as semantic rules, e.g., as rules of analytic or conceptual entailment, if reference were made to codes in the original sense, where codes are (substitutional) languages. But the codes Prinz is talking about do not - as I have argued - constitute a semantic domain, and for this reason we do not have any semantic rules here, so that the metaphor of a "mental language" is more misleading than illuminating. On the other hand, the construction rules in question cannot readily be understood as lawlike, causal generalizations. For we cannot know any such generalizations unless we know a domain of entities with clearly identified, causally relevant, nomic properties. And it does not look as if the perception and action codes mentioned in this volume will ever live up to this requirement. To sum up: Prinz' construction rules cannot be semantic rules, because the codes to which they apply are too far removed from the domain of intentions, conventions, and language proper; neither can they be causal generalizations, for codes are entities too abstract and elusive to instantiate natural laws.

I want to close this commentary by reviewing briefly three moves that might be made to defend the coding approach against this sort of skepticism. First, it might be said that the key idea of the approach can just as well be couched in terms of mental *representations*, and it might be hoped that operating with this concept would make things clearer and easier. In fact, the concept of representing and representations appears occasionally in the texts, and where it does, it is closely related to the notion of codes. Prinz, for example, describes the "common coding approach" as the claim that "percept codes and act codes are formed in the same format and are stored or maintained in a common representational medium" (p. 171), and his "construction rules" are said to "link event codes to each other within a common representational medium" (p. 172). And MacKay treats "coding" and "representing" as practically synonymous, as can be seen from such phrases as: "The nodes .... are known as *content nodes*, because they represent the form or content of an action or perception" (p. 287); "a committed node that consistently codes a particular cognitive content" (p. 285); "the node representing those concepts" (p. 280); "a node ... can no longer code any other content" (p. 287). But are we really any better off with representations than with codes?

Conventional Representations

It is very doubtful. There are two ways to read the term "representation", and neither one can remove the problem I am insisting on. According to the first one, representations presuppose conventions and a use which is based on intentions and agreement. Maps are an example. Certain arrangements of lines and colors can be taken as representing buildings, streets, mountains, or lakes because the community of users has agreed on a set of conventions. It is not any natural facts about the lines and the color patches that bestow on them the status of representations. This status is due to the existence, the use, of a conventional mapping system. It is the system which creates reference and - if you like - "meaning" for certain items on the map, making them semantic items. Now it is easy to see that the perception and action codes talked about in this volume could not be viewed as representations in this sense. The considerations are similar to those regarding the original concept of a code: Just as perception and action codes do not qualify as codes proper because they are located beyond or below any level of conventions and intentional influence, they do not qualify as representations in this first sense, either, because there does not exist anything resembling a conventional mapping system. And to the extent to which the semantic features of representations are due to conventional facts, perception and action codes cannot be said to be representations with a semantic content.

The Homunculus Trap

Here we may add a point which similarly applies to the earlier discussion about codes: Suppose we treat perception and action codes as representations on a par with, for example, symbols on a map. Then we must look for someone to use these representations in accordance with the assumed conventional rules that guide them. But the required subjects are not ordinary perceiving and acting persons, for those do not know of any conventions which guide their perception and action codes. Unless they are psychologists in the coding tradition, they do not even know of the existence of such codes. Thus, we are forced to postulate an additional subject within an ordinary person to fulfill this function. But to be a user of representations and a follower of conventions, this internal subject will have to possess an intentional profile very much like that of ordinary people. In other words, it will have to be a homunculus. Obviously, this hypothetical result amounts to a reductio ad absurdum of the idea that perception and action codes might be viewed as ordinary, conventionally guided representations. I am sure that any advocate of mental codes and mental representations wants to avoid walking into this trap. I have mentioned the trap in order to show that avoiding it means avoiding the most seductive concept of representation.

Natural Representations

Is there another concept? We are used to saying that certain bodily phenomena are "symptoms" in the sense that they "indicate" a disease, e.g., the presence of a virus. Suppose we rephrase this idea of indication by saying that a symptom "represents" a pathological phenomenon in the body. We then get the concept of a representation in the sense of a *causal concomitant* of that which it represents. There is room here for both representations as effects and representations as causes. We may say both that thunder represents preceding lightning and that lightning represents ensuing thunder. The crucial point in both cases is that the representational status of certain phenomena derives from causal relations, i.e, from natural facts. Instead of a conventional mapping system it is the system of natural laws which creates representations in this second sense. Is this concept of representing suited to elucidating the notion of internal codes and of an autonomous level of explanation?

Once again, there are reasons for skepticism. They emerge when we look, as an example, at Prinz' conception of a "common coding" for perception and action. Prinz takes the content of afferent and efferent codes to be the environmental or bodily events to which they refer. If we assume that the homunculus trap is avoided, we will interpret this content and this reference in nonconventional, causal terms, and this means that we are talking about representations in our second sense: Certain internal items have content and reference in the sense that they are causal concomitants of certain perceptual stimuli and actions. Given this naturalistic perspective on codes, what does the issue of "common" versus "separate" coding amount to? Prinz writes: "An act code can be the same as (or, similar to) a percept code with respect to some of the representational dimensions, and different or dissimilar with respect to some others" (p. 172). Now if codes are causally interpreted representations, we can at once make out a "representational dimension" with respect to which perception codes necessarily differ very much from action codes: the dimension of content or reference. There is nothing here which might be "common" to both: The tokens of a perception code refer to sensory stimulations or perceptual scenes, and they are effects of their representational targets; the tokens of an action code refer to muscular activity or larger action patterns, and they are causes of their representational targets. In terms of my earlier reflections on identity and difference of codes, we must say that what we have here are two strictly different codes, because the coded (represented) elements are different, both intrinsically and with respect to their causal role.

What, then, might be the "representational dimension" with respect to which we could, nevertheless, say that perception and action codes "are formed in the same format and maintained in a common representational medium," as Prinz claims (p. 171)? I suspect strongly that what he has in mind are *semantic* features which tokens of the two codes might have in common - some sort of "meaning" which can be the same even if all content or reference is totally different. However, unless my reflections on the concept of representations were completely off the mark, precisely this idea will not work. For when talking about perception and action codes, we are moving on a level of representations which just do not possess any semantic features besides those which are straightforward causal features, like causal reference. This result has a consequence which no one in the coding camp will like but which strikes me as inevitable: There *is* no genuine issue of "common" versus "separate" coding. For insofar as perception and action codes can be said to consist of representations, they are, clearly and trivially, separate codes. Beyond this there are, indeed, questions about the representational medium. However, these questions cannot, for conceptual reasons, concern any semantic characteristics but most concern the purely causal, i.e., physical characteristics of the representations involved.

I cannot help thinking that the term "representation" is here creating an illusion similar to the illusion created by the term "code": Since there does exist a sense in which talk about representations opens up the logical space of semantic features and may render possible new forms of analysis and explanation beyond causal explanation, it may seem as if we are dealing with an autonomous level of psychological facts and questions, once we have couched the issue of the perception-action interface in terms of representations. But when we realize how austere and thoroughly causal the admissible concept of representation is, the autonomy of this level appears to dwindle, and we are left with the need for ordinary causal accounts of the transition from perception to action. And look, incidentally, at Prinz' reasoning from the functional advantage of matching over mapping to the existence of a common code: Isn't this a straightforwardly causal reasoning? And might it not just as well terminate in the notion of particularly "easy" or "smooth" causal chains as in the notion of a "common representational medium"?

Information Processing

Earlier I indicated that I have two more moves in mind which might be made to ward off the skepticism I am expressing. One of them would be to say that doubts about an autonomous style of psychological explanation can be settled if only we take seriously enough the idea of *information processing*. Since I do not have space left for an adequate dialectical treatment of this notion, I shall simply (and somewhat dogmatically) state what I believe is problematic about it. We are, I take it, talking about information in the sense of certain contents, and not simply about abstract quantities which measure the reduction of possibilities. How do these informational contents arise? Through nomic relations of their carriers to other items in the world.[5] If this is the right answer, it follows immediately that an information-processing system as such is a system which deals with representations in the naturalistic sense sketched above. Now, my central problem is that I just do not understand what it is to "process" information. Of course, I am just as used to the term as anyone else. But strictly speaking, I don't know what is being talked about. However, I think I know the reason for my lack of understanding. It is that there are only two sorts of processes here which strike me as intelligible, and neither one fits the theoretical role that the notion in question is supposed to play. First, I can think of the operation of transmitting semantic, propositional information along inferential pathways. Syllogistic reasoning would be a case in point. But this cannot be the model, for we are talking about a level clearly below any level of propositional information. Second, we are free to call any causal, e.g., physiological, process a "processing." But this is once again the wrong level. This time it is too low, not abstract enough, for the point of talking about information processing is precisely that no detailed physiological, causal knowledge is required on this level. What, then, is a positive, sober account of the operations involved? This much seems clear: They are operations on representations in the second, austere sense of the term. But what sort of relations between these representations is being exploited,if it is true that neither genuinely semantic nor straightforwardly causal relations are the target? And what, given this want of clearness, can it mean that (perceptual) information "controls" action?

The Instrumentalistic Move

This worry leads over to the last move I want to mention. It might be called the "instrumentalistic" move, and it would amount to a gambit. The theoretical framework of information processing, of codes and of representations, it might be said, must not be mistaken for a realistic, potentially true description of internal processes. Rather, these notions serve as provisional instruments for organizing observed material, and they help to frame predictions at a time when genuine theory is still lacking. No one in this volume advocates this position explicitly. But some papers do flirt with it occasionally, as when Prinz (as pointed out earlier on) qualifies crucial notions in his account as "metaphors." (p. 170; footnote 7). The instrumentalistic move could meet my skeptical questions with nonchalance: Why worry about the theoretical purity and perfect coherence of ideas which have, after all, no more than a heuristic value? But the charm of this attitude has a price: If the information-processing and coding accounts of perception, action and their interface were nothing more than provisional, nonrealistic stories, this would mean that our ignorance concerning this entire topic is a lot greater than this volume suggests.

---

[5] This answer has been developed in detail by Dretske (1981).

References

Dennett, D.C. (1978). Brainstorms. Philosophical essays on mind and psychology. Hassocks: Harvester.

Dennett, D.C. (1981a). True believers: The intentional strategy and why it works. In A.F. Heath (Ed.), Scientific explanation. Oxford: Clarendon.

Dennett, D.C. (1981b). Three kinds of intentional psychology. In R. Healy (Ed.). Reduction, time and reality. Cambridge: Cambridge University Press.

Dretske, F.I. (1981). Knowledge and the flow of information. Oxford: Blackwell.

Hacker, P. (1987). Languages, minds and brains. In: C. Blakemore & S. Greenfield (Eds.). Mindwaves. Oxford: Blackwell.

Sellars, W. (1963). Some reflections on language games. In: Science, perception and reality. London: Routledge & Kegan Paul.

Relations Between Perception and Action: Unity in Diversity[1]

E. SCHEERER

CONTENTS

Bieri on Perception and Action: A Brief Metacommentary

Professor Bieri has no story of his own to tell about perception and action. Nor do I. What is more, I do not even have an axe to grind. Having been raised in the information-processing approach myself, I saw nothing objectionable in its adoption by some of the contributors to this book. But now that Bieri has alerted us to its dangers and insufficiencies, I believe that a few words are in order, not in defense of the overall approach, but on some specific points raised by my co-commentator.

Bieri criticizes the use of the term 'code' in this volume because it "expresses a concept which is very different from the original or ordinary one," the original (cryptographic) meaning of the term being connected to language and suggesting the ideas of concealing and being based on convention. I do not want to quibble with Bieri's analysis of the "original or ordinary meaning" of the concept, though I believe it is based, in a onesided fashion, on its cryptographic meaning. (How about legal codes or behavior codes? Are these concealed from somebody? And inherently related to language?) My point is a trivial one: when sciences borrow words from ordinary language, or from other sciences, as a rule their original meaning gets watered down. And this happened, of course, with the word 'code'. All that the "genetic code" retains from the original meaning is the idea of transcription; but would we, on this account, expunge the term from genetics? In the neurosciences and in cognitive science the term 'code' is routinely used in a fashion even more remote from the original meaning. Still, I believe that a reasonably accurate definition can be given: 'Code' refers to the *format*, as opposed to the

[1] Preparation of this chapter was helped by a fellowship from the *Stiftung Volkswagenwerk*.

Relationships Between Perception and Action
Edited by O. Neumann and W. Prinz

content, of inner representations (see, for instance, Roitblat, 1982). Thus, for instance, we can speak of 'physical codes' and 'name codes' of letters, or of 'imaginal' and 'propositional' codes of geometrical figures. This use of the term is equivalent to the expressions 'binary' or 'hexadecimal code', occurring in the information sciences, which were the source of the coding concept in the life sciences. I have not checked whether the authors of this volume do in fact use this definition consistently, though I concede that occasionally it may have been muddled. For instance, Prinz insists that "percept codes and act codes are formed in the same format and are stored or maintained in the same medium," which is not compatible with the definition. But is his actual use of the term consistent with his insistence on a common format? Probably not. He says that the distinction "is not correlated with the nature of the coded events," and that percept and motor codes stand for "events-to-be-represented" and "events-to-be-effected," respectively; he then goes on to give examples where both codes stand for one and the same event. What, then, is their difference, other than a difference in format?

Let us concede, for the sake of argument, that in this book 'code' has sometimes been used as a synonym of 'representation'. What, then, would be so dangerous about it? Back to the roots! In classical Latin, *repraesentare* (the verb from which the noun is derived) meant 'to bring to mind, to visualize'. (*Repraesentatio* meant 'cash payment', but this does not buy us a solution to our problem.) Thus, the original meaning of the term is tied to awareness and to mental pictures, and to those not accustomed to the slang of cognitive sciences it still carries these connotations. 'Code' is more neutral, it can be applied to neural as well as to mental states, and 'motor code' does not imply an "image" of a movement, so it may in fact be preferable to 'motor representation'. Etymology, so it seems, cuts both ways. We had better turn to substantial issues.

A fundamental objection Bieri raises is that the authors neglect the role of "intentional mediation," i.e., of "beliefs and desires, of volitions and intentions in the narrow sense of the word," and that practical reasoning does not occur in their analyses. Granted, intention is mentioned only obliquely in the texts. But perhaps there may be a good reason for it: the book is concerned, not with action in general, but with the relations between perception and action. More specifically, it is entirely clear from the selection of topics included in the volume that the *execution*, rather than the *formation*, of intentions is the basic focus of the authors. Now I think a good point can be made that once an intention has been formed, the details of its execution are no longer under the control of the intention, but are jointly controlled by perceptually represented environmental information and by the structure and dynamics of the motor apparatus (including, of course, its central sections). Grasping an object, for instance, certainly is an intentional, rather than a merely reflex-like action. And yet its execution is to all intents and purposes cognitively impenetrable. We open the hand while we are reaching for the object, and we adjust the opening of the hand in anticipation of, rather than as a consequence of, the actual encounter of the object. This basic pattern is already present in infants younger than 2 years of age (von Hofsten & Rönnqvist, 1988).

The autonomy of the executive, sensorimotor aspect of intentional action has been recognized by philosophers of all ages. For Aristotle, the premises of the practical syllogism were statements about beliefs and desires, but its conclusions were *the actions themselves*, rather than beliefs about actions; thus, actions were cognitively impenetrable. His little book *Movement of Animals* (where he expounds the practical syllogism) deals almost exclusively with actions rather than with their cognitive or motivational antecedents. Thomas Aquinas constructed an elaborate hierarchical scheme of volitional action, encompassing judgement, will in the narrow sense, intention, deliberation, consent, choice, and even some kind of anticipated reinforcement ("fruition"), but at the bottom of the hierarchy stands a simple command (*imperium*) to the body, which then utilizes this command on its own account. William James, of course, knew it; witness his chapters on "Habit" and on "The Will." Perhaps the editors of the volume should have made it clear that the book is a sequel to Aristotle's *Movement of Animals* (and his *Locomotion of Animals*) rather than to his *Nicomachean Ethics*; but the mere fact that the two fields of enquiry can be pursued relatively independent from each other is as true today as it was in Aristotle's times.

Bieri complains that the control of action by beliefs and desires - a cornerstone of the commonsense interpretation of volitional action - does not figure in the book. But is this really the case? Consider the act of picking apples, so dear to the psychology of attention. Commonsense psychology may quote various beliefs and desires antecedent to picking an apple; we pick an apple only if we need one, and we pick apples rather than leaves because we believe that certain visible things are apples. But now refer to *Neumann's* chapter. It does not contain the apple picking example. But if applied to it, his theory offers an elaborate analysis of what is needed in order to arrive at the belief that a certain object in front of us is an apple, and to act on the relevant belief. We need to form and update an internal representation of the visual environment; many objects (apple-like and non-apple-like) compete for the control of our action; however, the picking of apples intends not the apple type, but a concrete apple token. Thus, the properties defining apple-ness (e.g., those fixating our perceptual belief that there is an apple in front of us, in philosophical parlance) are not sufficient and must be supplemented by the (contingent) properties of one apple in front of us, and these properties are used for the specification of our concrete apple-picking act. And so on.

What is the relation between this type of task-related analysis, resulting in a definition of meaningful functional problems, and the belief-desire perspective of commonsense psychology? I think they are apples and oranges; that is, while they are different, they are both attractive and legitimate objects of our desire for knowledge. But there is no need to confront them, only to spell our their difference clearly. It seems to me that commonsense psychology is interested, predominantly or exclusively, in *indicational information*: beliefs tell us that an object is an apple, and desires tell us to pick one. But commonsense psychology is silent about *specificational information*. It is scarcely interested in how we recognize an apple, and it is totally uninterested in how we manage to pick one. These things are simply taken for granted by common sense but scientific psychology (and physiology, for that matter) cannot take them for granted.

To sum up: Bieri may have been right in reminding us that there is more to perception and action than the "how" of it. But a *complete* "story," if it is ever to be told, certainly will have to include the kind of problems that are dealt with in this volume. And while scientists are glad to listen to a philosopher, they also have the right to expect a philosopher to listen to them.

Diversity as Explained by History

My task is not to comment on Bieri's commentary, but to provide my own commentary on the main body of this book. The transition is most easily made by registering an additional disagreement with Bieri. In criticizing the information-processing approach presumably underlying those chapters of the book that are not "outright physiological," he seems to work from the presupposition that all relevant chapters share the same theoretical platform. I got a somewhat different impression, perhaps because I read the chapters with the eyes of an amateur historian. Historians tend to suffer from their own variety of skepticism, which is epitomized in the old saying, "There is nothing new under the sun." I appreciate that this book is meant to provide reports on the current state of the art, and that this goal has been realized in an admirable way. Nevertheless, when I first read it I was haunted by some diffuse *déjà vu* experiences.

Eventually, they crystallized in the conviction that the theoretical homogeneity of the information-processing approach is more apparent than real, and that it tends to hide important continuities with earlier approaches and controversies in the history of science and philosophy. Let me briefly point out some of these continuities.

Perception and Action: Separate or Common Codes?

While the 'coding' terminology is new, the problem itself goes back at least to the nineteenth century and ultimately is part of the European cultural heritage. There can be no doubt that in the Western tradition cognition (perception) and action (movement) were always kept apart. Usually, the historical "culprit" is taken to be Descartes (e.g., MacKay, 1987), but the tendency can be traced back to Aristotle, who assigned the movements of animals to the *orektikon* (striving, desire), as opposed to the *logistikon* (the rational faculty exercised in cognition). There was also the almost unanimous tendency to subordinate action to cognition, as evidenced by the traditional definition of volition as 'rational desire'. As an outgrowth of this (ultimately ethical) "intellectualism," cognitive processes including perception were assiduously studied by philosophers and scientists, while the executive aspect of action received little interest. To sum up, the dominant European perspective on perception and action was "separate and unequal," where 'unequal' also implied a difference in rank. The founders of scientific psychology were deeply immersed in this tradition; Boring could write a book on "Sensation and Perception in the History of Experimental Psychology," but he could never have written one about "Perception and Action".

Nevertheless, there were also some countercurrents against the prevailing "separate-and-unequal" mainstream. Throughout the nineteenth century we find a tradition of "motor theorizing," maintaining that the spatial organization of perception is derived from eye movement and more generally from explorative movements of the organism (Scheerer, 1984). Then there was the concept of "ideo-motor action," implying that at least some movements are directly guided by "ideas of movement" corresponding to them (Prinz, 1987). While the two countertrends had a common pioneer in the person of Lotze (1852), they were never merged into a "common-coding" theory of perception and action. As far as I can see, there were two reasons for this. One of them was the sweeping success of the reflex-arc idea, with its implication of a clear and insurmountable functional differentiation between afference and efference. The other was the introspectionist bias of classical experimental psychology, which required the discovery of sensory contents everywhere in mental life and had the result that "movement sensations" (kinesthesis) and "ideas of movement" were substituted for the actual movement; because movement, in itself, was not accessible to introspection, it was denied the status of an element of mental life. Perhaps classical psychology needed a term like 'code', defined in purely functional terms and thus not restricted to the sensory or quasi-sensory (imaginal) texture of awareness as revealed by introspection.

It is therefore no surprise that advances toward a common-coding conception of perception and action were made by developmental psychologists and neurologists, that is, in fields where even a defender of the introspective method would not consider it applicable. Prinz himself has acknowledged that Piaget's theory of imitation is a nice instance of the common-coding approach; I concur with this evaluation and want to add only that through some complicated filiations the theory is a descendant of the motor theories in vogue around the turn of the century (Scheerer, 1984). Perhaps the entire Piagetian conception of sensorimotor intelligence could be subsumed under the common-coding approach, but this depends on the definition of a common code. Probably unknown outside of the German-speaking countries, the neurologist Viktor von Weizsäcker (1973) has developed a theory maintaining the essential unity of perception and movement. Its name - "Gestalt-Circle Theory" (*Gestaltkreislehre*) - embodies its main thought: one and the same "psychophysical" (i.e., mind-brain) state can subserve either perception of movement - but apparently not both at the same time, contrary to Prinz's theory. According to von Weizsäcker, the unity of perception and movement is so compact as to render it unamenable to scientific analysis; it can be revealed only when it no longer exists, that is, when the "circle" is broken under pathological conditions.

Had von Weizsäcker not insisted on the unanalyzability of the intact Gestalt Circle, his theory would have had considerable scientific value, for it was directed against the classical identification of 'afferent' with 'sensory' (or 'perceptual') and 'efferent' with 'motor'. Subsequently, J.J. Gibson made the point with greater success, when he stated that "there is an output to perceptual systems and an input to motor systems" (Gibson, 1966, p. 56; see Reed, 1982). By virtue of this principle, Gibson's theory is akin to the common-coding idea, though to call it a "common-coding theory" would provoke the protests of Gibson's followers, in view of their rejection of the concept of inner representation.

Given that Gestalt theory seems to be compatible with the common-coding idea and Gestalt psychologists did not develop an explicit common-coding theory only because they were not interested in perception-action relations (but see Prinz on Koffka's common-coding conception of imitation), the counter-current against the classical "separate-and-unequal" conception seems much stronger than it appears at first sight. In fact, we may very well ask who still clings to the classical conception? In order to answer this question, we must first find out how we can diagnose a separate-coding theorist.

The safest criterion has been mentioned already. Separate-coding theorists identify 'afferent' with 'sensory' and 'efferent' with 'motor'. The stimulus-response (or input-output) terminology, when applied to the processes occurring inside the organism and coupled with the idea of linear, unidirectional causality, is usually sufficient to recognize a separate-coding theorist. According to this criterion, neobehaviorism is the classical example of a separate-coding approach. In fact, the neobehaviorists were so obsessed with the stimulus-response scheme that they applied it even when they were on the brink of discovering common codes; I am referring to the typical 'r-s' couplings postulated by Clark Hull and his followers, which were at once (internal, unobservable but inferrable) responses *and* stimuli. Of course, neobehaviorism is supposed to be dead by now, but I shall show below that it survives in modern clothes, and has its representative in this volume.

However, separate-coding theory also can occur in the context of an antibehaviorist theoretical perspective. For instance, if it is assumed that a motor program "runs off" independently of extero- or proprioceptive afference, then the motor system is endowed with a "code" that has no counterpart in the perceptual system. Thus, we are led to a second dichotomy that is logically independent of (though historically somewhat correlated with) the separate vs common-coding dichotomy. Before leaving the present topic, a brief comment on the historical background of MacKay's chapter is needed. In his recent book, MacKay (1987, p. 7) praises Lashley as forcefully opposing the "assumption that components for perception and action are completely separate and unshared." MacKay is able to quote textual evidence supporting this claim, but his quotation from Lashley (1951) pertains to speech perception and production only, not to the general question of perception-action relations. With respect to he latter, Lashley would seem to have pioneered the motor program concept and, by implication, to have belonged to the camp of the separate-coding theorists. The solution of the puzzle may be that speech is special, a point to which I shall return in the last section of this epilogue.

Bottom-up or Top-down Processing?

This somewhat overworked alternative does not figure too heavily in the chapters, but I need it because it provides a convenient vantage point for comparing the chapters by, say, Massaro and by Neumann.

Obviously, their style of theorizing is very different. Massaro works his way from the stimulus, through successive processing stages, to the response. Neumann starts with an intended action and asks what kind of information is needed for its planning and execution. Massaro theorizes bottom-up, Neumann top-down. Are there historical antecedents for these two styles?

The stage analysis of reaction time has a long history going back to Donders, Exner, and Wundt (Brozek, 1970). Near the end of the nineteenth century it fell into disrepute; typical arguments were that the stages overlapped, or that one and the same stimulus parameter could have an effect on more than one stage. Thus, the approach had vanished long before the behavioristic revolution. Its revival by Posner and Sternberg in the early 1960s marks the victory of the information-processing approach over neobehaviorism. Accordingly, Bieri is astonished that allusions to intentional mediation are so rare in texts "that were written long after the behaviorist interruption of psychology was terminated." But is this so astonishing? After all, the task of the new approach was to trace the "vicissitudes" of stimulus information while it is being subjected to various processing operations. To this extent, it retains certain characteristics of neobehaviorism and most notably its overall program, viz, to fill in the O in the S-O-R formula (Scheerer, 1988). Occasionally, even the terminology of neobehaviorism has been retained, for instance in the expression "stimulus-response compatibility." It should also be remembered that one pioneer text of the information-processing approach (Broadbent, 1958) contains a lengthy chapter written from the standpoint of Hull's behavior system. And finally, the very assumption of an "information transmission channel" of limited capacity, taken over from Shannon's mathematical theory of communication, retained the linear causality typical for neobehaviorism, except that 'stimulus' and 'response' were replaced by 'input' and 'output'.

In Germany, the fate of the reaction-time fractionation method was different from that in the United States. Even after the criticism of the fractionation method, reaction times were still collected in the psychology laboratories, but under the heading of *volition* rather than *cognition* - or, to use the terminology of this book, pertaining to *action* rather than to *perception*. This trend was inaugurated by Narziss Ach (1905), whose monumental work *Analysis of the Will* (1935) apparently remained unknown in the United States. Ach reinterpreted the classical reaction-time experiments in a systematic top-down fashion; the stimulus was taken not as the *cause* of an internal chain of processes, but as the *occasion* for the execution of a movement that arises from an impulse of the will, which in turn is controlled by a system of "determinations" set up by the instructions and activated by the stimulus. Incidentally, despite his acknowledgement of the role of intentions in voluntary action, Ach analyzed the *execution* of intentions without reference to them;

once a certain task has been adopted by the subject, the selection and control of subsidiary mental processes is performed at an unconscious level.

Wilhelm Wundt, who in his younger years had enthusiastically embraced the reaction-time fractionation method, became more cautious at the end of his carreer as an experimental psychologist (Wundt, 1911, p. 428), under the influence of the criticism by Ach and others, and interpreted the results of mental chronometry as indicators of the complexity of the underlying mental process - in surprisingly close anticipation of the relevant arguments of Pylyshyn (1984, chap. 5). But his real importance in the present context is his combination of the bottom-up and the top-down approaches to psychological theorizing, where the bottom-up direction is represented by the *associative*, and the top-down direction by the *apperceptive* level of mental functioning. (This applies, at least, to *active* apperception; according to Wundt, there was also a passive form of apperception.) Being the prototype of an (internal, mental) act of volition, apperception was related to the "action" rather than to the "perception" side of mental activity, and it had the property "to produce, in any given moment, *one* action only" (Wundt, 1911, p. 548). "Capacity limitations," thus we can conclude with Wundt - and Neumann - result from the necessities of action control, rather than from any "inborn" limit of the brain (or the mind, or of consciousness) to process information in parallel.

To be sure, science is international. But there still seem to be some national (or continental) traditions. How else should we explain the different styles of theorizing to be found in the chapters by Massaro and Neumann? (Neumann, of course, is not alone; van der Heijden and Prinz belong in the same camp.) This book will have served an important purpose if it succeeds in familiarizing Americans with perception and action, European style.

Things are not that simple, though. Again, the problem arises of where to put MacKay. In turning to him, we travel from the world of codes to the world of nodes. Of course, he is most heavily indebted to Lashley. But behind his network of nodes lurks, again, Wilhelm Wundt. Not the Wundt of psychological voluntarism, but the Wundt of association vs apperception. Priming is association, activation is apperception, of the passive type. The details are MacKay's own property, but the general outline seems familiar to a Wundtian. Perhaps this does not come as a surprise, because in his earlier work MacKay (1973) had acknowledged a Wundtian influence.

Peripheral or Central Control of Behavior?

This is approximately the equivalent, in physiological terms, of the bottom-up vs top-down dichotomy, which is stated in functional terms. Accordingly, I use it as a frame of reference for discussing the physiological papers. Central pattern generators and motor programs (Cruse, Dean, Heuer, and Schmidt) are examples of motor control with a predominantly central component. In addition, the advance use of sensory information (Cruse et al.), if its function is the selection of motor programs, belongs to this category, as do the 'planned movements' discussed by the developmental psychologists (Vinter; Hauert, Zanone and Mounoud). And finally, though it is used in the service of perception rather than action, the efference copy (Bridgeman) also qualifies as an example of central control.

Today, the relative contribution of periphery and center to the control of behavior is a factual question that is to be decided by experimentation and modeling. But the problem used to have definite ideological overtones. Around the turn of the century, reflex physiology reigned supreme, and as a consequence the initiation of impulses by the central nervous system alone was inadmissible. Sustained forms of activity, such as seemingly autonomous rhythms, were explained by the circular arrangement of reflexes (see Dean's chapter). This 'peripheralism' was not restricted to physiology but also pervaded psychology. Within classical experimental psychology, innervation sensations (Scheerer, 1987) and centrally aroused images were no longer tenable around 1900. The peripheralist style of theorizing was inherited rather than originated by Watson's behaviorism. Nevertheless, behaviorism was the main force behind the perpetuation of peripheralist thinking until the middle of this century, when authors such as Lashley, Hebb, and Sperry began to acknowledge the existence of centrally originated neural processes.

In Germany, a strong attack on reflex physiology and thus on peripheralism was made by the biologist Jacob von Uexküll (1973) whose *Theoretical Biology* first appeared in 1920. Von Uexküll is not mentioned in this volume, though his theory of movement should be highly relevant to some of the chapters. To him, all movements of an organism are the expression of the inherent 'planfulness' (*Planmäßigkeit*) of nature. Planfulness is of two kinds. One is immanent in the organic structure itself, the other is the result of the intervention of a central agency called, in a metaphorical way, the 'management' (*Betriebsleitung*) of the organism. (For purposes of exposition, I shall call these 'structural' and 'functional' planfulness, respectively). Von Uexküll distinguished between seven types of movement according to their "control mechanism," e.g., what type of planfulness they instantiate in the four components making up every movement: the receptors, the "registering organ" (*Merkorgan*), the "action organ" (*Wirkorgan*), and the effectors. At one extreme are reflex actions; they are structurally planful at all levels. At the other extreme are "controlled actions," e.g., the maintenance of a bird's singing ability by listening to its own song. They depend on the interaction between the functionally planful activity of the registering and the action organs. To von Uexküll, "interaction" was embodied in a loop structure - the functional circle (*Funktionskreis*) - where efferent processes act back on afferent processes. He envisaged various kinds of such loops, going through the environment, through the periphery, and remaining entirely central.

Unfortunately, von Uexküll's concept of planfulness was marred by vitalism; also, he engaged in a purely theoretical analysis and did not dispose of a methodology for the empirical investigation of the various types of control mechanisms. Another zoologist, Erich von Holst, made up for these deficits and therefore is rightly regarded as having brought about the demise of reflex physiology, and thus of peripheralism. Recognizing, in 1939, that rhythmic stimuli are not necessary to provoke rhythmic motor behavior, he rehabilitated the notion of spontaneous activity of the central nervous system and introduced the concept of central neural oscillators and their interaction. And, of course, in his seminal paper with Mittelstaedt (von Holst & Mittelstaedt, 1950) he pioneered the analysis of motor behavior in terms of an interaction between efferent signals and their reafferent consequences. The optokinetic reaction of the fly, to which the concept of the efference copy was first applied, does not strike us as an example of behavior controlled by volition. But when von Holst and Mittelstaedt extended their analysis to the problem of spatial localization constancy under eye movement, the efference copy became a scientifically respectable substitute for what von Helmholtz had said about the "effort of the will" as a reference for the localization of retinal images; respectable because it was defined, not in terms of introspective awareness, but in functional terms, and because it was amenable to mathematical treatment as a signal exactly cancelling the reafferent component of the afferent signal.

Although von Holst (as far as the "reafference principle" was concerned) stressed the *interaction* between central and peripheral signals, his approach carried the danger of replacing an earlier, dogmatic, peripheralism by a similarly dogmatic centralism. Such a move would not have lacked empirical foundations, in view of the fact that deafferentation is often found to have only minor effects on motor performance. However, as Cruse et al. show, distortion of a signal can have drastic effects though eliminating the same signal has little effect. Moreover, this is not a contradiction, because both cases can be explained by one and the same design principle of parallel open-loop and closed-loop control. The moral of their lucid analysis, in the present context, is twofold. First, one should not exclude a potential source of motor control signals on the basis of one experimental paradigm alone. And second, what does and what does not count as a signal cannot be decided by experiment alone, but is dependent on the particular model through which we approach our data.

Dean offers a rich choice of models for the interactions between efferent and afferent information in motor control. These interactions occur at many different levels in the nervous system. Using his chapter as a guide to the current state of the art, we notice that today the theoretical debate no longer focuses on peripheralism vs centralism, but on heterarchy vs hierarchy. If we describe neural function in information-processing terms, we are more likely to arrive at a hierarchy, while the actual structure of neural networks is more compatible with the notion of a heterarchy. Surprisingly, the notion of a heterarchy has generated a "new peripheralism" of sorts, which focuses on the important role played by the mechanical properties of the peripheral motor apparatus, both for the actual execution of movements and for the sensory feedback generated by them. However, this new, motor peripheralism seems to have little or no historical continuity with the classical, sensory peripheralism exemplified by reflex physiology.

Unity Despite Diversity: Some Common Themes

Sketchy and incomplete as they are, my historical observations may have exaggerated somewhat the diversity among the approaches represented in the book. When I began to look for some common themes running through all or most of the chapters, it became clear to me that it was easier to describe them in negative than in positive terms. What the authors do *not* say is at least as important as what they in fact are saying. The following is a brief, commented list of such assertions defined, in part, through negation or simple silence on part of the authors.

The Locus of Perception-Action Relations Is the Organism as a Biological System, and More Specifically the Central Nervous System. In itself, this statement may seem ridiculously trivial, but it must be contrasted with two other loci of perception-action relations that are favored in current approaches to perception and action.

One of them is the environment, or the relations of the organism to the environment. This approach is taken by Neo-Gibsonianism, or "ecological realism," and I think it is obvious that none of the authors of this book share it. This is not to say that they are totally uninfluenced by it. Topics originating from the "Gibson-Bernstein" approach to motor behavior are discussed in the chapters by Cruse et al. and by Dean, and a typically Gibsonian concept, the "specification" of actions or movements by environmental information, figures frequently in the pages of the book. Perhaps Neumann's chapter is closest to the Gibsonian perspective. But even he is not willing to follow the "direct" realists in their rejection of the concept of mental representation - as far as I can see, because he feels that perception not only is in the service of immediate action control, but also subserves functions with a longer time perspective. Nor are the developmental psychologists (Vinter, Hauert et al.) content to describe development in terms of mere changes in information pickup; what changes, according to them, is the organization of action, that is, the temporal and functional relations between perception and action. We also note that, apart from Dean's sympathy for Turvey's ideas about heterarchical control, the type of physiological and functional analysis of motor control offered in the relevant chapters does not readily square with Gibsonian pronouncements on the proper physiological basis of the ecological approach. Perhaps it is safe to say that most of the authors would go along with the "ecological" part of the Gibsonian program, as far as it relates to the specification of movement by higher-order environmental invariances, but that all of them reject the notion of a "direct" effect of behaviorally relevant environmental properties on movement and action. Hatfield (1988) recently has argued that ecological realism is compatible with the notion of representation, if the latter is not conceived in terms of cognitive operations and formal symbolic structures. I think that most contributors to this volume would agree with him.

Another, currently much-favored "locus" for perception-action relations does not occur in nature but is what the old philosophers used to call an *ens rationis*, a product of our reason. I am referring to the computer metaphor. If I am not mistaken, it is mentioned only once in the entire book, by Cruse et al. in the context of motor programs. But even there it is made clear that motor programs could as well be coded in an analog format. At present, the computer metaphor is no longer a unitary one. If my historical observations are not entirely off the mark, then it is no wonder that the classical, von Neumann architecture does not appeal to most of the authors, on account of the strict seriality and limited capacity of the Central Processing Unit. In fact, both Neumann and van der Heijden argue strongly against these presuppositions. Their arguments are cast in psychological terms and apply to the field of attention only. However, attention has always been the most favorite locus for central capacity limitations, and therefore I think their arguments can be generalized to show that the von Neumann computer metaphor is inappropriate in the *entire* domain of perception-action relations.

How about more recent development, such as parallel distributed processing (PDP) or connectionism? In view of the year of its publication, marked as it is by an unprecedented enthusiasm for these approaches, it is remarkable that the book contains relatively few allusions to them. The reason, I suspect, is that the authors prefer to look at real neurons and neuronal networks rather than at idealized models of them. This is quite obvious in the physiological sections. The psychologists are a little more favorably disposed to the idea of a "conceptual nervous system." Van der Heijden amends current PDP models by the notion of lateral inhibition between neural modules, a concept taken directly from physiology. The most outspoken instance of quasi-neural modeling (with an emphasis on 'quasi') is found in MacKay's chapter. His nodes behave somewhat like neurons (priming being analogous to presynaptic (local) potentials, activation to action potentials, self-inhibition to the refractory period, etc.) though they are certainly not instantiated by single neurons. It is interesting to note that both van der Heijden's lateral inhibition mechanism and MacKay's activation mechanism are introduced to construe certain apparent capacity limitations as emergent properties of systems of networks; in this sense, they are attractive alternatives to most current network modeling efforts, which as a rule are *totally* capacity free and thus tend to throw out the baby with the bathwater. On the other hand, one central feature of connectionist networks, their capacity for adaptive learning, scarcely shows up in the book. Bridgeman makes it clear that the adaptive tuning of sensory systems to the environment does not occur in the sensory systems alone, but critically depends on the organism's interaction with the environment.

Perhaps this gives us a clue as to why the computer is not likely to provide a complete model of perception-action relations. After all, computers do not move around freely in their environment. Artificial intelligence *might* occur in a programmed computer, just as natural intelligence *might* occur in a "brain in the vat," but artificial locomotion requires more than this. To simulate it (as opposed to merely modeling it) we need a robot, equipped with sensors, effectors, and capacities such as sensorimotor integration and the planning of actions. Robotics, not just computer science, is the technological field from which we might expect insights into the functional problems associated with perception-action relations. It is regrettable that it was not included in the book. I appreciate the author's emphasis on organisms as biological systems. But apart from the sheer difference in complexity, I did not find a principled argument against the possibility of robots instantiating the kind of "intentionally impenetrable" perception-action relations that are studied in the book. Or is is that robots do not dispose of "intentional profiles" (Bieri)? If so, I think the reason is that they do not have a *social* environment; and indeed, the social shaping and maintenance of action is totally neglected in the book.

Perception Controls Action in Many Ways, Not Just in One or Two. Historically, the life sciences have had a tendency to reduce the effects of perception on action (or of sensory signals on movements, or of stimuli on responses; I am lumping these together for ease of exposition) to one or at best two principles. 'Elicitation' (or 'triggering', or simply 'causation') was the only catchword for reflex physiology and early behaviorism, but soon dichotomous classifications became more fashionable and are to be found in all theoretical quarters. To name just a few: excitatory vs inhibitory effects (Pavlov); elicitation vs "occasioning" (Skinner); discrimination vs reinforcement (passim among the behaviorists); feedforward vs feedback (see the recent discussion by Gardner and Gardner, 1988); and a more recent arrival, indicational vs specificational information (telling what to do and telling how to do it; Pattee, 1977). Often, but not always, such dichotomies are based on dichotomous classifications of responses or of temporal and functional relations between responses and stimuli.

The human mind may have a penchant for dichotomization, but I, for one, was pleasantly surprised that in the present volume it has been largely avoided. Instead, the authors draw a rich and intricate picture of what perception does for action. The best example is the chapter by Cruse et al. First of all, they replace a customary dichotomization by a threefold categorization (intermittent, continuous, and advance use of sensory signals), and then they introduce various subcategories under each heading. For instance, advance use (i.e., 'elicitation' according to the usual scheme) has the additional functions of selecting motor programs and specifying parameters. Neumann's classification is less detailed, and there is only partial overlap with the scheme put forward by Cruse et al. But this is as it should be. He is interested in the functional relations between the various uses of sensory information. More importantly, he is concerned with the analysis of actions rather than with single movements. Small wonder, then that his list contains uses of sensory information such as "updating of the internal representation of the visual environment" and "effector recruitment," which do not figure in the account given by the motor control specialists.

Still other action-related functions of sensory information are recorded in the book, but there is no use in trying to be exhaustive. The list would not be complete, anyway. The book is concerned with the *immediate*, "real-time" functions of perception for action; the use of perception in the service of delayed action control is rarely mentioned. In order to be useful for purposes of orientation and locomotion, mental models or cognitive maps must contain other information than the type of information that is used in immediate action control. Observer-centered descriptions will have to be replaced by object-centered descriptions, optical flow patterns and the like will have to be converted into more static representations. Categorization, mentioned in the book as an important problem by Massaro only, will be needed for developing symbolic representations out of the analog "codes" making up the perceptual level. The emergence of perceptual awareness, as opposed to the mere use of sensory information, may be related to the necessities of delayed action control. The Gibsonian "direct" approach may be appropriate as long as perception is literally direct, i.e., subserves action occurring here and now, but it is likely to fail when confronted with the problems of delayed action control. Given that the authors are not Gibsonians, I would have expected them to be more explicit about perception for action at a temporal distance.

No Basic Unit, no Fixed Hierarchy: The Organization of Action Must Be Studied at a Multitude of Levels. Is there a basic, irreducible unit of action? At the beginning of the century, the answer was of course 'yes, the reflex,' though Sherrington already recognized that the reflex was a largely hypothetical entity. Subsequently, various other basic units were introduced, ranging from Skinner's operant, over the fixed action pattern of the ethologists and the TOTE unit of Miller, Galanter and Pribram (1960), to the "basic actions" of analytical philosophy (Danto, 1965). Compared with the reflex, these units were often more inclusive and had a definite 'top-down' flavor. A related development was the emergence of hierarchical models for sensorimotor integration and coordination (Weiss, 1941; see Gallistel, 1980).

One message of the present book is that a *universal* functional organization of perception-action relations simply does not exist. As a result, it is futile to look for basic units or fixed hierarchical arrangements. Dean points out that not even the most obvious, anatomical unit, the single nerve cell, acts in the uniform way that one might expect from the textbooks: for many neurons there is no strict separation of input and output synapses (thus, the reflex arc scheme breaks down); graded potentials have a much more important function than previously assumed; individual neurons may be differentiated into functionally separate regions; and so on. Things get even more complicated when we move from the cellular to the circuit level and consider neurons in terms of the functions they assume in neuronal networks. For instance, "command neurons" do not have a unitary function; a distinction must be made between gating, modulatory, and triggering functions. Working back from functions to neural networks, we get a similarly complicated picture. Perceptual (i.e., feature detection and object recognition) and motor functions are determined by the same genes and probably utilize many of the same neural circuits. Finally, one and the same action system may be realized by widely different network architectures, as Dean shows with respect to swimming in marine snails and aquatic worms.

Well, so one might argue, it is not surprising that the organization of action is affected by phylogenetic differences; moreover, when speaking of the functional organization of action, we are not interested in how the nervous system instantiates perception-action relations but want to stay at a purely functional level of analysis.

Now look at the chapters on ontogenetic development, however, with respect to action control in infants, Vinter argues for a developmental sequence beginning with a phase where behavior is "completely planned without any guidance of movement being possible" and leading to the full integration of preprogrammed and guided components of action control. Hauert et al., while favoring a hierarchical model of action control, nevertheless point out that perceptuo-motor coordination is reorganized in childhood, again involving a shift from pure preprogramming to a continuous use of visual information. Thus, ontogeny reinforces the lesson taught by phylogeny: the phenotypical identity (or similarity) of actions does not guarantee that they are mediated by identical functional organizations.

The conclusion is that the organization of action must be studied at different levels, where 'level' refers on the one hand to the organism's phylogenetic and ontogenetic status and on the other hand to the "grain size" of the analysis. However, here a problem arises. Would it not be possible, and preferable, to reduce, in a principled fashion, the number of levels needed for the purposes of analysis? Two dichotomies suggest themselves, one related to subject matter, the other to methodology. The first is embodied in two correlated pairs of concepts: 'sensation/perception' and 'movement/action'. A minimal criterion for establishing these two levels would be the reference to proximal vs distal stimuli. 'Sensation' covers the effects of proximal stimuli and 'perception' the representation of distal stimuli. Similarly, we speak of 'movement' in referring to muscular events as such, while 'action' refers to the effects of movements on the environment. As far as I can see, this useful descriptive terminology has been followed throughout the book. But I wonder whether it is sufficient to establish two basic, unitary levels of analysis. As mentioned earlier, the perception/action level introduces complications which are absent at the sensation/movement level, but the mere fact that all actions (except purely mental acts) must be realized by movements would render it unwise to exclude problems of movement control from the analysis of action in the strict sense. Analogously, while the physiologist may directly apply proximal stimuli or even use direct stimulation of the nervous system, outside of the laboratory movements are ultimately under environmental control. There are also theoretical approaches that deny the usefulness of the distinction altogether, but these apparently have no followers in the present volume.

The other, methodological, dichotomy confronts functional analysis, or what Dennett (1981, p. 237-9) has called the "design stance," with structural analysis, or Dennett's "physical stance," the difference being that "knowledge or assumptions about the system's design" are sufficient to make predictions in the first case, while "examination of the innards of the particular object," coupled with "knowledge of the laws of nature," is required in the second case. Dennett believes that the physical stance is needed only when we want to explain "instances of breakdown" or malfunction, while the design stance is sufficient for explaining the behavior of systems as long as they are functioning properly. (In the present context I am not concerned with Dennett's "intentional stance," which - if we follow Bieri - has not been adopted in sufficient measure by the contributors).

Undoubtedly, the chapters of the book can be roughly grouped according to the predominant adoption of either the physical or the design stance, and such a grouping would by and large coincidence with the physiology/psychology distinction. But a closer look reveals that the two stances are "mixed" in virtually all chapters. The physiologists freely employ the design stance: see Fig. 5 in Dean's chapter, which contains no reference at all to neural structures. And the psychologists use the physical stance: Neumann explains the necessity of central selection by the anatomical structure of the retina, Vinter uses biochemical considerations to account for the peculiarities of infant locomotion, reaching, and imitation. Neither of these uses has to do with malfunction, nor are the physiologists, with their predominantly physical stance, particularly interested in it.

Philosophers may frown on this lack of a principled division of labor. But this is how science (as opposed to technology) works. Dennett's "stance stance" (Dretske, 1988) was developed to explain how we understand computers, not animals or people. In a computer, the "design" of a program is known beforehand. In the life sciences, typically the design (i.e., the functional organization) has to be found out in the first place, before we can make any predictions on the basis of our knowledge of it. For this purpose, we often must take into account purely physical (mechanical, anatomical) constraints on the functioning of the system. Or we need to dissect the system, either physically or metaphorically, by severing functional links and observing the effects. Real or counterfactually imagined malfunction, thus, informs us about function. On the one hand, the functional "meaning" of anatomically or physiologically defined subsystems (such as neuronal circuits) will become clear only when we adopt the design stance. The design stance and the physical stance presuppose each other, rather than being mutually exclusive. They do not serve to demarcate two clearly separated levels of analysis.

Given that the organization of action needs to be studied at a variety of levels, one might wish to know what these levels are. To this question, though, the book does not give a ready answer. Yet one point seems to emerge clearly: the levels adopted by the investigator should not be confused with the levels which are functional for the system itself. In this connection, it is useful to return to Dean's observation that the design stance (what he calls the information-processing structure) results in a more hierarchical conception than the physical stance (the actual structure of networks). A hierarchical scheme like the one proposed by Weiss (1941) - motor unit, coordination of motor units comprising a muscle, muscle groups, organ, organ system, simple acts of the organism as a whole - is useful for analytical purposes, as a classificatory device. But we should not presuppose that the hierarchy is of the "connective" sort (Dawkins, 1976), where superior levels invariably control inferior levels (and only adjacent ones, at that) in the manner of a military command structure.

The Taxonomy of Action: Function is Not Perfectly Correlated with Mechanism, but Gives Clues to it. The taxonomy of action has been a central preoccupation of philosophers and scientists. Perhaps because of its ethical implications, from Aristotle on, philosophers - including Bieri - have been interested mostly in the basic dichotomy between voluntary and involuntary actions. At the end of the nineteenth century, a virtual consensus had been reached that the "involuntary" category should be split up into reflexes, instincts (primary automatisms), and habits (secondary automatisms), while the voluntary category remained "undivided" (though, of course, its mechanism was hotly disputed). Obviously, this classification is based on at least three criteria (presence vs absence of conscious control, innate vs acquired, simple vs complex). Meanwhile, evolutionary biology had introduced a new viewpoint, the classification of actions in terms of their adaptive value with respect to basic needs of the organism, such as reproduction or nutrition. Yet another fresh approach was taken by von Uexküll (see above), whose classification of movements was based on their internal organization. I propose to call these two approaches, the Darwinian and the Uexküllian, "functional" and "organizational" (the latter term is meant to include both abstract design features and their concrete, structural instantiation), respectively, and in what follows I shall be concerned with their interrelations.

In the contemporary literature, there is a tendency to assign mutually exclusive rather than complementary roles to the two viewpoints. For instance, Reed (1982) urges exclusive adoption of the functional viewpoint, which he uses to develop a taxonomy of action systems, and one of the criticisms he levels at the organizational viewpoint is that a consistent taxonomy of actions on the basis of the "structure of the processes that generated the behavior" (Gallistel, 1980, as quoted by Reed, 1982, p. 121) is impossible, *inter alia* because discoveries about organization would necessitate continuous revisions of the taxonomy. Nor is a taxonomy combining the two aspects possible, because functionally different action systems draw on the same neural mechanism.

The present volume is not organized according to an explicit taxonomical scheme of action systems. Reed would not be astonished at this; the authors are interested predominantly in the mechanism of action rather than in its adaptive function. He would also note with satisfaction that traces of a functional taxonomy are nevertheless present, both in single chapters and in the chapter outline of the book. For instance, Vinter deals with imitation, reaching, and locomotion, and the fact that the organization of these action systems changes with development does not induce her to reclassify them. Also the subject matter of some chapters (e.g., Neumann) is clearly defined on the basis of function. On the other hand, some chapters (e.g., Cruse et al., Dean) are explicitly concerned with a taxonomy of mechanisms. In a sense, these taxonomies substantiate Reed's claims, because they do not result in a classification of actions in terms of the Gibsonian affordances they realize. But I found them particularly revealing precisely because they cut across obvious functional categories such as the locomotor vs the "performatory" system, and thus draw our attention to the possible involvement of common mechanisms (such as, for instance, neural oscillators) in types of action that are so diverse that one would scarcely expect them to display such commonalities at the level of mechanism.

Why not leave it at that and accept a legitimate coexistence of the functional and the organizational viewpoint? And why should we be interested in taxonomy at all? I must here confess a personal motive. When I began to work at this "epilogue," I was literally dazzled by the sheer variety of models and mechanisms that are proposed and discussed in the book. I then thought that I could reduce the variety by explaining mechanisms in terms of functions. To achieve this, I had planned to extract two taxonomies, one based on function and the other on mechanism, and to arrange them in a double-entry table. My hope was that only the main diagonal of the table would be filled, perhaps with some scatter around it. But this hope did not materialize. Mechanism is not that closely correlated with function. Consider, for instance, functions as diverse as locomotion and attention (as a part of the investigatory system). To be sure, there will be mechanisms specific to each of them; neural oscillators are likely to be involved in locomotion and unlikely to be involved in attention. Object selection, so important in an attentional context, plays a minor role in locomotion, but the selection of texture properties is important. Locomotion is predominantly controlled by ambient vision and attention by focal vision, but ambient vision may be involved in "distributed attention" and focal vision in mountain climbing. In fact, perception subserves *all* action systems, it is multifunctional; this is the main reason why it can be legitimately studied in its own right.

Nevertheless, I do think that functions may give clues to underlying mechanisms, in the way that a Chinese character does not specify how a word is pronounced but merely hints at how it might be pronounced. It all depends on how 'function' is defined. Broad biological categories, conceived in terms of selection pressures, may be useful for other purposes, but they leave mechanism underdetermined. A more promising approach is to analyze *task requirements*, what Marr (1982) somewhat misleadingly called the "computational" level of analysis. Precisely defined task requirements may still be fulfilled by more than one "algorithm," but they certainly constrain the number of possible algorithms. Task analysis is most appropriate when subjects in an experimental setting are confronted with actual tasks, but it can also be applied in more naturalistic settings. At least an implicit task analysis is always presupposed when the functional organization of actions is described by means of block or flow diagrams.

While I am convinced that a taxonomy and analysis at the task level would reveal significant correlations with a taxonomy of mechanisms, the scope of the present epilogue prevents such an undertaking. Instead, I want to venture a tentative suggestion as to how task analysis could be used to reconciliate two conflicting viewpoints expressed in the book.

One of them is the common vs separate coding issue. The common-coding viewpoint is promoted most explicitly by Prinz and by MacKay, and I wonder whether this can be explained by the tasks which constituted the initial evidence for their theorizing. If I am not mistaken, Prinz's initial focus was on imitation and MacKay's was on the immediate coupling of speech perception and speech production in tasks like shadowing and speaking under delayed auditory feedback. Now, these tasks are special in that their perfect performance requires not only isomorphism between perception and action, but at a certain abstract level literal *identity* between components of perception and of action, respectively. This is most clearly the case in shadowing, where it is required that the phonemes in the shadowed message are exactly reproduced by the shadowing subject. A similar point can be made for those cases of imitation that qualify as copying, although it is more difficult to define the perception/action components in this case. At any rate, I do not believe that the special status of imitation in the strict sense (which comprises shadowing) should be watered down by appealing to the notion of specification. The latter encompasses a wide variety of perception-action relations, ranging from identity, over isomorphism (e.g., grapheme-phoneme correspondences in visual word recognition), a "key-lock" relation such as in grasping, the specification of parameters in reaching or catching, to a mere triggering function (e.g., the critical value of the *tau* parameter with respect to the gannet's wing-folding.) Just how likely is it that all of these cases are subserved by Prinzian common codes?

To return to Prinz and MacKay, I think that even the specific versions of their common-coding theories can be explained by resorting to their initial evidential basis. For instance, MacKay's insistence on the dualism of sequential and content nodes (which has no counterpart in Prinz's theory) is a very obvious reflection of his preoccupation with the speech perception/speech production problem. Prinz, on the other hand, is concerned more with common coding in the spatial domain, an obvious choice in view of the importance of spatial dimensions in copying nonverbal behavior. Of course, everybody is free to apply an analysis initially based on special cases to more general problems, but I think both authors will agree that the degree to which such an extension is legitimate is an open question, to be debated without theoretical blindfolds. Prinz's "mapping tasks" are a natural candidate for separate coding, though he himself makes a valiant effort to assimilate them to his common-coding "framework." But I think this effort must fail, because in most reaction-time tasks of the mapping variety the "response" and its latency is used as a mere indicator for studying the functional organization of inner processes. The fact that verbal or nonverbal responses of the "yes/no" and "same/different" variety can be given to any kind of stimulus class ensures that for them, at least, a separate motor code must be available.

Massaro draws almost exclusively on experiments of the "indicator response" sort, and so it is not astonishing that his stage model encompasses several "response" (i.e., motor) stages which are clearly separate from the "perception" stage. The mere notion of a linear arrangement of stages sets his chapter apart from almost all other contributions to the book, which on the whole lean more toward the notion of continuous information flow. Are we to accept this as a theoretical dichotomy pertaining to the entire perception/action field? I think not. Rather, a certain (general) task structure must be responsible for the emergence of a stage structure. The problem is only: what task structure?

One rather easy solution would be to dismiss the entire stage structure as an artifactual result of arbitrary tasks lacking ecological validity being imposed on the subject. The typical reaction-time experiment has a discontinuous temporal structure consisting of discrete events, and perhaps the system is thrown into a non-natural state of functioning. The reflex, it was once said, is the silly answer of the organism to a silly question, and the same might be true for tachistoscopic and reaction-time experiments. But I wonder just how sound the ecological validity argument is, in a technological world full of flashing signals and flickering video terminals; it is not accidental that the information-processing approach first made its appearance in an ergonomic context. Also, a system biologically attuned to a continuous flow of environmental information would still respond in its customary "smooth" ways to discrete information from the outside world. Thus, we must look for a way to understand the stage structure as an emergent property of a smoothly, continuously functioning system.

I think that this structure necessarily emerges under the conditions of the reaction-time experiment. First of all, we must separate the stage idea from the notion of serial processing, a move that Massaro himself considers legitimate as long as the independence of parallel processes is maintained. Second, we notice that reaction-time experiments are specifically designed to reveal the *temporal* structure of processing; unless we are interested in speed-accuracy trade-off, errors are considered a nuisance. To attain errorless or near errorless performance, the system is forced to rely on the "final," asymptotic "product" of processes rather than on intermediate stations on the way. Third, a temporal ordering of the "final products" is to be expected for at least two general reasons. One of them is that parallel processes will vary in complexity and thus in the time needed for their completion. The second is that often one process is logically dependent on the results of some other process, and thus is forced to "wait" for its completion. Seriality will then emerge - not in virtue of any intrinsic property of the brain or the mind, but as the outcome of a "logical" task structure superimposed on the temporal dynamics of a smoothly functioning processing system.

Epilogue of an Epilogue

I do not know whether the contributors to this volume agree with the interpretations I have offered for their positions and with my attempts at reconciling conflicting viewpoints, and thus I want to apologize for any faults that I may have committed. In writing this epilogue I have concentrated on the relations between chapters rather than on individual contributions. Perhaps my task would have been easier and more rewarding if I had chosen the second alternative. All chapters are highly informative, and the variety of viewpoints to be found in the book increases rather than diminishes its information value. Also, my conclusion regarding the lack of a universal structure of perception-action relations was not meant to convey a negative evaluation. As I have tried to make clear, this follows from the complexity of the subject matter itself, not from any deficiencies in theorizing about it. Given that perception-action relations are polymorphous, all we can do is to apply the analytical tools and conceptual structures provided in this book to the study of individual instances of sensorimotor coordination.

References

Ach, N. (1905). *Über die Willenstätigkeit und das Denken*. Göttingen: Vandenhoeck & Ruprecht.

Ach, N. (1935). *Analyse des Willens*. Berlin: Urban & Schwarzenberg.

Broadbent, D.E. (1958). *Perception and communication*. London: Pergamon.

Brozek, J. (1979). F.C. Donders (1818-1889) and the timing of mental operations. *Psychological Reports*, 26, 563-569.

Danto, A.C. (1965). Basic actions. *American Philosophical Quarterly*, *2*, 141-148

Dawkins, R. (1976). Hierarchical organization: A candidate principle for ethology. In: P.P.G. Bateson & R.A. Hinde (Eds.). *Growing points in ethology*. Cambridge: Cambridge University Press.

Dennett, D.C. (1981). *Brainstorms: Philosophical essays on mind and psychology*. Cambridge, MA: MIT Press.

Dretske, F. (1988). The stance stance. *Behavioral and Brain Sciences*, *11*, 511-512

Gallistel, C.R. (1980). *The organization of action: A new synthesis*. Hillsdale, NJ.: Erlbaum.

Gardner, A.R. & Gardner, B.T. (1988). Feedforward versus feedbackward: An ethological alternative to the law of effect. *Behavioral and Brain Sciences*, *11*, 429-494.

Gibson, J.J. (1966). *The senses considered as perceptual systems*. Boston: Houghton-Mifflin.

Hatfield, G. (1988). Representation and content in some (actual) theories of perception. *Studies in the History and Philosophy of Science*, *19*, 175-214.

Lashley, K.S. (1951). The problem of serial order in behavior. In: L.A. Jeffress (Ed.) *Cerebral mechanisms in behavior*, (pp. 112-146). New York: Wiley

Lotze, R.H. (1852). *Medicinische Psychologie oder Physiologie der Seele*. Leipzig: Weidmannsche Buchhandlung.

MacKay, D.G. (1973). Aspects of the theory of comprehension, memory and attention. *Quarterly Journal of Experimental Psychology*, *25*, 22-40.

MacKay, D.G. (1987). *The organization of perception and action: A theory for language and other cognitive skills*. Berlin, Heidelberg, New York, Tokyo: Springer.

Marr, D. (1982). *Vision*. San Francisco: Freeman.

Miller, G.A., Galanter, E., & Pribram, K.H. (1960). *Plans and the structure of behavior*. New York: Holt, Rinehart & Winston.

Pattee, H.H. (1977). Dynamic and linguistic modes of complex systems. *International Journal of General Systems*, *3*, 259-266.

Prinz, W. (1987). Ideo-motor action. In: H. Heuer & A.F. Sanders (Eds.), *Perspectives on Perception and Action*, (pp. 47-76) Hillsdale, NJ.: Erlbaum.

Pylyshyn, Z.M. (1984). *Computation and cognition: Toward a foundation for cognitive science*. Cambridge, MA: MIT Press.

Reed, E.S. (1982). An outline of a theory of action systems. *Journal of Motor Behavior*, *14*, 98-134.

Roitblat, H.L. (1982). The meaning of representation in animal memory. *Behavioral and Brain Sciences*, *5*, 353-406.

Scheerer, E. (1984). Motor theories of cognitive structure: A historical review. In: W. Prinz & A.F. Sanders (Eds.) *Cognition and motor processes*. (pp. 77-98). Berlin, Heidelberg, New York, Tokyo: Springer.

Scheerer, E. (1988). Muscle sense and innervation feelings: A chapter in the history of perception and action. In: H. Heuer & A.F. Sanders (Eds.) *Perspectives on perception and action* (pp. 171-194). Hillsdale, NJ.: Erlbaum.

Scheerer, E. (1988). Towards a history of cognitive science. *International Social Science Journal*, *40*(1), 7-20.

Von Hofsten, C., & Rönnqvist, L. (1988). Preparation for grasping an object: a developmental study. *Journal of Experimental Psychology: Human Perception and Performance*, *14*, 610-621.

Von Holst, E. (1939). Die relative Koordination als Phänomen und als Methode zentralnervöser Funktionsanalyse. *Ergebnisse der Physiologie*, *42*, 228-306.

Von Holst, E., & Mittelstaedt, H. (1950). Das Reafferenzprinzip (Wechselwirkungen zwischen Zentralnervensystem und Peripherie). *Naturwissenschaften*, *37*, 464-476.

Von Uexküll, J. (1973). *Theoretische Biologie*, Frankfurt: Suhrkamp (original work published 1920).

Von Weizsäcker, V. (1973). *Der Gestaltkreis*. Frankfurt/M.: Suhrkamp (original work published 1940).

Weiss, P. (1941). Self-differentiation of the basic patterns of coordination. *Comparative Psychology Monographs*, *17*(4).

Wundt, W. (1911). *Grundzüge der physiologischen Psychologie* (6th ed.), Vol. 3. Leipzig: Engelmann.

Author Index

Page numbers in *italics* refer to the References

Gazzaniga MS 243, *264*

261, 239

328, *341*

Subject Index

MIX
Papier aus verantwortungsvollen Quellen
Paper from responsible sources
FSC® C105338
FSC
www.fsc.org

If you have any concerns about our products,
you can contact us on
ProductSafety@springernature.com

In case Publisher is established outside the EU,
the EU authorized representative is:
Springer Nature Customer Service Center GmbH
Europaplatz 3, 69115 Heidelberg, Germany

Printed by Libri Plureos GmbH
in Hamburg, Germany